D1224925

# Methods in Enzymology

Volume 318
RNA–LIGAND INTERACTIONS
Part B
Molecular Biology Methods

# METHODS IN ENZYMOLOGY

EDITORS-IN-CHIEF

## John N. Abelson    Melvin I. Simon

DIVISION OF BIOLOGY
CALIFORNIA INSTITUTE OF TECHNOLOGY
PASADENA, CALIFORNIA

FOUNDING EDITORS

## Sidney P. Colowick and Nathan O. Kaplan

*Methods in Enzymology*

*Volume 318*

# RNA–Ligand Interactions

*Part B*
*Molecular Biology Methods*

EDITED BY

## Daniel W. Celander

LOYOLA UNIVERSITY
CHICAGO, ILLINOIS

## John N. Abelson

CALIFORNIA INSTITUTE OF TECHNOLOGY
PASADENA, CALIFORNIA

**ACADEMIC PRESS**

San Diego   London   Boston   New York   Sydney   Tokyo   Toronto

Academic Press
*A Harcourt Science and Technology Company*
525 B Street, Suite 1900, San Diego, California 92101-4495, USA
http://www.academicpress.com

Academic Press Limited
32 Jamestown Road, London NW1 7BY, UK
http://www.hbuk.co.uk/ap/

International Standard Book Number: 0-12-182219-2

PRINTED IN THE UNITED STATES OF AMERICA
00  01  02  03  04  05  06  MM  9  8  7  6  5  4  3  2  1

# Table of Contents

## Section I. Solution Probe Methods

## Section II. Tethered-Probe Methodologies

### A. Photochemical Reagents

# Section IV. Genetic Methodologies for Detecting RNA–Protein Interactions

## A. Bacterial Systems

## B. Eukaryotic Systems

# Section V. Protein Engineering Methodologies Useful for RNA–Protein Interaction Studies

## Section VI. Cell Biology Methods

# Contributors to Volume 318

Article numbers are in parentheses following the names of contributors.
Affiliations listed are current.

REBECCA W. ALEXANDER (9), *The Skaggs Institute for Chemical Biology, The Scripps Research Institute, La Jolla, California 92037*

MANUEL ARES, JR. (32), *Center for Molecular Biology of RNA, University of California, Santa Cruz, California 95064*

JEFFREY E. BARRICK (19), *Division of Chemistry and Chemical Engineering, California Institute of Technology, Pasadena, California 91125*

JOEL G. BELASCO (21), *Skirball Institute and Department of Microbiology, New York University School of Medicine, New York, New York 10016*

KRISTINE A. BENNETT (22), *Department of Microbiology and College of Medicine, University of Illinois at Urbana-Champaign, Urbana, Illinois 61801*

EDOUARD BERTRAND (33), *Institut de Genetique Moléculaire de Montpellier, CNRS, 34033 Montpellier, France*

CHRISTINE BRUNEL (1), *UPR 9002 du CNRS, Institut de Biologie Moléculaire et Cellulaire, 67084 Strasbourg, France*

YURI BUKHTIYAROV (9), *DuPont Pharmaceuticals Co., Wilmington, Delaware 19880*

DANIEL W. CELANDER (22), *Department of Chemistry, Loyola University Chicago, Chicago, Illinois 60626*

PASCAL CHARTRAND (33), *Department of Anatomy and Structural Biology, Albert Einstein College of Medicine, Bronx, New York 10461*

JIUNN-LIANG CHEN (10), *Department of Plant and Microbial Biology, University of California, Berkeley, California 94720-3102*

LILY CHEN (28), *Center for Biomedical Laboratory Science, San Francisco, California 94132*

BARRY S. COOPERMAN (9), *Department of Chemistry, University of Pennsylvania, Philadelphia, Pennsylvania 19104-6323*

GLORIA M. CULVER (30, 31), *Department of Biochemistry, Biophysics and Molecular Biology, Iowa State University, Ames, Iowa 50011*

ZHANNA DRUZINA (9), *Department of Chemistry, University of Pennsylvania, Philadelphia, Pennsylvania 19104-6323*

ANDREW D. ELLINGTON (14), *Institute for Molecular and Cellular Biology, University of Texas, Austin, Texas 78712*

BRICE FELDEN (11), *Biochimie, Université de Rennes 1, Faculté des Sciences Pharmaceutiques et Biologiques, 35043 Rennes Cedex, France*

STANLEY FIELD (27), *Departments of Genetics and Medical Genetics, Howard Hughes Medical Institute, University of Washington, Seattle, Washington 98195-1700*

DERRICK E. FOUTS (22), *Department of Microbiology and College of Medicine, University of Illinois at Urbana-Champaign, Urbana, Illinois 61801*

ALAN D. FRANKEL (20, 23, 28), *Department of Biochemistry and Biophysics, University of California, San Francisco, California 94143-0448*

RICHARD GIEGÉ (11), *UPR 9002 Structure de Macromolécules Biologiques et Mécanismes de Reconnaissance, Institut de Biologie Moléculaire et Cellulaire du CNRS, 67084 Strasbourg Cedex, France*

KAZUO HARADA (20), *Department of Material Life Sciences, Tokyo Gakugei University, Tokyo 184-8501, Japan*

ix

HANSJÖRG HAUSER (24), *Department of Gene Regulation and Differentiation, GBF— German Research Center for Biotechnology, D-38124 Braunschweig, Germany*

MATTHIAS W. HENTZE (25), *Gene Expression Program, European Molecular Biology Laboratory, D-69117 Heidelberg, Germany*

THOMAS HERMANN (3), *Institut de Biologie Moléculaire et Cellulaire du CNRS, F-67084 Strasbourg, France*

HERMANN HEUMANN (3), *Max-Planck-Institut für Biochemie, D-82152 Martinsried, Germany*

JOHN M. X. HUGHES (32), *Center for Molecular Biology of RNA, University of California, Santa Cruz, California 95064*

A. HALLER IGEL (32), *Center for Molecular Biology of RNA, University of California, Santa Cruz, California 95064*

CHAITANYA JAIN (21), *Skirball Institute and Department of Microbiology, New York University School of Medicine, New York, New York 10016*

SIMPSON JOSEPH (13), *Department of Chemistry and Biochemistry, University of California San Diego, La Jolla, California 92093*

ALEXEI V. KAZANTSEV (10), *Department of Plant and Microbial Biology, University of California, Berkeley, California 94720-3102*

TAD H. KOCH (7), *Department of Chemistry and Biochemistry, University of Colorado, Boulder, Colorado 80309-0215*

HEIKE KOLLMUS (24), *Department of Gene Regulation and Differentiation, GBR— German Research Center for Biotechnology, D-38124 Braunschweig, Germany*

BRIAN KRAEMER (27), *Department of Biochemistry, University of Wisconsin, Madison, Wisconsin 53706*

STEPHEN G. LANDT (23), *Department of Biochemistry and Biophysics, University of California, San Francisco, California 94143-0448*

FENYONG LIU (17), *Program in Infectious Diseases and Immunity, School of Public Health, University of California, Berkeley, California 94720*

RIHE LIU (19), *Department of Molecular Biology, Massachusetts General Hospital, Boston, Massachusetts 02114*

ZHI-REN LIU (2), *Department of Animal and Dairy Science, Auburn University, Auburn, Alabama 36849-5415*

ROY M. LONG (33), *Department of Microbiology and Molecular Genetics, Medical College of Wisconsin, Milwaukee, Wisconsin 53226-0509*

KRISTIN A. MARSHALL (14), *Institute for Molecular and Cellular Biology, University of Texas, Austin, Texas 78712*

KRISTEN M. MEISENHEIMER (7), *Department of Chemistry, Angelo State University, San Angelo, Texas 76909*

PONCHO L. MEISENHEIMER (7), *Department of Chemistry, Angelo State University, San Angelo, Texas 76909*

NIELS ERIK MØLLEGAARD (4), *Center for Biomolecular Recognition, Department of Biochemistry and Genetics, The Panum Institute, University of Copenhagen, DK-2200 Copenhagen N, Denmark*

DANIEL P. MORSE (5), *Department of Biochemistry, University of Utah, Salt Lake City, Utah 84132*

DMITRI MUNDIS (8), *Magellan Labs, Research Triangle Park, North Carolina 27709*

KNUD H. NIERHAUS (18), *Max-Planck-Institut für Molekulare Genetik, D-14195 Berlin, Germany*

PETER E. NIELSEN (4), *Center for Biomolecular Recognition, Department of Medical Biochemistry and Genetics, The Panum Institute, University of Copenhagen, DK-2200 Copenhagen N, Denmark*

HARRY F. NOLLER (13, 30, 31), *Center for Molecular Biology of RNA, Sinsheimer Laboratories, University of California, Santa Cruz, California 95064*

NORMAN R. PACE (10), *Department of Molecular, Cellular and Developmental Biology, University of Colorado, Boulder, Colorado 80309-0347*

EFROSYNI PARASKEVA (25), *Zentrum für Molekulare Biologie, Universität Heidelberg, D-69120 Heidelberg, Germany*

HADAS PELED-ZEHAVI (20), *Department of Biochemistry and Biophysics, University of California, San Francisco, California 94143-0448*

RICHARD W. ROBERTS (19), *Division of Chemistry and Chemical Engineering, California Institute of Technology, Pasadena, California 91125*

PASCALE ROMBY (1), *UPR 9002 du CNRS, Institut de Biologie Moléculaire et Cellulaire, 67084 Strasbourg, France*

BRUNO SARGUEIL (2), *Centre de Genetique Moléculaire-CNRS, 91198 Gif-sur-Yvette Cedex, France*

RENEÉ SCHROEDER (15), *Institute of Microbiology and Genetics, University of Vienna, A-1030 Vienna, Austria*

DHRUBA SENGUPTA (27), *Department of Biochemistry, University of Wisconsin, Madison, Wisconsin 53706*

SNORRI TH. SIGURDSSON (12) *Department of Chemistry, University of Washington, Seattle, Washington 98195-1700*

VLADIMIR N. SIL'NIKOV (11), *Institute of Bioorganic Chemistry, Siberian Division of the Russian Academy of Sciences, Novosibirsk 630090, Russia*

ROBERT H. SINGER (33), *Department of Anatomy and Structural Biology, Albert Einstein College of Medicine, Bronx, New York 10461*

CHRISTOPHER W. J. SMITH (2), *Department of Biochemistry, University of Cambridge, CB2 1GA Cambridge, United Kingdom*

COLIN A. SMITH (20, 28), *Department of Biochemistry and Biophysics, University of California, San Francisco, California 94143-0448*

DREW SMITH (16), *Somalogic, University of Colorado, Boulder, Colorado 80303*

CHRISTIAN M. T. SPAHN (18), *Howard Hughes Medical Institute, State University of New York, Albany, New York 12201-0509*

ERICA A. STEITZ (22), *Department of Microbiology and College of Medicine, University of Illinois at Urbana-Champaign, Urbana, Illinois 61801*

ULRICH STELZL (18), *Max-Planck-Institut für Molekulare Genetik, D-14195 Berlin, Germany*

SCOTT W. STEVENS (26), *Division of Biology, California Institute of Technology, Pasadena, California 91125*

JACK W. SZOSTAK (19), *Department of Molecular Biology, Massachusetts General Hospital, Boston, Massachusetts 02114*

RUOYING TAN (23), *Incyte Pharmaceuticals, Inc., Palo Alto, California 94304*

BERND THIEDE (29), *Max-Delbruck-Centrum für Molekulare Medizin, D-13122 Berlin, Germany*

BRIAN C. THOMAS (10), *Department of Plant and Microbial Biology, University of California, Berkeley, California 94720-3102*

PHONG TRANG (17), *Program in Infectious Diseases and Immunity, School of Public Health, University of California, Berkeley, California 94720*

HEATHER L. TRUE (22), *Department of Microbiology and College of Medicine, University of Illinois at Urbana-Champaign, Urbana, Illinois 61801*

SERGUEI N. VLADIMIROV (9), *Department of Chemistry, University of Pennsylvania, Philadelphia, Pennsylvania 19104-6323*

VALENTIN V. VLASSOV (11), *Institute of Bioorganic Chemistry, Siberian Division of the Russian Academy of Sciences, Novosibirsk 630090, Russia*

SCOT T. WALLACE (15), *Intercell Biotechnologies, Vienna, Austria*

JUN WANG (17), *Program in Infectious Diseases and Immunity, School of Public Health, University of California, Berkeley, California 94720*

RUO WANG (9), *Schering-Plough Research Institute, Kenilworth, New Jersey 07033-0539*

MATT WECKER (16), *NeXstar Pharmaceuticals, Boulder, Colorado 80301*

SANDRA E. WELLS (32), *Center for Molecular Biology of RNA, University of California, Santa Cruz, California 95064*

MARVIN WICKENS (27), *Department of Biochemistry, University of Wisconsin, Madison, Wisconsin 53706*

BRIGITTE WITTMANN-LIEBOLD (29), *Max-Delbruck-Centrum für Molekulare Medizin, D-13122 Berlin, Germany*

PAUL WOLLENZIEN (8), *Department of Biochemistry, North Carolina State University, Raleigh, North Carolina 27695*

YI-TAO YU (6), *Department of Biochemistry and Biophysics, School of Medicine and Dentistry, University of Rochester, Rochester, New York 14642*

MARINA A. ZENKOVA (11), *Institute of Bioorganic Chemistry, Siberian Division of the Russian Academy of Sciences, Novosibirsk 630090, Russia*

BEILIN ZHANG (27), *Department of Biochemistry, University of Wisconsin, Madison, Wisconsin 53706*

NORA ZUÑO (9), *Department of Chemistry, University of Pennsylvania, Philadelphia, Pennsylvania 19104-6323*

# Preface

A decade has passed since *Methods in Enzymology* addressed methods and techniques used in RNA processing. As has been evident since its inception, research in RNA processing progresses at a rapid pace. Its expansion into new areas of investigation has been phenomenal with novel discoveries being made in a variety of subspecialty areas. The subfield of RNA–ligand interactions concerns research problems in RNA structure, in the molecular recognition of structured RNA by diverse ligands, and in the mechanistic details of RNA's functional role following ligand binding. At the beginning of this new millennium, we celebrate the explosive development of exciting new tools and procedures whereby investigators explore RNA structure and function from the perspective of understanding RNA–ligand interactions.

New insights into RNA processing are accompanied with improvements in older techniques as well as the development of entirely new methods. Previous *Methods in Enzymology* volumes in RNA processing have focused on basic methods generally employed in all RNA processing systems (Volume 180) or on techniques whose applications might be considerably more specific to a particular system (Volume 181). RNA–Ligand Interactions, Volumes 317 and 318, showcase many new methods that have led to significant advances in this subfield. The types of ligands described in these volumes certainly include proteins; however, ligands composed of RNA, antibiotics, other small molecules, and even chemical elements are also found in nature and have been the focus of much research work. Given the great diversity of RNA–ligand interactions described in these volumes, we have assembled the contributions according to whether they pertain to structural biology methods (Volume 317) or to biochemistry and molecular biology techniques (Volume 318). Aside from the particular systems for which these techniques have been developed, we consider it likely that the methods described will enjoy uses that extend beyond RNA–ligand interactions to include other areas of RNA processing.

This endeavor has been fraught with many difficult decisions regarding the selection of topics for these volumes. We were delighted with the number of chapters received. The authors have taken great care and dedication to present their contributions in clear language. Their willingness to share with others the techniques used in their laboratories is apparent from the quality of their comprehensive contributions. We thank them for their effort and appreciate their patience as the volumes were assembled.

DANIEL W. CELANDER
JOHN N. ABELSON

# METHODS IN ENZYMOLOGY

VOLUME LV. Biomembranes (Part F: Bioenergetics)
*Edited by* SIDNEY FLEISCHER AND LESTER PACKER

VOLUME LVI. Biomembranes (Part G: Bioenergetics)
*Edited by* SIDNEY FLEISCHER AND LESTER PACKER

VOLUME LVII. Bioluminescence and Chemiluminescence
*Edited by* MARLENE A. DeLUCA

VOLUME LVIII. Cell Culture
*Edited by* WILLIAM B. JAKOBY AND IRA PASTAN

VOLUME LIX. Nucleic Acids and Protein Synthesis (Part G)
*Edited by* KIVIE MOLDAVE AND LAWRENCE GROSSMAN

VOLUME LX. Nucleic Acids and Protein Synthesis (Part H)
*Edited by* KIVIE MOLDAVE AND LAWRENCE GROSSMAN

VOLUME 61. Enzyme Structure (Part H)
*Edited by* C. H. W. HIRS AND SERGE N. TIMASHEFF

VOLUME 62. Vitamins and Coenzymes (Part D)
*Edited by* DONALD B. McCORMICK AND LEMUEL D. WRIGHT

VOLUME 63. Enzyme Kinetics and Mechanism (Part A: Initial Rate and Inhibitor Methods)
*Edited by* DANIEL L. PURICH

VOLUME 64. Enzyme Kinetics and Mechanism (Part B: Isotopic Probes and Complex Enzyme Systems)
*Edited by* DANIEL L. PURICH

VOLUME 65. Nucleic Acids (Part I)
*Edited by* LAWRENCE GROSSMAN AND KIVIE MOLDAVE

VOLUME 66. Vitamins and Coenzymes (Part E)
*Edited by* DONALD B. McCORMICK AND LEMUEL D. WRIGHT

VOLUME 67. Vitamins and Coenzymes (Part F)
*Edited by* DONALD B. McCORMICK AND LEMUEL D. WRIGHT

VOLUME 68. Recombinant DNA
*Edited by* RAY WU

VOLUME 69. Photosynthesis and Nitrogen Fixation (Part C)
*Edited by* ANTHONY SAN PIETRO

VOLUME 70. Immunochemical Techniques (Part A)
*Edited by* HELEN VAN VUNAKIS AND JOHN J. LANGONE

VOLUME 71. Lipids (Part C)
*Edited by* JOHN M. LOWENSTEIN

VOLUME 72. Lipids (Part D)
*Edited by* JOHN M. LOWENSTEIN

VOLUME 91. Enzyme Structure (Part I)
*Edited by* C. H. W. HIRS AND SERGE N. TIMASHEFF

VOLUME 92. Immunochemical Techniques (Part E: Monoclonal Antibodies and General Immunoassay Methods)
*Edited by* JOHN J. LANGONE AND HELEN VAN VUNAKIS

VOLUME 93. Immunochemical Techniques (Part F: Conventional Antibodies, Fc Receptors, and Cytotoxicity)
*Edited by* JOHN J. LANGONE AND HELEN VAN VUNAKIS

VOLUME 94. Polyamines
*Edited by* HERBERT TABOR AND CELIA WHITE TABOR

VOLUME 95. Cumulative Subject Index Volumes 61–74, 76–80
*Edited by* EDWARD A. DENNIS AND MARTHA G. DENNIS

VOLUME 96. Biomembranes [Part J: Membrane Biogenesis: Assembly and Targeting (General Methods; Eukaryotes)]
*Edited by* SIDNEY FLEISCHER AND BECCA FLEISCHER

VOLUME 97. Biomembranes [Part K: Membrane Biogenesis: Assembly and Targeting (Prokaryotes, Mitochondria, and Chloroplasts)]
*Edited by* SIDNEY FLEISCHER AND BECCA FLEISCHER

VOLUME 98. Biomembranes (Part L: Membrane Biogenesis: Processing and Recycling)
*Edited by* SIDNEY FLEISCHER AND BECCA FLEISCHER

VOLUME 99. Hormone Action (Part F: Protein Kinases)
*Edited by* JACKIE D. CORBIN AND JOEL G. HARDMAN

VOLUME 100. Recombinant DNA (Part B)
*Edited by* RAY WU, LAWRENCE GROSSMAN, AND KIVIE MOLDAVE

VOLUME 101. Recombinant DNA (Part C)
*Edited by* RAY WU, LAWRENCE GROSSMAN, AND KIVIE MOLDAVE

VOLUME 102. Hormone Action (Part G: Calmodulin and Calcium-Binding Proteins)
*Edited by* ANTHONY R. MEANS AND BERT W. O'MALLEY

VOLUME 103. Hormone Action (Part H: Neuroendocrine Peptides)
*Edited by* P. MICHAEL CONN

VOLUME 104. Enzyme Purification and Related Techniques (Part C)
*Edited by* WILLIAM B. JAKOBY

VOLUME 105. Oxygen Radicals in Biological Systems
*Edited by* LESTER PACKER

VOLUME 106. Posttranslational Modifications (Part A)
*Edited by* FINN WOLD AND KIVIE MOLDAVE

VOLUME 107. Posttranslational Modifications (Part B)
*Edited by* FINN WOLD AND KIVIE MOLDAVE

VOLUME 141. Cellular Regulators (Part B: Calcium and Lipids)
*Edited by* P. MICHAEL CONN AND ANTHONY R. MEANS

VOLUME 142. Metabolism of Aromatic Amino Acids and Amines
*Edited by* SEYMOUR KAUFMAN

VOLUME 143. Sulfur and Sulfur Amino Acids
*Edited by* WILLIAM B. JAKOBY AND OWEN GRIFFITH

VOLUME 144. Structural and Contractile Proteins (Part D: Extracellular Matrix)
*Edited by* LEON W. CUNNINGHAM

VOLUME 145. Structural and Contractile Proteins (Part E: Extracellular Matrix)
*Edited by* LEON W. CUNNINGHAM

VOLUME 146. Peptide Growth Factors (Part A)
*Edited by* DAVID BARNES AND DAVID A. SIRBASKU

VOLUME 147. Peptide Growth Factors (Part B)
*Edited by* DAVID BARNES AND DAVID A. SIRBASKU

VOLUME 148. Plant Cell Membranes
*Edited by* LESTER PACKER AND ROLAND DOUCE

VOLUME 149. Drug and Enzyme Targeting (Part B)
*Edited by* RALPH GREEN AND KENNETH J. WIDDER

VOLUME 150. Immunochemical Techniques (Part K: *In Vitro* Models of B and T Cell Functions and Lymphoid Cell Receptors)
*Edited by* GIOVANNI DI SABATO

VOLUME 151. Molecular Genetics of Mammalian Cells
*Edited by* MICHAEL M. GOTTESMAN

VOLUME 152. Guide to Molecular Cloning Techniques
*Edited by* SHELBY L. BERGER AND ALAN R. KIMMEL

VOLUME 153. Recombinant DNA (Part D)
*Edited by* RAY WU AND LAWRENCE GROSSMAN

VOLUME 154. Recombinant DNA (Part E)
*Edited by* RAY WU AND LAWRENCE GROSSMAN

VOLUME 155. Recombinant DNA (Part F)
*Edited by* RAY WU

VOLUME 156. Biomembranes (Part P: ATP-Driven Pumps and Related Transport: The Na,K-Pump)
*Edited by* SIDNEY FLEISCHER AND BECCA FLEISCHER

VOLUME 157. Biomembranes (Part Q: ATP-Driven Pumps and Related Transport: Calcium, Proton, and Potassium Pumps)
*Edited by* SIDNEY FLEISCHER AND BECCA FLEISCHER

VOLUME 158. Metalloproteins (Part A)
*Edited by* JAMES F. RIORDAN AND BERT L. VALLEE

VOLUME 245. Extracellular Matrix Components
*Edited by* E. RUOSLAHTI AND E. ENGVALL

VOLUME 246. Biochemical Spectroscopy
*Edited by* KENNETH SAUER

VOLUME 247. Neoglycoconjugates (Part B: Biomedical Applications)
*Edited by* Y. C. LEE AND REIKO T. LEE

VOLUME 248. Proteolytic Enzymes: Aspartic and Metallo Peptidases
*Edited by* ALAN J. BARRETT

VOLUME 249. Enzyme Kinetics and Mechanism (Part D: Developments in Enzyme Dynamics)
*Edited by* DANIEL L. PURICH

VOLUME 250. Lipid Modifications of Proteins
*Edited by* PATRICK J. CASEY AND JANICE E. BUSS

VOLUME 251. Biothiols (Part A: Monothiols and Dithiols, Protein Thiols, and Thiyl Radicals)
*Edited by* LESTER PACKER

VOLUME 252. Biothiols (Part B: Glutathione and Thioredoxin; Thiols in Signal Transduction and Gene Regulation)
*Edited by* LESTER PACKER

VOLUME 253. Adhesion of Microbial Pathogens
*Edited by* RON J. DOYLE AND ITZHAK OFEK

VOLUME 254. Oncogene Techniques
*Edited by* PETER K. VOGT AND INDER M. VERMA

VOLUME 255. Small GTPases and Their Regulators (Part A: Ras Family)
*Edited by* W. E. BALCH, CHANNING J. DER, AND ALAN HALL

VOLUME 256. Small GTPases and Their Regulators (Part B: Rho Family)
*Edited by* W. E. BALCH, CHANNING J. DER, AND ALAN HALL

VOLUME 257. Small GTPases and Their Regulators (Part C: Proteins Involved in Transport)
*Edited by* W. E. BALCH, CHANNING J. DER, AND ALAN HALL

VOLUME 258. Redox-Active Amino Acids in Biology
*Edited by* JUDITH P. KLINMAN

VOLUME 259. Energetics of Biological Macromolecules
*Edited by* MICHAEL L. JOHNSON AND GARY K. ACKERS

VOLUME 260. Mitochondrial Biogenesis and Genetics (Part A)
*Edited by* GIUSEPPE M. ATTARDI AND ANNE CHOMYN

VOLUME 261. Nuclear Magnetic Resonance and Nucleic Acids
*Edited by* THOMAS L. JAMES

VOLUME 262. DNA Replication
*Edited by* JUDITH L. CAMPBELL

VOLUME 263. Plasma Lipoproteins (Part C: Quantitation)
*Edited by* WILLIAM A. BRADLEY, SANDRA H. GIANTURCO, AND JERE P. SEGREST

VOLUME 264. Mitochondrial Biogenesis and Genetics (Part B)
*Edited by* GIUSEPPE M. ATTARDI AND ANNE CHOMYN

VOLUME 265. Cumulative Subject Index Volumes 228, 230–262

VOLUME 266. Computer Methods for Macromolecular Sequence Analysis
*Edited by* RUSSELL F. DOOLITTLE

VOLUME 267. Combinatorial Chemistry
*Edited by* JOHN N. ABELSON

VOLUME 268. Nitric Oxide (Part A: Sources and Detection of NO; NO Synthase)
*Edited by* LESTER PACKER

VOLUME 269. Nitric Oxide (Part B: Physiological and Pathological Processes)
*Edited by* LESTER PACKER

VOLUME 270. High Resolution Separation and Analysis of Biological Macromolecules (Part A: Fundamentals)
*Edited by* BARRY L. KARGER AND WILLIAM S. HANCOCK

VOLUME 271. High Resolution Separation and Analysis of Biological Macromolecules (Part B: Applications)
*Edited by* BARRY L. KARGER AND WILLIAM S. HANCOCK

VOLUME 272. Cytochrome P450 (Part B)
*Edited by* ERIC F. JOHNSON AND MICHAEL R. WATERMAN

VOLUME 273. RNA Polymerase and Associated Factors (Part A)
*Edited by* SANKAR ADHYA

VOLUME 274. RNA Polymerase and Associated Factors (Part B)
*Edited by* SANKAR ADHYA

VOLUME 275. Viral Polymerases and Related Proteins
*Edited by* LAWRENCE C. KUO, DAVID B. OLSEN, AND STEVEN S. CARROLL

VOLUME 276. Macromolecular Crystallography (Part A)
*Edited by* CHARLES W. CARTER, JR., AND ROBERT M. SWEET

VOLUME 277. Macromolecular Crystallography (Part B)
*Edited by* CHARLES W. CARTER, JR., AND ROBERT M. SWEET

VOLUME 278. Fluorescence Spectroscopy
*Edited by* LUDWIG BRAND AND MICHAEL L. JOHNSON

VOLUME 279. Vitamins and Coenzymes (Part I)
*Edited by* DONALD B. MCCORMICK, JOHN W. SUTTIE, AND CONRAD WAGNER

VOLUME 280. Vitamins and Coenzymes (Part J)
*Edited by* DONALD B. MCCORMICK, JOHN W. SUTTIE, AND CONRAD WAGNER

VOLUME 281. Vitamins and Coenzymes (Part K)
*Edited by* DONALD B. MCCORMICK, JOHN W. SUTTIE, AND CONRAD WAGNER

# Section I

# Solution Probe Methods

# [1] Probing RNA Structure and RNA–Ligand Complexes with Chemical Probes

*By* CHRISTINE BRUNEL *and* PASCALE ROMBY

## Introduction

The diversity of RNA functions demands that these molecules be capable of adopting different conformations that will provide diverse points of contact for the selective recognition of ligands. Over the years, the determination of RNA structure has provided complex challenges in different experimental areas (X-ray crystallography, nuclear magnetic resonance (NMR), biochemical approaches, prediction computer algorithms). Among these approaches, chemical and enzymatic probing, coupled with reverse transcription, has been largely used for mapping the conformation of RNA molecules of any size and for delimiting a ligand-binding site. The method takes into account the versatile nature of RNA and yields secondary structure models that reflect a defined state of the RNA under the conditions of the experiments. The aim of this article is to list the most commonly used probes together with an experimental guide. Some clues will be provided for the interpretation of the probing data in light of recent correlations observed between chemical reactivity of nucleotides within RNAs and X-ray crystallographic structures.

## Probes and Their Target Sites

Structure probing in solution is based on the reactivity of RNA molecules that are free or complexed with ligands toward chemicals or enzymes that have a specific target on RNA. The probes are used under statistical conditions where less than one cleavage or modification occurs per molecule. Identification of the cleavages or modifications can be done by two different methodologies depending on the length of the RNA molecule and the nature of the nucleotide positions to be probed. The first path, which uses end-labeled RNA, only detects scissions in RNA and is limited to molecules containing less than 200 nucleotides. The second approach, using primer extension, detects stops of reverse transcription at modified or cleaved nucleotides and therefore can be applied to RNA of any size. Table I lists structure-specific probes for RNA that are found in the litera-

## TABLE I
### Structure-Specific Probes for RNAs

| Probes[a] | Size[b] | Target | Detection[c] | | | Uses[d] | |
|---|---|---|---|---|---|---|---|
| | | | Direct | RT | Structure | Footprint | Interference |
| **Electrophiles and alkylating reagents** | | | | | | | |
| DMS | + | A(N1) | − | + | II, III | + | + |
| | | C(N3) | s | + | II, III | + | + |
| | | G(N7) | s | s | II, III | + | + |
| DEPC | + | A(N7) | s | + | II, III | − | + |
| ENU | + | Phosphate | s | s | II, III | + | + |
| Kethoxal | + | G(N1–N2) | e | + | II, III | + | + |
| CMCT | + | G(N1); U(N3) | − | + | II, III | + | + |
| **Radical generators/oxidants** | | | | | | | |
| Op-Cu | ++ | Binding pocket | + | + | III | | − |
| $Rh(phen)_2phi^{3+}$ | ++ | Tertiary interactions | s | s | III | | − |
| $Rh(DIP)_3^{3+}$ | ++ | G–U base pair | + | + | II | | − |
| $Fe^{2+}/EDTA/H_2O_2$ | + | Ribose (C1', C4') | + | + | III | + | − |
| $Fe^{2+}/MPE/H_2O_2$ | ++ | Paired N | s | s | II, III | + | − |
| KONOO | + | Ribose (C1', C4') | + | + | II, III | + | − |
| X-rays | + | Ribose (C1', C4') | + | + | III | + | − |
| Fe-bleomycin | ++ | Specific sites, loops | + | + | II, III | | − |
| NiCR and derivatives/ KHSO$_5$ | ++ | G(N7); G–U base pair | s | s | II, III | + | |
| Isoalloxazine | ++ | G–U base pair | s | s | II, III | | |
| **Hydrolytic cleavages and nuclease mimicks** | | | | | | | |
| $Mg^{2+}, Ca^{2+}, Zn^{2+}, Fe^{2+}$ | + | Specific binding sites | + | + | III | − | − |
| $Pb^{2+}$ | + | Specific binding sites; dynamic regions | + | + | II, III | + | − |
| Spermine-imidazole | ++ | Unpaired Py-A bond | s | + | II, III | + | − |
| Phosphorothioates | + | Phosphate | s | s | | + | + |
| **Biological nucleases** | | | | | | | |
| V1 RNase | +++ | Paired or stacked N | + | + | II, III | + | − |
| S1 nuclease | +++ | Unpaired N | + | + | II, III | + | − |
| *N. crassa* nuclease | +++ | Unpaired N | + | + | II, III | + | − |
| T1 RNase | +++ | Unpaired G | + | + | II, III | + | − |
| U2 RNase | +++ | Unpaired A > G ≫ C > U | + | + | II, III | + | − |
| T2 RNase | +++ | Unpaired A > C, U, G | + | + | II, III | + | − |
| CL3 RNase | +++ | Unpaired C ≫ A > U | + | + | II, III | + | − |

[a] DMS, dimethyl sulfate; DEPC, diethyl pyrocarbonate; ENU, ethylnitrosourea; kethoxal, $\beta$-ethoxy-$\alpha$-ketobutyraldehyde; CMCT, 1-cyclohexyl-3-(2-morpholinoethyl)carbodiimide metho-p-toluene sulfonate; Op-Cu, bis(1,10-phenanthroline)copper(I); [Rh(phen)$_2$(phi)$^{3+}$], bis(phenanthroline) (phenanthrene quinonediimine)rhodium(III); [Rh(DIP)$_3^{3+}$] tris(4,7-diphenyl-1,10-phenanthroline)rhodium(III); $Fe^{2+}/MPE/H_2O_2$, methidiumpropyl-EDTA-Fe(II); KONOO, potassium peroxonitrite; NiCR, (2,12-dimethyl-3,7,11,17-tetraazabicyclo[11.3.1]heptadeca-1 (2,11,13,15,17-pentaenato)nickel(II) perchlorate.

[b] Below 10 Å (+), 10–100 Å (++), and above 100 Å (+++).

[c] Direct: detection of cleavages on end-labeled RNA molecule. RT: detection by primer extension with reverse transcriptase. +, the corresponding detection method can be used; s, a chemical treatment is necessary to split the ribose-phosphate chain prior to the detection; e, RNase T1 hydrolysis can be used after kethoxal modification when end-labeled RNA is used. In that case, modification of guanine at N-1, N-2 will prevent RNase T1 hydrolysis [H. Swerdlow and C. Guthrie, *J. Biol. Chem.* **259,** 5197 (1984)].

[d] Probes useful for footprint or chemical interference are denoted by a plus sign, whereas probes that cannot be used for these purposes are denoted by a minus sign. II and III: probes that can be used to map the secondary (II) and tertiary (III) structure. Adapted from Giegé *et al.*[3]

ture. The mechanism of action for some of them has been described previously.[1-3]

*Enzymes*

Most RNases induce cleavage within unpaired regions of the RNA.[4,5] This is the case for RNases T1, U2, S1, and CL3 and nuclease from *Neurospora crassa*. In contrast, RNase V1 from cobra venom is the only probe that provides positive evidence for the existence of a helical structure. These enzymes are easy to use and provide information on single-stranded and double-stranded regions, which help to identify secondary structure RNA elements. However, because of their size, they are sensitive to steric hindrance and therefore cannot be used to define a ligand-binding site precisely. Particular caution has also to be taken as the cleavages may induce conformational rearrangements in RNA that potentially provide new targets (secondary cuts) to the RNase.

*Base-Specific and Ribose–Phosphate-Specific Probes*

Base-specific reagents have been largely used to define RNA secondary structure models. Indeed, the combination of dimethyl sulfate (DMS), 1-cyclohexyl-3-(2-morpholinoethyl)carbodiimide metho-*p*-toluene sulfonate (CMCT), and β-ethoxy-α-ketobutyraldehyde (kethoxal) allows probing the four bases at one of their Watson–Crick positions (Table I). DMS methylates position N-1 of adenines and, to a lower extent, N-3 of cytosines. CMCT modifies position N-3 of uridine and, to a weaker degree, N-1 of guanines. Kethoxal reacts with guanine, giving a cyclic adduct between positions N-1 and N-2 of the guanine and its two carbonyls. Reactivity or the nonreactivity of bases toward these probes identify the paired and unpaired nucleotides.

Position N-7 of purines, which can be involved in Hoogsteen or reverse Hoogsteen interactions, can be probed by diethyl pyrocarbonate (DEPC), DMS, or nickel complex (Table I). In contrast to DMS, nickel complex[6]

[1] C. Ehresmann, F. Baudin, M. Mougel, P. Romby, J. P. Ebel, and B. Ehresmann, *Nucleic Acids Res.* **15**, 9109 (1987).

[2] H. Moine, B. Ehresmann, C. Ehresmann, and P. Romby, *in* "RNA structure and function" (R. W. Simons and M. Grunberg-Manago, eds.), p. 77. Cold Spring Harbor Laboratory Press, Cold Spring Harbor Laboratory, NY, 1998.

[3] R. Giegé, M. Helm, and C. Florentz, *in* "Comprehensive Natural Products Chemistry" (D. Söll and S. Nishimura, eds.), Vol. 6, p. 63. Pergamon Elsevier Science, NY, 1999.

[4] H. Donis-Keller, A. M. Maxam, and W. Gilbert, *Nucleic Acid Res.* **4**, 2527 (1977).

[5] G. Knapp, *Methods Enzymol.* **180**, 192 (1989).

[6] Chen, S. A. Woodson, C. J. Burrows, and S. E. Rokita, *Biochemistry* **32**, 7610 (1993).

appears to be strictly dependent on the solvent exposure of guanines at position N-7. DEPC is very sensitive to the stacking of base rings and therefore N-7 of adenines within a helix are never reactive except if the deep groove of the helix is widened.[7]

Another class of probes encompassing ethylnitrosourea (ENU) and hydroxyl radicals attacks the ribose–phosphate backbone. ENU is an alkylating reagent that ethylates phosphates. The resulting ethyl phosphotriesters are unstable and can be cleaved easily by a mild alkaline treatment.[8] Hydroxyl radicals are generated by the reaction of the Fe(II)–EDTA complex with hydrogen peroxide, and they attack hydrogens at positions C-1′ and C-4′ of the ribose.[9] Studies performed on tRNA whose crystallographic structure is known revealed that the nonreactivity of a particular phosphate or ribose reflects its involvement in hydrogen bonding with a nucleotide (base or ribose) or its coordination with cations.[10–12] Hydroxyl radicals can also be produced by potassium peroxonitrite via transiently formed peroxonitrous acid.[13] A novel method based on the radiolysis of water with a synchrotron X-ray beam allows sufficient production of hydroxyl radicals in the millisecond range. This time-resolved probing is useful in determining the pathway by which large RNAs fold into their native conformation and also in obtaining information on transitory RNA–RNA or RNA–protein interactions.[14]

*Chemical Nucleases*

Divalent metal ions are required for RNA folding and, under special circumtances, can promote cleavages in RNA.[15] This catalytic activity was first discovered with $Pb^{2+}$ ions and later on with many other di- and trivalent cations (Table I). Two types of cleavages have been described: (1) strong cleavage resulting from a tight divalent metal ion-binding site and appro-

[7] K. M. Weeks and D. M. Crothers, *Science* **261,** 1574 (1993).
[8] B. Singer, *Nature* **264,** 333 (1976).
[9] R. P. Hertzberg and P. B. Dervan, *Biochemistry* **23,** 3934 (1984).
[10] V. V. Vlassov, R. Giegé, and J. P. Ebel, *Eur. J. Biochem.* **119,** 51 (1981).
[11] P. Romby, D. Moras, B. Bergdoll, P. Dumas, V. V. Vlassov, E. Westhof, J. P. Ebel, and R. Giegé, *J. Mol. Biol.* **184,** 455 (1985).
[12] J. A. Latham and T. R. Cech, *Science* **245,** 276 (1989).
[13] M. Götte, R. Marquet, C. Isel, V. E. Anderson, G. Keith, H. J. Gross, C. Ehresmann, B. Ehresmann, and H. Heumann, *FEBS Lett.* **390,** 226 (1996).
[14] B. Sclavi, S. Woodson, M. Sullivan, M. R. Chance, and M. Brenowitz, *J. Mol. Biol.* **266,** 144 (1996).
[15] T. Pan, D. M. Long, and O. C. Uhlenbeck, *in* "The RNA World" (R. F. Gesteland and J. F. Atkins, eds.), p. 271. Cold Spring Harbor Laboratory Press, Cold Spring Harbor Laboratory, NY, 1993.

priate stereochemistry of the cleaved phosphodiester[15] and (2) low-intensity cleavages at multiple sites in flexible regions[16] (usually interhelical or loop regions and bulged nucleotides). Despite the fact that Pb(II)-induced cleavages are not always easy to interpret, this probe can detect subtle conformational changes on ligand binding and determine structural changes between different mutant species of the same RNA molecule.[17]

Metal coordination complexes have been developed in order to recognize special features of RNAs (see Table I for reviews[18,19]). For example, methidiumpropyl-Fe(II)-EDTA displays intercalating properties that stimulate cleavage efficiency around the intercalation site in double-stranded or stacked regions of RNA.[9] Bis(phenanthroline)(phenanthrenequinonediimine)rhodium(III) [Rh(phen)$_2$(phi)$^{3+}$] only binds by intercalation in a distorted deep groove of RNA due to base tilting or propeller twisting.[20] The rhodium complex,[21] tris(4,7-diphenyl-1,10-phenanthroline)rhodium(III) [Rh(DIP)$_3^{3+}$], and isoalloxazine derivatives[22] present a remarkable selectivity for G–U base pairs in RNA. The porphyrin cation photochemical method probes the solvent accessibility of guanines and monitors the folding of coaxially stacked helices in RNAs.[23]

Several chemical probes have been designed to mimic the active site of RNase A (Table I). Two imidazole residues conjugated to an intercalating phenazine dye by linkers of variable length induced cleavages at the Py-A sequence located in flexible regions of tRNA.[24] Identical cleavages were obtained with a spermine–imidazole construct,[25] with hydrolysis being induced by the addition of a second imidazole residue from the buffer.

*Applications of Nucleotide Modifications*

The elaboration of secondary and tertiary structure models requires the use of a large panel of probes with different specificities. Base-specific reagents such as DMS, CMCT, and kethoxal defined the reactivity of the

[16] P. Gornicki, F. Baudin, P. Romby, M. Wiewiorowski, W. Kryzosiak, J. P. Ebel, C. Ehresmann, and B. Ehresmann, *J. Biomol. Struct. Dyn.* **6,** 971 (1989).
[17] L. S. Behlen, J. R. Sampson, A. B. DiRenzo, and O. C. Uhlenbeck, *Biochemistry* **29,** 2515 (1990).
[18] P. W. Huber, *FASEB J.* **7,** 1367 (1993).
[19] J. R. Morrow, *Adv. Inorg. Biochem.* **9,** 41 (1994).
[20] C. S. Chow, L. S. Behlen, O. C. Uhlenbeck, and J. K. Barton, *Biochemistry* **31,** 972 (1992).
[21] C. S. Chow and J. K. Barton, *Biochemistry* **31,** 5423 (1992).
[22] P. Burgstaller, T. Hermann, C. Huber, E. Westhof, and M. Famulok, *Nucleic Acids Res.* **25,** 4018 (1997).
[23] D. W. Celander and J. M. Nussbaum, *Biochemistry* **35,** 12061 (1996).
[24] M. A. Podyminogin, V. V. Vlassov, and R. Giegé, *Nucleic Acids Res.* **21,** 5950 (1993).
[25] V. V. Vlassov, G. Zuber, B. Felden, J. P. Behr, and R. Giegé, *Nucleic Acids Res.* **23,** 3161 (1995).

four bases at one of their Watson–Crick positions. Together with RNases, these probes are appropriate to define single-stranded regions and helical domains. Furthermore, chemical reactions can be conducted under a variety of conditions. For instance, the influence of monovalent or divalent ions (such as magnesium) on the folding of the RNA can be tested, and the thermal transition of RNA molecules can be followed by varying the temperature. Such experiments provide information concerning the stability of the different secondary structure domains. They also allow the identification of tertiary elements, as these interactions are the first to break in a cooperative manner during the melting process of an RNA structure. Noncanonical base pairs that involve base protonation such as A + C can also be identified by the pH dependence of DMS and DEPC modification (Table II). Various chemicals can be used to map the RNA structure *in vivo* under different cell growth conditions.[26,27] DMS is used extensively because it passes through the cell membranes readily and the reaction can be quenched by 2-mercaptoethanol. Because only a limited number of probe can be used *in vivo*, comparison between *in vivo* and *in vitro* probing provides complementary data for determining the functional RNA structure.

Enzymes and chemicals are used extensively to map the binding site of a specific ligand (antibiotic, RNA, proteins) and to follow the structural rearrangements that occur on binding. A list of the most appropriate chemicals is given in Table I. Caution needs to be taken when using chemicals because they can modify the protein moiety. Because of their small size and their insensitivity to the secondary structure, hydroxyl radicals generated by the reaction of the Fe(II)–EDTA complex with hydrogen peroxide or by potassium peroxonitrite and ENU represent ideal probes to map the footprint of a protein.

Chemical interference defines a set of nucleotides that have lost the capability to interact with a ligand when they are modified by a chemical probe. The analysis involves random modification of atomic positions in a given RNA. Using different screening procedures, such as gel filtration, gel retardation, and affinity chromatography, the RNA molecules that are competent for a specific function are then separated from those that have lost their function. After purification of the different RNA species, strand scission is induced at the modified base or phosphate. The modification can be detected either by primer extension or by using end-labeled RNAs.

[26] A. Méreau, R. Fournier, A. Grégoire, A. Mougin, P. Fabrizio, R. Lührmann, and C. Branlant, *J. Mol. Biol.* **273,** 552 (1997).
[27] E. Bertrand, M. Fromont-Racine, R. Pictet, and T. Grange, *Proc. Natl. Acad. Sci. U.S.A.* **90,** 3496 (1993).

TABLE II
USING STRUCTURE-SPECIFIC PROBES[a]

| Probes | Molecular weight | Target | Product | Special considerations/buffers, pH, temperature, etc. |
|---|---|---|---|---|
| **Chemicals** | | | | |
| DMS | 126 | A(N-1) | N1-CH$_3$ | Reactive at pH ranging from 4.5 to 10 and temperature from 4° to 90°. Tris buffers should be avoided as DMS reacts with amine groups |
| | | C(N-3) | N3-CH$_3$ | Reactive at pH ranging from 4.5 to 10 and temperature from 4° to 90°. Tris buffers should be avoided as DMS reacts with amine groups |
| | | G(N-7) | N7-CH$_3$ | Reactive at pH ranging from 4.5 to 10 and temperature from 4° to 90°. Tris buffers should be avoided as DMS reacts with amine groups |
| DEPC | 174 | A(N-7) | N7-CO$_2$H$_2$ | Reactive at pH ranging from 4.5 to 10 and temperature from 4° to 90°. Tris buffers should be avoided as DEPC reacts with amine groups |
| Kethoxal | 148 | G(N1-1–N-2) | N1-CHOH / N2-CROH | Slightly acidic pH and borate ions stabilize the guanine-kethoxal adduct |
| CMCT | 424 | G(N-1) | N1-C=N-R / NH-R' | Optimal reactivity at pH 8 and over a wide range of temperature. CMCT still soluble up to 300 mg/ml in water |
| | | U(N-3) | N3-C=N-R / NH-R' | Optimal reactivity at pH 8 and over a wide range of temperature. CMCT still soluble up to 300 mg/ml in water |
| ENU | 117 | Phosphate | -O-CH$_2$-CH$_2$ | Optimal reactivity at pH 8 and at temperature ranging from 4° to 90°. Tris buffers should be avoided as ENU reacts with amine groups |
| Fe$^{2+}$/EDTA/H$_2$O$_2$ | 17 | Ribose (C1', C4') | | Reactivity relatively insensitive to buffer components, pH, temperature, sulfhydryl compound, and cation concentration. Only sodium phosphate buffer should be avoided. Glycerol concentration as low as 0.5% impairs the reaction |
| **Biological nucleases** | | | | |
| T1 RNase | 11,000 | Unpaired G | ...Gp (3'p) | Active under a wide range of conditions (e.g., temperature between 4° and 55°, with or without magnesium ion and salt, in urea) |
| T2 RNase | 36,000 | Unpaired A > C, U, G | ...Ap (3'p) | Active under a wide range of conditions (e.g., temperature between 4° and 55°, with or without magnesium ion and salt) |
| U2 RNase | 12,490 | Unpaired A > G ≫ C > U | ...Np (3'p) | Specificity for A best at acidic pH (pH 3.5). Some cleavage after G is observed at physiological pH. Active with or without magnesium |
| V1 RNase | 15,900 | Paired or stacked N | pN... (5'p) | Absolutely requires divalent cations. Active under a wide range of temperature (4–50°) |

[a] Molecular weight, specificity, and products generated by the probe action are indicated. Comments are given for the most commonly used probes.

A variety of reagents can be used provided that they do not induce scission of the molecule (see Table I). RNA can also be modified during *in vitro* transcription by the statistical insertion of phosphorothioates.[28] Metal ion coordination can be deduced from "rescue" experiments where the inhibition of RNA cleavage is strongly relieved by the presence of more thiophilic cations such as $Mn^{2+}$ or $Cd^{2+}$.

Chemical probes can also be used to induce site-specific cleavage of a proximal RNA providing topographical information on ligand–RNA complexes. Coordination complexes such as phenanthroline–Cu(II)[29] and Fe(II)–EDTA[30] have been tethered to tRNA in order to map the ribosomal environment of the acceptor end of tRNA. Iron(II) can be chelated by an EDTA linker [1-(*p*-bromoacetamidobenzyl)-EDTA] (BABE) to a specific cysteine residue located at the surface of a protein.[31]

Experimental Guide for Chemical Probing

Optimal reactivity conditions vary with the different probes, and the possibility exists that subtle conformational changes occur under different incubation conditions (Table II). Therefore, probing the conformation of free or complexed RNAs requires strictly defined buffer conditions (pH, ionic strength, magnesium concentration, temperature) and the probe:RNA ratio must be adapted. The experimental guide is based on the conditions used to probe the solution structure of 5S rRNA[32] and will be limited to the most commonly used chemical probes. Buffer conditions (e.g., pH, ionic strength) may vary depending on the RNA. Because the experiments have to be conducted under statistical conditions, it is advisable for the first experiment to perform a concentration or time scale dependence for the different probes. Other detailed protocols for the detection of cleavages and modifications have also been reported previously.[33–35]

[28] F. Eckstein, *Annu. Rev. Biochem.* **54,** 367 (1985).
[29] W. E. Hill, D. J. Bucklin, J. M. Bullard, A. L. Galbralth, N. V. Jammi, C. C. Rettberg, B. S. Sawyer, and M. A. van Waes, *Biochem. Cell. Biol.* **73,** 1033 (1995).
[30] S. Joseph and H. F. Noller, *EMBO J.* **15,** 910 (1996).
[31] K. S. Wilson and H. F. Noller, *Cell* **92,** 131 (1998).
[32] C. Brunel, P. Romby, E. Westhof, C. Ehresmann, and B. Ehresmann, *J. Mol. Biol.* **221,** 293 (1991).
[33] S. Stern, D. Moazed, and H. F. Noller, *Methods Enzymol.* **164,** 481 (1988).
[34] J. Christiansen, J. Egebjerg, N. Larsen, and R. A. Garrett, *Methods Enzymol.* **164,** 456 (1988).
[35] A. Krol and P. Carbon, *Methods Enzymol* **180,** 212 (1989).

*Equipment and Reagents*

*Equipment.* Electrophoresis instrument for sequencing gels.

*Chemicals and Enzymes.* Aniline, CMCT, and Pb(II) acetate are from Merck (Nogent Sur Marne, France); DMS is from Aldrich Chemicals Co. (Milwaukee, WI); DEPC and hydrazine are from Sigma (St. Louis, MO); kethoxal is from U.S. Biochemical Co. (Cleveland, OH); calf intestinal phosphatase and T4 RNA ligase are from Roche Molecular Biochemicals (Meylan, France); avian myeloblastosis virus reverse transcriptase is from Life Science (St. Petersburg, FL); and T4 polynucleotide kinase, [$\gamma$-$^{32}$P]ATP (3200 Ci/mmol), and [5'-$^{32}$P]pCp (3000 Ci/mmol) are from Amersham (Orsay, France).

*Safety Rules Using Chemicals.* Most of the chemical reagents are hazardous and should be used with caution. Dispense all chemicals in a fume hood while wearing protective gloves. Discard DMS, DEPC, and aniline wastes in 1 $M$ sodium hydroxide, hydrazine waste in 2 $M$ ferric chloride, and CMCT wastes in 10% acetic acid.

*Buffers.* Buffer N1: 25 m$M$ sodium HEPES, pH 7.5, 5 m$M$ magnesium acetate, 100 m$M$ potassium acetate. Buffer N2: 50 m$M$ sodium cacodylate, pH 7.5, 5 m$M$ MgCl$_2$, 100 m$M$ KCl. Buffer D2: 50 m$M$ sodium cacodylate, pH 7.5, 1 m$M$ EDTA. Buffer N3: 50 m$M$ sodium borate, pH 8, 5 m$M$ MgCl$_2$, 100 m$M$ KCl. Buffer D3: 50 m$M$ sodium borate, pH 8, 1 m$M$ EDTA. Buffer N4: 50 m$M$ sodium borate, pH 7.5, 5 m$M$ MgCl$_2$, 100 m$M$ KCl. Buffer D2: 50 m$M$ sodium borate, pH 7.5, 1 m$M$ EDTA. RNA loading buffer: 8 $M$ urea, 0.02% xylene cyanol, 0.02% bromphenol blue. DNA loading buffer: 80% deionized formamide, 0.02% xylene cyanol, 0.02% bromphenol blue. RT buffer: 50 m$M$ Tris–HCl, pH 7.5, 20 m$M$ MgCl$_2$, 50 m$M$ KCl. TBE buffer: 0.09 $M$ Tris–borate, pH 8.3, 1 m$M$ EDTA.

*Protocol 1: Direct Detection of Chemical Modifications or Cleavages on End-Labeled Molecules*

This strategy, developed initially by Peattie and Gilbert[36] to probe the conformation of tRNA$^{Phe}$ in solution, is restricted to the detection of cleavages in the RNA: nuclease cuts or modifications that allow subsequent strand scission by an appropriate treatment (see Table I).

*Step 1. End Labeling of 5S rRNA*

For 5' end labeling, the RNA has been dephosphorylated previously at its 5' end[37] and then labeled using [$\gamma$-$^{32}$P]ATP and T4 polynucleotide

[36] D. A. Peattie and W. Gilbert, *Proc. Natl. Acad. Sci. U.S.A.* **77**, 4679 (1980).
[37] M. Shinagawa and R. Padmanabhan, *Anal. Biochem.* **95**, 458 (1979).

kinase.[38] The 3' end labeling is performed with [5'-$^{32}$P]pCp and T4 RNA ligase.[39]

Labeled RNAs are repurified by electrophoresis on a 10% polyacrylamide (0.5% bisacrylamide):8 $M$ urea slab gels. Molecules are stored in the gels at $-20°$ in order to minimize the degradation of the RNA. Before each experiment, the RNA is eluted, precipitated twice with ethanol, and resuspended in water (to get around 50,000 cpm/$\mu$l).

### Step 2. Denaturation and Renaturation of RNA

It is important to verify that a conformationally homogeneous population of molecules is studied, as the RNA is often in contact with denaturing reagents during its purification. Thus, it is worth carrying out a renaturation process before probing the experiments. However, chemical probing can be used to map alternative structures, providing the fact that the conformers have different electrophoretic mobilities. After chemical modification, the coexisting structures are separated on a native polyacrylamide gel, and the modification sites for each conformer are then identified.[40] Renaturation of "native" 5S rRNA is as follows: the RNA is preincubated for 2 min at 90° in doubly distilled water, cooled quickly on ice (2 min), and brought back slowly (20 min) at room temperature in the appropriate buffer.

### Step 3. Chemical Modifications

Chemical probing and lead-induced cleavages are performed on end-labeled RNA (50,000 cpm) supplemented with carrier tRNA in order to get 2 $\mu$g of RNA. Incubation controls that detect nonspecific cleavages in RNA are performed for all conditions. For RNA–protein footprinting experiments, the complex has been formed previously in an appropriate buffer. Usually, some reducing agents such as DTT or 2-mercaptoethanol are added to the buffers.

LEAD-INDUCED HYDROLYSIS. Lead(II) acetate is dissolved in water just before use. Labeled 5S rRNA is incubated for 5 min at 20° in 20 $\mu$l of buffer N1 in the presence of 0.1–30 m$M$ lead(II) acetate. The reaction is stopped by adding 5 $\mu$l of 0.25 $M$ EDTA.

DMS MODIFICATION. Native conditions: 5S rRNA is incubated at 20° for 10–20 min in 20 $\mu$l of buffer N2 in the presence of 1 $\mu$l of DMS (diluted one-eighth in ethanol just before use). Semidenaturing conditions: same procedure as for native conditions, but in buffer D2. Denaturing conditions: reaction is performed at 90° for 1 min in buffer D2.

[38] M. Silberklang, A. M. Gillam, and U. L. RajBhandary, *Nucleic Acids Res.* **4,** 4091 (1977).
[39] A. G. Bruce and O. C. Uhlenbeck, *Nucleic Acids Res.* **4,** 2527 (1978).
[40] A. R. W. Schröder, T. Baumstark, and D. Riesner, *Nucleic Acids Res.* **26,** 3449 (1998).

DEPC MODIFICATION. Native conditions: same procedure as for DMS but in the presence of 2 $\mu$l of DEPC for 15 or 30 min at 20° with occasional stirring. Semidenaturing conditions: same procedure as for native conditions, but in buffer D2. Denaturing conditions: reaction is performed at 90° for 5 min in buffer D2.

ENU ALKYLATION. Native conditions: reaction is performed in 20 $\mu$l of buffer N3 in the presence of 5 $\mu$l of ENU. ENU is prepared extemporaneously as a saturated solution in ethanol by adding some crystals until the solution becomes yellow and saturated. The reaction is performed at 20° for 2 hr or at 37° for 30 min. Semidenaturing conditions: same conditions but in buffer D3. Denaturing conditions: reactions are done at 90° for 2 min in buffer D3.

All the chemical reactions are stopped by ethanol precipitation in the presence of 0.3 $M$ sodium acetate, pH 6.0. In RNA–protein footprinting experiments, the protein is removed by phenol extraction. The RNA is precipitated twice, washed with 80% ethanol, vacuum dried, and redissolved in the appropriate buffer. Modifications of C(N-3) and G(N-7) toward DMS, of A(N-7) toward DEPC, and of phosphates toward ENU are then detected after the subsequent chemical treatment.[10,36]

CLEAVAGES AT G(N-7). The RNA is dissolved in 10 $\mu$l of 1 $M$ Tris–HCl, pH 8, and incubated at 4° for 5 min in the dark in the presence of freshly prepared 10 $\mu$l 200 m$M$ NaBH$_4$. Strand scission is then performed in 10 $\mu$l of 1 $M$ aniline–acetate buffer, pH 4.5 (10 $\mu$l aniline, 93 $\mu$l H$_2$O, and 6 $\mu$l acetic acid), for 10 min in the dark at 60°. The reaction is stopped by adding 100 $\mu$l of 0.1 $M$ sodium acetate, pH 6.0, and precipitation of the RNA by 300 $\mu$l of ethanol. The RNA is precipitated twice, washed with 80% ethanol, and vacuum dried.

CLEAVAGES AT C(N-3). The RNA is dissolved in 10 $\mu$l of 10% hydrazine for 5 min at 4°. Reactions are stopped by ethanol precipitation in the presence of 0.3 $M$ sodium acetate, pH 6.0. The RNA is precipitated twice, washed with 80% ethanol, vacuum dried, and treated with aniline as described earlier.

CLEAVAGES AT POSITION A(N-7). After DEPC modification, the samples are treated with aniline as described earlier.

CLEAVAGE AT PHOSPHOTRIESTER BOND. Cleavages are performed at 50° for 10 min in 10 $\mu$l of 100 m$M$ Tris–HCl, pH 9. Reactions are stopped by ethanol precipitation as described earlier.

### Step 4. Fractionation on Denaturing Polyacrylamide Gels

RNA fragments are resuspended on 6 $\mu$l of RNA loading buffer and sized by electrophoresis on 10% polyacrylamide (0.5% bisacrylamide) or

15% polyacrylamide (0.75% bisacrylamide)–8 $M$ urea gel in TBE 1×. To obtain information on the very first nucleotides adjacent to the end label, a 15% gel is better. Gels should be prerun and run warm to avoid band compression. The migration conditions must be adapted to the length of the RNA, knowing that on a 15% gel, xylene cyanol migrates to 39 nucleotides and bromphenol blue to 9 nucleotides and that on a 10% gel, to 55 and 12 nucleotides, respectively. The cleavage positions are identified by running RNase T1 and formamide ladders in parallel.[4]

RNASE T1 LADDER. End-labeled 5S rRNA (25,000 cpm) is preincubated at 50° for 5 min in 5 $\mu$l containing 1 $\mu$g total tRNA, 20 m$M$ sodium citrate, pH 5, 10 $M$ urea, 0.02% xylene cyanol, and 0.02% bromphenol blue. The reaction is then done with a 0.005 unit of RNase T1 for 5 min at 50°.

FORMAMIDE LADDER. End-labeled 5S rRNA (50,000 cpm) is incubated at 90° for 25 min in the presence of total tRNA (1 $\mu$g) in 3 $\mu$l of formamide.

At the end of the run, the 10% gel is fixed for 5 min in a 10% ethanol, 6% acetic acid solution, transferred to Whatman (Clifton, NJ) 3MM paper, dried, and autoradiographed. The 15% gel is transferred directly on an old autoradiography, covered with Saran wrap plastic film, and exposed at −80° using an intensifying screen. An example of the ENU experiment is shown in Fig. 1A.


### Protocol 2. Detection of Modification and/or Cleavages Using Primer Extension

Primer extension has been developed originally by HuQu et al.[41] for probing the structure of large RNA molecules. Reverse transcription stops dNTP incorporation at the residue preceding a cleavage or a modification at a Watson–Crick position. While carbethoxylation of A(N-7) by DEPC stops reverse transcriptase, DMS methylation of G(N-7) does not stop the enzyme and a subsequent treatment is necessary to induce a cleavage at the site of modification (Table I). Alkylation of RNA by ENU has also to be followed by an alkaline treatment to induce cleavage at the phosphotriester bond (see earlier discussion).


### Step 1. Choice of Primer and 5' End Labeling

The length of the primers usually varies from 10 to 18 nucleotides. This provides sufficient specificity even if the primers are used on a mixture of RNAs. Because natural RNA can present posttranscriptional modifications (such as $m^2G$, $m^6_2A$), which may interfere with reverse transcription, the

[41] L. HuQu, B. Michot, and J. P. Bachellerie, *Nucleic Acids Res.* **11**, 5903 (1983).

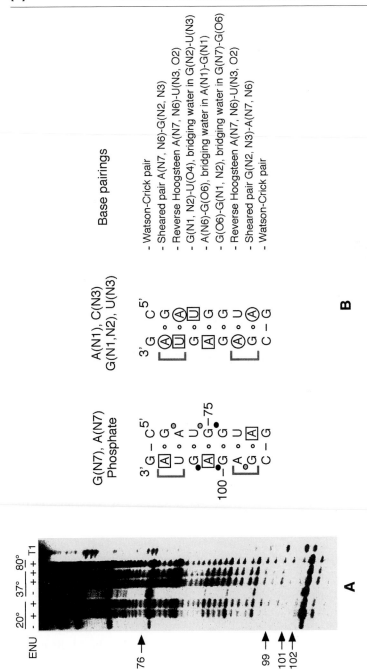

FIG. 1. Correlation between chemical reactivity of bases and phosphates in *E. coli* loop E of 5S rRNA and its X-ray structure. (A) Gel electrophoresis fractionation of products resulting from ENU modification using 5′ end-labeled 5S rRNA. Reactions have been performed on free 5S rRNA under native conditions in the absence (−) or presence (+) of ENU at 20° for 2 hr and at 37° for 30 min. Lane T1 corresponds to the RNase T1 ladder. (B) Reactivity of Watson–Crick positions, of purines at N-7, and of phosphates toward chemical probes taken from Brunel *et al.*[32] Base pairings found in the crystal structure of loop E are from Correll *et al.*[53]

primers have to be chosen accordingly. For long RNA, primers are selected every 200 nucleotides due to gel resolution. Before probing the RNA structure, it is wise to select (1) the concentration of the RNA and the primer and (2) the conditions of hybridization to be used.

The protocols for 5' end labeling and purification of the labeled primer are identical to the procedure described for 5S rRNA. For this RNA, two DNA primers complementary to nucleotides 112–120 (ATGCCTGGC) and 102–120 (ATGCCTGGCAGTTCCCTAC) are used.

### Step 2. Chemical Modifications

For 5S rRNA, chemical probing is performed on 2 $\mu$g of unlabeled RNA. A control of an unmodified RNA is run in parallel in order to discriminate between stops induced specifically by modification and other stops due to stable secondary structures or to spontaneous cleavages at Py-A sequences.

DMS, DEPC, AND ENU MODIFICATIONS. Same conditions as described previously for end-labeled 5S rRNA.

CMCT MODIFICATION. Native conditions: 5S rRNA is incubated at 20° for 20 and 40 min in 20 $\mu$l of buffer N3 in the presence of 5 $\mu$l of CMCT (42 mg/ml in water just before use). Semidenaturing conditions: same dure as for native conditions, but in 20 $\mu$l of buffer D3 for 10 and 15 min.

KETHOXAL MODIFICATION. Native conditions: 5S rRNA is incubated at 20° for 15 or 30 min in 20 $\mu$l of buffer N4 in the presence of 2 $\mu$l of kethoxal (20 mg/ml in 20% ethanol). Semidenaturing conditions: same procedure as for native conditions, but in 20 $\mu$l of buffer D4 for 10 or 20 min. To stabilize the kethoxal adduct, 20 $\mu$l of 0.05 $M$ potassium borate, pH 7.0, is added.

All the reactions are stopped by ethanol precipitation in the presence of 0.3 $M$ sodium acetate, pH 6.0. For RNA–protein footprinting ments, the protein is removed by phenol extraction. The RNA is precipitated twice, washed with 80% ethanol, vacuum dried, and resuspended in the appropriate buffer.

### Step 3. Detection and Identification of Cleavages and Modified Nucleotides

Primer annealing conditions are selected in order to maximize the unfolding of the probed RNA and to minimize RNA degradation. Hybridization of the end-labeled DNA primer (around 100,000 cpm) to the modified 5S rRNA is performed in a total volume of 6 $\mu$l. The mixture is heated at 90° for 1 min and is cooled quickly on ice, followed by 20 min of annealing at room temperature in RT buffer. Primer extension is performed in the presence of 2 units of reverse transcriptase and 2.5 m$M$ of each of the triphosphate deoxyribonucleotides in 15 $\mu$l of RT buffer for 30 min at 37°. The RNA template is then hydrolyzed in 20 $\mu$l of 50 m$M$ Tris–HCl, pH 7.5, 7.5 m$M$ EDTA, and 0.5% SDS in the presence of 3.5 $\mu$l of 3 $M$ KOH at 90° for 3 min followed by an incubation at 37° for 1 hr. The reaction is stopped by adding 6 $\mu$l of 3 $M$ acetic acid, 100 $\mu$l of 0.3 $M$ sodium acetate, and 300 $\mu$l of ethanol. The pellets are washed twice with 80% ethanol and vacuum dried. The samples are resuspended in 10 $\mu$l of the DNA loading buffer, mixed carefully, denatured for 2 min at 90°, and fractionated on a 8% polyacrylamide (0.4% bisacrylamide)–8 $M$ urea gel in TBE 1×. As described earlier, gels should be prerun and run warm to avoid band compression. Migration conditions must be adapted to the length of the RNA to be analyzed, knowing that on an 8% gel, xylene cyanol migrates to 81 nucleotides and bromophenol blue to 19 nucleotides. The cleavage positions are identified by running a sequencing reaction in parallel.[42] The elongation step is performed as described earlier except in the presence of one dideoxyribonucleotide ddXTP (2.5 $\mu M$), the corresponding deoxyribonucleotide dXTP (25 $\mu M$), and the three other deoxyribonucleotides (100 $\mu M$). After running, the gels are dried and autoradiographed at −80° with an intensifying screen for 12 hr.

## Use of Probing Data to Study RNA Structure

Chemical probing had become a useful tool to study RNA folding and the effect of mutations on the RNA structure, to investigate RNA–ligand interactions, and to monitor conformational changes of RNA. The correlation between X-ray structure and chemical modification of different RNAs can be used to unravel the existence of particular structural features in RNA molecules and certain noncanonical base pairs (sheared purine base pairs and Hoogsteen reverse A–U base pair). This knowledge has been used to discover the tRNA structural features in different RNAs such as the 3' domain of turnip yellow mosaic virus,[43,44] the 5' leader region of *Escherichia coli thrS* mRNA,[45] and *E. coli* mtRNA.[46]

[42] F. S. Sanger, S. Nicklen, and A. R. Coulson, *Proc. Natl. Acad. Sci. U.S.A.* **74,** 5463 (1977).
[43] C. Florentz, J. P. Briand, P. Romby, L. Hirth, J. P. Ebel, and R. Giegé, *EMBO J.* **1,** 269 (1982).
[44] K. Rietveld, R. Van Poelgeest, C. W. A. Pleij, J. H. van Boom, and L. Bosch, *Nucleic Acids Res.* **10,** 1929 (1982).
[45] H. Moine, P. Romby, M. Springer, M. Grunberg-Manago, J. P. Ebel, B. Ehresmann, and C. Ehresmann, *J. Mol. Biol.* **216,** 299 (1990).
[46] B. Felden, H. Himeno, A. Muto, J. P. McCutcheon, J. F. Atkins, and R. F. Gesteland, *RNA* **3,** 89 (1997).

The case of 5S rRNA illustrates how chemical probing combined with phylogenetic approaches has provided the basis for the elaboration of secondary and tertiary structure models allowing selection of subdomains for NMR and X-ray studies.

Based on probing experiments, models of 5S rRNA from different origins have been proposed using computer graphic modeling.[32,47] Of particular interest is 5S rRNA containing an internal purine-rich loop E that is part of the site of ribosomal protein L25 in *E. coli*,[48] as well as TFIIIA in *Xenopus laevis*.[49] In both 5S rRNA, the nonreactivity of several Watson–Crick and position N-7 of purines in loop E were interpreted as due to an unusual secondary structure.[32,49] Probing experiments further demonstrated a participation of the magnesium ion in the folding of loop E, as enhanced reactivities to base- and phosphate-specific probes were observed in the absence of magnesium[32,49] (Fig. 1). Structure probing of several *X. laevis* 5S rRNA mutants has also revealed that the intrinsic conformation of loop E is strictly sequence dependent and that the stability of a noncanonical interaction (mispair, base-phosphate interaction, etc.) is very sensitive to nearest-neighbor effects.[50]

Fragments containing *E. coli* loop E have been studied by NMR spectroscopy[51,52] and X-ray crystallography.[53] These studies confirmed that the internal loop has a closed conformation mediated by noncanonical interactions, although the nature of these interactions was not strictly identical to those proposed from chemical probing. Indeed, three novel water-mediated noncanonical base pairs identified in the crystal structure of *E. coli* loop E 5S rRNA could not be diagnosed from chemical probing by itself. However, the chemical reactivity pattern is well correlated with the crystal structure of *E. coli* 5S rRNA, providing some clues that can be used for RNAs for which no X-ray data are available (Fig. 1).

In the crystal structure, loop E of *E. coli* 5S rRNA is closed by two identical motifs, which consist of a Watson–Crick G–C base pair followed by a sheared A–G base pair and a reverse Hoogsteen A–U base pair. This motif was called "a cross-strand A stack" due to the fact that the two

[47] E. Westhof, P. Romby, P. Romaniuk, J. P. Ebel, C. Ehresmann, and B. Ehresmann, *J. Mol. Biol.* **207,** 417 (1989).

[48] S. Douthwaite, A. Christensen, and R. A. Garrett, *Biochemistry* **21,** 2313 (1982).

[49] P. J. Romaniuk, I. leal de Stevenson, C. Ehresmann, P. Romby, and B. Ehresmann, *Nucleic Acids Res.* **16,** 2295 (1988).

[50] I. Leal de Stevenson, P. Romby, F. Baudin, C. Brunel, E. Westhof, C. Ehresmann, B. Ehresmann, and P. J. Romaniuk, *J. Mol. Biol.* **219,** 243 (1991).

[51] B. Wimberly, G. Varani, and I. Tinoco, Jr., *Biochemistry* **32,** 1978 (1993).

[52] A. Dallas and P. B. Moore, *Structure* **15,** 1639 (1997).

[53] C. C. Correll, B. Freeborn, P. B. Moore, and T. A. Steitz, *Cell* **91,** 705 (1997).

adenines, which come from opposite strands, are stacked on each other, inducing a severe kink in the backbone of the A of the reverse Hoogsteen base pair.[53] The reactivity pattern of these residues is unusual, as position N-1 of adenines is highly reactive toward DMS, whereas position N-7 is not. Furthermore, position N-3 of the uridine involved in the reverse-Hoosgteen pair and position N-1 of the guanine in the sheared base pair are not reactive (Fig. 1). In addition, the crystal structure reveals the presence of five metal ions that bind to the major groove of loop E and make contact primarily with purines and nonbridging phosphate oxygens.[53] The presence of these magnesium ions fits with the protection of several phosphates toward ENU (Fig. 1). In particular, two metal ions bridge the phosphoryl groups of G75 and U74 on one strand and of A99 on the other strand explaining their nonreactivity toward ENU. It is interesting to note that analogous phosphates were protected against ENU not only in chloroplast,[54] but also in *X. laevis*[49] 5S rRNAs, indicating the conservation of the tertiary folding of loop E across phylogenetic groups. Using this knowledge, Leontis and Westhof[55] have reinterpretated chemical data obtained on chloroplastic 5S rRNA[54] and have shown that the loop E can adopt a structure similar to *E. coli* 5S rRNA loop E. Furthermore, they have identified possibly similar loop E submotifs in other RNAs that have been probed experimentally, such as 16S rRNA[56] and 4.5S rRNA.[57]

The two noncanonical base pairs, the sheared A–G base pair, and the reverse Hoogsteen A–U base pair are widespread in RNA[58] and both can be detected easily by chemical probing. For example, a sheared A–G base pair was proposed to occur in the GAGA hairpin loop of *X. laevis* rRNA, essentially based on chemical probing and graphic modeling.[47,50] This unusual structure was later confirmed by NMR[59] and crystallographic studies.[60,61] The existence of a tandem of sheared G–A/A–G base pairs was also shown by chemical probing coupled to mutagenesis in the SECIS

[54] P. Romby, E. Westhof, R. Toukifimpa, R. Mache, J. P. Ebel, C. Ehresmann, and B. Ehresmann, *Biochemistry* **27**, 4721 (1988).
[55] N. Leontis and E. Westhof, *J. Mol. Biol.* **283**, 571 (1998).
[56] D. Moazed, S. Stern, and H. F. Noller, *J. Mol. Biol.* **187**, 399 (1986).
[57] G. Lentzen, H. Moine, C. Ehresmann, B. Ehresmann, and W. Wintermeyer, *RNA* **2**, 244 (1996).
[58] S. R. Holbrook, *in* "RNA Structure and Function" (R. W. Simons and M. Grunberg-Manago, eds.), p. 147. Cold Spring Harbor Laboratory Press, Cold Spring Harbor Laboratory, NY, 1998.
[59] H. Heus and A. Pardi, *Science* **253**, 191 (1991).
[60] H. W. Pley, K. M. Flaherty, and D. B. McKay, *Nature* **372**, 68 (1994).
[61] M. Perbandt, A. Nolte, S. Lorenz, R. Bald, C. Betzel, and V. A. Erdmann, *FEBS Lett.* **429**, 211 (1998).

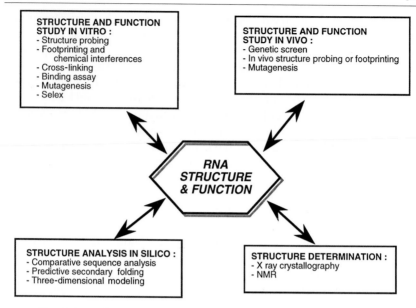

FIG. 2. Strategies used to study RNA structure and relation to its function.

element of the glutathione peroxidase mRNA that mediates selenoprotein translation.[62]

Chemical and enzymatic probing may also provide some indication of
the relative orientation of helices. The presence of RNase V1 overlapping
two contiguous helices or the stacking of G residues at branch point as
indicated by the nonreactivity of their N-7 can help define the coaxiality
of two helices.[23,63] Determination of the region where the phosphate–ribose
backbone is inside or outside of the molecule (using ENU or hydroxyl
radicals) provides additional constraints for RNA folding.[12]

In summary, in the absence of X-ray data, the most successful strategy
(1) to determine the structure of a large RNA in solution and (2) to establish
the essential link between structure and function lies essentially on the
interconnection of various approaches (Fig. 2). By combining probing data
and other information resulting from diverse sources, such as computational
analyses and mutagenesis, the three-dimensional model of RNA can be
derived by graphic modeling. Whatever its degree of complexity and refinement, a model represents a heuristic tool to understand the molecular

[62] R. Walczak, P. Carbon, and A. Krol, *RNA* **4,** 74 (1998).
[63] A. Krol, E. Westhof, R. Bach, R. Luhrmann, J. P. Ebel, and P. Carbon, *Nucleic Acids Res.*
**18,** 3803 (1990).

basis of a given function. Challenging the model can be achieved by testing the effect of site-directed mutagenesis on selected positions critical for the tertiary folding or the RNA function (Fig. 2).

## Perspectives and Conclusion

A growing list of examples underscores the roles that RNA motifs play in many cellular regulatory functions. Furthermore, methodologies have been developed to select new functional RNA targets. These strategies include genomic selex (Gold *et al.*[64]; Brunel and McKeown, personal communication, 2000), genetic methodologies for detecting specific RNA–protein interactions (see elsewhere in this volume), and computational analyses (search for conserved RNA motif in the data bank). A deeper understanding of the RNA structure in solution requires finding new structural features, such as noncanonical base pairs or tertiary interactions, and approaches to the dynamic of RNA folding (kinetic aspect of RNA folding pathway, alternative functional structure, role of metal ion and water). Rules that dictate the folding of a large RNA into a unique functional three-dimensional structure still have to be defined. It might be expected that thermodynamically stable alternative structures or transitory RNA–RNA or RNA–protein interactions may influence the RNA function. This will require the development of new chemical probes that react rapidly (in the order of milliseconds) in order to follow the folding of a large RNA molecule and to study kinetically the footprint of a specific RNA ligand. The development of new artificial RNases will also be another challenge. These reagents may combine a structure-specific domain coupled to a reactive group that will modify or cleave the RNA.

## Acknowledgments

We are grateful to B. Ehresmann and C. Ehresmann for their constant support and for critical reading of the manuscript. We also thank H. Moine, E. Westhof, and R. Giegé for helpful discussions.

[64] B. S. Singer, T. Shattland, D. Brown, and L. Gold, *Nucleic Acids Res.* **25,** 781 (1997).

# [2] Methylene Blue-Mediated Cross-Linking of Proteins to Double-Stranded RNA

*By* ZHI-REN LIU, BRUNO SARGUEIL, and CHRISTOPHER W. J. SMITH

## Introduction

Photochemical cross-linking methods have proven invaluable for identifying proteins that interact with a particular target RNA and for mapping the precise sites of contact between proteins and RNA.[1–3] A typical cross-linking experiment to identify RNA-binding proteins involves incubation of a radiolabeled RNA with a crude cell extract or a purified protein, followed by irradiation under conditions that induce covalent protein–RNA cross-links. The RNA is then degraded and proteins that have become radiolabeled, because they were bound to RNA initially, are identified by SDS–PAGE and autoradiography. The commonest approach is to induce cross-linking with shortwave ultraviolet (UV) light (~254 nm). Various refinements to this basic approach include the use of modified photoreactive base analogs, such as 4-thiouridine, which can be photoactivated at longer wavelength UV, avoiding the generalized damage to RNA bases and amino acids induced by shortwave UV. Site-specific incorporation of radiolabel, often in conjunction with a single photoreactive base, allows the analysis only of proteins that interact at a specific part of an RNA molecule.

In general, UV cross-linking methods are efficient at detecting proteins that interact with single-stranded RNA. In addition, 4-thio U-labeled RNAs have been used to detect interactions between different RNA strands (e.g., Refs. 4, 5). However, UV cross-linking is not efficient at inducing cross-links between proteins and extensively base-paired double-stranded RNA (dsRNA). The reasons for this are not entirely clear, but are probably connected with the fact that dsRNA takes up the A-form helical configuration, with its poorly accessible major groove.[6] Therefore, protein side chains are unlikely to be positioned favorably for cross-linking to the RNA bases.

[1] M. J. Moore and C. C. Query, "RNA : Protein Interactions: A Practical Approach" (C. W. J. Smith, ed.), p. 75. Oxford Univ. Press, Oxford, 1998.

[2] M. M. Hanna, *Methods Enzymol.* **180**, 383 (1989).

[3] J. Rinke-Appel and R. Brimacombe, *in* "RNA : Protein Interactions: A Practical Approach" (C. W. J. Smith, ed.), p. 255. Oxford Univ. Press, Oxford, 1998.

[4] D. A. Wassarman and J. A. Steitz, *Science* **257**, 1918 (1992).

[5] A. Newman, S. Teigelkamp, and J. D. Beggs, *RNA* **1**, 968 (1995).

[6] K. M. Weeks and D. M. Crothers, *Science* **261**, 1574 (1993).

Many of the more interesting and functional RNAs in the cell—snRNAs, rRNAs, RNase P, and self-splicing introns—have secondary structures intermediate between the two extremes of unbase-paired and fully Watson–Crick base-paired A-form dsRNA. These RNAs adopt complex tertiary folds, much like globular proteins. While many of the bases are involved in normal Watson–Crick base pairs, others are involved in tertiary interactions.[7] Therefore, as with fully dsRNA, there may be a number of regions within these highly structured RNAs that are unable to UV cross-link to bound proteins.

In an effort to develop a method that could be used to detect proteins that interact with dsRNA, we turned to the phenothiazinium dye, methylene blue (MB, Fig. 1A). MB is highly photoreactive in response to visible light irradiation and can mediate damage to bases (mainly guanine) as well as many other macromolecules via the production of radicals.[8] Biophysical studies had shown that MB binds to double-stranded DNA at low stoichiometry and with high affinity principally by intercalation.[8] At higher stoichiometries, it can also bind in the grooves of the double helix and via electrostatic interaction with the phosphodiester backbone (phenothiazinium dyes are cationic, Fig. 1A). Moreover, intercalated MB had been shown to mediate protein–DNA cross-links in chromatin, as assayed by the loss of 260 nm absorbance after the phenol extraction of MB and visible light-treated chromatin.[9,10] We reasoned that MB might similarly induce protein–dsRNA cross-links and therefore prove useful in identifying proteins that bind to dsRNA. Consistent with these expectations, we found that irradiation with visible light in the presence of MB led to the formation of RNA–protein cross-links with a marked preference for dsRNA. Frequently, these interactions were undetectable by UV cross-linking.[11–13] MB cross-linking was inefficient at inducing protein–ssRNA or RNA–RNA cross-links between Watson–Crick base-paired strands.[11] The preference for dsRNA makes MB a useful investigative tool that is complementary in specificity to UV. Although we employed known dsRNA-binding proteins to establish the MB cross-linking method, we have subsequently used the approach to identify novel proteins in nuclear extracts that interact at specific intra- and intermolecular RNA duplexes.[12,13] The basic MB cross-

[7] G. Varani and A. Pardi, in "RNA-Protein Interactions" (K. Nagai and I. Mattaj, eds.), p. 1. Oxford Univ. Press, Oxford, 1994.
[8] E. M. Tuite and J. M. Kelly, J. Photochem. Photobiol. B 21, 103 (1993).
[9] R. Lalwani, S. Maiti, and S. Mukherji, J. Photochem. Photobiol. B 7, 57 (1990).
[10] R. Lalwani, S. Maiti, and S. Mukherji, J. Photochem. Photobiol. B 27, 117 (1995).
[11] Z. R. Liu, A. M. Wilkie, M. J. Clemens, and C. W. J. Smith, RNA 2, 611 (1996).
[12] Z. R. Liu, B. Laggerbauer, R. Luhrmann, and C. W. J. Smith, RNA 3, 1207 (1997).
[13] Z. R. Liu, B. Sargueil, and C. W. J. Smith, Mol. Cell. Biol. 18, 6910 (1998).

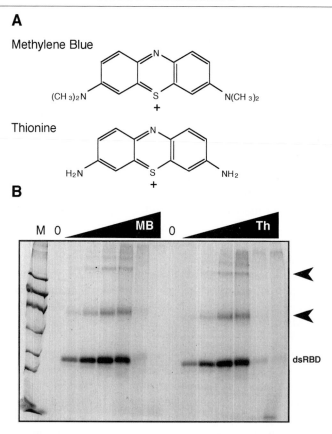

FIG. 1. dsRNA–protein cross-linking mediated by phenothiazinium dyes. (A) Structure of methylene blue and thionine. (B) Cross-linking of labeled adenovirus VAI RNA to a GST fusion protein containing a double-stranded RNA-binding domain from the staufen protein. Each binding reaction contained 1.5 ng labeled RNA (2.5 n$M$) and 300 ng recombinant protein (0.8 $\mu M$). Methylene blue (MB) or thionine (Th) was add to 0, 0.1, 0.3, 1, 3, 10, and 30 ng/ $\mu$l. The position of the cross-linked protein is indicated by "dsRBD" to the right of the gel. Arrowheads indicate the presence of radiolabeled proteins whose size corresponds precisely to dimers and trimers of the GST-dsRBD protein. Note that cross-linking with both MB and Th in this experiment is optimal between 0.3 and 3 ng/$\mu$l.

linking technique is similar to UV cross-linking both in its simplicity and in the propensity to induce damage to both nucleic acids and proteins. We have introduced some refinements to the method to try and reduce undesirable side reactions. This article details the basic experimental details for MB cross-linking followed by a commentary on the technical aspects and limitations of the approach and a discussion of applications.

## Materials and Reagents

Light: a 1.2-m-long 60-W fluorescent tube light

Standard apparatus for SDS–PAGE, autoradiography, and/or phosphorimaging

Microtiter plates

Stock aqueous solution of methylene blue ~2.5 mg/ml. The precise concentration is determined by absorbance at 665 nm; $\varepsilon_{665nm} = 81{,}600 \text{ cm}^{-1} \cdot M^{-1}$. Store in ~1-ml aliquots at $-20°$ in light-tight tubes.

RNA labeled to high specific activity using either one or more $[\alpha\text{-}^{32}\text{P}]$NTPs for body-labeled RNAs, $[\gamma\text{-}^{32}\text{P}]$ATP for 5′ end-labeled or site specifically labeled RNAs produced by oligonucleotide-mediated ligation (1), or $[5'\text{-}^{32}\text{P}]$pCp for 3′ end-labeled RNAs.

10× binding buffer: The composition of binding buffer depends on the RNA and protein under investigation. We have found that MB cross-linking is compatible with the monovalent salt and $MgCl_2$ concentrations commonly used in such buffers (unaffected by 150 m$M$ NaCl or 10 m$M$ MgCl$_2$[11]). It may be advisable to avoid EDTA in binding buffers as this reduces the excited triplet state of MB.[8] This possibly could lead to reduced cross-linking yields and/or nonspecific damage by the generation of radicals. As a *rough* guide, a 1× binding buffer might typically contain 10 m$M$ Tris–HCl, pH 7.5, 20–100 m$M$ KCl, 2 m$M$ MgCl$_2$, and 0.5 m$M$ dithiothreitol (DTT). This will vary according to the interaction under investigation.

Ribonucleases: RNase A (10 mg/ml), RNase T1 (100,000 U/ml), and RNase V1 (700 U/ml)

5× SDS sample buffer: 10% (w/v) SDS, 1 $M$ Tris–HCl, pH 6.8, 50% (v/v) glycerol, 0.03% (w/v) bromphenol blue. Store at room temperature. Add 2-mercaptoethanol to 5% (v/v) 2-mercaptoethanol immediately before use.

Ascorbic acid: Prepared as a small volume of stock 1 $M$ aqueous solution by dissolving directly in RNase-free water. The pH of the solution is not adjusted as aqueous ascorbate is oxidized rapidly at higher pH. The solution is either made fresh each time or stored at $-70°$ for short periods (less than a month).

## Basic Method

1. The binding reaction is typically assembled in a volume of ~10 $\mu$l in the wells of a microtiter plate. The required quantities of RNA and protein will vary according to the affinity of the RNA–protein interaction. For instance, using a recombinant dsRNA-binding domain from staufen, we typically use 0.4 $\mu M$ protein with 0.01 $\mu M$

RNA.[11] Excess protein is required to maximize signal, as the estimated $K_d$ for binding is ~0.1 $\mu M$. With crude extracts, we typically use 10 $\mu g$ total protein. The incubation may be only 1–2 min at room temperature for a simple bimolecular association. Longer incubations may be necessary for the assembly of larger complexes such as spliceosomes.

2. Place the microtiter plate on a bed of ice that is covered with a sheet of aluminum foil (to maximize the subsequent illumination). Alternatively, use an aluminum foil-covered adjustable platform in a cold room. Add MB to a final concentration of ~0.1 to 3 ng/$\mu$l from a 20× stock solution. Mix the MB into the binding reaction by pipetting up and down. Then add ascorbic acid or another quenching reagent to an appropriate concentration, if desired (see later). Add ascorbic acid along with a twofold molar excess of Tris–HCl to maintain the pH of the sample. Mix the quenching reagent into the binding reaction by pipetting up and down.

3. Place the microtiter plate 1–5 cm below the light source and illuminate for 5–20 min.

4. During illumination, prepare a 10× RNase mix (5 $\mu g/\mu$l RNase A, 3 U/$\mu$l RNase T1, 0.35 U/$\mu$l RNase VI). Add 1 $\mu$l of RNase mix and incubate at 37° for 30 min. This and subsequent steps can be carried out in the microtiter plate or after transferring the reactions to 0.5-ml microcentrifuge tubes.

5. Add 4 $\mu$l of 5× SDS sample buffer to each sample. Samples can be stored temporarily at −20° before electrophoresis. Denature samples immediately prior to electrophoresis by incubating in a boiling water bath for 3 min or in a heating block or oven at 85° for 5 min. Although not essential, it is often useful to stain the gel after electrophoresis with Coomassie blue and then partially destain. This allows a visual inspection for the equal loading of samples and also for any signs of protein–protein cross-linking induced by the MB treatment. Finally, expose the dried gel to a phosphorimager screen or to X-ray film.

### Variables in Basic Protocol and Properties of MB Cross-Linking

*MB Concentration.* We typically use a final concentration in the range of 0.1–3 ng $\mu l^{-1}$. The optimal concentration may vary considerably according to whether purified protein or a crude cell extract is being used. Figure 1B shows titrations of both MB and the related phenothiazinium dye thionine into cross-linking reactions containing a recombinant dsRNA-binding domain. Note that cross-linking of the ~38-kDa protein is optimal at 0.3–3 ng/$\mu$l. Note also that additional bands that correspond precisely to dimers and trimers of the protein appear at higher concentrations. These presumably result from direct protein–protein cross-linking as they are detected even with RNA probes that are only large enough to bind a single protein. At higher concentrations ($\geq$10 ng $\mu$l$^{-1}$) the RNA–protein cross-linking yield diminishes due to the inhibition of RNA–protein interactions, presumably by high levels of intercalated MB (Ref. 11, Fig. 1B). The optimal concentration of MB is therefore often a compromise between the maximal detection of the interaction of interest and the need to reduce unwanted reactions.

*Suppressing Side Reactions.* MB is highly photoreactive, and although its ability to induce RNA–protein cross-links shows a high preference for dsRNA, it also induces many unwanted side reactions, such as the protein dimers and trimers seen in Fig. 1B. We have also observed that the MB concentrations used in our cross-linking reactions lead to substantial damage to proteins in cell extracts. This is indicated by the complete loss of some high molecular weight protein bands from Coomassie blue-stained gels. Likewise, in cross-linking reactions using radiolabeled RNA, we have often observed substantial amounts of radiolabeled material remaining in the wells of the SDS gel. For example, Fig. 2 (lane 1) shows an experiment in which a novel 65-kDa protein was detected that cross-links at the duplex formed between U1 snRNA and a radiolabeled pre-mRNA.[13] Note that in addition to the p65 band there is a substantial amount of radiolabeled material stuck in the well. This presumably arises from large aggregates of cross-linked protein and RNA. Because dsRNA–protein cross-linking is mediated by intercalated MB, whereas many of the other photochemical reactions are presumably mediated by solution phase dye, we reasoned that it may be possible to screen for compounds that preferentially suppress the unwanted side reactions. The compounds we have tested include ascorbic acid, histidine, semicarbazide, and 1,4-diazabicyclo[2.2.2]octane (DABCO). Semicarbazide (Fig. 2, lanes 12–16) and DABCO, which quench singlet oxygen, inhibited both specific dsRNA–protein cross-linking and the side reactions at the same concentrations. However, histidine (Fig. 2, lanes 7–11) and, to a greater extent, ascorbic acid (Fig. 2, lanes 2–6) suppressed the formation of aggregated material and, at optimal concentrations, led to an increase in the specific signal, although at the highest concentrations they also inhibited RNA–protein cross-linking. Over a similar concentration range, Coomassie blue staining of gels showed that general damage and cross-linking of proteins were also reduced. Note also that the bands at ~30–35 kDa show a higher degree of sensitivity to ascorbate and histidine than p65. For various reasons, we suspect that these are SR proteins—ssRNA-binding proteins involved in splicing. These are detected with much higher efficiency by UV cross-linking. We suspect that their

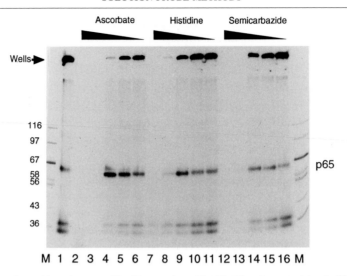

FIG. 2. Quenching of nonspecific side reactions. The 5′ splice site-containing [$\alpha$-$^{32}$P]GTP-labeled RNA GC + DX/XhoI was incubated with 30% HeLa nuclear extract for 15 min before being subjected to MB cross-linking. Under these conditions, a 65-kDa protein is detected that interacts with the duplex formed between the 5′ splice site and U1 snRNA (lane 1).[13] Ascorbate, histidine, and semicarbazide were added to 50, 10, 2, 0.4, and 0.08 m$M$ in lanes 2–6, 7–11, and 12–16, respectively. Lanes M, protein markers, sizes in kDa shown to the left. Note that in lane 1 there is a significant amount of radiolabeled material stuck in the wells of the gel. We attribute this to the formation of large cross-linked RNA–protein aggregates. Note that ascorbate preferentially reduces the amount of radioactive material in the wells while the amount of cross-linked p65 increases (lanes 4–6). At the highest concentrations ($\geq$10 m$M$), all of the tested compounds fully inhibited cross-linking. In experiments containing HeLa nuclear extracts, the optimal ascorbate concentration was usually in the region of 2 m$M$.

detection by MB may be due to the purine-rich nature of their RNA-binding sites. It is possible that MB may be able to mediate some ssRNA–protein cross-linking by stacking between purine bases. Because ascorbate causes complete inhibition of dsRNA–protein cross-linking at higher concentrations (~10 m$M$), it is again advisable to carry out a preliminary titration to find the most effective dose. While ascorbate may not be a complete panacea for all unwanted side reactions, it appears to be a useful additive to many reactions.

*Light Source and Duration of Illumination.* The light source is probably not crucial. We simply use the standard domestic fluorescent light tubes that are mounted at the back of laboratory benches. According to the output of the light, the duration of illumination may need to be adjusted. We usually find that cross-linking plateaus within 5 min.

*Choice of Labeled Nucleotide and RNases.* For long, purely double-stranded RNAs, the choice of labeled nucleotide and RNases is probably not very significant. However, for short specific duplexes it is worth considering the optimal combination of labeled bases and RNases. The ability to detect a cross-linked protein depends on the ability of nucleotides in close proximity to the protein to mediate photochemical cross-linking and whether a labeled phosphate will be retained with the cross-linked fragment after RNase digestion. MB cross-linking appears to show no major base specificity (see later) so the choice of radiolabel and RNase should be the only important consideration. For example, the p65 protein that cross-links to the short intermolecular duplex formed between pre-mRNA 5' splice sites and the 5' end of U1 snRNA (Fig. 2) could be detected with RNAs labeled with $[\alpha\text{-}^{32}P]$GTP or ATP, but not UTP or CTP. It is possible that cross-linking could have been observed with the labeled UTP or CTP if we had used a cocktail of RNases that did not cleave as frequently. Our usual cocktail contains RNase A (cleaves ssRNA to leave Pyp 3'), RNase T1 (ssRNA to leave Gp 3'), and RNase V1 (dsRNA to leave 5' pNoh).

*MB Cross-Linking of Purified Complexes.* UV cross-linking has proven useful in identifying the higher order complexes within which certain RNA–protein cross-links occur. For instance, UV cross-linking can be carried out easily on complexes fractionated by glycerol gradient or gel filtration (e.g., Ref. 14). It has even been carried out on complexes separated by native gel electrophoresis by irradiating the gel slice excised from the native gel.[15] We have successfully carried out MB cross-linking of glycerol gradient-separated complexes.[13] The main technical obstacle here was that the glycerol gradient-separated fractions were in a larger volume (0.4 ml) and, being more dilute, required concentration before electrophoresis. The samples could not be accommodated within the wells of a microtiter plate: they were placed on the inverted lid of a microtiter plate after the addition of MB to 2 ng/$\mu$l. After cross-linking, dilute samples need to be concentrated. This can be achieved by a number of methods and the concentration step can be carried out before or after ribonuclease digestion. With glycerol gradient fractions, we digested with ribonucleases first and then added (in a 15-ml polypropylene tube) 4 volumes of methanol (1.6 ml), 1 volume of chloroform (0.4 ml), and 3 volumes of water (1.2 ml) sequentially. The mixture was vortexed thoroughly after each addition. After bench-top centrifugation for 5 min, the upper organic layer was discarded, with care being taken to retain the protein-containing interface and the small volume of

[14] D. Staknis and R. Reed, *Mol. Cell. Biol.* **14,** 7670 (1994).
[15] D. L. Black, R. Chan, H. Min, J. Wang, and L. Bell, *in* "RNA:Protein Interactions: A Practical Approach" (C. W. J. Smith, ed.), p. 109. Oxford Univ. Press, Oxford, 1998.

aqueous phase. Finally, 3 volumes (1.2 ml) of methanol was added and the protein precipitate was pelleted by microcentrifugation for 10 min. The protein pellet can be resuspended in SDS–PAGE loading buffer. Alternative methods for concentration would be to use centrifugal microconcentrators or ethanol precipitation followed by ribonuclease treatment.

In principle, MB cross-linking of gel-filtered complexes also should not be difficult. In contrast, combining native gel electrophoresis with MB cross-linking may be more challenging. One would need to infiltrate the dye into the acrylamide gel slice prior to illumination.

*Immunoprecipitation of MB Cross-Linked Proteins.* Identification of cross-linked proteins by immunoprecipitation with candidate antibodies is a commonly used approach. We have found that this approach can be applied to MB cross-linked proteins in the same way as it is commonly applied to UV cross-linking reactions.[12,13] Although we cannot rule out the possibility that MB may cause damage to the epitopes recognized by antibodies, in the cases that we have investigated, any such damage must have been well below 100%.

*Other Phenothiazinium Dyes.* MB is a member of a family of related dyes that differ only in the number of methyl substituents. We have tested thionine, the fully unmethylated member of the family (Fig. 1A), on the basis that it is reported to show the lowest activity in inducing base damage to DNA.[8] We found that thionine behaved indistinguishably from MB in RNA–protein cross-linking, inducing specific cross-links as well as protein–protein cross-linking (Fig. 1B). Nevertheless, if an attempt is going to be made to map sites of cross-linking by, for instance, reverse transcription using the cross-linked RNA–protein complex as substrate, thionine might be preferable because there should be fewer nonspecific reverse transcriptase arrests due to base damage.

*Mechanism of MB Cross-Linking: Lack of Base Specificity.* Much evidence has pointed toward guanine as being the major target of MB-mediated photoreactive damage in DNA (reviewed in Ref. 8). If dsRNA–protein cross-linking were mediated similarly in a highly base-specific fashion, the ability to detect interactions of proteins with some short RNA duplexes of specific defined sequences might be restricted severely. To address this issue, a number of cross-linking experiments were carried out using model dsRNA-binding proteins (RED-1 editing enzyme and the recombinant staufen GST-dsRBD) and a model RNA hairpin in which one arm of the stems was composed purely of pyrimidines and the complementary arm only of purines. We found that the efficiency of cross-linking was comparable (10–20%) when the stem was labeled with any of the four individual [$\alpha$-$^{32}$P]NTPs. Likewise, cross-linking of the GST-dsRBD was comparable to a hairpin consisting solely of A–U base pairs as to a pure G–C base pair hairpin. Thus, cross-linking chemistry appears to show no strong base specificity, which implies that the base composition of any particular RNA substrate should not restrict the ability to detect RNA–protein interactions.

In general, it appears that the more base paired the RNA, the greater is the likelihood that MB will detect proteins that are invisible to UV. Nevertheless, the mechanism of MB-mediated cross-linking, and thus the range of RNA–ligand interactions that are amenable to detection using this method, remains unclear. For instance, it is possible that efficient cross-linking requires contact of the protein with the minor groove of the RNA. In limited support of this conjecture, nuclear magnetic resonance analysis of the staufen dsRBD, which we have used in developing the MB cross-linking method, in complex with an RNA hairpin, demonstrates that this domain has a direct contact with the minor groove via a conserved amino acid loop (A. Ramus, D. St. Johnston, and G. Varani, personal communication, 1998). In contrast, attempts to cross-link the bacteriophage R17 coat protein to a high-affinity RNA ligand were unsuccessful, despite control filter-binding experiments showing that methylene blue itself was not disrupting the RNA–protein interaction. The structure of this RNA–protein complex at 2.7 $\text{Å}$ [16] shows no interactions between the protein and the minor groove of the short RNA stem–loop. Possibly, detection of RNA–protein contacts by MB cross-linking may require a particular apposition of amino acid side chains with the minor groove of the structured RNA.

### Applications of MB Cross-Linking

To date, we have applied MB cross-linking to two separate problems. The first application, for which we originally devised the procedure, was to identify components of spliceosomes that had been blocked prior to step 2 of pre-mRNA splicing by obstructing hairpin structures. In some mammalian alternatively spliced introns the branch point is located hundreds of nucleotides from the 3′ splice site. The 3′ splice site is invariably the first AG downstream of the branch point and appears to be located by a scanning type of mechanism. The analogy with translational scanning extends to the effects of stable hairpin structures between the branch point and the 3′ splice site, which prevent step 2 of splicing. We deployed MB cross-linking to detect components of spliceosomes that had been blocked by such structures. We successfully identified the 116-kDa component of U5 snRNP in contact with the hairpins.[12] This shows that MB cross-linking can be used to detect a protein that is not necessarily stably bound to a

[16] K. Valegard, J. B. Murray, N. J. Stonehouse, S. van den Worm, P. G. Stockley, and L. Liljas, *J. Mol. Biol.* **270,** 724 (1997).

specific RNA, but that is simply in close proximity. In a second series of experiments we detected a 65-kDa protein that interacts at the duplex between a pre-mRNA 5' splice site and U1 snRNA, which forms transiently during splicing.[13] This is a short intermolecular duplex, and the protein detected had not been observed in previous UV cross-linking experiments, despite the fact that interactions at the 5' splice site had been investigated intensively over the course of several years. MB cross-linking could also be used to look for proteins in cell extracts that interact with more conventional dsRNA ligands. In many cases, it is likely that many of the same proteins may be detected by MB as by UV cross-linking. This may especially be the case with RNAs that are highly structured yet contain some unpaired bases. For instance, MB cross-linking has been used to detect the interaction of cellular proteins with highly structured poliovirus internal ribosome entry segments. Although a number of apparently identical proteins were identified by MB and UV cross-linking, some proteins were only detected using MB (T. A. A. Pöyry and R. J. Jackson, personal communication). In principle, MB cross-linking might be able to detect the interaction of ligands other than proteins with RNA. We showed previously that MB does not induce cross-links between Watson–Crick base-paired RNA strands.[11] We have subsequently attempted to induce cross-links between different RNA segments that are known to have tertiary contacts via minor groove docking interactions (e.g., in group I and group II introns). However, to date, we have not been able to detect any such tertiary RNA–RNA interactions via MB cross-linking.

## Concluding Remarks

The major advantages and disadvantages of MB cross-linking are similar to those of shortwave UV cross-linking. MB treatment can be relatively nonspecific, damaging both amino acid side chains and nucleic acid bases that are not involved directly in the RNA–protein interaction of interest. The use of quenchers such as ascorbic acid and/or other phenothiazinium dyes such as thionine may help reduce such side reactions. The main advantages of the MB cross-linking procedure are its simplicity and the fact that it detects a complementary array of interactions to UV cross-linking, sometimes detecting interactions that are invisible to UV. Many experimental elaborations that are commonly used with UV cross-linking (site-specific radiolabeling, immunoprecipitation of cross-linked proteins, cross-linking of glycerol gradient-purified fractions[12,13]) can also be applied to MB cross-linking. In fact, any RNA ligand that has been synthesized for UV cross-linking could also be tested by MB cross-linking.

## Acknowledgments

B.S. is a member of the CNRS and was the recipient of a Marie Curie Research Training Fellowship from the European Union and a NATO Fellowship during the course of this investigation. This work was supported by a grant from the Wellcome Trust (040375) to C.W.J.S.

# [3] Structure and Distance Determination in RNA with Copper Phenanthroline Probing

*By* THOMAS HERMANN and HERMANN HEUMANN

## Introduction

Complexes of redox-active metals serve as chemical nucleases for probing secondary and tertiary structure of RNA molecules.[1-4] Among these metal complexes, 1,10-phenanthroline-copper (OP-Cu) is especially useful, as it cleaves RNA with high specificity in ordered single-stranded regions.[5-8] The cleavage reaction proceeds in an oxidative attack of OP-Cu on the ribose moiety of nucleotides followed by strand scission.[1-4,9,10] The character of the reactive species has not been finally established.[4] However, there is evidence that OP-Cu may generate diffusible hydroxyl radicals that attack riboses of nucleic acid by H-abstraction.[4,11-14] The structure specificity in OP-Cu-mediated cleavage of RNA originates from specific interactions of

[1] D. S. Sigman and C. B. Chen, *Annu. Rev. Biochem.* **59,** 207 (1990).

[2] C. S. Chow and J. K. Barton, *J. Am. Chem. Soc.* **112,** 2839 (1990).

[3] D. M. Perrin, A. Mazumder, and D. S. Sigman, *Progr. Nucleic Acid Res. Mol. Biol.* **52,** 23 (1996).

[4] W. K. Pogozelski and T. D. Tullius, *Chem. Rev.* **98,** 1089 (1998).

[5] G. J. Murakawa, C. B. Chen, W. D. Kuwabara, D. P. Nierlich, and D. S. Sigman, *Nucleic Acids Res.* **17,** 5361 (1989).

[6] Y.-H. Wang, S. R. Sczekan, and E. C. Theil, *Nucleic Acids. Res.* **18,** 4463 (1990).

[7] A. Mazumder, C. B. Chen, R. Gaynor, and D. S. Sigman, *Biochem. Biophys. Res. Commun.* **187,** 1503 (1992).

[8] T. Hermann and H. Heumann, *RNA* **1,** 1009 (1995).

[9] T. E. Goyne and D. S. Sigman, *J. Am. Chem. Soc.* **109,** 2846 (1987).

[10] O. Zelenko, J. Gallagher, Y. Xu, and D. S. Sigman, *Inorg. Chem.* **37,** 2198 (1998).

[11] L. M. Pope, K. A. Reich, D. R. Graham, and D. S. Sigman, *J. Biol. Chem.* **257,** 12121 (1982).

[12] H. R. Drew and A. A. Travers, *Cell* **37,** 491 (1984).

[13] T. B. Thederahn, M. D. Kuwabara, T. A. Larsen, and D. S. Sigman, *J. Am. Chem. Soc.* **111,** 4941 (1989).

[14] M. Dizdaroglu, O. I. Aruoma, and B. Haliwell, *Biochemistry* **29,** 8447 (1990).

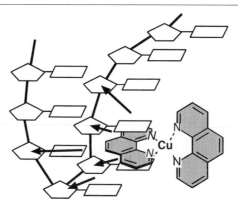

FIG. 1. The tetrahedral metal complex OP-Cu binds to nucleic acid and cleaves nucleotides surrounding the binding site (arrows). For RNA, the cleavage specificity of OP-Cu for single-stranded stacked regions was attributed to a "bookmarking" binding mode in which a single phenanthroline ligand of the metal complex partially intercalates into a stack of bases.

the metal complex with the nucleic acid.[8,15] For double-stranded helical conformations, the differences in reactivity found between A form RNA and B form DNA suggest that OP-Cu binds in the minor groove of the helix. While double-stranded DNA in B form is cleaved with high efficiency, A form nucleic acid, RNA as well as DNA, is inert to OP-Cu, probably due to the shallow shape of the minor groove in the A conformation, which prevents OP-Cu binding.[7,16]

Single-stranded RNA often forms highly ordered structures with extensive stacking interactions between neighboring bases. OP-Cu cleaves preferably within stacks of single-stranded nucleotides, as revealed in cases where the three-dimensional structure of the RNA is known from X-ray crystallography or nuclear magnetic resonance spectroscopy.[6,8] It has been proposed that OP-Cu binds to nucleotide stacks in RNA, like a bookmark, by partial intercalation of a single phenanthroline ligand between two adjacent bases (Fig. 1).[8] This binding mode is characterized by two strong cuts flanked by bands of decreasing intensity observed frequently in RNA cleavage patterns. The pattern is most likely generated by diffusible hydroxyl radicals, which radiate from the bound OP-Cu and cleave the RNA backbone. Because the radicals are "diluted" with increasing distance from the radical source, the cleavage intensity contains distance information. The observed signal intensity at a cleaved nucleotide roughly displays an inverse proportionality to the distance ($1/r$) between the radical source and the attacked

[15] D. S. Sigman, A. Spassky, S. Rimsky, and H. Buc, *Biopolymers* **24,** 183 (1985).
[16] G. J. Murakawa and D. P. Nierlich, *Biochemistry* **28,** 8067 (1989).

ribose. In the model system of the tRNA$^{Phe}$ anticodon loop, it has been shown that the $1/r$ dependence of the cleavage efficiency on the distance $r$ of the cleaved nucleotide to the OP-Cu-binding site holds quantitatively.[8]

Based on the binding specificity of OP-Cu for stacked single-stranded nucleotides, OP-Cu probing provides information on the location of such regions in RNA. In addition, the $1/r$ dependence of the OP-Cu cleaving pattern can be used to determine relative distances between cleaved nucleotides. However, this is only feasible in cases where the cleavage pattern is generated from a single OP-Cu-binding site. Multiple overlapping OP-Cu-binding sites give rise to cleavage patterns that cannot be deconvoluted.

OP-Cu has been used successfully for secondary structure probing of RNA,[5-8] footprinting investigations on RNA/protein complexes,[16-19] and sequence-specific cleavage by oligonucleotide-linked OP-Cu.[19-21] This article describes the use of OP-Cu as a probe for RNA secondary structure and, if applicable, for determining relative distances between nucleotides within RNA folds.

## Materials

### Enzymes and Reagents

T4 RNA ligase and RNase T1 are from Roche, Penzberg, [5'-$^{32}$P]pCp is from Amersham Pharmacia, Freiburg

tRNA carrier is commercially available yeast tRNA$^{Phe}$ from Boehringer Mannheim

XC/BPB dye mix: 0.02% (w/v) xylene cyanol, 0.02% (w/v) bromphenol blue, 50% (v/v) glycerol

### Buffers

3'-end-labeling buffer: 50 m$M$ $N$-2-hydroxyethylpiperazine-$N'$-2-ethanesulfonic acid (HEPES) (pH 7.5), 10 m$M$ MgCl$_2$, 10% (v/v) dimethyl sulfoxide (DMSO), 3 m$M$ dithioerythritol (DTE)

1× TBE: 90 m$M$ Trizma base, 90 m$M$ boric acid, 1 m$M$ EDTA; pH 8.3

[17] P. Darsillo and P. W. Huber, *J. Biol. Chem.* **266**, 21075 (1991).
[18] L. Pearson, C. B. Chen, R. P. Gaynor, and D. S. Sigman, *Nucleic Acids Res.* **22**, 2255 (1994).
[19] D. J. Bucklin, M. A. van Waes, J. M. Bullard, and W. E. Hill, *Biochemistry* **36**, 7951 (1997).
[20] J. Sun, J.-C. François, R. Lavery, T. Saison-Behmoaras, T. Montenay-Garestier, N. T. Thuong, and C. Hélène, *Biochemistry* **27**, 6039 (1988).
[21] C. B. Chen and D. S. Sigman, *J. Am. Chem. Soc.* **110**, 6570 (1988).

MG extraction buffer: 500 m$M$ ammonium acetate, 10 m$M$ magnesium acetate, 0.1 m$M$ EDTA, 0.1% (w/v) sodium dodecyl sulfate (SDS)
Hydroxide hydrolysis buffer: 50 m$M$ NaHCO$_3$/Na$_2$CO$_3$, 1 m$M$ EDTA; pH 9.2
RNase T1 digestion buffer: 20 m$M$ sodium acetate, 7 $M$ urea, 1 m$M$ EDTA; pH 5.0
OP-Cu probing buffer: 50 m$M$ Tris–HCl, 10 m$M$ MgCl$_2$, 50 m$M$ NaCl; pH 8.0

## Methods

### Principle of Method

The 3'-end-labeled RNA is treated with the OP-Cu reagent under "single hit" conditions where each RNA molecule in the sample receives statistically less than one cut. The resulting RNA fragments are fractionated electrophoretically by size on polyacrylamide gels. The length of the separated fragments corresponds to the distance of the strand scission from the terminal radioactive label. Comparison of the fragment sizes with RNA "ladders," obtained by partial digestion with G-specific RNase T1 and by partial alkaline hydrolysis, allows to identify the positions of the cleaved nucleotides in the RNA molecule. Thus, stacked single-stranded regions in the RNA can be located directly by OP-Cu probing.

In cases where a cleavage signal can be attributed to a single OP-Cu-binding site, further quantification of band intensities may be used to determine relative distances between the cleaved nucleotides. Band intensities of the cleaved fragments are obtained by the integration of photometric scans of the X-ray autoradiograph or determined directly from PhosphoImager scans. Relative internucleotide distances $r$ are calculated from the band intensities $I$ using a simple inverse relation with a scalable proportionality factor $q$: $I = q/r$.

### Probing Experiment

*1. 3'-End-Labeling of RNA.* Purified RNA is 3'-end-labeled with [$^{32}$P]pCp according to England *et al.*[22] The RNA precipitate is resuspended in water at a concentration of 1 mg/ml and kept on wet ice. For the labeling reaction, 1 $\mu$g of RNA (1 $\mu$l) is added to 30 $\mu$l labeling buffer containing 100 $\mu M$ adenosine triphosphate (ATP), 10 $\mu$g/ml bovine serum albumin (BSA), 5 U of T4 RNA ligase, and 100 $\mu$Ci of [5'-$^{32}$P]pCp (10 mCi/

[22] T. E. England, A. G. Bruce, and O. C. Uhlenbeck, *Methods Enzymol.* **65,** 65 (1980).

ml ~ 3000 Ci/mmol). After incubation at 15° for 6 hr (time may need to be adapated for different RNAs), RNA is precipitated by adding 300 m$M$ sodium acetate (pH 5.4), 5 $\mu$g carrier tRNA, followed by 3 volumes of ice-cold ethanol. To complete precipitation, the mixture is cooled on dry ice for 5 min and centrifuged at 12,000–15,000$g$ for 5 min.

2. *Purification of Labeled RNA.* The pellet of [32]P-labeled RNA from ethanol precipitation is rinsed with 70% (v/v) ethanol, air dried, and resuspended by vortexing in 5 $\mu$l water. After adding 5 $\mu$l of xylene cyanol (XC)/bromphenol blue (BPB) dye mix solution, the sample is loaded onto a polyacrylamide–8 $M$ urea gel[23] (40 × 20 cm; thickness, 0.5 mm). Depending on the size of the RNA, gels of 10 or 20% (w/v) polyacrylamide [19% (w/v) acrylamide and 1% (w/v) $N,N'$-methylenebisacrylamide] in 1× TBE buffer are used. Electrophoresis is carried out at 500–600 V in 1× TBE buffer until the XC dye has moved to ~10 cm from the bottom of the gel (5–8 hr). An estimation can be made, considering that BPB and XC dyes migrate on 10% gels with RNAs of ~10 and ~60 nucleotides (nt), on 20% gels with RNAs of ~5 and ~25 nt, respectively.[24]

The band containing the labeled RNA is identified by autoradiography, cut out from the gel, and transferred to a microfuge tube. The gel slice is cut into small pieces and shaken at room temperature for at least 10 hr suspended in 300 $\mu$l of MG extraction buffer. The supernatant is transferred to a fresh tube and 10 $\mu$g of carrier tRNA is added, followed by 800 $\mu$l of ice-cold ethanol. After vortexing, the tube is cooled on dry ice for 5 min and centrifuged at 12,000–15,000$g$ for 5 min. The pellet is washed at 4° with 70% (v/v) ethanol and resuspended in 50 m$M$ Tris–HCl buffer (pH 7.5). Aliquots are removed, corresponding to 50,000 cpm for the following OP-Cu probing reaction (see later), 20,000 cpm for the alkaline hydrolysis sequence ladder, and 10,000 cpm for the RNase T1 digestion. The remaining RNA is ethanol precipitated and stored at −20°.

3. *Partial Alkaline Hydrolysis of RNA.* An aliquot of ~20,000 cpm [32]P-labeled RNA is mixed with 10 $\mu$g carrier tRNA and 10 $\mu$l of hydroxide hydrolysis buffer. The sample is incubated at 90° for 5 min and then immediately put on ice for further use in producing a sequence ladder in polyacrylamide gel electrophoresis of the RNA fragments from OP-Cu cleavage.

4. *G-Specific RNase T1 Partial Digestion of RNA.* To an aliquot of ~10,000 cpm [32]P-labeled RNA, 2 $\mu$g carrier tRNA, 10 $\mu$l digestion buffer, and 0.02 U of RNase T1 are added. The reaction mixture is incubated at 55° for 10 min and then kept on ice for later use in producing a G ladder in gel electrophoresis.

[23] A. M. Maxam and W. Gilbert, *Methods Enzymol.* **65,** 499 (1980).
[24] A. Krol and P. Carbon, *Methods Enzymol.* **180,** 212 (1989).

5. *OP-Cu Probing Reaction.* A solution of ~50,000 cpm [32]P-labeled RNA is mixed with 100 ng carrier tRNA and 20 $\mu$l of probing buffer. After addition of the OP-Cu reagent solution, the final reaction volume should be ~50 $\mu$l. The OP-Cu reagent is freshly prepared from 1,10-phenanthroline and copper(II) sulfate (2:1) to a final concentration of 40 and 20 $\mu M$, respectively (concerning the concentration of OP-Cu, see the remark on the evaluation of probing data for distance determination). To the reaction mixture, hydrogen peroxide is added to a final concentration of 7 m$M$. The reaction is started by the addition of 5 m$M$ of the reducing agent 3-mercaptopropionic acid (MPA). In order to achieve rapid mixing of the reactants, it is convenient to carefully set droplets of $H_2O_2$ and MPA at the wall of the microfuge tube containing the mixture of RNA and OP-Cu reagent. The reaction is then started by centrifuging briefly to spin down the droplets. The probing reaction is allowed to proceed for 2–5 min at 37°. The optimal reaction time must be adapted for each RNA target, depending on the achieved cleavage rate. The probing reaction is quenched by adding an excess of 2,9-dimethyl-1,10-phenanthroline (neocuproine) solution to a final concentration of 1 m$M$. Neocuproine is a strong chelating agent that forms oxidatively stable complexes with Cu(II).[25]

The cleaved RNA is precipitated by adding 3 volumes of ethanol in the presence of 300 m$M$ sodium acetate (pH 5.4). After centrifuging and reprecipitating from 100 $\mu$l of 300 m$M$ sodium acetate, the pellet is rinsed with 100 $\mu$l of ice-cold 70% (v/v) ethanol and air dried.

6. *Analysis of Cleaved RNA.* For the electrophoretic fractionation of cleavage fragments, the RNA pellet is dissolved in 5 $\mu$l of water and 5 $\mu$l of XC/BPB dye mix is added. The sample is loaded onto a polyacrylamide–8 $M$ urea sequencing gel (length, 40 cm; thickness, 0.5 mm) made in 1× TBE buffer. Depending on the length of the intact RNA, gels of 10–15% (w/v) polyacrylamide [19% (w/v) acrylamide and 1% (w/v) *N,N'*-methylenebis-acrylamide] are used. Gels of higher percentage (BPB dye runs halfway from the start) are suitable to resolve the short RNA fragments resulting from cleavage close to the 3' end label. Gels are prerun at 40 W in order to improve band resolution by both warming the gel and equilibrating ionic differences between the gel and the buffer reservoirs. Prerunning is done for 1–2 hr until the BPB in a preloaded dye sample has moved halfway down from the start. Electrophoresis of the RNA fragments is carried out at 40–70 W for times adapted to the size of the RNA to be analyzed. For an estimation, it can be considered that XC dye migrates to ~60 nt on 10% gels.[24]

[25] D. R. Graham, L. E. Marshall, K. A. Reich, and D. S. Sigman, *J. Am. Chem. Soc.* **102,** 5419 (1980).

In order to obtain reference sequence ladders, reaction mixtures of partial alkaline hydrolysis and partial RNase T1 digestion are electrophoresed together with the cleavage fragments. For the reference lanes, 4 $\mu$l of each reaction mixture is loaded onto the gel. Sequence reading is facilitated by applying two lanes of partial alkaline hydrolysis flanking the OP-Cu probing reactions and partial T1 digestion. It is recommended to run at least one control lane where RNA is loaded, which was kept in probing buffer without the added OP-Cu reagent. Workup of the control sample by RNA precipitation is identical to the OP-Cu-treated reaction mixture. If the OP-Cu cleavage pattern is quantified for distance determination (see later), the control lane is used for background subtraction and should thus contain the same amount of $^{32}$P-labeled RNA as is loaded for the probing experiment.

At the end of electrophoresis, the gel is fixed in 10% (v/v) acetic acid and dried *in vacuo*. An autoradiograph of the the dried gel is recorded either by PhosphoImaging or by exposition to X-ray film using an intensifying screen.

## Evaluation of Probing Data

*1. Structural Information.* Nucleotides susceptible to OP-Cu scission are identified on the autoradiograph by comparing the fragments from OP-Cu probing with sequence ladders. When assigning bands of the different cleavage reactions, the nature of the fragments, thus their electrophoretic mobility, must be considered. Both alkaline hydrolysis and RNase T1 digestion produce 5'-OH ends at 3'-end-labeled fragments. RNase T1 cleaves GpN phosphodiester bonds 3' to unpaired guanines, leaving the phosphate at the unlabeled 5' fragment. Strand scission by OP-Cu proceeds by elimination of the susceptible nucleoside, leaving phosphate groups at both the 3' and the 5' fragment.[9] Therefore, both cleavage reactions result in products that differ slightly in their electrophoretic mobility.

Cleavage of RNA by OP-Cu typically yields clusters of fragments that display decaying intensity centered around one or two strong bands (Fig. 2A). The example of tRNA$^{Phe}$, for which a high-resolution crystal structure is available,[26] shows that unpaired nucleotides are cleaved preferentially in regions where base stacking occurs.[5,8] Molecular modeling studies have suggested that OP-Cu binds to stacked single-stranded regions within RNA folds by partial intercalation of a single phenanthroline ligand between two adjacent bases (Fig. 1).[8] Two major cleavage sites are observed in tRNA$^{Phe}$, namely in the D/T loop junction around G18/G57 and in the anticodon

[26] E. Westhof, P. Dumas, and D. Moras, *Acta Crystallogr. A* **44,** 112 (1988).

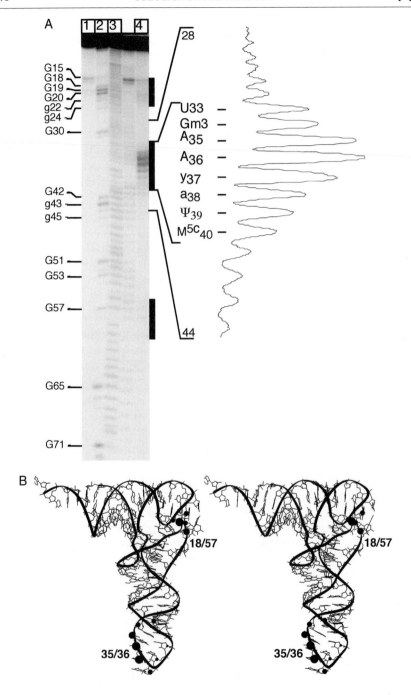

loop around A35/A36 (Fig. 2).[5,8] The three-dimensional structure reveals extensive stacking interactions of the bases in both loop regions (Fig. 2B).[26] In contrast, low reactivity toward OP-Cu cleavage is found in the single-stranded variable loop where less stacking occurs.

2. *Distance Information.* Relative internucleotide distances in RNA folds may be determined using the OP-Cu cleavage pattern, which are likely to originate from single nonoverlapping OP-Cu-binding sites. Such sites are recognized by a well-defined intensity maximum of one to three bands flanked on both sides by nucleotides cleaved with gradually decreasing efficiency.[8] Multiple and unspecific binding of OP-Cu to RNA is reduced by applying low concentrations of the metal complex in the probing experiment. The concentration of OP-Cu given earlier, being 10-fold lower than concentrations occasionally reported in the literature, yielded an optimal single-site probing pattern for tRNA$^{Phe}$.[8] However, the OP-Cu concentration applied in the probing experiment is a parameter that must be optimized for each RNA target, especially if probing data are to be quantified. If OP-Cu is used exclusively for nonquantitative probing of the RNA secondary structure, up to 10-fold higher concentrations of the reagent may be applied.

Cleavage pattern are quantified either directly from PhosphoImager data or, if X-ray film is used, by photodensitometric scanning of autoradiographs. For photodensitometry, lightly exposed films are desirable on which the bands of interest are within the nonsaturating density range of the X-ray film (Fig. 2A). The lane(s) of the OP-Cu probing reaction and a control lane with uncleaved RNA are scanned. Densities of OP-Cu-cleaved fragment bands are corrected for background by subtracting the scanned densities in the same region of the control lane. Numerical integration of the peak areas under the density scan is performed to calculate the probability of cleavage at the reactive nucleotides. Peak area integration can be done directly in the analysis software of the PhosphoImager or photodensitome-

---

FIG. 2. (A) Autoradiograph of the cleavage fragments obtained by OP-Cu probing of 3'-end-labeled tRNA$^{Phe}$. Lane 1, control with untreated tRNA$^{Phe}$; lane 2, G-specific RNase T1 digestion; lane 3, alkaline hydrolysis; lane 4, OP-Cu cleavage; unlabeled lane, control with Fe-EDTA-generated OH-radicals. The regions of enhanced cleavage around G18/G57 and A35/A36 are marked with bars. A low exposition time was chosen here in order to yield a nonsaturating signal suitable for quantification for the A35/A36 cleavage site. As a consequence, the G18/G57 signals are weak on this autoradiograph. On the right, a densitometric scan of the OP-Cu cleavage reaction (lane 5) is shown for the nucleotides C28–A44. (Adapted from Hermann and Heumann.[8]) (B) Stereo view of the three-dimensional structure of tRNA$^{Phe}$ with spheres marking nucleotides cleaved by OP-Cu. Sphere size indicates cleavage probability at the two major cleavage sites designated 18/57 and 35/36.

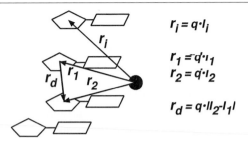

$$r_i = q \cdot I_i$$

$$r_1 = q \cdot I_1$$

$$r_2 = q \cdot I_2$$

$$r_d = q \cdot |I_2 - I_1|$$

FIG. 3. Principle of distance determination using OP-Cu cleavage pattern (see also Fig. 1). The proportionality of the cleavage probability $I$ at a nucleotide $i$ and the distance $r_i$ between the nucleotide and the OP-Cu reagent (black circle) allows the calculation of relative distances between nucleotides even without knowing the proportionality factor $q$.

ter or with the help of common data analysis programs (CricketGraph, DeltaGraph, Igor, KaleidaGraph, etc.).

Distances $r_i$ between the bound OP-Cu and the cleaved nucleotides $i$ are obtained from peak areas $I$ of the corresponding RNA fragment bands according to the proportionality $r_i = qI_i$. Because the proportionality factor $q$ is not known *a priori,* relative internucleotide distances can be obtained by forming differences between pairs of nucleotides (Fig. 3). For a rough estimation of absolute distances, it can be assumed that if OP-Cu binds by partial intercalation, the Cu(II) atom of OP-Cu is located approximately 5–6 Å away from the center of the nearest nucleotide sugars immediately flanking the OP-Cu-binding site. Thus, $q$ may be calculated from the cleavage probability determined by the intensity of the strongest cut at an assumed OP-Cu-binding site. Other absolute nucleotide distances are then calculated using this $q$ value.

Relative internucleotide distances obtained from OP-Cu probing data will at first yield information whether the probed RNA region deviates from a straight conformation.[8] If nucleotides are bending away from an OP-Cu-binding site, their distances to the reactive Cu(II) center and thus the probing signal will change more markedly than in a straight assembly. Second, internucleotide distances may be used in modeling of the RNA three-dimensional structure. RNA structure modeling calls for a combination of data from different theoretical and experimental approaches, such as comparative phylogeny, chemical and enzymatic probing, cross-linking, mutational analysis, and ligand-binding studies.[27–29] While the few distance

[27] E. Westhof and F. Michel, in "RNA-Protein Interactions" (K. Nagai and I. W. Mattaj, eds.), pp. 25–51. IRL Oxford Univ. Press, Oxford, 1994.
[28] P. Romby, this volume.
[29] T. Hermann and E. Westhof, *Curr. Opin. Biotech.* **9,** 66 (1998).

constraints obtained from OP-Cu probing are not sufficient to fully determine the three-dimensional structure of an RNA target, they provide useful
information on the local folding of the nucleic acid. This knowledge is
especially valuable considering that the RNA three-dimensional structure
is assembled from modular units.[30,31]

### Acknowledgments

Matthias Götte is thanked for providing probing data. T.H. was supported by an EMBO
long-term fellowship. H.H. thanks the Deutsche Forschungsgemeinschaft for support (He
1285/8-3).

[30] F. Michel and E. Westhof, *Science* **273,** 1676 (1996).
[31] E. Westhof, B. Masquida, and L. Jaeger, *Fold. Des.* **1,** 78 (1996).

# [4] Applications of Uranyl Cleavage Mapping
# of RNA Structure

*By* Niels Erik Møllegaard and Peter E. Nielsen

### Introduction

The uranyl(VI) ion, $UO_2^{2+}$, binds very strongly to DNA and, on irradiation with long wavelength UV light, induces single-strand breaks.[1,2] The
excited state of $UO_2^{2+}$ is a very strong oxidant, and DNA cleavage patterns
support a mechanism in which a uranyl ion coordinated to a phosphate
group oxidizes proximal sugars via a direct electron-transfer mechanism.[2]
    Thus uranyl cleavage of DNA reflects phosphate accessibility as well
as affinity for the uranyl ion. Therefore, uranyl photoprobing can be employed to study protein(ligand)–DNA interactions in terms of phosphate–
protein contacts[1,3–5] as well as conformational variations in the DNA double

[1] C. Jeppesen, O. Buchardt, and P. E. Nielsen. *FEBS Lett.* **235,** 122 (1988).
[2] P. E. Nielsen, C. Hiort, O. Buchardt, O. Dahl, S. H. Sönnichsen and B. Nordén, *J. Amer.
Chem. Soc.* **114,** 4967 (1992).
[3] C. Jeppesen and P. E. Nielsen, *Nucleic Acids Res.* **17,** 4947 (1989).
[4] N. E. Møllegaard, P. E. Rasmussen, P. Valentin-Hansen, and P. E. Nielsen, *J. Biol. Chem.*
**268,** 17471 (1993).
[5] N. E. Møllegaard and P. E. Nielsen, *Methods Mol. Biol.* **90,** 43 (1997).

helix.[6–8] In this way the uranyl probing complements the Fe/EDTA method and is significantly more sensitive to helix conformations, e.g., regarding minor groove width, than Fe/EDTA probing. Furthermore, uranyl photocleavage is not temperature dependent in a biologically relevant range (0–60°) and is not sensitive to radical quenchers, such as glycerol.[2] It is, however, quenched by phosphate and the efficiency decreases dramatically above pH 7–8.[2]

In the presence of a chelator, such as citrate, high-affinity binding sites for uranyl and, by inference, biologically relevant ions such as $Mg^{2+}$ may be identified in protein–DNA complexes [5] and in folded nucleic acids, such as the four-way junction.[9]

Three studies on the photocleavage of RNA by uranyl have been published.[10–12] These studies indicate that some structural information may be extracted from the cleavage result, but, more importantly, the cleavage pattern can be used to identify possible binding sites for divalent metal ions (e.g., $Mg^{2+}$) that are important in folding and function (e.g., ribozyme catalysis).[11,12]

Uranyl Photocleavage

*Materials*

RNA: labeled at either the 3′ or the 5′ end
Uranyl nitrate [$UO_2(NO_3)_2$]: 100 m$M$ stock solution in water (this is stable for several years at room temperature)
Uranyl cleavage buffer: For instance, 50 m$M$ Tris–HCl, pH 7.2
Sodium citrate: 100 m$M$ stock solution
Buffer for precipitation of RNA: 0.5 $M$ sodium acetate, pH 4.5. The pH is important because uranyl does not coprecipitate at low pH[10]
Gel running buffer: 1× TBE (90 m$M$ Tris–borate, 1 m$M$ EDTA, pH 8.3)
Loading buffer (80% formamide, 1× TBE, 0.1% bromphenol blue, 0.1% xylene cyanol)

[6] P. E. Nielsen, N. E. Møllegaard, and C. Jeppesen, *Nucleic Acids Res.* **18,** 3847 (1990).
[7] S. H. Sönnichsen and P. E. Nielsen, *J. Mol. Recogn.* **9,** 219 (1996).
[8] C. Bailly, N. E. Møllegaard, P. E. Nielsen, and M. Waring, *EMBO J.* **14,** 2121 (1995).
[9] N. E. Møllegaard, A. I. H. Murchie, D. Lilley, and P. E. Nielsen, *EMBO J.* **13,** 1508 (1994).
[10] R. Gaynor, E. Soultanakis, M. Kuwabara, J. Garcia, and D. S. Sigman, *Proc. Natl. Acad. Sci. U.S.A.* **74,** 4832 (1989).
[11] P. E. Nielsen and N. E. Møllegaard, *J. Mol. Recogn.* **9,** 228 (1996).
[12] G. Bassi, N. E. Møllegaard, A. I. H. Murchie, E. von Kitzing, and D. Lilley, *Nature Struct. Biol.* **2,** 45 (1995).

Polyacrylamide gels: 8–15%, 0.3 bisacrylamide, 7 $M$ urea, 1× TBE
Light source: Philips TL 40-W/03 fluorescent light tube

## Methods

End labeling of RNA is done by standard techniques, e.g., using either 5' labeling with T4 polynucleotide kinase and [$\gamma$-$^{32}$P]ATP or 3' labeling with RNA T4 ligase and [$^{32}$P]pCp. Because the quality of the RNA is important, it is recommended purifying the labeled RNA on a denaturing gel to ensure a full-length homogeneous RNA preparation. The labeled RNA is purified on 6–10% denaturing polyacrylamide gels. The choice of gel depends on the length of the RNA.

A typical uranyl cleavage reaction is performed by adding 10 $\mu$l of a 10 m$M$ uranyl nitrate solution to the 90-$\mu$l reaction mixture containing the RNA (e.g., in 50 m$M$ Tris–HCl, pH 7.0), reaching a final concentration of 1 m$M$ uranyl (it is possible to use a lower concentration of uranyl). The 10 m$M$ uranyl solution is made from a 100 m$M$ uranyl nitrate stock solution. Immediately after the addition of uranyl, the samples are placed with open lids under a Philips TL 40-W/03 fluorescent light tube of wavelength 420 mm. For obtaining maximal light exposure, it is important to place the tube as close as possible to the light source. Following 20–30 min of irradiation, 20 $\mu$l of 0.5 $M$ sodium acetate, pH 4.5, is added to prevent precipitation of the uranyl, which will cause hydrolysis upon heating prior to gel loading.[11] The RNA is precipitated by adding 2 volumes of 96% (v/v) ethanol followed by centrifugation. The pellet is washed with 70% (v/v) ethanol, dried, and redissolved in 5 $\mu$l of 80% formamide containing xylene cyanol and bromphenol blue. The RNA samples are analyzed by standard polyacrylamide gel electrophoresis (10–15% polyacrylamide, 7 m$M$ urea, 1× TBE) and visualized by autoradiography using intensifying screens or by phosphorimager.

### Probing of Metal Ion Binding Sites in RNA

Appropriate controls have to be included in the experiment to ensure that the cleavage signals are due to uranyl photocleavage. A typical uranyl photoprobing of metal ion binding sites should include the following samples.

1. One sample irradiated without uranyl to show that the light source is without a direct damaging effect on the RNA.
2. One sample with the amount of uranyl used in the photocleavage reaction incubated without irradiation (e.g., wrapped in aluminum foil) to monitor thermal cleavage.

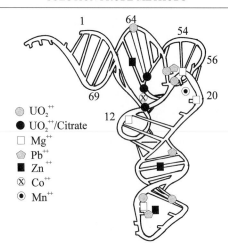

Fig. 1. Tertiary structure of yeast tRNA[Phe]. Sites of both uranyl and uranyl/citrate hyperreactivity are shown together with binding sites of other divalent metal ions found in the crystal.[16]

3. Several samples all with 1 m$M$ uranyl but with varying concentrations of citrate. It is recommended to include at least five concentrations (0, 0.25, 0.5, 0.75, 1.0, and 1.25). By increasing the citrate concentration, the general uranyl cleavage of the RNA will be suppressed, while hypersensitive sites (i.e., metal ion binding sites) will remain the same. At the highest citrate concentrations the hypersensitive sites are suppressed as well, dependent on the affinity for metal ions. The uranyl/citrate solution has to be mixed before adding it to the RNA sample.
4. Samples with the addition of an increasing amount of a competitive ion, e.g., magnesium. This will demonstrate the affinity for other metal ions in addition to uranyl. The competitive ion has to be added before uranyl.

Discussion

The use of uranyl photocleavage of DNA for footprinting, conformational analysis, and (divalent) metal ion binding site studies is now well documented.[1,9] However, only few studies with RNA have been published.[10–12]

Nonetheless, these studies clearly demonstrate that uranyl may indeed become a very useful probe for studying RNA structure, folding, and function, especially regarding the influence and role of divalent metal ions. It has already been demonstrated that uranyl hypersensitive sites that persist in the presence of citrate correlate well with the putative binding site for $Mg^{2+}$ in the hammerhead ribozyme[12] and the hyperreactive cleavage site for the yeast tRNA$^{Phe}$ cluster around the T$\Psi$C arm cavity and in the anticodon loop,[11] areas that also harbor binding sites for a variety of other divalent metal ions (Fig. 1).[13-17] Moreover, citrate persistent sites are very pronounced and are located exclusively in the T$\Psi$C arm cavity[11] (Fig. 1).

The localization of metal ion binding sites with uranyl should be comparable to studies using $Fe^{2+}$ cleavage. In contrast to $Fe^{2+}$ cleavage, however, the uranyl cleavage is light induced and temperature independent. Thus low temperature studies are possible and, more importantly, RNA structures may be folded in the presence of uranyl (and other metal ions) and cleavage will, in contrast to the $Fe^{2+}$ method, not occur until the sample is irradiated. Therefore kinetic studies of RNA folding processes are also possible using the uranyl technique. Once again, however, we wish to emphasize that some limited thermal cleavage of RNA (in contrast to DNA) does occur with uranyl and it is therefore crucial to include proper nonirradiated controls.

The modulation of the cleavage pattern of duplex DNA observed at acidic pH (pH $\leq$ 6.5) has been correlated with a minor groove width.[6,7] With RNA the modulation is more complicated to interpret due to RNA three-dimensional structure, and at this stage we cannot say if the modulation in the absence of a chelator (citrate) also reflects RNA helix conformation to some extent.

In conclusion, we believe that the uranyl photoprobing technique has an unexplored potential in RNA structure–function studies.

[13] A. Rich and U. L. RajBhandary, *Annu. Rev. Biochem.* **45,** 806 (1976).

[14] R. S. Brown, J. C. Dewan, and A. Klug, *Biochemistry* **24,** 10163 (1985).

[15] G. J. Quigley, M. M. Teeter, and A. Rich, *Proc. Natl. Acad. Sci. U.S.A.* **75,** 64 (1978).

[16] J. R. Rubin, J. Wang, and M. Sundaralingam, *Biochim. Biophys. Acta* **756,** 111 (1983).

[17] P. Tao, D. M. Long, and O. C. Uhlenbeck, *in* "The RNA World" (R. F. Gesteland and J. F. Atkins, eds.), p. 271. Cold Spring Harbor Laboratory Press, Cold Spring Harbor, NY, 1993.

# [5] Identification of Messenger RNAs That Contain Inosine

*By* Daniel P. Morse

## Introduction

Adenosine deaminases that act on RNA (ADARs) convert adenosines to inosines within double-stranded regions of RNA.[1,2] During the time since the discovery of this family of RNA-editing enzymes, several viral and cellular ADAR substrates have been identified.[2] These substrates were discovered by noticing discrepancies between genomic and cDNA sequences. In each case, adenosines within genomic DNA appeared as guanosines in the corresponding cDNA. Such A to G changes are diagnostic of A to I changes at the RNA level because inosine pairs with cytosine during cDNA synthesis.

What are the biological consequences of A to I conversion in RNA? A partial answer to this question is provided by some of the known ADAR substrates. Because inosine is read as guanosine by the translational machinery, ADARs can produce codon changes in messenger RNAs (mRNAs), resulting in the synthesis of multiple protein isoforms with distinct functional properties. The best studied examples of this type of editing are found in several glutamate receptor (GluR) pre-mRNAs,[3] $5HT_{2C}$ serotonin receptor pre-mRNA,[4] and the antigenome of hepatitis delta virus.[5] ADAR activity has been observed in every metazoan examined,[2] and surprisingly high levels of inosine have been detected in mRNA derived from multiple rat tissues.[6] Thus, there are probably many substrates and biological roles for ADARs remaining to be discovered.

Toward the goal of understanding the full scope of ADAR function, we developed a systematic method to identify new substrates.[7,8] Our approach was to look for mRNAs that contain inosine. The protocol evolved

---

[1] B. L. Bass, K. Nishikura, W. Keller, P. H. Seeburg, R. B. Emeson, and M. A. O'Connell, *RNA* **3**, 947 (1997).

[2] B. L. Bass, *Trends Biochem. Sci.* **22**, 157 (1997).

[3] P. H. Seeburg, *J. Neurochem.* **66**, 1 (1996).

[4] C. M. Burns, H. Chu, S. M. Rueter, L. K. Hutchinson, H. Canton, and E. Sanders-Bush, *Nature* **387**, 303 (1997).

[5] A. G. Polson, B. L. Bass, and J. L. Casey, *Nature* **380**, 454 (1996).

[6] M. S. Paul and B. L. Bass, *EMBO J* **17**, 1120 (1998).

[7] D. P. Morse and B. L. Bass, *Biochemistry* **36**, 8429 (1997).

[8] D. P. Morse and B. L. Bass, *Proc. Natl. Acad. Sci. U.S.A.* **96**, 6048 (1999).

in three stages. First, we developed a method that allowed specific cleavage of phosphodiester bonds 3' to inosines.[7] We next devised an amplification strategy that, when the RNA sequence was known, could confirm the presence of inosine in a candidate ADAR substrate.[7] Finally, by adapting the differential display method, we modified the amplification strategy in order to identify cleaved RNAs in the absence of sequence information.[8] It is this modified strategy that allows the discovery of new ADAR substrates.

This article provides detailed protocols for inosine-specific cleavage and for the two amplification strategies. The notes following many of the protocols provide helpful hints, describe expected results, and explain the rationale and nuances of some of the steps. These should be read carefully.

## Methods

### General Considerations

Unless stated otherwise, phenol–chloroform extractions are performed with 1 volume of phenol–chloroform–isoamyl alcohol (25:24:1, v/v) and nucleic acids are precipitated with 1/10 volume of 3 $M$ sodium acetate and 2.5 volumes of ethanol. To ensure reproducibility when performing multiple reactions, master mixes of common reagents should be used.

### Preparation of RNA

*Synthesis of Control RNA.* The use of a synthetic RNA containing a single inosine was critical to the development of most of the procedures in this article.[7] As described later in the relevant sections, this RNA (subsequently referred to as the control RNA) is important for both optimizing the methods and monitoring the RNA recovery after each step. The control RNA is synthesized by joining two half-molecules using the oligo-bridged ligation method of Moore and Sharp.[9] In this method, two RNA halves are annealed to a complementary DNA oligonucleotide that spans the ligation junction. The nick in the resulting RNA/DNA duplex is sealed with T4 DNA ligase. The two half-molecules are produced by *in vitro* transcription from either linearized plasmids or polymerase chain reaction (PCR) products that contain phage promoters upstream of the desired sequences. Inosine is placed at the 5' end of the 3' half-molecule by initiating transcription with an IpG dinucleotide using a 5:1 molar ratio of dinucleotide to GTP in the transcription reaction. Because IpG is not available commercially, we synthesize it on an Applied Biosystems 394 DNA/RNA

[9] M. J. Moore and P. A. Sharp, *Science* **256**, 992 (1992).

synthesizer. The 3' half-molecule is 5' end labeled using T4 polynucleotide kinase and [$\gamma$-$^{32}$P]ATP and is purified on a denaturing polyacrylamide gel.

Ligation reactions (20 $\mu$l) contain 75 pmol 3' half; 100 pmol 5' half; 80 pmol bridging oligodeoxynucleotide; 1× ligase buffer [50 m$M$ Tris–HCl (pH 7.6), 10 m$M$ MgCl$_2$, 1 m$M$ ATP, 1 m$M$ dithiothreitol (DTT), 5% polyethylene glycol (PEG) 8000]; and 10 U T4 DNA ligase. The two RNAs and the bridging DNA are first mixed in water, heated to 100° for 1 min, and placed at 42°. Buffer is added, and the mixture is kept at 42° for 20 min to anneal the bridge to the RNA molecules. After cooling to room temperature, DNA ligase is added, and the reaction is allowed to proceed at room temperature for 4 hr. The reaction is diluted to 100 $\mu$l with water, extracted once with phenol–chloroform, and nucleic acids are precipitated. The ligation product is purified on a denaturing polyacrylamide gel and quantified by liquid scintillation counting.

*Notes.* (1) It is not necessary to gel purify the 5' half because any prematurely terminated transcripts will not be ligated to the 3' half. (2) The bridge must have a C in the position that will pair with inosine. (3) The specific activity of the ligation product will be equal to the specific activity of the [$\gamma$-$^{32}$P]ATP because only phosphorylated 3' half-molecules participate in the ligation reaction. (4) In our hands, this protocol typically gives 10 to 40% ligation efficiency. The efficiency drops as the sizes of the RNA halves increase. (5) The control RNA should be as large as possible while maintaining a reasonable ligation efficiency. We have had good success with ligating RNAs on the order of 200 nucleotides in length. For RNAs in this size range, we use a 40 nucleotide bridging oligonucleotide with approximately 20 nucleotides of complementarity to both RNAs. (6) The sequence of the control RNA is not important.

*Purification of Cellular RNA.* To detect inosine at candidate ADAR-editing sites in RNAs of known sequence, one can use total RNA as the starting material, but the sensitivity of detection is improved by starting with poly(A)$^+$ RNA. Many methods are available for preparing total and poly(A)$^+$ RNA. The choice depends on several factors: the source of RNA, the desired yield and purity, and cost.

Identification of new ADAR substrates requires starting material that is highly enriched for poly(A)$^+$ RNA and nearly free of ribosomal RNA. In our hands, no commercially available kit yields RNA of the required quality. For this application, we purify poly(A)$^+$ RNA from total RNA using the method of Bantle *et al.*[10] In this method, total RNA is subjected to two (optionally three) rounds of poly(A) selection on an oligo(dT) cellulose column. Prior to the second round, the RNA is heated in dimethyl

[10] J. A. Bantle, I. H. Maxwell, and W. E. Hahn, *Anal. Biochem.* **72,** 413 (1976).

sulfoxide (DMSO) and buffered LiCl, which dissociates mRNA–rRNA complexes. This treatment results in poly(A)$^+$ RNA with little, if any, detectable rRNA contamination.

We prepare total RNA using the guanidinium thiocyanate–phenol method.[11] Ten milligrams of total RNA is loaded onto a 1-ml oligo(dT) cellulose column. Following the standard protocol for binding, washing, and elution,[12] the first round of selection typically yields 100 to 200 $\mu$g of poly(A)$^+$ RNA in a 3- to 4-ml volume. The RNA is precipitated, dissolved in 400 $\mu$l of water, reprecipitated, and dissolved in 20 $\mu$l of 10 m$M$ Tris–HCl (pH 7.5). We perform the first precipitation in 8 to 10 separate microfuge tubes, dissolve each sample in 40 to 50 $\mu$l of water, and combine the 8 samples into one tube for the second precipitation. Next, 180 $\mu$l of DMSO and 20 $\mu$l of buffered LiCl [1 $M$ LiCl, 50 m$M$ EDTA, 2% SDS, 10 m$M$ Tris–HCl (pH 6.5)] are added. The RNA is heated at 55° for 5 min, diluted with 2 ml loading buffer, and loaded onto the same column used for the first round of selection. The second round typically yields 50 to 100 $\mu$g of RNA. The RNA is precipitated twice as before and dissolved to a final concentration of 2 $\mu$g/$\mu$l. A 5-$\mu$g sample of the RNA is run on a formaldehyde gel. If significant amounts of ribosomal RNA are still visible by ethidium bromide staining, then the procedure followed for the second round can be repeated in a third round of selection. A small amount of rRNA contamination is acceptable but will increase the background of false positives as described in the section on identifying new ADAR substrates.

*Oxidation of 3'-Hydroxyl Groups.* This method is used only when following the protocol for identifying new ADAR substrates (see later). To detect inosine in a candidate substrate of a known sequence, skip this step and proceed to inosine-specific cleavage.

Reactions contain 10 $\mu$l (20 $\mu$g) poly(A)$^+$ RNA; 3.3 $\mu$l of 0.5 $M$ sodium acetate (pH 5.5); and 3.3 $\mu$l of 50 m$M$ sodium periodate (freshly dissolved). After 1 hr in the dark at room temperature, 16.6 $\mu$l of 2% (v/v) ethylene glycol is added and the incubation is continued for another 10 min. The reaction is diluted to 400 $\mu$l with water and the RNA is precipitated. The RNA is dissolved in 174 $\mu$l of water, which is sufficient for four glyoxal reactions containing 5 $\mu$g of RNA each (see later).

*Note.* We have not rigorously tested whether this step is required. Its purpose is to prevent elongation of the original poly(A) tails, which could

[11] F. M. Ausubel, R. Brent, R. E. Kingston, D. D. Moore, J. G. Seidman, and J. A. Smith, "Current Protocols in Molecular Biology," p. 4.2.1. Wiley, New York, 1987.

[12] J. Sambrook, E. F. Fritsch, and T. Maniatis, "Molecular Cloning: A Laboratory Manual," 2nd ed., p. 7.26. Cold Spring Harbor Press, Cold Spring Harbor, NY, 1989.

interfere with tailing of the cleavage sites and reverse transcription (see later).

## Inosine-Specific Cleavage of RNA

*Principle of Method.* Ribonuclease T1 (RNase T1) cleaves RNA 3′ of both guanosine and inosine. The strategy for cleaving RNA specifically after inosine (Fig. 1) exploits the ability of the reagent glyoxal to discriminate between these structurally similar nucleotides. In the presence of borate ions, glyoxal forms a stable adduct with guanosine, but not with

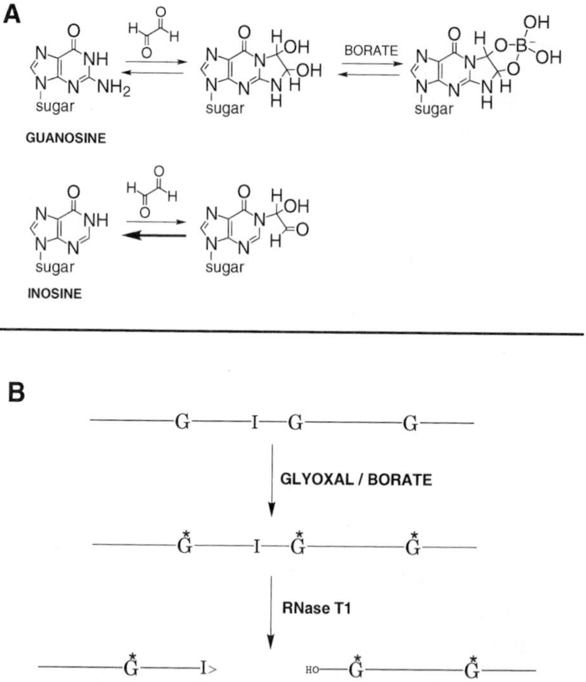

Fig. 1. Strategy for inosine-specific cleavage of RNA. (A) The diagram shows the reaction of glyoxal ($C_2H_2O_2$) with guanosine and inosine and stabilization of the guanosine adduct with borate. As shown, inosine does not react stably with glyoxal. (B) Scheme for inosine-specific RNase T1 cleavage of glyoxalated RNA. Asterisks mark sites of glyoxalated guanosines that are resistant to RNase T1. The RNase T1 fragment that contains the inosine is shown with a 2′,3′-cyclic phosphate (>) because the RNase T1 reaction with inosine frequently does not go to completion. Reprinted with permission from Morse and Bass.[7] Copyright 1997 American Chemical Society.

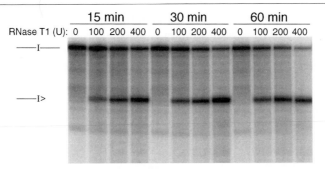

FIG. 2. Optimization of inosine-specific cleavage using control RNA. For each lane, 1 fmol of RNA containing a single inosine was spiked into 5 µg of total yeast RNA and treated with glyoxal and borate as described in the text. The glyoxalated RNA was incubated with increasing amounts (U) of RNase T1 and time, as indicated. The PhosphorImager image shows electrophoretically separated starting material and a single product of the size expected for cleavage after the inosine, which accumulated with increasing time and RNase T1. Positions of full-length and cleaved RNAs (determined in control experiments) are shown on the left. In this experiment, treatment with 400 units of RNase T1 for 30 min was optimal. Treatment with 400 units for 60 min resulted in an unacceptably high level of nonspecific cleavage as indicated by the decrease in the intensity of the specific cleavage product. Reprinted with permission from Morse and Bass.[7] Copyright 1997 American Chemical Society.

inosine,[13] and glyoxalated guanosines are resistant to RNase T1.[14] Therefore, an inosine-containing RNA that is modified stably by glyoxal is cleaved by RNase T1 only after inosine.[7]

*Optimization.* The control RNA is used to determine the optimal conditions for the RNase T1 reaction (Fig. 2). For each condition to be tested, 1 fmol of the labeled control RNA is added to 5 µg of carrier RNA (e.g., total yeast RNA) and subjected to the inosine-specific cleavage method described later. Following the RNase T1 reaction, the products are separated on a 6% denaturing polyacrylamide gel. Because the control RNA is labeled immediately 5' of the single inosine, only two discreet bands should be visible (see Fig. 2). The two bands correspond to uncleaved RNA and the 5' half of the inosine-specific cleavage product. Optimal conditions are those that give the maximum amount of inosine-specific cleavage and the minimum amount of nonspecific cleavage. Nonspecific cleavage appears as an increase in the intensity of the background smear (compared to no enzyme control) coupled with a decrease in the intensities of the two discreet bands. We find that about 80% of the input control RNA can be cleaved specifically after inosine before nonspecific cleavage becomes

[13] N. E. Broude and E. I. Budowsky, *Biochim. Biophys. Acta* **254,** 380 (1971).
[14] P. R. Whitfeld and H. Witzel, *Biochim. Biophys. Acta* **72,** 338 (1963).

significant. Optimization should be repeated with each new lot of enzyme or whenever a new reagent solution is made.

*General Considerations.* Each reaction is performed in duplicate to assess reproducibility. Therefore, four 5-$\mu$g samples of RNA are processed in parallel: two samples are treated with RNase T1 and two samples are not. Reactions that contain glyoxalated RNA should be performed in siliconized tubes to facilitate dissolving of the RNA following precipitation. Although tubes can be siliconized in the laboratory, better results are seen using commercially available presiliconized tubes such as those available from Phenix.

*Glyoxal Reaction.* Reactions contain 43.5 $\mu$l (5 $\mu$g) of poly(A)$^+$ RNA diluted in water; 1 $\mu$l (1 fmol) of $^{32}$P-labeled control RNA diluted in water; 4 $\mu$l of 250 m$M$ sodium phosphate (pH 7); 50 $\mu$l of DMSO; and 1.5 $\mu$l of 40% deionized glyoxal. After 45 min at 37°, 100 $\mu$l of 1 $M$ sodium borate (pH 7.5) is added, and the reaction products are precipitated with 500 $\mu$l of ethanol (no sodium acetate is added) and dissolved in 15 $\mu$l of Tris–borate [10 m$M$ Tris–HCl (pH 7.8), 1 $M$ sodium borate (pH 7.5)].

*Notes.* (1) The 40% glyoxal stock solution (Sigma) is deionized with AG 501-X8 (Bio-Rad, Hercules, CA) until pH >5. (2) We prepare solutions containing 1 $M$ borate from a stock solution of 1.4 $M$ boric acid that has been adjusted to pH 7.5 with NaOH. The boric acid will not dissolve completely until the pH approaches neutrality. (3) At 37°, the pH of the Tris–HCl in the Tris–borate solution changes to 7.5, which is optimal for the RNase T1 reaction. (4) On addition of ethanol, the solution becomes very cloudy, and the resulting pellet is huge. This is normal but the precipitate should not be cooled excessively or spun for a long time as there will be an even larger pellet. We usually cool at −70° for about 2 min (cooling is probably unnecessary) and spin at room temperature for 10 min. (5) It takes about 20 min at room temperature to dissolve the pellet. There is no need to vortex until after the 20-min incubation. The labeled control RNA is useful at this stage (and every subsequent precipitation) for confirming that the RNA is dissolved completely.

*Digestion with RNase T1.* One microliter of water (for no RNase T1 controls) or 1 $\mu$l of RNase T1 containing the previously determined optimal number of units (typically 100 to 400 U of BRL enzyme) is added to the redissolved glyoxalated RNA. After incubating at 37° for the previously determined optimal time (typically 30 min), RNase T1 is inactivated as follows: (1) Add 0.6 $\mu$l of 15 $\mu$g/$\mu$l proteinase K (Boehringer Mannheim) and incubate for 20 min at 37°. (2) Add 100 $\mu$l of phenol–chloroform and vortex 30 sec. (3) Add 85 $\mu$l of water and vortex 30 sec. (4) Transfer the aqueous phase to a new tube and extract again with 100 $\mu$l of phenol–chloroform. The RNA is diluted to 400 $\mu$l with water, precipitated, and dissolved in 43 $\mu$l of water (or dissolved in 10 $\mu$l of water if the reaction products are to be run on a gel).

*Notes.* (1) RNase T1 is diluted in water immediately prior to use. (2) Treatment with both proteinase K and phenol–chloroform ensures that RNase T1 is completely inactivated. (3) The first phenol–chloroform extraction is done before diluting the reaction so that glyoxal does not begin to dissociate prior to inactivation of the enzyme. (4) The RNA is diluted to 400 $\mu$l prior to precipitation to prevent the coprecipitation of excess borate. Excess borate will inhibit the subsequent removal of glyoxal (see later).

*Postcleavage Processing of RNA*

*Removal of 3'-and 2',3'-Cyclic Phosphates.* RNase T1 normally produces 3'-phosphates via of 2',3'-cyclic phosphate intermediate.[14] Under our conditions, the reaction with inosine does not go to completion, resulting in a mixture of 3'- and cyclic phosphates. Removal of phosphates, which is required for the first step of each of the amplification protocols, is accomplished with T4 polynucleotide kinase. In addition to its more well-known activity, this enzyme has both 3'-phosphatase[15] and 2',3'-cyclic phosphodiesterase[16] activities.

Kinase reactions contain 43 $\mu$l of RNA (from the RNase T1 digestion); 5 $\mu$l of 10× buffer [200 m$M$ Tris–HCl (pH 8), 100 m$M$ MgCl$_2$]; and 2 $\mu$l of 30 U/$\mu$l T4 polynucleotide kinase (USB). After 1 hr at 37°, 0.6 $\mu$l of 15 $\mu$g/$\mu$l proteinase K (Boehringer Mannheim) is added, and the incubation is continued for another 20 min. The RNA is diluted to 100 $\mu$l with water, extracted twice with phenol–chloroform, diluted to 400 $\mu$l with water, precipitated, and dissolved in 46 $\mu$l of water. The RNA may not dissolve completely in water at this stage, but it will dissolve once DMSO is added in the next step.

*Notes.* (1) The reaction is optimal at pH 8 and requires a high concentration of enzyme ($\sim$1 U/$\mu$l). ATP should not be added. (2) We found that without proteinase K treatment, much of the RNA is pulled into the organic phase during the phenol–chloroform extraction. This is likely due to the large amount of enzyme in the reaction. (3) Because borate stabilizes the glyoxal adducts, it is important to remove the residual borate before attempting to remove glyoxal in the next step. This is the purpose of diluting the RNA to 400 $\mu$l prior to precipitation at this (and the previous) step.

*Removal of Glyoxal Adducts.* Glyoxal must be removed from RNA prior to first-strand cDNA synthesis because reverse transcriptase terminates at glyoxalated guanosines. Glyoxal adducts are unstable at alkaline pH, but

[15] V. Cameron and O. C. Uhlenbeck, *Biochemistry* **16**, 5120 (1977).
[16] C. L. Greer and O. C. Uhlenbeck, personal communication (1996).

the use of high pH results in unacceptable levels of RNA degradation. Instead we remove glyoxal by incubation at neutral pH and high temperature. These conditions result in only a negligible amount of RNA hydrolysis.

Reactions contain 46 $\mu$l of RNA (from the kinase reaction); 4 $\mu$l of 250 m$M$ sodium phosphate (pH 7); and 50 $\mu$l of DMSO. After 3 hr at 60°, the RNA is diluted to 400 $\mu$l with water, precipitated, and dissolved in either 9 or 13.5 $\mu$l of water, depending on which protocol is followed next.

*Notes.* (1) The buffer is the same as that used for the glyoxal reaction (minus glyoxal, of course). (2) The RNA is diluted prior to precipitation to prevent coprecipitation of glyoxal. The RNA pellet should dissolve readily now that glyoxal is removed. (3) The control RNA can be used to confirm that glyoxal has been removed because glyoxalated RNA migrates more slowly in a gel than unmodified RNA.

### Detection of Inosine at Candidate Editing Sites

*Principle of Method.* A candidate ADAR substrate is an RNA for which A-to-G changes have been observed when comparing genomic and cDNA sequences. If an RNA is an ADAR substrate, it will be cleaved efficiently by RNase T1 3' of positions corresponding to the A-to-G discrepancies. Figure 3A outlines an amplification strategy for determining where (or if) cleavage has occurred within a candidate substrate. The method entails ligation of an oligonucleotide (the anchor) to both the original and newly created 3' ends to provide priming sites for reverse transcription and subsequent PCR. In order to preferentially amplify cDNA derived from RNA molecules that had inosine at their 3' ends, we use one of four discriminating anchor primers (DAPs) for the downstream PCR primer (see later). The second priming site for PCR is provided by the known sequence of the RNA upstream of the candidate editing site. Thus, an RNase T1-dependent PCR product of a predictable size should be produced. The sequence of the product is determined to confirm that the candidate RNA was cleaved at the expected site. Figure 3B shows the results of applying this method to detect cleavage of the control RNA.

*Ligation of Anchor (Fig. 3A, Step 1).* Under the appropriate conditions, T4 RNA ligase can attach a DNA oligonucleotide to the 3' end of an RNA molecule. The reaction joins a 5'-phosphate to a 3'-hydroxyl. The 3'-hydroxyls at the RNA cleavage sites are produced by the kinase reaction described earlier. (The original 3' ends naturally have 3'-hydroxyls.) A 5'-phosphate is added to the anchor with T4 polynucleotide kinase or it is incorporated during synthesis of the oligonucleotide. To prevent multimerization, the 3'-hydroxyl on the anchor must be blocked. This can be accomplished by incorporating a 3'-deoxynucleotide at the 3' end. For example,

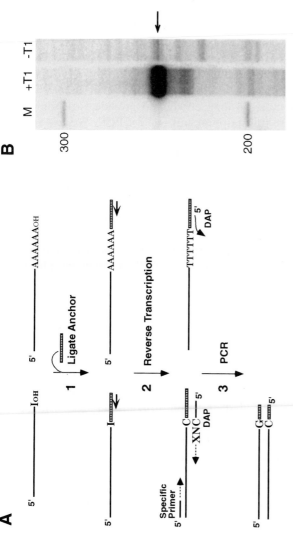

FIG. 3. Detection of inosine in candidate ADAR substrates. (A) Strategy for specific amplification of RNase T1 cleavage products that have inosine at their 3' ends. Starting material is poly(A)⁺ RNA that has been subjected to inosine-specific cleavage and postcleavage processing. Step 1: Ligation of anchor to all 3' hydroxyls. Step 2: Synthesis of first-strand cDNA using a primer that is complementary to the anchor. Step 3: Amplification of the inosine-containing fragments using a sequence-specific upstream primer and one of four possible DAPs (X in 3' terminal CNX sequence is G, A, C, or T; see text for description of DAPs). The curved arrow indicates that DAPs do not prime at that site. Note that the cDNA templates are drawn 3' to 5'. (B) Detection of inosine in control RNA; 0.5 fmol of the control RNA was spiked into 5 μg of yeast RNA and subjected to inosine-specific cleavage using 0 units (−T1) or 400 units (+T1) of RNase T1. The cleaved RNA was then subjected to the protocol of A. For this experiment, X = C at the 3' end of the DAP because there was a G two nucleotides upstream of the inosine in the control RNA. The arrow points to an RNase T1-dependent band whose sequence confirmed that it derived from the control RNA cleaved precisely 3' to its single inosine. M, 100-bp ladder. Figure 3B adapted with permission from Morse and Bass.[7] Copyright 1997 American Chemical Society.

we incorporate cordycepin (3′-deoxyadenosine) during synthesis of the anchor. Alternatively, a dideoxynucleotide can be added postsynthetically using terminal transferase.

Anchor ligation is performed as in Tessier et al.[17] with minor modifications. Reactions contain 9 $\mu$l of RNA (from the glyoxal removal step); 4 $\mu$l of 10 pmol/$\mu$l anchor oligonucleotide; 5 $\mu$l of 10× ligation buffer [500 m$M$ Tris–HCl (pH 8), 100 m$M$ MgCl$_2$, 100 $\mu$g/ml bovine serum albumin (BSA)]; 2.5 $\mu$l of 20 m$M$ hexamminecobalt chloride; 4 $\mu$l of 250 $\mu M$ ATP; 25 $\mu$l of 25% PEG 8000; and 0.5 $\mu$l of 20 U/$\mu$l T4 RNA ligase (BRL, Gathersburg, MD). After 2 hr at room temperature, the RNA is diluted to 400 $\mu$l with water, precipitated, and dissolved in 22.5 $\mu$l of water.

*Notes.* (1) The anchor must be long enough to provide two primer-binding sites: one for reverse transcription and one for PCR. We use a 36 nucleotide anchor. (2) The sequence of the anchor is unimportant except it should not be self-complementary and it should not be present in the RNA being tested. (3) We do not use phenol–chloroform to inactivate the enzyme because the RNA is extracted into the organic phase. This seems to be due to the hexamminecobalt chloride, but this reagent is required for ligation. (4) Ligation to the cleaved RNA depends on the success of the previous phosphate removal step as well as the ligation reaction itself. (5) In our hands, the reaction is very inefficient (about 5% of the anchor is ligated) but adequate. (6) Ligation to the control RNA can be visualized on a gel. Products of ligation to both cleaved and uncleaved RNA should be visible as slightly larger bands. (7) The RNA pellet is difficult to dissolve after this step. We simply do the best we can and accept the losses. (8) Despite the inefficient ligation and losses during dissolving, the subsequent amplification works well.

*First-Strand cDNA Synthesis (Fig. 3A, Step 2).* Reactions contain 22.5 $\mu$l of RNA (from the anchor ligation step); 10 $\mu$l of 5× MLV RT buffer [250 m$M$ Tris–HCl (pH 8.3), 375 m$M$ KCl, 15 m$M$ MgCl$_2$]; 5 $\mu$l of 10 m$M$ DTT; 2.5 $\mu$l of 10 m$M$ dNTPs; 8 $\mu$l of 10 pmol/$\mu$l primer; and 2 $\mu$l of 200 U/$\mu$l MLV reverse transcriptase (BRL). Prior to adding enzyme, the reaction is incubated for 5 min at 65° (to denature the RNA) and cooled to 50°. The enzyme is added, and the 50° incubation is continued for 2 hr. The resulting cDNA is used directly for PCR without further processing.

*Notes.* (1) The primer is complementary to the 3′ end of the anchor sequence. (2) The reaction temperature will depend on the melting temperature of the primer and the thermal stability of the reverse transcriptase. We use a 50° reaction temperature to maximize specificity. (3) Because the anchor is attached to the newly created 3′ ends as well as the original

[17] D. C. Tessier, R. Brousseau, and T. Vernet, *Anal. Biochem.* **158,** 171 (1986).

ends, cDNA will be made from all of the RNA in the sample. Additional specificity is provided by the PCR in the next step.

*Specific Amplification of Cleavage Products (Fig. 3A, Step 3).* The PCR is performed with a specific upstream primer and a discriminating anchor primer (DAP; Fig. 3A, step 3). The specific primer is designed based on the known sequence of the RNA being analyzed. To maximize sensitivity during the PCR, we typically use a priming site that is no more than 200 to 300 nucleotides upstream of the candidate editing site. The DAP is complementary to the extreme 5' end of the anchor sequence and has three extra nucleotides, of the form CNX, at its 3' end. The extra C pairs with G at the RNase T1 cleavage site. (Inosine in the RNA becomes guanosine in the cDNA.) N is a randomized position and X is chosen to pair with the nucleotide that is two bases upstream of the cleavage site. In control experiments, we found that the extra nucleotides provided the necessary specificity to detect cleavage at known editing sites.[7]

Reactions contain 1 $\mu$l of cDNA; 2.5 $\mu$l of 10× buffer [100 mM Tris–HCl (pH 8.3), 500 mM KCl, 15 mM MgCl$_2$, 0.1% gelatin]; 2 $\mu$l of 250 $\mu$M dNTPs; 2 $\mu$l of 10 pmol/$\mu$l specific primer; 2 $\mu$l of 10 pmol/$\mu$l DAP; 0.5 $\mu$l of 10 $\mu$Ci/$\mu$l [$\alpha$-$^{32}$P]dCTP (3000 Ci/mmol); 14.5 $\mu$l of water; and 0.5 $\mu$l of 5 U/$\mu$l *Taq* DNA polymerase. The optimal cycling conditions depend on the particular primer pair used. We typically perform 37 cycles with 1-min denaturing and annealing steps and 30-sec extensions. Three microliters of each PCR product is added to 2 $\mu$l of formamide loading buffer (95% formamide, 20 mM EDTA, 0.05% bromphenol blue, 0.05% xylene cyanol), heated to 100° for 1 min, and loaded onto a 6% sequencing gel. A 100-bp ladder that has been 5' end labeled with $^{32}$P serves as a size marker. The gel is run at 32 mA until the xylene cyanol is near the bottom. Fluorescent markers for aligning the gel with the autoradiogram [we use "glogos" (Stratagene La Jolla, CA)] are placed on the dried gel and an X-ray film is exposed overnight. An RNase T1-dependent band of the expected size should be visible on the autoradiogram.

*Notes.* (1) The low dNTP concentration produces high specific activity products. (2) The success of the entire procedure can be monitored by amplifying the control RNA using an upstream primer specific for its sequence (see Fig. 3B).

*Elution, Reamplification, Cloning, and Sequencing of PCR Products.* To confirm that the observed band is the expected PCR product, the band is eluted from the gel, reamplified, cloned and sequenced. Because the anchor is ligated directly to the RNase T1 cleavage site, the sequence reveals both the identity of the RNA and the location of the cleavage site.

The exposed X-ray film is placed on top of the dried gel and aligned with the fluorescent markers (e.g., glogos). A needle is used to punch holes

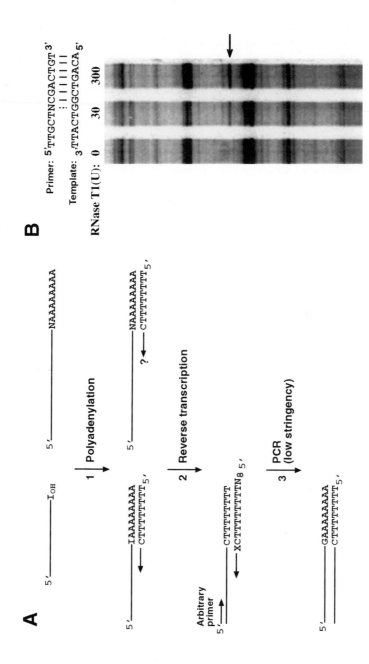

through the film into the gel on either side of the band of interest. The band is cut out of the gel with a razor blade using the holes as a guide. A second X-ray film or a phosphorimager screen is used to confirm that the correct band has been excised. The gel slice is soaked in 100 $\mu$l of water for 10 min, heated at 100° for 15 min, and cooled to room temperature. The gel slice is spun at top speed for 2 min in a microcentrifuge, and the water containing the eluted DNA is removed to a fresh tube. Three microliters of the eluted DNA is reamplified in a 50-$\mu$l PCR. The reaction components and cycling conditions are identical to the original PCR except the final dNTP concentration is 200 $\mu M$ and no isotope is added. Ten microliters of the reaction is run on an agarose gel with a DNA mass ladder (BRL) to estimate the yield, and the rest of the reaction is extracted once with phenol–chloroform and precipitated. The DNA is cloned using a T-vector system (see note) and sequenced using vector primers. The sequence should be that of the candidate RNA, and the anchor should be attached directly to the anticipated editing site.

*Note.* A significant fraction of a PCR product made with a nonproofreading enzyme such as *Taq* has a single additional adenosine at each 3' end. Such products can be ligated directly to a T-vector, which is a linearized plasmid with single additional thymidines at its 3' ends. There are several commercially available T-vectors; we use pGem-T vector (Promega, Madison, WI).

## Identification of New ADAR Substrates

*Principle of Method.* The protocol outlined in Fig. 4A is used to find new ADAR substrates in organisms whose genome sequencing projects

---

FIG. 4. Identification of new inosine-containing RNAs. (A) Differential display strategy. The starting material is periodate-oxidized, poly(A)[+] RNA that has been subjected to inosine-specific cleaveage and postcleavage processing. Step 1: A poly(A) tail is added 3' to the inosine at the cleavage site to create a primer binding site for first-strand cDNA synthesis. Step 2: Reverse transcription with a $T_{12}C$ primer. The question mark indicates that the primer will extend on uncleaved RNA only if N = G. Step 3: Low-stringency PCR. X at the 3' end of the downstream primer is G, A, C, TG, TA, TC, or TT. $N_8$ is the 8-bp extension on the 5' end of the downstream primer. Note that the cDNA template is drawn 3' to 5'. (B) Detection of control RNA using the differential display strategy; 0.5 fmol of the control RNA was spiked into 5 $\mu$g of yeast RNA and subjected to inosine-specific cleavage using the indicated amounts of RNase T1. The cleaved RNA was then subjected to the protocol of A. The complementarity between the upstream arbitrary PCR primer and its priming site is shown above the gel; N represents a randomized position. For the downstream PCR primer, X = G because the nucleotide on the 5' side of the inosine in the control RNA was a C. The arrow points to an RNase T1-dependent band whose sequence confirmed that it derived from the control RNA cleaved precisely 3' to its single inosine. Reprinted with permission from Morse and Bass.[8] Copyright 1999 National Academy of Sciences, U.S.A.

are nearing completion. A genomic DNA sequence is required to screen out false positives as described later. For this reason, we have confined our search to *Caenorhabditis elegans,* but other eukaryotic genome sequences (including human) will be available in the near future.

Prior to inosine-specific cleavage of poly(A)$^+$ RNA, the 3'-hydroxyls are oxidized to prevent elongation of the original poly(A) tails (the oxidation protocol has been described earlier). Following inosine-specific cleavage and postcleavage processing, a poly(A) tail is added to the cleavage sites. The RNA is then amplified by arbitrarily primed RT/PCR as in the differential display method.[18] After first-strand cDNA synthesis, small aliquots of the cDNA are amplified in numerous low-stringency PCRs. Each reaction is performed with one of seven different downstream primers and one of a large collection of arbitrary upstream primers. The downstream primers are of the form GAGACCAGT$_{12}$CX, where X is one of G, A, C, TG, TA, TC or TT. The upstream primers are 13-mers whose sequences are chosen "arbitrarily." (See notes following section on arbitrarily primed PCR for further discussion of primers and differential display.) Each PCR amplifies a subset of the cDNA population and the products are "displayed" on a sequencing gel. Candidate ADAR substrates are identified as RNase T1-dependent bands. True ADAR substrates are distinguished from false positives by comparing the cDNA and genomic sequences for each candidate. Figure 4B shows the results of applying this method to detect cleavage of the control RNA. Figure 5 shows a typical differential display gel produced with *C. elegans* poly(A)$^+$ RNA.

*Polyadenylation (Fig. 4A, Step 1).* Reactions contain 13.5 μl of RNA (from the glyoxal removal step); 4 μl of 5× poly(A) polymerase buffer [100 m*M* Tris–HCl (pH 7.9), 250 m*M* KCl, 3.5 m*M* MnCl$_2$, 1 m*M* EDTA, 500 μg/ml BSA, 50% glycerol]; 1 μl of 10 m*M* ATP; 1 μl of 1 m*M* cordycepin triphosphate; and 0.5 μl of 500 U/μl poly(A) polymerase (USB). After 1 hr at 30°, the enzyme is inactivated by adding 0.5 μl of 15 μg/μl proteinase K (Boehringer Mannheim) and incubating at 37° for 20 min. The RNA is diluted to 100 μl with water, extracted twice with phenol–chloroform, and precipitated. The RNA is dissolved in 11.4 μl of water.

*Notes.* (1) We include cordycepin (3'-deoxyadenosine) triphosphate in the reaction to limit the lengths of the poly(A) tails. (2) Control RNA can be used to visualize the lengths of the added tails and to monitor the efficiency of the oxidation reaction. If oxidation was successful, only cleaved RNA will be tailed.

*First-Strand cDNA Synthesis (Fig. 4A, Step 2).* Reactions contain 11.4 μl of RNA (from the polyadenylation step); 4 μl of 5× MLV RT buffer

[18] P. Liang and A. B. Pardee, *Science* **257,** 967 (1992).

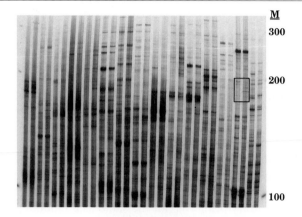

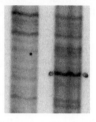

FIG. 5. A differential display gel from the analysis of *C. elegans* poly(A)$^+$ RNA is shown, with a region containing a T1-dependent band (boxed) enlarged below. Each pair of lanes corresponds to a different primer pair, minus (left lane) or plus (right lane) RNase T1. Dots on each side of the band are pinholes used to mark its position for excision and elution of the DNA. M, 100-bp ladder. Reprinted with permission from Morse and Bass.[8] Copyright 1999 National Academy of Sciences, U.S.A.

[250 m$M$ Tris–HCl (pH 8.3), 375 m$M$ KCl, 15 m$M$ MgCl$_2$]; 1 $\mu$l of 10 m$M$ DTT; 1.6 $\mu$l of 250 $\mu M$ dNTPs; 1 $\mu$l of 10 pmol/$\mu$l T$_{12}$C primer; and 1 $\mu$l of 200 U/$\mu$l MLV RT (BRL). Prior to adding the enzyme the mixture is incubated at 65° for 5 min (to denature the RNA) and cooled to 37°. The enzyme is added, and the reaction is incubated at 37° for 1 hr. The RNA is hydrolyzed by adding 2.5 $\mu$l of 1 $M$ NaOH and incubating at 50° for 30 min. The mixture is neutralized with 2.5 $\mu$l of 1 $M$ HCl, and the cDNA is diluted to 1 ml with water.

*Notes.* (1) To enrich for cleaved molecules, the reverse transcription primer has a 3′ terminal C that will pair with inosine (or guanosine) at the RNase T1 cleavage site (see Fig. 4A). The primer will be extended efficiently from the poly(A) tails on uncleaved RNA only when the nucleotide immediately 5′ of the tail (N in Fig. 4A) is a G. (2) We find that hydrolysis of the RNA improves the sensitivity of the procedure.

*Arbitrarily Primed PCR (Fig. 4A, Step 3).* Reactions contain 2 μl of cDNA (from first-strand synthesis); 1 μl of 10× PCR buffer [100 m$M$ Tris–HCl (pH 8.3), 500 m$M$ KCl, 0.1% gelatin]; 1 μl of 25 m$M$ MgCl$_2$; 0.8 μl of 250 μ$M$ dNTPs; 1 μl of 10 pmol/μl downstream primer; 1 μl of 10 pmol/μl arbitrary primer; 0.2 μl of 10 μCi/μl [α-$^{33}$P]dATP (2000–4000 Ci/mmol); 2.8 μl of water; and 0.25 μl of 5 U/μl Amplitaq Gold (Perkin-Elmer, Norwalk, CT). Cycling conditions are 94° for 9 min (to activate the enzyme); 50 cycles of 94° for 1 min, 40° for 2 min, and 72° for 1 min; and 1 cycle of 72° for 5 min.

*Notes.* (1) Only one pair of cDNA samples (plus and minus T1) is used at this stage. The duplicate cDNA samples are used to verify the reproducibility of any RNase T1-dependent PCR products detected (see later). (2) An arbitrary primer is not the same as a random primer. It has a single fixed sequence that is chosen "arbitrarily." (3) Because PCR is performed with a low annealing temperature (40°), each arbitrary primer can be extended from multiple poorly matched priming sites. Each combination of primers produces a unique pattern of bands that represent a subset of the molecules in the original population (see Fig. 5). By using a sufficient number of primer pairs, one can, in theory, sample the entire population. Typically, an arbitrary primer can produce a PCR product if only six to eight of its 3′-most nucleotides are complementary to the template. (4) We found that by randomizing the eighth position from the 3′ end of each arbitrary primer, we could improve sensitivity significantly (see Fig. 4B). (5) The one or two extra nucleotides on the 3′ end of the downstream primers result in each primer amplifying only a subset of the cleaved RNA molecules. This keeps the number of bands produced in each reaction within a manageable range and improves sensitivity. We found that a downstream primer with a single extra T was extended inefficiently. This problem was overcome by adding a second extra 3′ nucleotide (X = TG, TA, TC, or TT; see Fig. 4A). (6) The eight extra nucleotides at the 5′ end of the downstream primers (GAGACCAG) improve the efficiency of reamplifying RNase T1-dependent bands (see later). The 5′ extension can be any sequence but it is important to avoid palindromes (such as restriction sites) because it results in the production of dimeric molecules in which two PCR products are joined tail to tail. (7) We have found that Amplitaq Gold from Perkin-Elmer, a heat-activated enzyme, improves the sensitivity of this procedure greatly (compared to the original Amplitaq). (8) Because only 2 μl of cDNA is used in each PCR, the 1 ml of cDNA produced from 5 μg of RNA is sufficient for 500 reactions. This corresponds to using about 70 arbitrary primers coupled with the seven different downstream primers (70 × 7 = 490). (9) The control RNA can be used to monitor the success of the entire procedure. To mimic typical differential display conditions, we use an upstream primer that is complementary to the control RNA at only its last seven nucleotides. The eighth position from the 3' end of the primer is randomized as in all our arbitrary primers (see Fig. 4B).

*Identification and Processing of Candidates.* Six microliters of each PCR is added to 4 $\mu$l of formamide loading buffer (95% formamide, 20 m$M$ EDTA, 0.05% bromphenol blue, 0.05% xylene cyanol) and loaded onto a 6% sequencing gel. Samples that differ only in whether or not they were treated with RNase T1 are loaded in adjacent lanes (see Fig. 5). The gel is run at 35 mA until the xylene cyanol is about two-thirds of the way down the gel. We use a 5'-end-labeled 100-bp ladder as a molecular weight marker. Fluorescent markers (e.g., glogos) are placed on the dried gel for later alignment with the X-ray film. An X-ray film is placed on the gel and exposed for about 36 hr. The developed film is examined for the presence of bands that are more intense in samples that were treated with RNase T1 (see Fig. 5). Each PCR that produces such RNase T1-dependent bands is repeated using the duplicate pair of cDNA samples. Bands that are reproducibly dependent on RNase T1 represent candidate ADAR substrates. Each candidate band is excised from the gel, and the DNA is reamplified, cloned, and sequenced as described earlier. Frequently, several different PCR products comigrate in the gel so we routinely sequence three clones generated from each excised band. If at least two of the three sequences are the same, the majority sequence is analyzed further.

*Notes.* (1) The gel in Fig. 5 shows that the PCR products range in size from very small to about 300 bp. We ignore bands that are smaller than 100 nucleotides because these may derive from contaminating tRNA. (2) Bands often appear as doublets (sometimes triplets). We cut these out as a single gel slice because they usually represent the same sequence. The multiple bands may be due to nontemplated nucleotides added to the 3' ends of the PCR products or to the two strands of the DNA migrating differently in the gel.

## Confirmation of Candidates

*Principle.* Guanosines in glyoxalated RNA are not completely resistant to RNase T1. Therefore, most of the RNase T1-dependent bands are due to RNAs that have been cleaved after guanosine (see statistics later). This is not a serious problem because true ADAR substrates (I cleavages) are distinguished easily from false positives (G cleavages) by examining genomic sequences. Because ADARs convert adenosines to inosines, T1 cleavage sites within true ADAR substrates appear as adenosines in the corresponding genomic sequences. There are several other characteristics that are typical of ADAR substrates (discussed later). When present in a candidate RNA, these characteristics provide additional evidence for its status as a true ADAR substrate.

*Screening out False Positives.* cDNA sequences obtained from the excised bands are used in a BLAST search to identify the corresponding genomic DNA sequences. Genes that contain an adenosine (rather than a guanosine) at the RNase T1 cleavage site are likely to encode ADAR substrates. The T1 cleavage sites are identified easily in the cDNA sequences because they are immediately 5' of the added poly(A) tail. If a genomic sequence is not identified by the BLAST search, the cDNA sequence can be used to design probes for screening genomic libraries. However, this approach becomes impractical as the number of genes to be cloned increases.

*Confirmation of A to G Changes.* It is important to independently confirm the sequences of the gene and cDNA for each of the candidates that remain after the BLAST search. This is to screen out false positives that could arise from two possible (although unlikely) sources: sequencing errors in the database or RNase T1 cleavage after adenosine in the RNA. To confirm the A-to-G changes, a region surrounding each candidate editing site is amplified from both genomic DNA and cDNA and their sequences are compared. Sequences found in the previous BLAST searches provide the information needed to design PCR primers. Uncloned PCR products can be sequenced directly or the sequences of multiple individual clones can be determined. These two approaches provide complementary information. The sequences of individual clones reveal the distributions of A-to-G changes within single molecules (see later). By sequencing the PCR products directly, one can estimate the fraction of molecules that are deaminated at each site. A very sensitive and accurate method to measure the efficiency of editing at one particular site is limited primer extension performed on the PCR products (see Melcher *et al.*[19] for an example).

*Other Characteristics of ADAR Substrates.* ADARs usually deaminate multiple adenosines within double-stranded regions of RNA. Thus, most candidate substrates should have the potential to fold into one or more stem–loop structures, and multiple A-to-G changes should be found within these potential structures. RNAs that are predicted to be almost completely double stranded should be deaminated at more sites than those whose structures are frequently interrupted by mismatches, bulges, and loops. ADARs have a 5'-nearest neighbor preference: A or U is preferred over C, which is preferred over G.[20] These preferences should be reflected in

[19] T. Melcher, S. Maas, M. Higuchi, W. Keller, and P. H. Seeburg, *J. Biol. Chem.* **270**, 8566 (1995).
[20] A. G. Polson and B. L. Bass, *EMBO J.* **13**, 5701 (1994).

the deamination patterns seen in candidate ADAR substrates. That is, adenosines in good context should be deaminated in a greater fraction of the population than those in poor context.

*Notes.* (1) A candidate ADAR substrate should not be considered a false positive due to the lack of a detectable secondary structure. It is possible that the required dsRNA is formed by intermolecular base pairing with an antisense transcript or the structure may be difficult to detect due to multiple mismatches or a large distance between inverted repeats (see Herb *et al.*[21] for an example of the latter). (2) Future experiments could reveal that the 5'-nearest neighbor preferences for ADARs from some organisms differ from those observed to date.

*Statistics.* We have applied the differential display strategy to search for new ADAR substrates in *C. elegans.* Our results so far provide some useful statistics. In 420 PCRs (60 arbitrary primers × 7 downstream primers) we identified 82 RNase T1-dependent bands. Forty-nine of the bands contained a majority sequence and 46 of these were found in the *C. elegans* database. Five of the 46 candidates contained an adenosine at the RNase T1 cleavage site. The remaining 41 were cleaved at a guanosine as expected for false positives. A-to-G changes for 4 of the 5 remaining candidates were confirmed by comparing genomic and cDNA sequences. Thus, it required an average of 15 arbitrary primers to find one true ADAR substrate. With a single person doing the work, all 4 substrates were found in a period of about 2 months.

### Acknowledgments

Work was done in the laboratory of Dr. Brenda L. Bass in the Department of Biochemistry and the Howard Hughes Medical Institute at the University of Utah. D. Morse was supported by an NIH training grant (CA 09602) and a postdoctoral fellowship from the American Cancer Society (PF 3891).

[21] A. Herb, M. Higuchi, R. Sprengel, P. H. Seeburg, *Proc. Natl. Acad. Sci. U.S.A.* **93,** 1875 (1996).

# Section II

# Tethered-Probe Methodologies

### A. Photochemical Reagents
*Articles 6 through 10*

### B. Chemical Reagents
*Articles 11 through 13*

# [6] Site-Specific 4-Thiouridine Incorporation into RNA Molecules

*By* YI-TAO YU

## Introduction

4-Thiouridine (s⁴U, ⁴SU, or 4-thioU) site-specific cross-linking is a powerful tool for probing the neighboring microenvironment of this residue in a complex macromolecular system. Since the early demonstration that 4-thioU could be used as a photoactivatable agent for site-specific cross-linking,[1,2] this technique has been applied to studies of protein translation and pre-mRNA splicing, providing detailed information as to how an mRNA makes close contacts with ribosomal constituents[3–5] and how the conserved elements in a pre-mRNA (the 5′ and 3′ splice sites and branch site) are recognized.[6–10]

The use of 4-thioU as a cross-linking probe has a number of advantages.[11] For instance, because 4-thioU is a uridine derivative with the oxygen atom at position 4 of the pyrimidine ring substituted by a sulfur atom (Fig. 1), no bulky moiety is introduced and no base-pairing properties are altered; 4-thioU can be activated on irradiation at wavelength above 300 nm, which holds nonspecific UV cross-linking at a minimum; 4-thioU cross-linking can effectively probe both RNA–RNA and RNA–protein interactions.

However, because the results of cross-linking are easiest to interpret when an RNA is substituted with 4-thioU at only one specific site, construction of such RNAs has been an important challenge. In previous studies, two methods were widely used. The first method involved *in vitro* phage RNA polymerase transcription of a short template composed almost entirely of A, G, and C residues.[3–5] For a long RNA molecule containing

[1] M. Yaniv, A. Favre, and B. G. Barrell, *Nature* **223,** 1331 (1969).
[2] J. Ninio, A. Favre, and M. Yaniv, *Nature* **223,** 1333 (1969).
[3] K. Stade, J. Rinke-Appel, and R. Brimacombe, *Nucleic Acids Res.* **17,** 9889 (1989).
[4] W. Tate, B. Greuer, and R. Brimacombe, *Nucleic Acids Res.* **18,** 6537 (1990).
[5] J. Rinke-Appel, N. Junke, K. Stade, and R. Brimacombe, *EMBO J.* **10,** 2195 (1991).
[6] M. J. Moore and P. A. Sharp, *Science* **256,** 992 (1992).
[7] J. R. Wyatt, E. J. Sontheimer, and J. A. Steitz, *Genes Dev.* **6,** 2542 (1992).
[8] E. J. Sontheimer and J. A. Steitz, *Science* **262,** 1989 (1993).
[9] R. K. Gaur, J. Valcarcel, and M. R. Green, *RNA* **1,** 407 (1995).
[10] A. J. Newman, S. Teigelkamp, and J. D. Beggs, *RNA* **1,** 968 (1995).
[11] A. Favre, *in* "Bioorganic Photochemistry: Photochemistry and the Nucleic Acids" (H. Morrison, ed.), p. 379. Wiley, New York, 1990.

METHODS IN ENZYMOLOGY, VOL. 318
0076-6879/00 $30.00

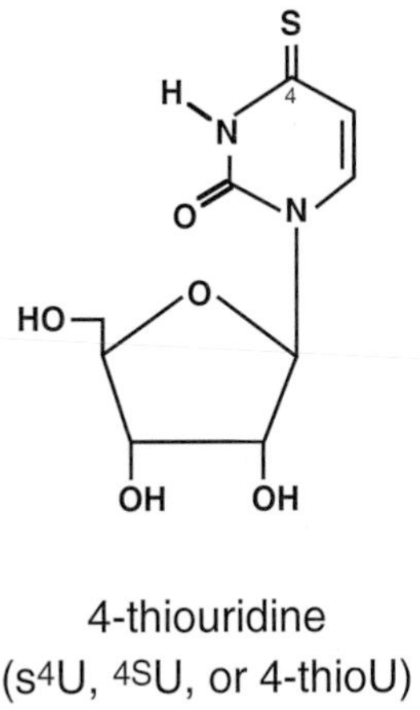

4-thiouridine
($s^4$U, $^4$SU, or 4-thioU)

Fig. 1. Structure of 4-thiouridine. The fourth position of the pyrimidine ring is indicated.

multiple uridines, however, site-specific incorporation of 4-thioUTP via *in vitro* transcription is not possible. In the second method, two half-RNAs, which in combination correspond to a full-length RNA, were prepared separately. 4-ThioUpG was used to initiate the transcription of the 3′-half RNA. Following 5′ phosphorylation (usually with [$\gamma$-$^{32}$P]ATP), the resultant 3′-half RNA and the 5′-half RNA were aligned with a bridging deoxyoligonucleotide and ligated with T4 DNA ligase.[6–10,12] Because phage RNA polymerases (T3, T7, and SP6) all require guanosine to efficiently initiate transcription,[13,14] priming of the 3′-half transcripts with 4-thioUpG is necessary. Although in principle this method is quite useful for introducing a single 4-thioU into a long RNA molecule, its requirement of 4-thioUpG priming constrains the sequence surrounding the site of 4-thioU substitution: the uridines that are to be substituted must be followed by guanosines. Clearly, many sites of interest within a substrate RNA do not meet this requirement.

An alternative approach to introducing 4-thioU into specific sites in a long RNA molecule has been developed in our laboratory.[15] This method involves four basic steps that are schematized in Fig. 2. First (I), 5′-half and 3′-half RNAs are synthesized via phage RNA polymerase transcription and/or RNase H site-specific cleavage directed by 2′-*O*-methyl

[12] E. J. Sontheimer, *Mol. Biol. Rep.* **20**, 35 (1994).
[13] J. F. Milligan and O. C. Uhlenbeck, *Methods Enzymol.* **180**, 51 (1989).
[14] C. Kang and C.-W. Wu, *Nucleic Acids Res.* **15**, 2279 (1987).
[15] Y.-T. Yu and J. A. Steitz, *RNA* **3**, 807 (1997).

RNA–DNA chimeras.[16] When combined, these two half-RNAs correspond to the sequence of the full-length RNA, with a single nucleotide gap at the junction that will be filled with a 4-thiouridylate. In the second step (II), $^{4S}$UpN (N can be any nucleotide), a 4-thioU-containing dinucleotide, is first phosphorylated (usually with radioactive phosphate) at its 5′ end and is then subjected to RNase A digestion, which results in the formation of 4-thioU 5′,3′-diphosphate (p$^{4S}$Up). In step 3 (III), a single p$^{4S}$Up is attached to the 3′ end of the 5′-half RNA with T4 RNA ligase. The 3′-phosphate of the ligated product is subsequently removed by calf intestinal alkaline phosphatase (CIP) to produce a 3′-hydroxyl (OH) group. In the final step (IV), the resulting 5′-half RNA and the 3′-half RNA with a 5′-phosphate group (which can also be radioactive) are aligned with a bridging deoxyoligonucleotide and ligated with T4 DNA ligase. The sequence of the joined RNA corresponds to that of the desired full-length RNA, with a single uridine being site-specifically replaced by 4-thioU. Because 4-thioUpG priming is eliminated, this approach has no flanking sequence requirement and therefore can be applied to virtually any RNA. Detailed procedures for this approach and its applications are described.

## Site-Specific 4-Thiouridine Incorporation

### Step I. Preparation of 5′- and 3′-Half RNAs

The 5′-half RNA can be synthesized by *in vitro* transcription with phage RNA polymerase. This is the simplest and most convenient way to produce large quantities of RNA. In some cases, however, phage RNA polymerases incorporate additional nucleotides at the 3′ end of a transcript, causing problems in the later bridging oligonucleotide-mediated two-piece ligation step. This difficulty can be circumvented by site-specific cleavage of a full-length RNA transcript with a hammerhead or hairpin ribozyme[17–19] or with RNase H and a 2′-*O*-methyl RNA–DNA chimera.[16] Because the cleavage occurs precisely at the desired site and is virtually sequence independent, this approach has tremendous advantages for creating a homogeneous 3′ end on the 5′-half RNA. Site-specific cleavage is also a good method for producing a 3′-half RNA with a clean 5′ end. More importantly, when the residue in the first or second position downstream of the target uridine is

[16] J. Lapham and D. M. Crothers, *RNA* **2**, 289 (1996).
[17] C. A. Grosshans and T. R. Cech, *Nucleic Acids Res.* **19**, 3875 (1991).
[18] M. Altschuler, R. Tritz, and A. Hampel, *Gene* **122**, 85 (1992).
[19] A. R. Ferré-D'Amaré and J. A. Doudna *Nucleic Acids Res.* **24**, 977 (1996).

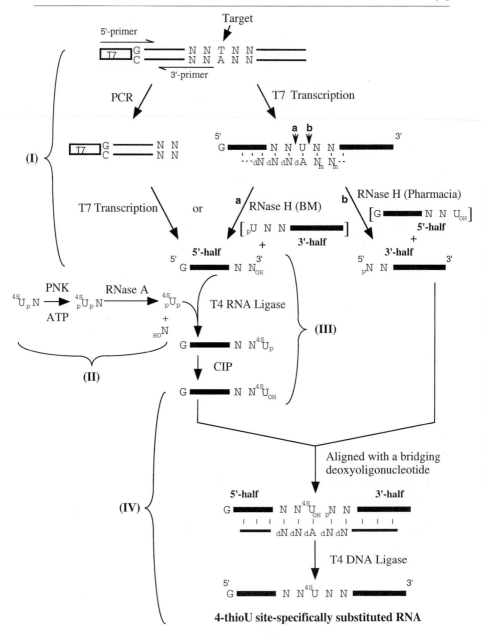

**4-thioU site-specifically substituted RNA**

not a guanosine, phage RNA polymerase transcription cannot be used to produce the 3'-half RNA. In these cases, sequence independent, site-specific cleavage appears to be the ideal approach for creating the desired 3'-half RNA.[15]

Because of space limitations, only *in vitro* transcription and site-specific RNase H cleavage directed by a 2'-O-methyl RNA–DNA chimera are detailed.

### A. Preparation of 5'- and 3'-half RNAs via in Vitro phage RNA Polymerase Transcription

#### Buffers, Reagents, and Equipment

1. For transcription:

   5× transcription buffer: 200 m$M$ Tris–HCl, pH 7.5, 30 m$M$ MgCl$_2$, 10 m$M$ spermidine

   T3, T7, or SP6 RNA polymerase, 45–69 units/$\mu$l (Pharmacia, Piscataway, NJ)

   RNase inhibitor, 40 units/$\mu$l (Boehringer Mannheim, Mannheim, Germany)

   Dithiothreitol (DTT), 0.5 $M$ (Sigma, St. Louis, MO)

   NTPs (Pharmacia): 10 m$M$ ATP, 10 m$M$ CTP, 10 m$M$ UTP, 10 m$M$ GTP

   Transcription primer (or initiator): 10–20 m$M$ G(5')ppp(5')G (or another cap analog) (Pharmacia), GMP (Pharmacia), or dinucleotide (Sigma) (if desired)

   Radioactive nucleotide: [$\alpha$-$^{32}$P]UTP (Amersham, Arlington Heights, IL)

   RNase-free autoclaved distilled water

   Polymerase chain reaction (PCR)-produced DNA template (or linearized plasmid containing the desired template) (see note a)

---

FIG. 2. Strategy for introducing a 4-thioU into a specific position in an RNA molecule. The strategy (overall scheme) involves four basic steps that are detailed in the text: (**I**) preparation of a 5'-half and a 3'-half RNA; (**II**) conversion of $^{4S}$UpN to p$^{4S}$Up; (**III**) attachment of 4-thioU to the 3' end of the 5'-half RNA; (**IV**) ligation of the 5'-half and the 3'-half RNAs. In step I, the T7 promoter, PCR primers (5'-primer and 3'-primer), and a 2'-O-methyl RNA–DNA chimera (---dNdNdNdANmNm--, where the dotted line and Nm represent 2'-O-methyl ribonucleotides and dN stands for deoxynucleotide) are depicted. When RNase H from Boehringer Mannheim (BM) is used, the RNA is cleaved at position a (indicated by arrow a), producing the desired 5'-half RNA. When RNase H from Pharmacia is used, cleavage occurs at position b (indicated by arrow b), yielding the desired 3'-half RNA (see text). In step II, PNK stands for polynucleotide kinase. In step III, CIP stands for calf intestinal alkaline phosphatase. In step IV, thinner lines indicate the bridging oligodeoxynucleotide used for two-piece ligation with T4 DNA ligase. RNA chains are represented by thick lines.

RQ1 RNase-free DNase, 1 unit/$\mu$l (Promega, Madison, WI)
2. For PCA extraction and ethanol precipitation:
   G50 buffer: 20 m$M$ Tris–HCl, pH 7.5, 300 m$M$ sodium acetate, 2 m$M$ EDTA, 0.25% sodium dodecyl sulfate (SDS)
   PCA: [Tris–HCl (pH 7.5) buffered phenol : chloroform : isoamyl alcohol (50 : 49 : 1)]
   Ethanol, 100%
3. For gel purification and elution:
   Formamide loading buffer: 95% formamide, 10 m$M$ EDTA, 0.1% xylene cyanol FF, 0.1% bromphenol blue
   Polyacrylamide–8 $M$ urea gel solution: Sequel NE (American Bioanalytical, Natick, MA)
   Electrophoresis apparatus
   Film (Eastman Kodak, Rochester, NY)
   G50 buffer, PCA, and ethanol as described earlier
   Glycogen, 10 mg/ml (Boehringer Mannheim)

*Procedure*

TRANSCRIPTION

1. At room temperature, mix together 14 $\mu$l of autoclaved distilled water, 20 $\mu$l of 5× transcription buffer, 10 $\mu$l of 0.1 $\mu$g/$\mu$l PCR-produced DNA template (or 10 $\mu$l of 1 $\mu$g/$\mu$l of linearized plasmid containing desired template) (see note a), 12 $\mu$l each of 10 m$M$ ATP, CTP, UTP, and GTP (if 5' cap is required, 12 $\mu$l 10 m$M$ GTP is replaced by a mixture of 2 $\mu$l of 10 m$M$ GTP and 10 $\mu$l of 10–20 m$M$ cap analog) (see notes b and c), 1 $\mu$l of 0.5 $\mu$Ci/$\mu$l [$\alpha$-$^{32}$P]UTP (see note d), 1 $\mu$l of 0.5 $M$ DTT, 1 $\mu$l of 40 units/$\mu$l RNase inhibitor, and 5 $\mu$l of 45–69 units/$\mu$l of phage RNA polymerase (T3, T7, or SP6, depending on the promoter sequence) in a 1.5-ml Eppendorf tube.
2. Remove 1 $\mu$l and determine cpm (see note d). Place the tube containing the rest of the reaction (99 $\mu$l) in a 37° water bath and incubate for 1–2 hr.
3. Add 2 $\mu$l of 1 unit/$\mu$l of RQ1 RNase-free DNase to the tube and incubate at 37° for another 30 min.

PCA EXTRACTION AND ETHANOL PRECIPITATION

4. Bring the volume of the just-described DNase-treated transcription reaction to ~250 $\mu$l with G50 buffer (need ~150 $\mu$l), add 2 volumes (~500 $\mu$l) of PCA, mix vigorously by vortexing, and microfuge for 2 min.

5. Remove the aqueous phase, mix thoroughly with 2.5 volume (~625 $\mu$l) of ethanol, and place on dry ice for 10 min. Microfuge at 14,000$g$ for 10–15 min. Remove and discard the supernatant promptly.

GEL PURIFICATION AND ELUTION

6. Resuspend the pellet in 10 $\mu$l of autoclaved distilled water, mix with 20 $\mu$l of formamide loading buffer, heat at 95° for 3 min, chill on ice immediately, and load the sample on a 5 or 6% polyacrylamide–8 $M$ urea gel.
7. Expose the gel marked with fluorescent ink to EK film for about 5–10 min.
8. Locate and excise the RNA transcript band and place it in a 1.5-ml Eppendorf tube.
9. Add 450 $\mu$l G50 buffer to the tube; place the tube on dry ice for 5 min and then transfer to room temperature for elution overnight (~16 hr).
10. Microfuge the tube containing the gel slice at 14,000$g$ for 15 min. Transfer the supernatant to a new 1.5-ml Eppendorf tube. Extract the supernatant with 500 $\mu$l of PCA and precipitate the gel-purified RNA with 1000 $\mu$l of ethanol (using 1 $\mu$l of 10 mg/ml glycogen as carrier), as described earlier.
11. Determine the total cpm incorporated and calculate the exact amount of RNA transcribed (see note d). Resuspend the RNA pellet in autoclaved-distilled water at 20 pmol/$\mu$l.

*Notes to Step I, A*

a. The DNA template for transcribing the 5′-half RNA, which is designed to end at the nucleotide preceding the target uridine [Fig. 2, (I)], can be either a linearized plasmid or a PCR product. In many cases, however, the template sequence in a plasmid contains no convenient restriction sites that allow linearization at a desired position. Hence, the use of plasmid templates is greatly limited. However, DNA produced by PCR can be designed to end at any desired site, favoring the use of a PCR product as the transcription template for the 5′-half RNA [Fig. 2, (I)]. The 5′ PCR primer contains a phage promoter sequence (T3, T7 or SP6), a guanylate (as the starting nucleotide), and the 5′-most 15–20 nucleotides of the desired RNA. It is critical that the first templated nucleotide be a guanylate, as phage RNA polymerases all require guanylate to efficiently initiate transcription.[13,14] The 3′ PCR primer is a 15 to 20-mer that is complementary to the sequence preceding the target uridine. Detailed descriptions of PCR techniques are presented elsewhere.[20]

b. For RNAs whose 5' ends need to be capped, the ratio of GTP:cap analog used is between 1:10 and 1:5. The excess of cap analog ensures efficient priming of the RNA. The 5' cap structure is important for many applications in which the RNA is to be incubated in a cellular environment (e.g., *in vivo* or *in vitro* splicing in *Xenopus* oocytes or nuclear extracts, respectively) where $5' \rightarrow 3'$-exonuclease activities would degrade an uncapped RNA transcript rapidly.

c. If the residue in the first or second position downstream of the target uridine is a guanosine, the 3'-half RNA may also be synthesized via phage RNA polymerase transcription. GMP or NpG (dinucleotide), respectively, is used to initiate transcription (the ratio of GTP:GMP or GTP:NpG is 1:10–1:5). Because a 5'-phosphate on the 3'-half RNA is required for the later two-piece ligation (see later), an RNA transcript initiated with NpG needs to be phosphorylated with ATP (or $[\gamma\text{-}^{32}P]ATP$) and polynucleotide kinase (the phosphorylation protocol is described in Step II).

d. A trace amount of radioactive nucleotide ($[\alpha\text{-}^{32}P]UTP$) is included in addition to the unlabeled nucleotide (UTP) to allow calculation of the percentage incorporation and thereby accurate determination of the amount of RNA transcribed. For templates with average transcription efficiency, a 100-$\mu$l reaction can produce 50–100 pmol of 5'-capped RNA.

*B. Synthesis of 5'-Half and 3'-Half RNAs via Site-Specific RNase H Cleavage Directed by 2'-O-Methyl RNA–DNA Chimeras*

*Buffers, Reagents, and Equipment*

Desired full-length RNA transcript (see note a)

2'-O-methyl RNA–DNA chimeric oligonucleotides (Keck Oligonucleotide Synthesis Facility at Yale University) (see note b)

2× RNase H buffer: 40 m$M$ Tris–HCl, pH 7.5, 20 m$M$ MgCl$_2$, 200 m$M$ KCl, 50 mM DTT, 10% sucrose

RNase inhibitor, 40 units/$\mu$l (Boehringer Mannheim)

RNase H, 1 unit/$\mu$l (Pharmacia or Boehringer Mannheim) (see note b)

Buffers, reagents, and equipment for PCA extraction, ethanol precipitation, and gel purification and elution are the same as described earlier

---

[20] R. A. Eeles and A. C. Stamps, "Polymerase Chain Reaction (PCR): The Technique and Its Applications." CRC Press, Boca Raton, FL, 1993.

*Procedure*

1. In a 1.5-ml Eppendorf tube, mix 2 $\mu$l of 20 pmol/$\mu$l trace-labeled full-length RNA transcript (see note a) with 2 $\mu$l of 35 pmol/$\mu$l 2'-O-methyl RNA–DNA chimera (see note b). Heat at 95° for 3 min and then anneal at 37° for 10 min (see note c).
2. Add to the annealing mixture 10 $\mu$l of 2× RNase H buffer, 1 $\mu$l of 40 units/$\mu$l RNase inhibitor, and 5 $\mu$l of 1 unit/$\mu$l RNase H (either Pharmacia or Boehringer Mannheim) (see note b). Incubate at 37° for 2 hr.
3. Terminate the reaction by adding 300 $\mu$l of G50 buffer. PCA extract and ethanol precipitate the sample as described earlier.
4. Gel purify the cleaved half-RNAs as described (also see note d). Resuspend the recovered half-RNAs in autoclaved distilled water to a final concentration of ~20 pmol/$\mu$l.

*Notes to Step I, B*

a. A full-length RNA synthesized by phage RNA polymerase is used as a substrate for site-specific RNase H cleavage. The transcription procedure is essentially the same as that for synthesizing a 5'-half RNA, except that the DNA template is a linearized plasmid containing an insert coding for the full-length RNA.
b. To obtain RNase H cleavage at the desired site, the correct design of a 2'-O-methyl RNA–DNA chimera is crucial. A 2'-O-methyl RNA–DNA chimera consisting of four deoxynucleotides flanked by 2'-O-methyl ribonucleotides directs RNase H cleavage of a complementary RNA at a specific site.[16] This cleavage reaction is sequence independent; however, depending on the source of the enzyme, the cleavage site shifts one nucleotide.[21,22] Both Sigma and Pharmacia RNase H cleave the RNA at the phosphodiester bond 3' to the ribonucleotide that is base paired with the 5'-most deoxynucleotide of the chimera[16,21] [Fig. 2, (I)]. In contrast, RNase H from Boehringer Mannheim (BM) cleaves one nucleotide upstream, 5' to the ribonucleotide that base pairs to the 5'-most deoxynucleotide[15,21,22] [Fig. 2, (I)]. Use of two different enzymes with only one chimeric oligonucleotide can therefore generate both a 5'-half RNA and a 3'-half RNA. The chimera should be designed in such a way that the 5'-most deoxynucleotide base pairs with the uridylate to be substituted in the RNA [Fig. 2, (I)]. Cleavage by RNase H from Boehringer Mannheim will

[21] J. Lapham, Y.-T. Yu, M.-D. Shu, J. A. Steitz, and D. M. Crothers, *RNA* **3**, 950 (1997).
[22] Y.-T. Yu, M.-D. Shu, and J. A. Steitz, *RNA* **3**, 324 (1997).

occur at the phosphodiester bond 5' to the target uridine, generating the 5'-half RNA [Fig. 2, (I), pathway a], whereas Pharmacia RNase H will cleave the phosphodiester bond 3' to the target uridine, yielding the 3'-half RNA [Fig. 2, (I), pathway b]. In rare cases, RNase H from Boehringer Mannheim may cleave the RNA at two adjacent positions. Primer-extension analysis of the 3'-half RNA to check the exact cleavage site(s) is recommended. If double cuts are observed, alternative approaches (e.g., designing two different chimeras for use with Pharmacia RNase H) should be considered. Another economic way to generate 5'-half and 3'-half RNAs is to use a mutant RNA substrate with the target uridylate deleted. Chimera-directed RNase H cleavage at the deletion site then yields both the 5'-half and the 3'-half RNAs in a single step.

c. For some RNAs, heating and gradual cooling of the RNA substrate mixed with the 2'-O-methyl RNA–DNA chimera in the absence of salt may not produce efficient hybridization. In these cases, KCl can be included in the hybridization mixture at a concentration of 500 m$M$. After heating and reannealing, a modified 2× RNase H buffer omitting salt can be used to enable the RNase H cleavage reaction to proceed.

d. Before loading on the gel, the sample is heated at 94° for 3 min, ensuring that all cleaved RNA/chimera duplexes become completely denatured. Because the RNA substrate is trace labeled, the cleaved RNAs are located conveniently on the gel by autoradiography and the yield can be calculated accurately. Usually, RNase H cleavage is very efficient (>95%), and recovery of cleaved RNA reaches at least 60% (>24 pmol) of the theoretical value (40 pmol).

e. RNase H cleavage yields a 3'-hydroxyl group on the 5'-half RNA, required for subsequent p$^{4S}$Up ligation with T4 RNA ligase (see Step III), and a phosphate group at the 5' end of the 3'-half RNA, ideal for the later bridging oligonucleotide-mediated two-piece ligation with T4 DNA ligase (see Step IV). However, if site-specific labeling of the phosphate 3' to the 4-thioU is desired, the 3'-half RNA can be subjected to dephosphorylation with calf intestinal alkaline phosphatase (CIP) followed by rephosphorylation with [$\gamma$-$^{32}$P]ATP and T4 polynucleotide kinase. Protocols for phosphorylation and dephosphorylation are detailed in Step II and Step III, respectively.

*Step II. Synthesis of 4-Thiouridine 5',3'-Diphosphate*
*(p$^{4S}$Up, or [$^{32}$P]p$^{4S}$Up)*

Although p$^{4S}$Up is not available commercially at present, it can be derived readily from 4-thioUpU ($^{4S}$UpU) or 4-thioUpN, where N can be any nucleotide (Sigma; Sierra Bioresearch, Tucson, AZ). The conversion of $^{4S}$UpU to p$^{4S}$Up involves two basic steps: 5'-phosphorylation of $^{4S}$UpU followed by RNase A digestion (RNase A cleaves 3' to pyrimidines, leaving 3' phosphate/5' hydroxyl termini) [Fig. 2, (II)].

### Buffers, Reagents, and Equipment

4-thioUpU ($^{4S}$UpU) or 4-thioUpN, 1 mM (Sigma; Sierra Bioresearch)
5× kinase buffer: 250 mM Tris–HCl, pH 7.5, 50 mM MgCl$_2$
ATP, 100 mM (Pharmacia), or [γ-$^{32}$P]ATP (6000 mCi/mmol, Du Pont/ NEN) (see note a)
Polynucleotide kinase, 7.9 units/μl (Pharmacia)
RNase A, 10 mg/ml (Sigma)
Proteinase K, 10 mg/ml (Boehringer Mannheim)
G50 buffer, PCA, ethanol, and RNase-free autoclaved distilled water (as described earlier)

### Procedure

PHOSPHORYLATION

1. In a 1.5-ml Eppendorf tube, mix together 4 μl of 5× kinase buffer, 10 μl of 1 mM of 4-thioUpU, 2 μl of 100 mM ATP (or 1.5–5 mCi of [γ-$^{32}$P]ATP, if desired) (see note a), and 4 μl of 7.9 units/μl polynucleotide kinase.
2. Place the tube at 37° and incubate for 1 hr.
3. Transfer to 65° for 10 min to terminate the reaction.

RNASE A DIGESTION AND PROTEINASE K TREATMENT

4. Add 4 μl of 10 mg/ml RNase A to the tube and incubate at 42° for 1 hr.
5. Bring the volume to 100–150 μl with G50 buffer, add 10 μl of 10 mg/ml proteinase K, and incubate at 42° for 1 hr (see note b).
6. PCA extract four times (each with 500 μl) (see note c). Remove the aqueous phase after each PCA extraction to a new 1.5-ml Eppendorf tube.
7. Add 2 μl of 100 mM ATP and 4 volumes (~400–600 μl) of ethanol to the aqueous phase of the final PCA extraction, place on dry ice for ~30 min, and then microfuge at 14,000g for 30 min (see note d).
8. Resuspend the pellet containing the product p$^{4S}$Up (see note d) in 3 μl of autoclaved distilled water.

### Notes to Step II

a. If site-specific labeling is desired, ATP can be replaced by 1.5–5 mCi of [γ-$^{32}$P]ATP (6000 mCi/mmol, Du Pont/NEN), which is roughly equivalent to 250–833 pmol. Because 5 mCi [$\gamma$-$^{32}$P]ATP is contained in a volume of ~33 $\mu$l, a larger reaction volume (~50 $\mu$l) is needed. For efficient phosphorylation, ATP must be in excess relative to $^{4S}$UpN; similarly, for site-specific labeling, the ratio of [$\gamma$-$^{32}$P]ATP/$^{4S}$UpN should be kept as high as possible. Because a large amount of radioactivity is involved, caution should be taken.

b. After RNase A digestion, it is necessary to completely inactivate this ribonuclease as it will degrade RNA and cause serious problems in later steps. Proteinase K treatment followed by PCA extraction is effective.

c. Extensive PCA extraction ensures complete removal of any residual ribonuclease.

d. Inclusion of ATP, complete freezing of the sample, and longer centrifugation can reduce the loss of p$^{4S}$Up during ethanol precipitation. Although the recovery may be less than complete, sufficient amounts of p$^{4S}$Up for subsequent ligations (see later) can be readily obtained in this way. Because $_{OH}$U is inert for ligation catalyzed by T4 RNA ligase, its residual presence in the mixture is not harmful; no additional purification steps are necessary.

*Step III. Ligation of p$^{4S}$Up (or [$^{32}$P]p$^{4S}$Up) to 3' End of 5'-Half RNA*

*Buffers, Reagents, and Equipment*

p$^{4S}$Up or [$^{32}$P]p$^{4S}$Up in 3 $\mu$l autoclaved distilled water (prepared in Step II)

5'-half RNA, 20–25 pmol/$\mu$l (prepared in Step I)

2× RNA ligation buffer: 100 m$M$ HEPES, pH 8.3, 10 $\mu M$ ATP, 20 m$M$ MgCl$_2$, 6.6 m$M$ DTT, 20% (v/v) dimethyl sulfoxide (DMSO), 30% (v/v) glycerol

RNase inhibitor, 40 units/$\mu$l (Boehringer Mannheim)

DTT, 50 m$M$

T4 RNA ligase, 6.9 units/$\mu$l (Pharmacia)

Sodium acetate, 3 $M$ (pH 5.2)

PCA, ethanol, and autoclaved distilled water (as described earlier)

10× dephosphorylation buffer: 500 m$M$ Tris–HCl, pH 8.5, 1 m$M$ EDTA

Calf intestinal alkaline phosphatase, 20 U/$\mu$l (Boehringer Mannheim)

*Procedure*

p$^{4S}$ Up Ligation

1. On ice, mix together 3 $\mu$l of p$^{4S}$Up (or [$^{32}$P]p$^{4S}$Up), 1 $\mu$l of 20–25 pmol/$\mu$l 5'-half RNA, 8 $\mu$l of 2× RNA ligation buffer, 1 $\mu$l of 40

U/$\mu$l RNase inhibitor, 1 $\mu$l of 50 m$M$ DTT, and 2 $\mu$l of 6.9 U/$\mu$l T4 RNA ligase (see note a).

2. Incubate at 4° (in the cold room or refrigerator) for ~16 hr (overnight).
3. Add 234 $\mu$l of autoclaved distilled water, PCA extract twice (500 $\mu$l each).
4. Add 25 $\mu$l of 3 $M$ sodium acetate (pH 5.2) and precipitate with 2.5 volumes of ethanol.
5. Resuspend the pellet in 8 $\mu$l of autoclaved distilled water.

DEPHOSPHORYLATION

6. Mix the 8 $\mu$l p$^{4S}$Up-ligated 5'-half RNA with 1 $\mu$l of 10× dephosphorylation buffer and 1 $\mu$l of 20 U/$\mu$l CIP.
7. Incubate at 50° for 1 hr.
8. Add 240 $\mu$l of autoclaved distilled water, PCA extract, and ethanol precipitate as described earlier.
9. Resuspend the pellet in 2 $\mu$l autoclaved distilled water (~10 pmol/$\mu$l).

*Notes to Step III*

a. The ligation reaction is essentially as described by England and Uhlenbeck.[23] The reaction volume is kept relatively small to concentrate the enzyme and its substrates, thereby achieving more efficient ligation. The efficiency obtained is usually comparable to that of conventional pCp ligation.[15]
b. After p$^{4S}$Up ligation, the 3' phosphate group of the 5'-half RNA must be removed by dephosphorylation as described earlier. The resultant 5'-half RNA containing a 4-thioU with a hydroxyl group at its 3' terminus can then be used for the subsequent two-piece ligation (see later).

*Step IV. Ligation of 5'-Half and 3'-Half RNAs with Bridging Deoxyoligonucleotide and T4 DNA Ligase*

*Buffers, Reagents, and Equipment*

4-thioU-containing 5'-half RNA, 10 pmol/$\mu$l (prepared in Step III)
3'-half RNA, 20 pmol/$\mu$l (prepared in Step I)
Bridging deoxyoligonucleotide, which is complementary to 15–20 nucleotides on each side of the ligation junction [Fig. 2, (IV)] (Keck Oligonucleotide Synthesis Facility at Yale University)

[23] T. E. England and O. C. Uhlenbeck, *Biochemistry* **17**, 2069 (1978).

5× ligation buffer: 100 m$M$ Tris–HCl, pH 7.5, 33 m$M$ MgCl$_2$, 1 m$M$ ATP, 2.5 mg/ml BSA

RNase inhibitor, 40 U/$\mu$l (Boehringer Mannheim)

DTT, 50 m$M$

T4 DNA ligase, 10 U/$\mu$l (USB)

Buffers, reagents, and equipment for PCA extraction, ethanol precipitation, and gel purification and elution are the same as described earlier.

*Procedure*

1. In a 1.5-ml Eppendorf tube, mix together 2 $\mu$l of 10 pmol/$\mu$l 5′-half RNA, 1.5 $\mu$l of 20 pmol/$\mu$l 3′-half RNA, and 1.5 $\mu$l of 20 pmol/$\mu$l bridging deoxyoligonucleotide.
2. Heat at 94° for 2 min and then incubate at room temperature for 10 min.
3. Microfuge briefly, and add 3 $\mu$l of 5× ligation buffer, 1 $\mu$l each of 40 U/$\mu$l RNase inhibitor and 50 m$M$ DTT, and 5 $\mu$l of 10 U/$\mu$l T4 DNA ligase (see note a).
4. Incubate at 37° for 2–4 hr (see note b).
5. Bring the volume to 250 $\mu$l with G50 buffer, PCA extract once, ethanol precipitate, and gel purify the ligated full-length RNA (see note c), as described.

*Notes to Step IV*

a. Ligation is conducted in a small volume (~15 $\mu$l), which concentrates the enzyme and its substrates to ensure efficient ligation.
b. For some RNAs, lengthening the ligation time can increase the efficiency considerably. After an initial 2-hr incubation at 37°, additional T4 DNA ligase (~2 $\mu$l) is added, and the ligation is allowed to proceed at room temperature for another ~16 hr (overnight). The efficiency of two-piece ligation for many RNAs can be greater than 90%.
c. Before resolution on a polyacrylamide–8 $M$ urea gel, it is important to heat (95° for 3 min) the sample adequately to denature any hybrids between the bridging oligonucleotide and substrate RNAs. Because the half-RNAs are labeled, the ligated full-length RNA can be visualized by autoradiography and eluted accordingly.

*Additional Comments*

Application of the methodology described in this article is not limited to synthetic RNAs. RNase H site-specific cleavage also permits site-specific

4-thioU incorporation into a cellularly derived RNA molecule, which may be otherwise modified posttranscriptionally. For instance, to replace a uridine with 4-thioU in a cellularly derived snRNA, RNase H site-specific cleavage immediately upstream and downstream of that U residue will produce the required 5′-half and 3′-half RNAs. After ligation of a 4-thioU to the 3′ end of the 5′-half RNA, the two half-RNAs can be rejoined via bridging oligonucleotide-mediated two-piece ligation.

It has been reported that radioactive labeling with $^{32}$P may result in dethiolation of 4-thioU.[24] It is therefore recommended that a freshly made RNA substrate containing a $^{32}$P-labeled 4-thioU be utilized immediately in a cross-linking experiment.

It should be noted that the method described here involves more steps (Steps I–IV) and is therefore more complex than the previously described approach[12] involving priming of the 3′-half RNA transcript with 4-thioUpG. However, this disadvantage is outweighed by its sequence independence. Use of the new method provides an ideal way to construct many 4-thioU site-specifically substituted RNAs that could not be obtained by the previous method.

Applications

The just-described method has been applied to both P120 pre-mRNA, which contains an AT-AC intron, and the adenovirus standard splicing substrate (Ad pre-mRNA). Specifically, a single 4-thioU has been introduced into three different positions in the P120 pre-mRNA—+2, +4, or +7 with respect to the conserved 5′ splice site[15,25]—and into one position in the Ad pre-mRNA—−3 relative to the 3′ splice site. None of these uridylates is followed by a guanylate. These 4-thioU-containing pre-mRNAs were then used in cross-linking experiments to dissect how splice sites are recognized during splicing in HeLa nuclear extract.

*4-ThioU Site-Specific Cross-Linking during Pre-mRNA Splicing*

*Procedure*

1. Set up a large *in vitro* splicing reaction with a 4-thioU-containing pre-mRNA (usually site-specifically labeled with $^{32}$P, ~$10^6$–$10^7$ cpm/ 20-$\mu$l reaction)[7,25] and place the reaction in a 30° water bath.

[24] G. Igloi, *Biochemistry* **27**, 3842 (1988).
[25] Y.-T. Yu and J. A. Steitz, *Proc. Natl. Acad. Sci. U.S.A.* **94**, 6030 (1997).

2. At various time points, transfer a portion of the reaction mixture as a drop onto a petri dish placed on ice.
3. Place a 365-nm UV light lamp (9815 Series, Cole Parmer Instrument Co., Chicago, IL) over the petri dish, 2–4 cm from the sample.
4. Irradiate for ~10 min.
5. Transfer the sample to a 1.5-ml Eppendorf tube.

*Assay for RNA–RNA Cross-Links*

6. Bring the volume of the irradiated sample to 250 $\mu$l with G50 buffer and add 10 $\mu$l of 10 mg/ml proteinase K.
7. Incubate at 42° for 15–30 min.
8. PCA extract twice and ethanol precipitate, as described earlier. Resuspend the pellet in 1–2 $\mu$l of autoclaved distilled water.
9. Add 2 volumes of formamide loading buffer, heat denature, and load the sample on a 4 or 5% polyacrylamide–8 $M$ urea gel, as described earlier.
10. Use autoradiography to visualize the cross-linked species.

*Assay for RNA–Protein Cross-Links*

11. To the irradiated sample (from step 5), add RNase A and/or other ribonucleases to ~0.5 $\mu$g/$\mu$l. [It is necessary to remove all but a small fragment of the RNA so that the protein migrates close to the real size on an SDS–PAGE (see later).]
12. Incubate at 37° or 42° for 30 min.
13. Add SDS loading buffer, heat at 90° for 5 min, and load the sample on an 10–12.5% SDS–polyacrylamide gel.
14. Dry the gel and use autoradiography to visualize the cross-linked protein partners. (Because the 4-thioU-containing RNA is usually site-specifically labeled with [32]P, after RNase digestion the label is transferred to the cross-linked proteins.)

The results of RNA–RNA cross-linking studies using the three 4-thioU-containing P120 pre-mRNAs have been published.[25] Briefly, upon irradiation, all three P120 substrates produced two early, ATP-independent RNA–RNA cross-links with similar kinetics. For one of the substrates, where the 4-thioU substitution occurred at the +2 position, a third cross-link that required ATP formed as the two early cross-links diminished. Subsequent analyses using a variety of approaches indicated that the two early cross-links involve U11, one of the small nuclear RNAs (snRNAs) required for the splicing of the AT-AC introns; the third cross-link contained U6atac, another AT-AC spliceosomal snRNA. These results strongly suggest that the 5' splice site is first recognized by U11 and then by U6atac, a mechanism

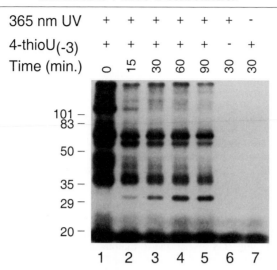

FIG. 3. A time course of RNA–protein cross-linking using Ad pre-mRNA substituted with $^{32}$P-labeled 4-thioU at position $-3$ (relative to the 3′ splice site). A large *in vitro* splicing reaction containing the 4-thioU-substituted Ad pre-mRNA was incubated at 30°. Portions were removed and irradiated with 365-nm UV light at the indicated times. The irradiated samples were subjected to RNase A digestion for 30 min at 37° and immediately subjected to 10% SDS–PAGE. Lanes 6 and 7 are 30-min controls where either 4-thioU was absent in the RNA or 365-nm UV irradiation was omitted. The positions of protein markers (in kDa) are indicated.

similar to that of conventional pre-mRNA splicing where the 5′ splice site is successively recognized by U1 and U6.[7,8,26]

RNA–protein cross-linking with the same 4-thioU-substituted pre-mRNAs has also been performed. Figure 3 reveals the proteins that become cross-linked to the Ad pre-mRNA containing a single 4-thioU at position $-3$ with respect to the 3′-splice site. While some of these cross-linked bands remain constant throughout the course of splicing, others change. Further characterization of these proteins using immunoprecipitation and Western blotting with available antibodies against splicing factors should provide more detailed information on the mechanism of 3′-splice site recognition.

It should be noted that the method described here is not limited to splicing substrates. Because there are no sequence constraints, this method can, in principle, be applied to any RNA that can subsequently participate in a biological reaction, e.g., synthetic or cellularly derived spliceosomal

[26] D. A. Wassarman and J. A. Steitz, *Science* **257**, 1918 (1992).

snRNAs that reconstitute pre-mRNA splicing in nuclear extracts[27–33] or in *Xenopus* oocytes,[34,35] snoRNAs that reconstitute rRNA modification in the *Xenopus* oocyte,[36] and ribozymes.[37]

## Acknowledgments

I am indebted to Joan Steitz for support, input, and encouragement during the development of the procedures described in this manuscript. I thank Joan Steitz, Lara Weinstein, and Tim McConnell for valuable comments on the manuscript. I was a fellow of the Cancer Research Fund of the Damon Runyon-Walter Winchell Foundation. This work was supported by Grant GM26154 from the U.S. Public Health Service to Joan Steitz.

[27] C. L. Will, S. Rumpler, J. Klein Gunnewiek, W. J. van Venrooij, and R. Lührmenn, *Nucleic Acids Res.* **24,** 4614 (1996).
[28] C. Wersig and A. Bindereif, *Mol. Cell. Biol.* **12,** 1460 (1992).
[29] V. Ségault, C. L. Will, B. S. Sproat, and R. Lührmann, *EMBO J.* **14,** 4010 (1995).
[30] Y.-T. Yu, P. A. Maroney, and T. W. Nilsen, *Cell* **75,** 1049 (1993).
[31] T. Wolff and A. Bindereif, *EMBO J.* **11,** 345 (1992).
[32] D. S. McPheeters, P. Fabrizio, and J. Abelson, *Genes Dev.* **3,** 2124 (1989).
[33] P. Fabrizio, D. S. McPheeters, and J. Abelson, *Genes Dev.* **3,** 2137 (1989).
[34] Z.-Q. Pan and C. Prives, *Genes Dev.* **3,** 1887 (1989).
[35] Y.-T. Yu, M.-D. Shu, and J. A. Steitz, *EMBO J.* **17,** 5783 (1998).
[36] K. T. Tycowski, C. M. Smith, M.-D. Shu, and J. A. Steitz, *Proc. Natl. Acad. Sci. U.S.A.* **93,** 14480 (1996).
[37] A. M. Pyle, *Science* **261,** 709 (1993).

# [7] Nucleoprotein Photo-Cross-Linking Using Halopyrimidine-Substituted RNAs

*By* KRISTEN M. MEISENHEIMER, PONCHO L. MEISENHEIMER, and TAD H. KOCH

Photo-cross-linking of nucleic acids to proteins is commonly used to establish residues of the nucleic acid in proximity with residues of the protein in a specific nucleoprotein complex. It can be accomplished by irradiation of an unmodified nucleoprotein complex using ultraviolet (UV) light in the region of 250–260 nm or by irradiation of a nucleoprotein complex in which either the nucleic acid or the protein has been chemically modified to enhance its photoreactivity in cross-linking. The most common modification of the protein is the addition of an azide functional group. Nucleic acids are modified by substitution of a hydrogen with an azide or halide (bromide or iodide) on the nucleic acid base or substitution of a carbonyl with a thiocarbonyl on the nucleic acid base. This article focuses on the use of 5-bromo- and 5-iodouracil and 5-iodocytosine bases in nucleo-protein photo-cross-linking. For more general coverage of the topic of nucleoprotein photo-cross-linking with emphasis on the nucleic acid that includes an extensive review of the halopyrimidines, the reader is referred to a recent review.[1]

Bromo- and iodo-substituted pyrimidines show specific photoreactivity with functional groups in proteins. The most reactive residues are the aromatic residues, phenylalanine, tyrosine, histidine, and tryptophan, and the sulfur-containing residues, cysteine, cystine, and methionine. Molecular structures for the cross-link are predicted from experiments with small molecules and are shown in Fig. 1.[2–5] To date, the most frequently reported residue in successful cross-linking experiments is tyrosine. Reactivity with only seven amino acid residues is not severely limiting because, in particular, the aromatic residues appear to be located with a $\pi$-stacking geometry at the nucleoprotein interface as indicated by cocrystal structures.[6,7]

Cross-linking appears to be initiated by either electron transfer from the amino acid residue to the halopyrimidine followed by loss of halide from the pyrimidine and radical combination or carbon halogen bond homolysis followed by radical addition to the aromatic ring of the amino acid residue. Experiments with model reactions support the former mechanism for cross-linking with the 5-bromouracil chromophore and the latter mechanism for cross-linking with 5-iodouracil and 5-iodocytosine chromophores.[8,9] The high specificity observed in nucleoprotein cross-linking with nucleic acids bearing halopyrimidines is more consistent with the electron transfer mech-anism for all of the chromophores and suggests that model reactions are an inadequate predictor of reactivity in a macromolecular complex. The major limitation of model studies is that the intimate association of halopy-rimidine and amino acid residue present in a nucleoprotein complex cannot be achieved. A schematic representation of a photoelectron transfer mecha-

[1] K. M. Meisenheimer and T. H. Koch, *Crit. Rev. Biochem. Mol. Biol.* **32,** 101 (1997).
[2] I. Saito, S. Ito, and T. Matsuura, *J. Am. Chem. Soc.* **100,** 2901 (1978).
[3] T. M. Dietz and T. H. Koch, *Photochem. Photobiol.* **46,** 971 (1987).
[4] T. M. Dietz and T. H. Koch, *Photochem. Photobiol.* **49,** 121 (1989).
[5] K. M. Meisenheimer, P. L. Meisenheimer, M. C. Willis, and T. H. Koch, *Nucleic Acids Res.* **24,** 981 (1996).
[6] C. Oubridge, N. Ito, P. R. Evans, C. H. Teo, and K. Nagai, *Nature* **372,** 432 (1994).
[7] K. Valegård, J. B. Murray, P. G. Stockley, N. J. Stonehouse, and L. Liljas, *Nature* **371,** 623 (1994).
[8] C. Norris, P. L. Meisenheimer, and T. H. Koch, *J. Am. Chem. Soc.* **118,** 5796 (1996).
[9] C. L. Norris, K. M. Meisenheimer, and T. H. Koch, *Photochem. Photobiol.* **65,** 201 (1997).

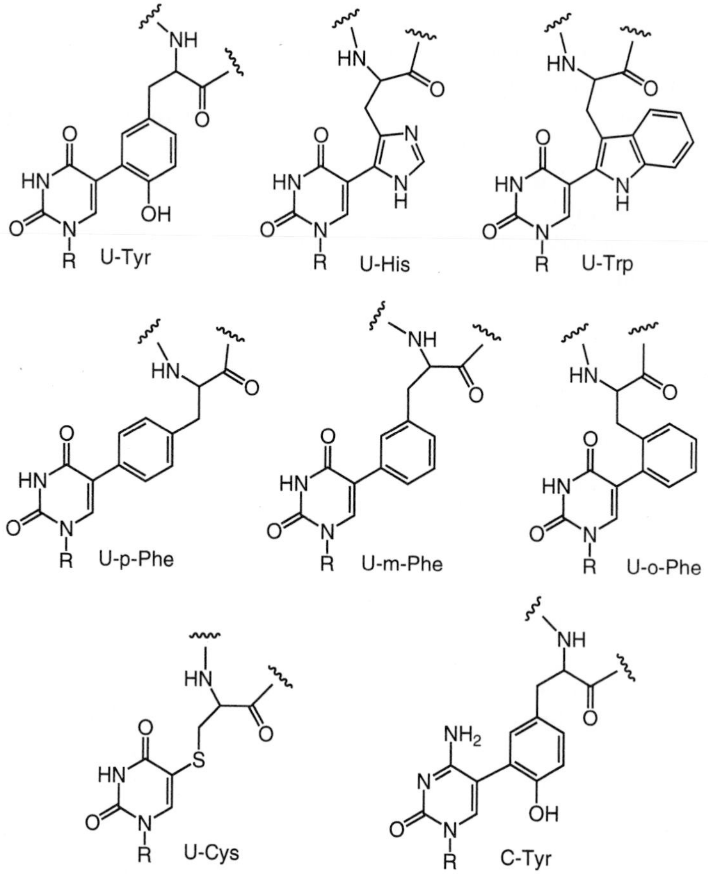

FIG. 1. The molecular nature of nucleoprotein cross-links resulting from irradiation of nucleic acids bearing 5-bromouracil, 5-iodouracil, or 5-iodocytosine chromophores adjacent to photoreactive amino acid residues. Predicted structures are based on the structural characterization of products of model reactions. Not all of the possible cross-links have been characterized by model reactions at this time.

nism with $\pi$-stacked chromophores in a nucleoprotein complex is shown in Fig. 2.

An important factor in the design of a successful cross-linking experiment is the choice of chromophores and excitation wavelength. The goal is to excite the cross-linking chromophore without exciting other chromophores in the molecule. UV absorption by the various chromophores present in halopyrimidine-substituted nucleoprotein complexes is shown in Fig. 3. The halopyrimidines all absorb at longer wavelengths than the natural

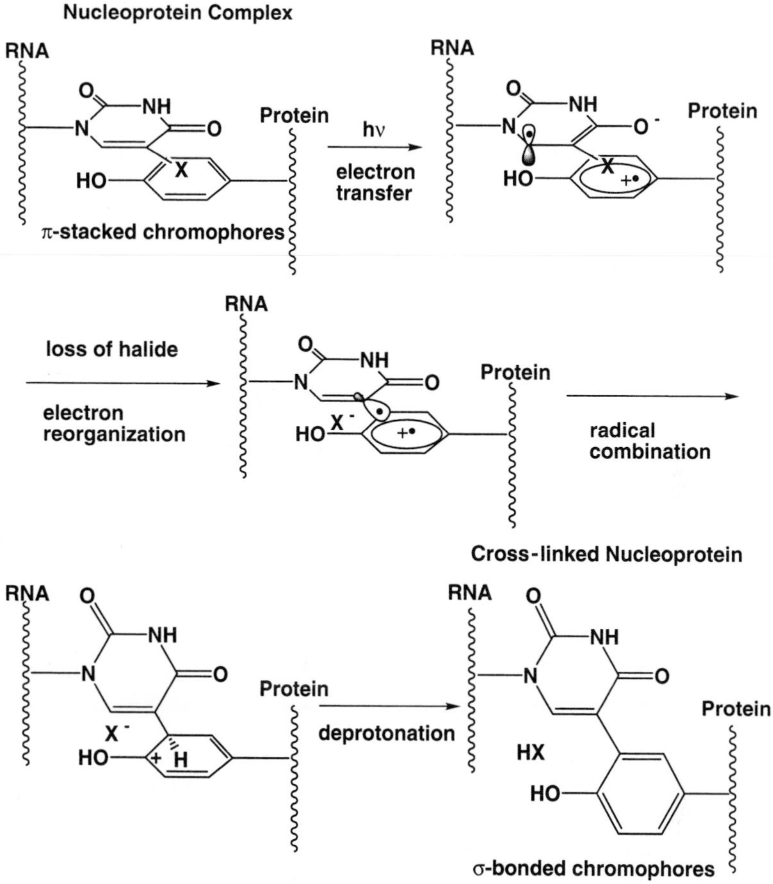

FIG. 2. A schematic diagram illustrating a possible photoelectron transfer mechanism for cross-linking of π-stacked chromophores. Note that ultimate formation of a σ bond requires a geometry change of the nucleoprotein.

bases and aromatic amino acid residues except tyrosine and tryptophan. UV absorption by tyrosine and tryptophan is compared with absorption by 5-bromouridine in Fig. 4. The separation of absorption bands is not large and, as a consequence, monochromatic light sources that excite the long wavelength tail of the halopyrimidine absorption bands are desirable. In this regard, the xenon chloride (XeCl) excimer laser emitting at 308 nm is optimum for the bromouracil chromophore, and the helium cadmium laser emitting at 325 nm is optimum for iodouracil and iodocytosine chromophores. In practice, the XeCl laser can be used to excite all of the halopy-

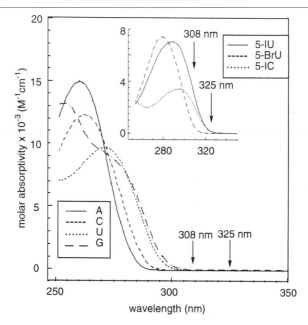

FIG. 3. UV absorption of the various chromophores present in halopyrimidine-substituted nucleic acids in water. Spectra are of the ribonucleosides.

rimidines. If the peak power per pulse per unit area of the XeCl laser beam is too high, undesirable multiphoton excitation will occur, but it can be eliminated with a defocusing quartz lens. Monochromatic light in the region 310–330 nm can also be achieved with a frequency-doubled rhodamine dye laser pumped with a frequency-doubled Nd : YAG laser. Although a YAG pumped dye laser is an expensive alternative to XeCl or HeCd lasers, it is often more available in academic institutions. A midrange transilluminator with maximum emission at 312 nm is also effective, but may give lower cross-linking yields because its emission is not as monochromatic as emission from lasers and will excite other chromophores. A polystyrene petri dish cover can sometimes help as a short wavelength filter.

Identification of which pyrimidine to substitute to achieve photo-cross-linking is initially guess work. If a secondary RNA structure is predictable, pyrimidines in loop or bulge positions are probably good candidates. If multiple substitution does not eliminate specific binding, all of the uracils or all of the cytosines can be substituted simultaneously during transcription. Single cross-links are still observed with multiple substituted RNAs because halopyrimidines not adjacent to reactive amino acid residues are more photostable than those adjacent to reactive amino acid residues. If all of

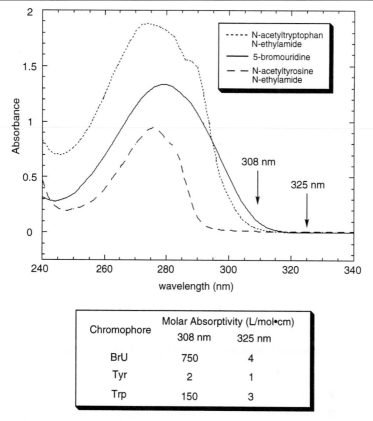

FIG. 4. UV absorption spectra of 0.18 m*M* 5-bromouridine, 0.73 m*M* *N*-acetyltyrosine *N*-ethylamide, and 0.32 m*M* *N*-acetyltryptophan *N*-ethylamide in 50 m*M*, pH 7, phosphate buffer. Molar absorptivities at 308 and 325 nm are compared in the tabulation below the spectra. Adapted from Ref. 9 with permission.

the uracils or cytosines are substituted, identification of the cross-linking site will require nucleic acid sequencing subsequent to cross-linking. Alternatively, single substitutions can be achieved through solid-state synthesis using the appropriate haloribonucleoside phosphoramidite or possibly the halodeoxyribonucleoside phosphoramidite.

Substitution of one or more pyrimidines for halopyrimidines can affect specific protein binding. Several examples of variously substituted RNAs in comparison with unsubstituted counterparts are summarized in Fig. 5. RNAs chosen for this presentation are stem–loop structures that model a small portion of the MS2 bacteriophage genome, which specifically binds to the coat protein. A cross-linking site that has been identified is the −5 position relative to the start codon for replicase synthesis.[10] The cross-linking site was established from measurements of binding and cross-linking as a function of RNA sequence. The substitution of halopyrimidines for pyrimidines has a variable effect on the dissociation constants ($K_d$ values). With single bromouridine or iodouridine substitution at the −5 position, binding to coat protein is actually better than with uridine at the −5 position.[10,11] However, multiple iodocytidine substitutions for cytidines decrease the binding significantly as indicated by a comparison of $K_d$ values for RNAs 4, 5, and 6.[5,10]

Measurement of the nucleoprotein-binding constant, commonly reported as a dissociation constant $K_d$, is recommended prior to a photo-cross-linking experiment because the $K_d$ is the protein concentration necessary to bind half of the RNA. If possible, a protein concentration an order of magnitude above the $K_d$ is desirable for maximum cross-linking yield. Such a protein concentration will minimize excitation of unbound RNA. The $K_d$ values in Fig. 5 were measured using a nitrocellulose filter-binding assay, and an experimental procedure for the measurement with RNAs 5 and 6 is given later.

Irradiations are best performed with the reaction mixture in disposable cuvettes but can also be performed with the reaction mixture in small centrifuge tubes, irradiating through the open top, or in microtiter plates. The actual irradiation time to maximum cross-linking yield is difficult to predict and can vary from seconds to hours as shown in Table I. It depends on the light source and the quantum yield of the actual cross-linking reaction. A significant variation in quantum yield has been observed and, although not understood, probably involves subtle differences in the orientation of the reacting groups in the nucleoprotein complex. Because of uncertainty in the irradiation time, one or more time course experiments are required where aliquots are removed at specific time points and analyzed for cross-linking. Another reason for running time course experiments is the photolability of some cross-linked nucleoproteins. Hence, samples can be irradiated too long. Cross-linking experiments are analyzed by denaturing polyacrylamide gel electrophoresis with $^{32}$P-labeled RNAs. Cross-linked nucleoproteins show a substantial gel shift relative to RNAs as shown in Figs. 6 and 7, which illustrate time course experiments with RNAs 3, 4, and 5. Gels are quantitated by phosphorimaging or by autoradiography. Maximum yields for photo-cross-linking of RNAs 2, 3, and 5 to MS2 coat protein and MS2 coat proteins mutated at the cross-linking site, Tyr-85,

[10] J. M. Gott, M. C. Willis, T. H. Koch, and O. C. Uhlenbeck, *Biochemistry* **30**, 6290 (1991).
[11] M. C. Willis, B. J. Hicke, O. C. Uhlenbeck, T. R. Cech, and T. H. Koch, *Science* **262**, 1255 (1993).

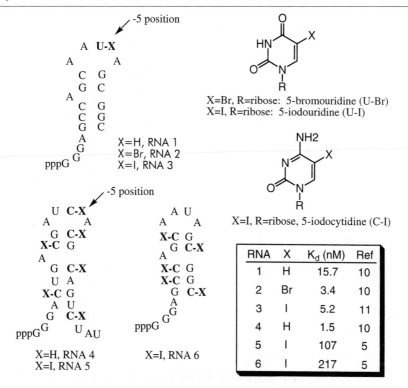

FIG. 5. Predicted secondary structures of RNAs discussed here and $K_d$ values for dissociation of their complexes with MS2 bacteriophage coat protein.

are summarized in Table I. In general, higher yields are obtained with iodouracil and iodocytosine substitution than with bromouracil substitution. The difference appears to result in part from a photoreactive Trp residue of MS2 not at the cross-linking site. Tryptophan is known to be the most photoreactive amino acid both in the presence and in the absence of dioxygen.[12] Further, as shown in Fig. 4, Trp will absorb competitively at 308 nm. In fact, a control experiment estalished that irradiation of MS2 coat protein alone at 308 nm lowered its binding to RNA.[10] If Trp were at the cross-linking site, exciting Trp in competition with exciting the halopyrimidine to which it cross-links is not predicted to be detrimental. This untested prediction is made because photoelectron transfer can be achieved by exciting either the electron donor or the electron acceptor. In fact, a model

[12] D. Creed, *Photochem. Photobiol.* **39,** 537 (1984).

TABLE I

MAXIMUM YIELDS AND IRRADIATION TIMES OBSERVED FOR PHOTO-CROSS-LINKING OF HALOPYRIMIDINE-SUBSTITUTED HAIRPIN RNAs TO BACTERIOPHAGE MS2 COAT PROTEIN[a]

| MS2 coat protein | Light source | Wavelength (nm) | % cross-linking with | | | Irradiation time (min) | Ref. |
|---|---|---|---|---|---|---|---|
| | | | RNA 2 | RNA 3 | RNA 5 | | |
| Tyr85 (wild type) | XeCl laser | 308 | 40 | — | — | 15 | 10 |
| | Transilluminator | 312 | 22 | — | — | 60 | 10 |
| | HeCd laser | 325 | 66 | — | — | 600 | 9 |
| | HeCd laser | 325 | — | 90 | — | 400 | 11 |
| | HeCd laser | 325 | — | — | 85 | 180 | 5 |
| Tyr85Ser | XeCl laser | 308 | 0 | — | — | 8 | 14 |
| Tyr85Phe | XeCl | 308 | 47 | 93 | — | 5 | — |
| | HeCd | 325 | — | 80 | — | 120 | 19 |
| Tyr85His | XeCl | 308 | 37 | — | — | 3 | 19 |
| | HeCd | 325 | — | 90 | — | 120 | 19 |
| Tyr85Cys | XeCl | 308 | — | 76 | — | 2 | — |
| | HeCd | 325 | — | 75 | — | 120 | 19 |

[a] As a function of coat protein mutation, pyrimidine substitution, and irradiation source. The light sources are those described in the text. The percentage cross-linking is the chemical yield of cross-linked nucleoprotein based on the amount of starting nucleic acid.

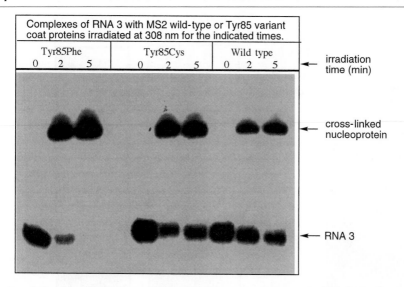

| Complexes of RNA 3 with MS2 wild-type or Tyr85 variant coat proteins irradiated at 308 nm for the indicated times. | | | | | | | | | |

Fig. 6. An Afga Studio II scan (1200 dpi) of an autoradiogram of a 20% denaturing gel showing RNA 3 photo-cross-linked to the MS2 wild-type (Tyr-85) and Tyr85Phe and Tyr85Cys variants of the coat protein. The irradiation was performed at 308 nm with samples removed from the reaction at 0, 2, and 5 min. Photo-cross-linking yields achieved after 5 min are shown in Table I.

study with a tyrosine derivative and 5-bromouridine showed predominant cross-linking from excitation of the tyrosine derivative.[9]

Two useful control experiments are illustrated in Fig. 7.[5] One is the irradiation of cross-linking RNA (such as RNA 5) in the absence of protein. The lack of a gel band at the point assigned to the cross-linked nucleoprotein establishes that cross-linking required protein. An alternate possibility is an RNA dimer resulting from an RNA–RNA cross-link. Another control is the irradiation in the presence of protein of an analogous RNA (such as RNA 6) missing the halopyrimidine at the presumed cross-linking site. The lack of a gel band at the point assigned to the cross-linked nucleoprotein establishes the cross-linking nucleotide. Figure 7 also shows some weak bands assigned to RNA cleavage. Little is known about this photoprocess in RNA, but extrapolation from DNA studies suggests that the process may be initiated by photoelectron transfer between halopyrimidines and adjacent purine bases.[13,14]

Once cross-linking has been established and irradiation conditions opti-

[13] I. Saito, H. Sugiyama, and Y. Tsutsumi, *J. Am. Chem. Soc.* **112,** 6720 (1990).
[14] G. P. Cook and M. M. Greenberg, *J. Am. Chem. Soc.* **118,** 10025 (1996).

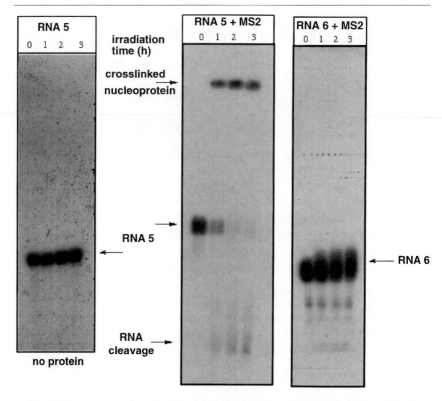

Fig. 7. Denaturing polyacrylamide gels (15%) showing cross-linked nucleoprotein from irradiation of RNA 5 in the presence of MS2 bacteriophage coat protein and the lack of cross-linking in the absence of coat protein. The photo-cross-linking yield is given in Table I. As a control, a band for cross-linked nucleoprotein is not observed for RNA 6 irradiated in the presence of coat protein; RNA 6 does not have an iodocytidine at the cross-linking site. Adapted from Ref. 5 with permission.

mized, larger amounts of cross-linked nucleoprotein can be prepared for protein sequencing. Identification of the cross-linking site on the protein has traditionally been accomplished by Edman sequencing of a fragment from enzymatic digestion. The fragment is separated by gel electrophoresis, and the cross-linked peptide is extracted from a gel slice or is transferred electrolytically to a membrane. This methodology with trypsin digestion was employed for the identification of Tyr-85 as the amino acid residue of MS2 coat protein at the cross-linking site.[15] As mass spectrometry develops,

[15] M. C. Willis, K. A. LeCuyer, K. M. Meisenheimer, O. C. Uhlenbeck, and T. H. Koch, *Nucleic Acids Res.* **22,** 4947 (1994).

it may become the preferred technique for sequencing. Measurement of the molecular mass of an intact photo-cross-linked nucleoprotein by electrospray mass spectrometry has been reported.[16]

The high yield and specificity observed in halopyrimidine-substituted nucleic acid cross-linking to associated proteins have led to another important application, photoSELEX. PhotoSELEX is a methodology for the identification of a nucleic acid aptamer that will bind with high affinity to a target protein and cross-link to the target in high yield. The procedure starts with a nucleic acid combinatorial library in which one of the pyrimidines is substituted with its halopyrimidine counterpart and selection is for both affinity and cross-linking yield. The first example employed 5-iodouridine in an RNA library with the target protein human immunodeficiency virus type-1 (HIV-1) Rev. A 32-mer bearing seven iodouridines that showed a $K_d$ of 1 n$M$ was identified, and it cross-linked to Rev in approximately 40% yield.[17]

## Materials, Equipment, and Procedures

### Materials

5-Bromouridine triphosphate (BrUTP), 5-iodouridine triphosphate (IUTP), 5-iodocytidine triphosphate (ICTP), and T7 RNA polymerase are all from Sigma Chemical Co. (St. Louis, MO). Radioactive [$\alpha$-$^{32}$P]ATP and [$\alpha$-$^{32}$P]CTP are available from Du Pont NEN Research Products (Boston, MA) and used as received. MS2 bacteriophage nonaggregating wild-type coat protein and nonaggregating coat protein were provided by Olke Uhlenbeck's laboratory (University of Colorado, Boulder, CO).[18,19] Synthetic DNA templates can be purchased from a number of commercial vendors, such as Integrated DNA Technologies (Coralville, IA), but for the experiments described here, they were supplied by Olke Uhlenbeck's laboratory. Distilled water is purified further to 18 M$\Omega$ cm with a Millipore Q water purification system (Millipore, Bedford, MA). Buffered and salt solutions are all sterilized in an autoclave prior to use. The 80% formamide loading buffer contains 80% formamide, 10 m$M$ EDTA (pH 8), 1 mg/ml xylene cyanol FF, and 1 mg/ml bromphenol blue.

[16] D. L. Wong, J. G. Pavlovich, and N. O. Reich, *Nucleic Acids Res.* **26,** 645 (1998).
[17] K. B. Jensen, B. L. Atkinson, M. C. Willis, T. H. Koch, and L. Gold, *Proc. Natl. Acad. Sci. U.S.A.* **92,** 12220 (1995).
[18] K. A. LeCuyer, L. S. Behlen, and O. C. Uhlenbeck, *Biochemistry* **34,** 10600 (1995).
[19] K. A. LeCuyer, L. S. Behlen, and O. C. Uhlenbeck, *EMBO J.* **15,** 6847 (1996).

## Equipment

Ultraviolet spectra are taken using a Hewlett-Packard Model 8452A UV-visible spectrometer. For the experiments described here, monochromatic light at 308 or 325 nm is produced by either a Lambda Physik EMG-101 xenon chloride pulsed excimer laser at a power of 40–55 mJ/pulse at 10 Hz (Fort Lauderdale, FL) or an Omnichrome helium–cadmium continuous wave laser emitting between 30 and 50 mW (Melles Griot, Carlsbad, CA), respectively. With the 308-nm excimer laser, approximately 15% of the beam is passed through a 7-mm-diameter circular beam mask. The laser beam diameter of the 325-nm helium–cadmium laser is 3 mm. Samples are irradiated in 1-cm Pyrex or quartz cuvettes or 4 mm × 1 cm methacrylic cuvettes placed in a cell holder thermostated with a refrigerated circulator. The power of both lasers is measured using a Scientech 360-001 disk calorimeter power meter (Boulder, CO) to assure reproducibility. An alternate less expensive 308-nm laser is an MPB PSX-100 xenon chloride pulsed excimer laser that operates at a power of 1 mJ/pulse at 50–60 Hz (Dorval, Québec, Canada) with a beam size of 4 × 4 mm. With the MPB laser the beam is used without attenuation with a beam mask. Photo-cross-linking is also accomplished with less expensive transilluminators such as a Spectroline (Spectronics Corp.) medium wavelength UV transilluminator (peak at 312 nm). Irradiation of samples in microtiter plates is simply accomplished with a transilluminator mounted at fixed distance (1 cm) with the microtiter plate cooled in ice. The irradiation time to maximum cross-linking yield is dependent on the light source and quantum yield of the specific cross-linking reaction and is determined from experimentation. With a transilluminator, reproducibility requires reproducible location of the sample with respect to the light bulb.

## Procedures

*Preparation of RNAs.* Halopyrimidine-substituted RNAs are prepared by *in vitro* transcription from synthetic DNA templates by T7 RNA polymerase. Transcription reactions contain 40 m$M$ Tris–HCl (pH 8.0 at 37°), 1 m$M$ spermidine, 5 m$M$ dithiothreitol (DTT), 50 $\mu$g/ml bovine serum albumin (BSA), 0.1% (v/v) Triton X-100, 80 mg/ml polyethylene glycol ($M_r$ 8000), 0.1 mg/ml T7 RNA polymerase, 25 m$M$ MgCl$_2$, and 1 $\mu$M DNA template. BrUTP or IUTP (2 m$M$) is used in place of UTP, or ICTP is used in place of CTP; depending on the radioactive nucleotide used, either CTP or ATP concentrations are kept low (0.2 m$M$) so that [$\alpha$-$^{32}$P]CTP or [$\alpha$-$^{32}$P]ATP can radiolabel the RNA effectively. The remainder of NTPs are 2 m$M$. The reaction is allowed to go for 1 hr at 37°. The RNA is then denatured by adding an equal volume of formamide loading buffer and is heated to 97° for 3 min. The RNA is loaded onto a 20% 7 $M$ urea denaturing polyacrylamide gel (29:1/acrylamide:bisacrylamide) and electrophoresed in 89 m$M$ Tris–borate/2 m$M$ EDTA (pH 8.0). The desired RNA is extracted from a slice of the gel overnight at 4° using 10 m$M$ Tris–HCl (pH 8.0), 2 m$M$ EDTA (pH 8.0), and 0.3 $M$ ammonium acetate. RNA is precipitated with 3 volumes of ethanol (0°, 2 hr) and is isolated by concentrating the RNA into a pellet via centrifugation. The pellet is dried and resuspended in 10 m$M$ Tris–HCl (pH 8.0) and 1 m$M$ EDTA.

*Irradiation at 308 nm of RNA 2 and MS2 Bacteriophage Wild-Type Coat Protein.* RNA 2 is heated in water to 95° for 3 min and cooled quickly on ice before use to ensure that the RNA is in a hairpin conformation.[20] Prior to irradiation, a 500-$\mu$l solution containing 10 n$M$ RNA 2, 100 n$M$ MS2 wild-type coat protein, 5 m$M$ DTT, 100 m$M$ Tris–HCl (pH 7.5 at 4°), 80 m$M$ KCl, 10 m$M$ magnesium acetate, and 80 $\mu$g/ml BSA is incubated on ice for approximately 30 min. The solution of the resulting nucleoprotein complex is then irradiated at 308 nm in a 4-mm-wide by 1-cm methacrylate cuvette at 4° with a Lambda Physik XeCl laser. Samples are removed from the reaction after 0, 2, 4, 6, 8, and 10 min of irradiation, heated to 95° for 5 min in formamide loading buffer, and electrophoresed through a 15% urea denaturing polyacrylamide gel in 89 m$M$ Tris–borate/2 m$M$ EDTA (pH 8.0). The percentage cross-link is quantitated by scanning the gel using a Molecular Dynamics PhosphorImager (Sunnyvale, CA), and the maximum cross-linking yield and irradiation time are reported in Table I.

*Irradiation at 308 nm of RNA 2 Complexed with Tyr85Ser, Tyr85His, or Tyr85Phe MS2 Coat Protein Variants.* Solutions of RNA 2/MS2 variant coat protein complexes are prepared in a manner analogous to the RNA 2/wild-type coat protein complex except for the concentrations of the RNAs and proteins, which are as follows: [RNA 2]/[Tyr85Ser mutant] 10 n$M$/2.1 $\mu$$M$; [RNA 2]/[Tyr85His mutant] 3.1 n$M$/13 $\mu$$M$; [RNA 2]/[Tyr85Phe mutant] 5n$M$/3 $\mu$$M$. Differences in protein concentration reflect differences in $K_d$ values.[19] Different time courses are followed for the irradiations of the complexes of the various protein mutants with RNA 2 because of differences in quantum yield. The nonaggregating wild-type protein (Tyr-85) and Tyr85Ser mutant samples are removed from the reaction at 0, 2, 4, 6, 8, and 10 min. The samples are taken from the Tyr85His and Tyr85Phe mutant irradiations with RNA 2 at 0, 1, 2, 3, 4, and 5 min. The percentage cross-linking for the nonaggregating wild-type (Tyr85) and Tyr85Ser is determined from a Molecular Dynamics PhosphorImager scan of the 15% denaturing polyacrylamide gel. The percentage cross-linking of the Tyr85His and Tyr85Phe mutants is determined by densitometry of an auto-

---

[20] D. R. Groebe and O. C. Uhlenbeck, *Nucleic Acids Res.* **16,** 11725 (1988).

radiogram of the 20% denaturing polyacrylamide gel using the software NIH Image, and the maximum value and the irradiation time to achieve it are reported in Table I.

*Irradiation at 325 nm of RNA 2 Complexed with MS2 Bacteriophage Coat Protein.* RNA 2 is renatured by heating to 97° in water for 5 min and cooling quickly on ice. Prior to irradiation, a 250-$\mu$l solution containing 5 n$M$ RNA, 200 n$M$ MS2 nonaggregating coat protein, 5 m$M$ DTT, 100 m$M$ Tris–HCl (pH 7.5 at 4°), 80 m$M$ KCl, 10 m$M$ magnesium acetate, and 100 $\mu$g/ml BSA is incubated on ice for 30 min. The 325-nm irradiation is carried out at 4° in a 4-mm-wide by 1-cm methacrylate cuvette. Samples are removed after 2, 4, 6, 8, 10, and 12 hr, heated to 97° for 5 min in formamide loading buffer, and electrophoresed through a 20% urea denaturing polyacrylamide gel in 89 m$M$ Tris–borate/2 m$M$ EDTA (pH 8.0). The percentage cross-linking is quantitated from a phosphorimager scan, and the maximum yield with irradiation time is reported in Table I. A long irradiation time is required because of the low optical density of the reacting chromophore, presumed to be Tyr-85 or a BrU(-5)-Tyr-85 $\pi$-complex in this case.[9]

*Irradiation at 308 nm of RNA 3 and Wild-Type MS2 Bacteriophage Coat Protein.* RNA 3 is heated in water to 95° for 3 min and cooled quickly on ice. Prior to irradiation, a 500-$\mu$l solution, which contained 6 n$M$ RNA 3, 35 n$M$ MS2 coat protein, 5 m$M$ DTT, 100 m$M$ Tris–HCl (pH 7.5 at 4°), 80 m$M$ KCl, 10 m$M$ magnesium acetate, and 80 $\mu$g/ml BSA, is incubated on ice for approximately 20 min. The solution of the resulting nucleoprotein complex is irradiated at 308 nm in a 4-mm-wide by 1-cm methacrylate cuvette at 4°. Samples are removed from the reaction after 0, 2, and 5 min of irradiation, heated to 95° for 3 min in 80% formamide loading buffer, and electrophoresed through a 20% urea denaturing polyacrylamide gel in 89 m$M$ Tris–borate/2 m$M$ EDTA (pH 8.0). The percentage cross-link is quantitated by scanning the gel using a Molecular Dynamics Phosphor-Imager, and the maximum yield, together with irradiation time, is reported in Table I.

*Irradiation at 308 nm of RNA 3 Complexed with MS2 Tyr85Phe or Tyr85Cys MS2 Coat Protein Variants.* Solutions of RNA 3 and the MS2 coat protein variants, Tyr85Phe and Tyr85Cys, are prepared and analyzed in a manner analogous to the RNA 3/wild-type complex, except for the concentration of the variant proteins in the reaction. Concentrations of the Tyr85Phe and Tyr85Cys variants are both 3.5 $\mu M$. The gel is shown in Fig. 6 and the maximum cross-linking yield is shown in Table I.

*Irradiation at 325 nm of RNA 3 Complexed with MS2 Tyr85His Variant Coat Protein.* The RNA 2/Tyr85His complex is prepared in a similar manner to the other MS2 wild-type and variant nucleoprotein complexes. However, the concentrations of RNA 3 and protein are 14 and 63 n$M$, respectively. The irradiation is carried out at 4° in a 4-mm-wide by 1-cm methacrylate cuvette using 325 nm. Samples from the irradiation are removed at 0, 30, and 60 min, diluted with an equal volume of 80% formamide loading buffer, and electrophoresed through a 20% 7 $M$ urea denaturing polyacrylamide gel in 89 m$M$ Tris–borate/2 m$M$ EDTA (pH 8.0). The percentage cross-linking is quantitated by densitometry from an Afga Studio II scan of the gel using the software NIH Image, and the maximum yield and irradiation time are reported in Table I.

*Binding Assays of RNA 5 and RNA 6 to the MS2 Bacteriophage Coat Protein.* Nitrocellulose filter-binding assays include 5 m$M$ DTT, 100 m$M$ Tris–HCl (pH 7.5 at 4°), 80 m$M$ KCl, 10 m$M$ magnesium acetate, and 80 $\mu$g/ml BSA and either trace RNA 5 or trace RNA 6 (assuming 1:1 stoichiometry with the lowest concentration of protein used in the assay). Protein concentrations range from 0.05 n$M$ to 1 $\mu M$. Equilibration of the RNA–MS2 protein complexes is done at 4° for 30 min. Prior to filtration, the 0.45-$\mu$m nitrocellulose membranes (Millipore Corp., Bedford, MA) are presoaked at 4° in the binding buffer. Filtration is carried out using a 12-port Millipore vacuum filtration-binding apparatus. The fraction of RNA bound to each membrane is determined by loading RNA, without protein, onto a membrane and subtracting those counts from each data point. The fraction of RNA bound to protein is determined from [32]P scintillation counting. Alternatively, filtration is carried out using the Convertible filtration manifold system (Gibco-BRL Products, Grand Island, NY). After filtration, the membrane is blotted with a KimWipe to remove excess buffer, and the fraction of RNA bound to protein is determined by quantitating a phosphor image scan of the membrane. For both methods, data points are fit to a retention efficiency and $K_d$ assuming bimolecular equilibrium using the program Kaleidagraph. The binding constants, reported in Fig. 5, are determined using the general curve fit for $y = M_0/[(M_2 + M_0)M_1]$, where $y$ is the percentage RNA bound, $M_0$ is the protein concentration, $M_2$ is $K_d$, and $M_1$ is the percentage RNA that is competent for binding.

*Irradiation at 325 nm of RNA 5 and RNA 6 with Wild-Type MS2 Bacteriophage Coat Protein.* RNA 5 is heated separately in water to 95° for 3 min and cooled quickly on ice. Prior to irradiation, a 250-$\mu$l solution containing 10 n$M$ RNA 5, 1 $\mu M$ MS2 nonaggregating mutant coat protein,[18] 5 m$M$ DTT, 100 m$M$ Tris–HCl (pH 7.5 at 4°), 80 m$M$ KCl, 10 m$M$ magnesium acetate, and 80 $\mu$g/ml BSA is incubated on ice for approximately 30 min. After equilibration, the solution is irradiated at 4° with 325-nm light. Samples are removed from the reaction after 0, 1, 2, and 3 hr of irradiation, heated to 95° for 5 min in an equal volume of formamide loading buffer, and electrophoresed through a 15% urea denaturing polyacrylamide gel in

89 m$M$ Tris–borate/2 m$M$ EDTA (pH 8.0). The RNA 6–MS2 bacteriophage coat protein complex is prepared and irradiated in an identical fashion to the RNA 5–MS2 bacteriophage coat protein complex. The percentage cross-linking is quantitated from a phosphor image scan of the gel shown in Fig. 7. The maximum yield for RNA 5 is reported in Table I. The gel of the irradiated RNA 6–MS2 complex does not show a cross-linking band because RNA 6 (Fig. 4) does not have an iodocytidine at the −5 position.

*Irradiation at 325 nm of RNA 5 in Absence of MS2 Coat Protein.* RNA 5 is also irradiated at 325 nm in the absence of coat protein to establish that the new band observed on irradiation in the presence of coat protein (Fig. 7) is protein dependent and does not simply represent an RNA–RNA cross-link. RNA 5 is heated in water at 97° for 3 min and cooled on ice. The 250-$\mu$l solution contains 10 n$M$ RNA 5, 5 m$M$ DTT, 100 m$M$ Tris–HCl (pH 7.5 at 4°), 80 m$M$ KCl, 10 m$M$ magnesium acetate, and 80 $\mu$g/ml BSA. After equilibration for 30 min on ice, the solution is irradiated at 4° with 325-nm monochromatic light for 3 hr. Samples are removed from the reaction after 0, 1, 2, and 3 hr of irradiation, heated to 95° for 5 min in formamide loading buffer, and electrophoresed through a 15% urea denaturing polyacrylamide gel using 89 m$M$ Tris–borate/2 m$M$ EDTA (pH 8.0) as the running buffer. The absence of cross-linking is shown in Fig. 7.

### Acknowledgments

We thank the Council for Tobacco Research, USA, the Colorado RNA Center at the University of Colorado, and NeXstar Pharmaceuticals, Inc. for financial assistance. We also thank Karen LeCuyer, Linda Behlen, and Olke Uhlenbeck for MS2 coat protein and variants of MS2 coat protein.

# [8] Structure Determination by Directed Photo-Cross-Linking in Large RNA Molecules with Site-Specific Psoralen

*By* DMITRI MUNDUS and PAUL WOLLENZIEN

## Introduction

Photo-cross-linking experiments can provide important information about intra- and intermolecular interactions in RNA and between RNA and proteins. A number of different types of reagents and strategies are being used for this, including ultraviolet (UV) light-induced cross-linking

of normal nucleosides[1] and naturally occurring modified nucleosides,[2] photoreactive nucleoside analogs incorporated into the RNA,[3,4] and photoreactive reagents that can be attached photochemically[5] or chemically[6-8] to
RNA. In the latter instances, two approaches can be taken. In the first,
probes are incorporated randomly into the RNA, either by incubation
of soluble reagents that depend on their interaction with the RNA by
intercalation or chemical reactivity or by random incorporation during *in
vitro* transcription. These are usually incorporated into the RNA at low to
moderate overall stoichiometries to not perturb the native RNA structure
and to avoid over-cross-linking the RNA. In this approach it is relative
easy to achieve incorporation and cross-linking; however, the resulting
cross-links may not occur in functionally interesting parts of the molecule
and may be difficult to analyze because of their heterogeneity. A second
type of approach is the placement of photoreagents at specific sites at as
high a stoichiometry as possible. The advantage of this is its specificity, it
simplifies the subsequent analysis of the cross-linking sites, and the structural and functional effects of the probe can be determined separately from
the cross-linking experiment. The main disadvantage is that a large amount
of work needed in preparing sufficient amounts of the derivatized RNA
sample before the experiment can be done and some site-directed strategies
leave internal breaks in the RNA sample. This article describes a site-
directed targeting method that uses psoralen photochemistry and overcomes some of these problems.

Psoralen photochemical cross-linking offers a number of features for
nucleic acids, including favorable absorbance properties, solubility, photochemical efficiencies, and specificity and reactivity to nucleic acids.[9,10] The

[1] S. Y. Wang, "Photochemistry and Photobiology of Nucleic Acids." Academic Press, New York, 1976.
[2] J. Ofengand, R. Liou, J. Kohut III, I. Schwartz, and R.A. Zimmermann, *Biochemistry* **18**, 4322 (1979).
[3] N. K. Tanner, M. M. Hanna, and J. Abelson, *Biochemistry* **27**, 8852 (1988).
[4] Y. Lemaigre-Dubreuil, A. Expert-Bezancon, and A. Favre, *Nucleic Acids Res.* **19**, 3653 (1991).
[5] S. T. Isaacs, C.-K. J. Shen, J. E. Hearst, and H. Rapoport, *Biochemistry* **16**, 1058 (1977).
[6] J. Ofengand, I. Schwartz, G. Chinali, S. S. Hixson, S. H. Hixson, *Methods Enzymol.* **46**, 683 (1977).
[7] T. Doring, P. Mitchell, M. Osswald, D. Bochkariov, and R. Brimacombe, *EMBO J.* **13**, 2677 (1994).
[8] M. E. Harris, A. V. Kazantsev, J.-L. Chen, and N. R. Pace, *RNA* **3**, 561 (1997).
[9] G. Cimino, H. B. Gamper, S. T. Isaacs, and J. E. Hearst, *Annu. Rev. Biochem.* **54**, 1151 (1985).
[10] P. Wollenzien, *in* "Psoralen-DNA Photochemistry, Photobiology and Phototherapies" (F. P. Gasparro, ed.), p. 57. CRC Press, New York, 1988.

method we describe depends on the use of a guide DNA complementary to the RNA target site,[11,12] which carries the psoralen molecule attached through a disulfide-containing linker in such a way that the psoralen can be phototransferred to the RNA in the DNA–psoralen–RNA complex (Fig. 1). The DNA is subsequently removed (after reduction of the disulfide bond) by gel electrophoresis and this step also selects the RNA that has reacted successfully with the psoralen. There are a number of attractive features in strategy. Usually it is possible to obtain good targeting to an individual nucleotide, or at least closely spaced nucleotides, even in a large and highly structured RNA.[13] Full-length native RNA can be used for the experiment, allowing refolding or function of the RNA if that depends on the natural posttranscriptional modifications. Because the addition is done without breaking and rejoining the RNA, no ligation step is needed in the construction of the modified RNA. Because psoralen is modest in molecular size, it should not induce significant perturbation of the native functional state of the RNA. Finally, the cross-links that psoralen makes are at an 8 Å distance, which provide strong distance constraints. The two constraints are that to avoid the possibility of unwinding of double-stranded regions by psoralen, it is preferable to choose nucleotides in single-stranded regions if the native secondary structure of the RNA is known and, in addition, only uridine residues are reactive in the targeting reaction.[10]

## Site-Specific Psoralens

Site-specific psoralen, 2-pyridyldithioethylmethylamidodiethoxyethanemethylaminomethyltrimethylpsoralen (SSP)[14] (Table I), is from Cerus Corp. (Concord, CA). SSP contains a linker of approximately 20 Å maximum extension between position C4′ and the sulfur atom and was the psoralen derivative with which the design of the targeting experiment was optimized. Two other versions of SSP with linkers of different lengths and chemical structures (Table I) subsequently have been prepared by the reaction of aminomethyltrimethyl psoralen (AMT, obtained from Calbiochem, La Jolla, CA) with the reagents SPDP and LC-SPDP (Pierce, Rockford, IL). These are important because after addition to the RNA, it is possible to add cross-linking or other reporter groups to the sulfur atom that will be at different distances from the RNA with the different SSP derivatives. For the syntheses of these, 1 μmol of AMT in 200 μl ethanol

[11] J. Teare and P. Wollenzien, *Nucleic Acids Res.* **17,** 3359 (1989).
[12] J. Teare and P. Wollenzien, *Nucleic Acids Res.* **18,** 855 (1990).
[13] D. Mundus and P. Wollenzien, *RNA* **4,** 1373 (1998).
[14] W. Saffran, M. Goldenberg, and C. Cantor, *Proc. Natl. Acad. Sci. U.S.A.* **79,** 4594 (1982).

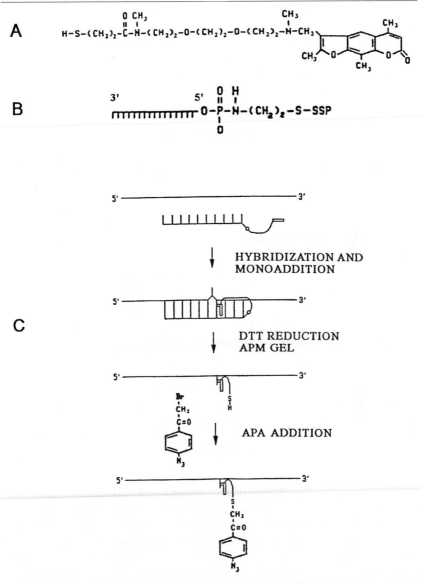

Fig. 1. Scheme for placement of site-specific psoralen in RNA. (A) Structure of site-specific psoralen (SSP20) in the reduced form. (B) Structure of the SSP–DNA attachment with the DNA polynucleotide shown as the line with vertical dashes. (C) Steps in hybridization, phototransfer, and purification of SSP-modified RNA. The last step indicates the modification of the sulfur of the SSP by the azidophenacylbromide reagent.

TABLE I
PSORALEN DERIVATIVES USED IN SITE-DIRECTED PLACEMENT REACTIONS

| Names | Formula[a] | Maximum distance[b] C4′–S (Å) |
|---|---|---|
| SSP (SSP20) | | 20 |
| LC-SPDP-AMT (SSP19) | | 19 |
| SPDP-AMT (SSP7) | | 7 |

[a] The reduced form of all three psoralen compounds is shown.
[b] Maximum distances are the extended lengths of side chains from the C4′ atom of the psoralen furan ring to the S atom. Distances are in angstrom units.

is mixed with 1.5 $\mu$mol SPDP in 200 $\mu$l methanol and reacted for 30 min at room temperature to give complete reaction or 1 $\mu$mol of AMT in 200 $\mu$l ethanol is mixed with 1.5 $\mu$mol LC-SPDP in 200 $\mu$l acetonitrile overnight at room temperature to give 40% reaction. Each reaction mixture is purified on a HPLC PRP-1 column (Hamilton, Reno, NV) using a 45–90% acetonitrile gradient for 30 min with 0.1 $M$ triethylammonium acetate buffer, pH 8.0, in each phase. The products are monitored at 290 nm and elute at about 70% acetonitrile. They are dried and redissolved in pyridine. The SPDP-AMT and LC-SPDP-AMT have linkers of about 7 and 19 Å and are named SSP7 and SSP19. These have been used for the addition to RNA with the DNA oligonucleotides designed for the SSP20 reagent and give similar results with respect to their addition to the RNA.

## DNA Sequence Design

The optimal sequence for the DNA oligonucleotides used for SSP targeting reactions was determined empirically[11] and was based on the fact that many reagents favor intercalation at bulge structures within base-paired regions.[15] To accomplish this, the DNA oligonucleotide, from its 3′ end, is complementary to the RNA for 12 to 20 nucleotides and then there is an omission of one nucleotide that would be complementary to the RNA, creating a bulge in the RNA adjacent to the nucleotide that is to react with the psoralen, and four more nucleotides complementary to the RNA are present at the 5′ end of the DNA (Fig. 1). This arrangement allows SSP attached to the 5′ end of the DNA oligonucleotide to reach the site adjacent to the bulge position. The orientation of the psoralen has not been determined, but irradiation of the complex results in SSP attachment as a mono-adduct to the RNA strand, so a large fraction of the psoralen must be in a favorable orientation for the desired photoreaction. If there are additional uridines adjacent to the target uridine, they are also reacted frequently[11] and if there are uridines within five to seven nucleotides of the intended target, these may sometimes be reacted, but this depends to a large extent on the specific sequence, the secondary structure, and the ionic conditions in which the psoralen phototransfer reaction is done.

## Synthesis of SSP–DNA

Synthesis of SSP–DNA is done in three steps (Fig. 1). First, cystamine is added to the 5′-phosphate of the DNA, which is followed by reduction of the cystamine and the addition of SSP. It is usually sufficient to purify DNA after synthesis by redissolving it in water, removing any insoluble material by low-speed centrifugation, and then ethanol precipitating the DNA and redissolving it in water. It is worthwhile to obtain the DNA molecule as a 5′-phosphorylated oligonucleotide during its chemical synthesis. Alternatively, 5′ phosphorylation can be done with up to 10 nmol of oligonucleotide in a 500-$\mu$l volume in a solution containing 50 m$M$ Tris–HCl, pH 8.0, 10 m$M$ MgCl$_2$, 10 m$M$ 2-mercaptoethanol with 5–10 $M$ excess of ATP, and 30 units of polynucleotide kinase at 37° for 2–4 hr. Phosphorylated DNA is purified by reversed-phase HPLC using conditions written later, dried, and redissolved in water before the addition of cystamine.

Cystamine (Aldrich, Milwaukee, WI) is added to the DNA in a reaction as described previously.[11,16] The following reaction is used for 100 nmol of

[15] J. M. Kean, S. A. White, and D. E. Draper, *Biochemistry* **24,** 5062 (1985).
[16] B. C. F. Chu and L. Orgel, *Nucleic Acids Res.* **16,** 3671 (1988).

DNA oligonucleotide: DNA in 250 $\mu$l $H_2O$, 100 $\mu$l 1 $M$ imidazole (titrated to pH 7.0 with HCl), 150 $\mu$l 1 $M$ 1-ethyl-3-(3-dimethylaminopropyl)carbodiimide, and 500 $\mu$l 1 $M$ cystamine (titrated to pH 7.0 with HCl) are mixed and incubated for 2.5 hr at 50°. The product is separated by reversed-phase HPLC according to the conditions written later. The cystamine–DNA is dried under vacuum and redissolved in TE (10 m$M$ Tris–HCl, pH 8.0, 1 m$M$ EDTA).

To accomplish addition of the SSP, the cystamine–DNA must be reduced first. The completeness of this reduction and the subsequent removal of excess dithiothreitol (DTT) are critical for good efficiency in the subsequent SSP addition step. For 20 nmol cystamine–DNA in 100 $\mu$l TE buffer, 1 $\mu$l 1 $M$ DTT is added and the mixture is incubated at 37° for 1 hr. This mixture is then pipetted onto 2 ml of a Sephadex G50-80 spin column freshly prepared in 50 m$M$ TEAC and 50 $\mu M$ DTT. TEAC (50 m$M$) is made from a 1 $M$ solution of triethylamine in water that has been brought to pH 7.5 by the bubbling of $CO_2$. The spin column is prepared in a 5-ml Econo-column (Bio-Rad, Hercules, CA) by pipetting a slurry of 70% (w/v) Sephadex G50-80 in 50 m$M$ TEAC, 50 $\mu M$ DTT buffer (4°) and centrifuging the column at 1000 rpm (approximately 300$g$) for 10 min in a clinical centrifuge with a swinging head at 4°. The 100 $\mu$l of mixture containing the reduced cystamine–DNA is then pipetted onto the top of the column and centrifuged at the same speed for 10 min. The sample is collected in a microfuge tube placed under the spin column and should be 100 $\mu$l in volume.

Eight microliters of SSP at 12 nmole/$\mu$l is added immediately to the reduced cystamine–DNA. All steps involving SSP should be done in reduced lighting: no fluorescent lights or sunlight should be allowed but incandescent lights are tolerated. The SSP–DNA mixture is incubated for 3 hr at 37°, and then the SSP–DNA product is isolated by HPLC chromatography as described later. For SSP20 and the other SSP compounds, a solution of $A_{250} = 1$ corresponds to a concentration of 100 nmol/ml SSP–DNA (for 24-mer DNA samples). The SSP–DNA is collected, dried, and redissolved in water at a concentration of about 100 pmol/$\mu$l.

## Chromatography Conditions

The purification of phosphorylated DNA, cystamine–DNA, and SSP–DNA is carried out by reversed-phase HPLC. Alternative techniques that use manual low-pressure reversed-phase columns were described earlier.[11] For HPLC purification of any of the DNA samples, volumes up to 1 ml in aqueous buffers are purified using a PRP-1 column (Hamilton, 0.41 × 15 cm) with two phases (phase A is 50 m$M$ triethylammonium acetate, pH

7.5, and phase B is 90% acetonitrile in water) applied in a linear gradient (8 to 32% phase B run from 5 to 35 min) with 1 ml per minute flow. In this system, DNA and phosphorylated DNA have retention times of 16–17 min, cystamine–DNA has a retention time of 16–20 min, and SSP–DNA has a retention time of 27–30 min. Figure 2 shows separations for the different species. The elution profile for the cystamine frequently looks like several species. We have not chemically characterized these further, but all of them are reactive in the addition of psoralen and result in the same final product. The chromatographic profiles were determined at 260 or 298 nm for test and preparative experiments.

## Synthesis of SSP on 3′ Terminus of DNA Oligonucleotides

In some instances it is desirable to add psoralen close to the 5′ end of an RNA molecule or close to the 3′ side of some feature, such as a posttranscriptionally modified nucleotide that prevents normal hybridization. In addition, there are some sites in ribosomal RNA that have sequences hybridized more readily on their 3′ side than on their 5′ side, and this may hold for other large structured RNA molecules. For these reasons it is sometimes desirable to synthesize the SSP onto the 3′ end of the guide DNA oligonucleotide for phototransfer to the RNA target site. This can be done with DNA chemically synthesized with its 3′ terminus phosphorylated. We have found that the DNA sequence should again be designed to create a single bulge in the RNA strand adjacent to the target site. The steps for cystamine addition are identical to the conditions for the 5′ modification, and the phototransfer reaction is similar in its efficiency.

## Addition of SSP to RNA

Addition of the SSP to the RNA is accomplished by forming an SSP–DNA hybrid with the RNA followed by irradiation of the complex to activate the psoralen for photochemical transfer to the RNA. The RNA is then separated from the DNA oligonucleotide, preferably in a way that selects RNA that has been reacted successfully in the phototransfer reaction.

## Hybridization of SSP–DNA and RNA: Phototransfer Reaction

Hybridization of SSP–DNA and the RNA can be done in several ways. For experiments involving ribosomal RNA, large amounts of derivatized RNA are needed, and the native 16S rRNA and the SSP–DNA are equally time consuming to prepare. Also, it is not desirable to have a large excess

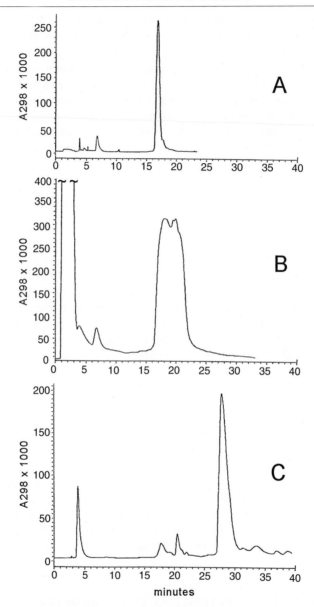

Fig. 2. Reversed-phase HPLC chromatographic behavior of DNA and modified DNA. (A) Phosphorylated DNA. (B) Cystamine–DNA. (C) SSP–DNA. All three of the samples shown were analyzed under the same chromatographic procedure described in the text. The vertical axis is the $A_{298}$ in a scale in which 1 OD = 1000. The DNA oligonucleotide used in these experiments is a 24-mer. The elution times indicated are typical, but there are some differences for DNA oligonucleotides of different lengths and base composition.

of SSP–DNA that might hybridize to other than the intended sites in the 16S rRNA. Therefore, the hybridization of SSP–oligonucleotide and 16S rRNA is performed at a 1 : 1 molar ratio. This is done in a buffer containing 1× TE (10 m$M$ Tris–HCl, pH 8.0, 1 m$M$ EDTA) and 0.1 $M$ NaCl for 10 min at 45°, followed by incubation on ice for 10 min. Usually, 4 mg (8 nmol) of 16S rRNA with 8 nmol SSP–oligonucleotide are hybridized in 600 $\mu$l total volume. The extent of hybridization in this condition is usually 30–50%, depending on the site chosen, but can be as low as 10% in some sites. For experiments in which a 500 nucleotide pre-mRNA is derivatized with psoralen, 50 $\mu$g RNA (about 300 pmol) is hybridized with equal molarity of SSP–DNA in 10 $\mu$l R-loop buffer (80% formamide, 400 m$M$ NaCl, 40 m$M$ PIPES, pH 6.8, 1 m$M$ EDTA). For that experiment, the hybridization procedure heats the mixture to 90° for 10 min, followed by slow cooling to 4°. This achieves about 30–80% hybridization, depending on the sites that are tested.

Irradiation of the hybridized samples is done in polypropylene microfuge tubes for 1.5 min at 4° using an irradiation device that produces 320–370 nm of light from mercury lamps and a circulating $Co(NO_3)_2$ solution to remove far-UV light below 320 nm.[5] This device has an estimated intensity of about 200 mW/cm$^2$. The short irradiation preferentially activates the 4′, 5′ furan-side reactive bond.[9] One alternative irradiation condition used 380 ± 20-nm light from a xenon light source with an emission monochromator at approximately 10 mW/cm$^2$.[11] In this case, the sample was irradiated in a quartz cuvette with stirring and nitrogen atmosphere for 1 hr. It is also possible to use a UV laser to accomplish the monoaddition reaction.[17]

## Purification of RNA

After the irradiation to add SSP covalently to the RNA, the sample is ethanol precipitated, redissolved in 400 $\mu$l of a buffer containing 1× TE and 7 $M$ urea, and reduced by adding DTT to a final concentration of 10 m$M$ and incubating for 15 min at 37°. The samples are then subjected to electrophoresis on a 4% polyacrylamide gel containing 1 mg/100 ml [(N-acryloylamino)phenyl]mercuric chloride (APM).[18] Electrophoresis is at 200 V (12.5 V/cm) at 45° for 5 hr in a thermostated apparatus. SSP-containing 16S rRNA is retarded greatly on such a gel because of the sulfhydryl group it contains and migrates only a few millimeters away from the well under these conditions. The RNA sample can be located by brief autoradiography

[17] J. W. Tessman, S. T. Isaacs, and J. E. Hearst, *Biochemistry* **24,** 1669 (1985).
[18] G. Igloi, *Biochemistry* **27,** 3842 (1988).

or exposure on a phosphorimager plate if part of it was [32]P labeled, and it can also be seen in the gel by its refractivity.

Gel material containing the RNA–SSP is cut out, and the RNA–SSP is recovered by ultracentrifugation through cushions containing 2 $M$ CsCl, 0.2 $M$ EDTA, pH 7.4, and 5 m$M$ DTT for 20 hr at 40,000 rpm,[19] resuspended in 100 $\mu$l of 5 m$M$ DTT, phenol extracted, and ethanol precipitated.

For RNAs smaller than 500 nucleotides, an alternative method can be used to isolate for reacted RNA. In this case, after the irradiation, reduction is *not* performed and the sample is added to a denaturing PAGE gel. The reacted RNA will be attached to the DNA oligonucleotide, which will produce a sufficient mobility shift to allow is isolation. The RNA–DNA sample is then purified by diffusion out of the gel material and then must be treated with DNase and rigorous reduction to remove the DNA.[10]

## Addition of Azidophenacylbromide to SSP–RNA

Because the SSP–RNA contains a reactive sulfhydryl group that is part of the SSP, it can be reacted with sulfhydryl-reactive reagents. One choice is to extend the range of the cross-linking reagent by adding another photo-reactive group to it. The procedures of Stade *et al.*[20] are followed for the addition of azidophenacylbromide (APAB). RNA is redissolved in 50 $\mu$M DTT and then 500 m$M$ Tris–HCl, pH 8.2, and 20 m$M$ APAB in methanol are added to a final concentration of 50 and 2 m$M$. After incubation for 1.5 hr at room temperature, 16S rRNA is ethanol precipitated, redissolved in 1$\times$ TE, extracted with phenol, and ethanol precipitated again. The RNA is redissolved in reconstitution buffer and incubated for 10 min at 37° prior to reconstitution. With APA added to the SSP molecule, there is a maximum physical distance of about 25 Å between the C4′ position of the psoralen and the reactive radical during the photochemical reaction; that compound is called SSP25APA. The other versions of the SSP with APA added are called SSP24APA and SSP12APA.

## RNA Cross-Linking, Purification, and Analysis of Specific Cross-Linked Molecules

Because psoralen absorbs in the near-UV, light sources can be used that activate it without inducing nonspecific far-UV cross-linking. The cobalt nitrate-filtered mercury light source described earlier is suitable for this. The amount of light needed to saturate the photochemical reaction is

[19] C. Wilms and P. Wollenzien, *Anal. Biochem.* **221,** 204 (1994).
[20] K. Stade, J. Rinke-Appel, and R. Brimacombe, *Nucleic Acids Res.* **17,** 9889 (1989).

reached in about 7–10 min. Because psoralen is a reagent that can cause a photodynamic effect, oxygen should be removed during the irradiation if possible. However, because there is a very low concentration of psoralen in the site-specific approach, this is not a significant problem. Therefore, a convenient method is to perform the cross-linking in closed microfuge tubes inserted in the thermostated holding cylinder of the irradiator.

In experiments with SSP-modified 16S rRNA, reconstitutions are done with total protein from 30S ribosomal subunits (TP30) according to the method of Krzyzosiak et al.[21] 16S rRNA without SSP modification is also reconstituted with TP30 and is used as a control to confirm that the irradiation does not produce any intramolecular cross-links. 30S ribosomal subunits that contain the SSP or SSP25APA are activated by incubating at 37° in activation buffer[22] and are irradiated in the same buffer at 4° at a concentration of 1.5 $\mu$g/$\mu$l in a microfuge tube. Irradiation is for 15 min in the device described previously.[5] The 16S rRNA is recovered from 30S ribosomal subunits by proteinase K digestion with 1% (w/v) sodium dodecyl sulfate (SDS) and 20 m$M$ EDTA, followed by phenol extraction and ethanol precipitation. Part of the sample (5 $\mu$g) is dephosphorylated with shrimp alkaline phosphatase for 15 min at 37° and purified by proteinase K digestion, phenol extraction, and ethanol precipitation. The 16S rRNA is then 5' end labeled with [$\gamma$-$^{32}$P]ATP by T4 polynucleotide kinase and mixed back with the unlabeled cross-linked sample.

An alternative to this is 3' end labeling using [$^{32}$P]pCp and RNA ligase according to the conditions of England and Uhlenbeck.[23] The advantage of 3' labeling is that there is less degradation of the RNA than in 5' labeling; however, the samples have somewhat less specific radioactivity than in the 5' labeling procedure.

### Analysis of RNA Cross-Linking Sites

Cross-linked 16S rRNA is separated by polyacrylamide gel electrophoresis using 3.6% acrylamide : bisacrylamide (40 : 1) with 8 $M$ urea and BTBE buffer (30 m$M$ Bis–Tris base, 30 m$M$ boric acid, 2 m$M$ EDTA, pH 6.8).[24] This method works because RNA molecules with cross-links between nucleotides distant in the primary sequence have a reduced electrophoretic mobility and are visualized as distinct bands in the gel electrophoresis

[21] W. Krzyzosiak, R. Denman, K. Nurse, W. Hellmann, M. Boublik, C. W. Gehrke, P. F. Agris, and J. Ofengand, *Biochemistry* **26**, 2353 (1987).
[22] A. Zamir, R. Miskin, A. Vogel, and D. Elson, *Methods Enzymol.* **30**, 406 (1973).
[23] T. E. England and O. C. Uhlenbeck, *Nature* **275**, 560 (1978).
[24] C. Wilms, J. W. Noah, D. Zhong, and P. Wollenzien, *RNA* **3**, 602 (1997).

pattern. For analytical separation, up to 2 μg RNA is loaded in each well (8 mm with a 0.38-mm spacer); for preparative separation, up to 40 μg RNA is loaded in a well 6.75 cm wide on a gel with the same spacer size. Electrophoresis is at 800 V (20 V/cm) at 45° for 17 hr in BTBE buffer in a thermostated gel electrophoresis apparatus (Hoeffer Scientific, San Francisco, CA). Figure 3 shows gel electrophoresis of 16S rRNA samples in which cross-linking was carried out with SSP at three different placement sites: U788/789,[132] U534/531, and U965. The gel electrophoresis patterns show the distinctively different cross-linking profiles for the different sam-

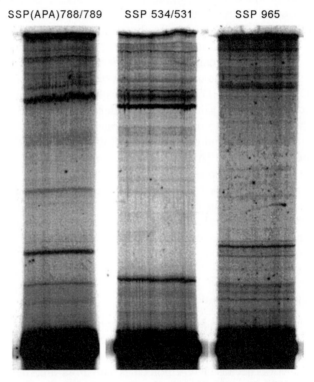

SSP(APA)788/789　　SSP 534/531　　SSP 965

FIG. 3. Examples of PAGE analysis of cross-linked 16S rRNA. 16S rRNA was derivatized with SSP20 at the nucleotide positions indicated at the top of the lanes. The SSP in sample 788/789 was reacted further with azidophenacylbromide. Derivatized RNAs were reconstituted, and 30S subunits were isolated and irradiated as described in the text. Forty-microgram samples were [32]P labeled and electrophoresed on 3.6% polyacrylamide gels as described in the text. Linear 16S rRNA has the mobility of the band at the bottom of the gel; cross-linked species appear as bands of reduced mobility. The pattern of cross-linking for the SSP(APA)788/789 sample has been reported in Ref. 13 in which the identity of 11 of the most frequent bands was identified by primer extension.

ples. The frequencies of the cross-linked species are up to 2% of the total RNA. If the gel separation is performed on 40 $\mu$g 16S rRNA, it is possible to obtain enough of cross-linked species present at 0.03% of the total for subsquent identification by primer extension analysis.

For smaller RNAs, conditions for polyacrylamide gel electrophoresis can be from 4 to 12% polyacrylamide, depending on the RNA size, in TBE, pH 8.3, with 8 $M$ urea,[12] as RNA smaller than 16S rRNA is usually less sensitive to temperature and pH-induced degradation.

The location of bands containing uncross-linked and cross-linked 16S rRNA is detected by autoradiography or phosphorimaging; they are cut out of the gel and eluted by ultracentrifugation through cushions containing 2 $M$ CsCl and 0.2 $M$ EDTA, pH 7.4, for 20 hr at 40,000 rpm.[15] To ensure recovery of the low amounts of RNA that may be present in some gel fractions, it is necessary to include tRNA carrier in the centrifugation step.[15] RNA pellets are redissolved in 100 $\mu$l $H_2O$, phenol extracted, and reprecipitated before further analysis.

## Determination of Cross-Linked Sites by Reverse Transcription Primer Arrest Assay

A convenient and usually successful way to determine the location of cross-links involves the use of primer extension on the cross-linked RNA fractions.[25] Nucleotides involved in photochemical cross-links are identified by the presence of an increased stop in the reverse transcription pattern in a particular fraction; for many types of cross-links, stops usually occur one nucleotide in the 3' direction of the cross-linked base.[4,8,26,27] For 16S rRNA (1542 nucleotides long), it is usually sufficient to use 10 DNA primers complementary to sites spaced throughout the 16S rRNA to allow reading of the 16S rRNA, except for the 3'-terminal 40 nucleotides. (There are nonreverse-transcribable dimethyladenine bases at A1518 and A1519.) It is usually a good idea to verify a cross-linking site with at least two primers to rule out reverse transcription and mispriming artifacts. Because the SSP monoadduct at the target site will be a complete stop for reverse transcriptase,[26] usually an additional primer, which begins reverse transcription just beyond the targeted site, should be used. Reverse transcription reactions are usually electrophoresed on 8% acrylamide : bisacrylamide (19:1) and 8 $M$ urea in TBE buffer.

[25] P. Wollenzien, *Methods Enzymol.* **164**, 319 (1988).
[26] G. Ericson and P. Wollenzien, *Anal. Biochem.* **174**, 215 (1998).
[27] R. Denman, J. Colgan, K. Nurse, and J. Ofengand, *Nucleic Acids Res.* **16**, 165 (1988).

Other Uses for SSP-Derivatized RNA

The methodology described here has dealt solely with cross-linking analysis of the RNA structure. However, SSP strategy is also amenable to other types of experiments in which reporter groups have been designed to be sulfur reactive. These include site-directed cleavage experiments with the reagents bromoacetamidobenzyl–EDTA–Fe(II)[28] or $o$-phenanthroline–Cu(II),[29] both of which have been shown to be valuable in establishing intermolecular distances.[30,31] A number of sulfhydryl-specific fluorescent probes are available that could also be attached to SSP. These types of experiments, together with cross-linking, provide important new tools in investigating RNA function and structure.

Acknowledgments

We thank John Tessmann of Cerus Corp. for providing SSP reagent. Mike Dolan and Tatjana Shapkina are thanked for critical comments on the manuscript. This work is supported by NIH Grant GM43237.

[28] T. Rana and C. Meares, *J. Am. Chem. Soc.* **112**, 2457 (1991).
[29] D. S. Sigman, M. D. Kuwabara, C. B. Chem, and T. C. Bruice, *Methods Enzymol.* **208**, 414 (1991).
[30] G. M. Heilek, R. Manusak, C. Meares, and H. F. Noller, *Proc. Natl. Acad. Sci. U.S.A.* **92**, 1113 (1995).
[31] D. J. Bucklin, M. A. van Waes, J. M. Bullard, and W. E. Hill, *Biochemistry* **36**, 7951 (1997).

# [9] Photolabile Derivatives of Oligonucleotides as Probes of Ribosomal Structure

*By* BARRY S. COOPERMAN, REBECCA W. ALEXANDER, YURI BUKHTIYAROV, SERGUEI N. VLADIMIROV, ZHANNA DRUZINA, RUO WANG, and NORA ZUÑO

The ribosome is the unique site of protein biosynthesis in all cells, and as such a detailed understanding of its structure and function is of fundamental importance to the more general understanding of cellular function at the molecular level. With the publications of a 9-Å crystallographic structure of the 50S ribosomal subunit[1] and of a 15-Å structure of

[1] N. Ban, B. Freeborn, P. Nissen, P. Penczek, R. A. Grassucci, R. Sweet, J. Frank, P. B. Moore, and T. A. Steitz, *Cell* **93**, 1105 (1998).

the 70S ribosome using electron microscopy data[2] and the potential for achieving even higher resolution for the 30S subunit,[3] prospects for a detailed structure of the ribosome, capable of explaining its function, have never been brighter. A major hurdle to achieving such a structure is the correct placement of ribosomal components: the accurate tracing of the RNA backbone and the orientation of ribosomal proteins alongside that backbone.

This article describes an approach toward overcoming this hurdle, which employs radioactive, photolabile derivatives of oligonucleotides (PHONTs) having sequences complementary to rRNA sequences. Such probes bind to their targeted sequences in intact ribosomal subunits and, on photolysis, incorporate into neighboring ribosomal components that can subsequently be identified. These photo-cross-links and related topographical data can be used as constraints for the construction of detailed models of ribosome structure.

The PHONT approach offers several important advantages for such an exercise. First, it allows the targeting of sequences of particular functional or structural significance throughout the ribosome structure. Second, it is readily applicable to the examination of the nature of the substantial conformational changes that the ribosome undergoes during the overall cycle of protein biosynthesis.[4] Third, the cross-links formed provide a defined upper limit distance for the separation of the linked components within the ribosome, given by the length of the tether arm. As the length of this arm can be varied, the approach can be used to identify components both immediately neighboring the target site or somewhat further away. Fourth, sample preparation, employing readily synthesized PHONTs and intact ribosomes, is straightforward, an important attribute given the large number of constraints required for model building.

Water Treatment

All of the aqueous solutions described are made up in diethyl pyrocarbonate (DEPC)-treated water, i.e., 0.1% diethyl pyrocarbonate incubated overnight at room temperature and autoclaved.

---

[2] A. Malhotra, P. Penczek, R. K. Agrawal, I. S. Gabashvili, R. A. Grassucci, R. Jünemann, N. Burkhardt, K. H. Nierhaus, and J. Frank, *J. Mol. Biol.* **280,** 103 (1998).

[3] A. Yonath, J. Harms, H. A. Hansen, A. Bashan, F. Schlunzen, I. Levin, I. Koelln, A. Tocilj, I. Agmon, M. Peretz, H. Bartels, W. S. Bennett, S. Krumbholz, S. Janell, S. Weinstein, T. Auerbach, H. Avila, M. Piolleti, S. Morlang, and F. Franceschi, *Acta Crystallogr. A* **54,** 945 (1998).

[4] K. S. Wilson and H. F. Noller, *Cell* **92,** 337 (1998).

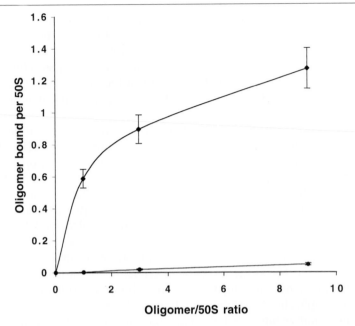

Fɪɢ. 1. Noncovalent binding of oligonucleotides to 50S subunits. 50S subunits (12.5–15 pmol) were incubated with various amounts of the indicated 5′ $^{32}$P-labeled nucleotides, complementary to 23S rRNA nucleotides 2604-2612, and filter binding was carried out as described. ♦, 2′-O-Me-RNA-p*2612-2604; ●, cDNA-p*2612-2604. Adapted from S. Vladimirov, Z. Druzina, R. Wang, and B. S. Cooperman, *Biochemistry* **39,** 183 (2000).

## Synthesis and Purification of Oligonucleotides

We use both 2′-O-methyl (2′-O-Me)-oligoRNAs and oligoDNAs, typically 9–11 nucleotides long, as PHONTs. 2′-O-Me-oligoRNAs offer the advantage of binding more tightly to their target sites[5] (Fig. 1). However, they do not generate an RNase H cleavage site on binding to the target site, which is useful both for demonstrating target site binding and for identifying labeled sequences by RNase H analysis (see later). We have shown that, as expected, a chimeric PHONT, made up of four consecutive DNA residues to generate an RNase H site[6] with the remainder of the residues being 2′-O-Me-RNAs, allows retention of the RNase H site while binding to its target with high affinity.

[5] H. Inoue, Y. Hayase, A. Imura, S. Iwai, K. Miura, and E. Ohtsuka, *Nucleic Acids Res.* **15,** 6131 (1987).
[6] H. Inoue, Y. Hayase, S. Iwai, and E. Ohtsuka, *Nucleic Acids Res. Symp. Ser.* **19,** 135 (1988).

OligoDNAs and 2'-O-Me-oligoRNAs are prepared using phosphor-amidite chemistry on a Milligen Biosearch Cyclone automated DNA synthesizer that carries out 3' to 5' solid-phase synthesis with reagents supplied by Glen Research (Sterling, VA). Amine or thiophosphoryl functionality is placed within oligonucleotides according to protocols also provided by Glen Research. Following deprotection, oligonucleotides are purified by RP-HPLC (ODS-silica column, SynChropak RP-P Ti 250 × 4.6 mm, MICRA Scientific, Inc., Northbrook, IL) using a linear gradient, 5–40% acetonitrile in 0.1 $M$ triethylammonium acetate, pH 7.6.

## Amine Functionality

3' Derivatives are made through the application of 3'-amino-modifier C7CPG (controlled pore glass), which places a 1° amine at the 3' terminus and is designed for use in automated synthesizers. This reagent introduces a branched chain containing a protected primary amine into the normal reagent used to initiate oligoDNA synthesis. Following oligoDNA synthesis and deprotection, the amino group is available for derivatization. Placement at the 3' end employs (1-dimethoxytrityloxy-3-fluorenylmethoxycarbon-ylaminohexane-2-methylsuccinoyl)-long chain alkylamino-CPG as the solid-phase support. Placement at the 5' end is accomplished by coupling the 5' terminus with $O$-($N$-(4-monomethoxytrityl)-6-aminohexyl)-$O'$-(2-cyanoethyl)-$N,N$-diisopropylphosphoramidite prior to deblocking. The tether size can be decreased or increased using commercially available reagents in which, for example, 2-aminoethyl or 12-aminododecyl replaces 6-aminohexyl.

## Phosphorothioate Functionality

Introduction of this functionality is accomplished by using the sulfurizing reagent 3$H$-1,2-benzodithiol-3-one 1,1-dioxide[7] in place of the normal oxidation reagent in the appropriate condensation cycle. For placement at the 5' end, the C3 spacer synthon 3-$O$-dimethoxytritylpropyl-1-[(2-(cyano-ethyl)-($N,N$-diisopropyl)]phosphoramidite is employed in the last condensation cycle and the C3 spacer is retained as a protecting group.

## Introduction of Photolability and Radioactivity

Photolability is introduced via the amine functionality by acylation with $N$-hydroxysuccinimidyl-4-azidobenzoate (HSAB, Pierce). Typically, 120

---

[7] R. P. Iyer, L. R. Phillips, W. Egan, J. B. Regan, and S. L. Beaucage, *J. Organ. Chem.* **55**, 4693 (1990).

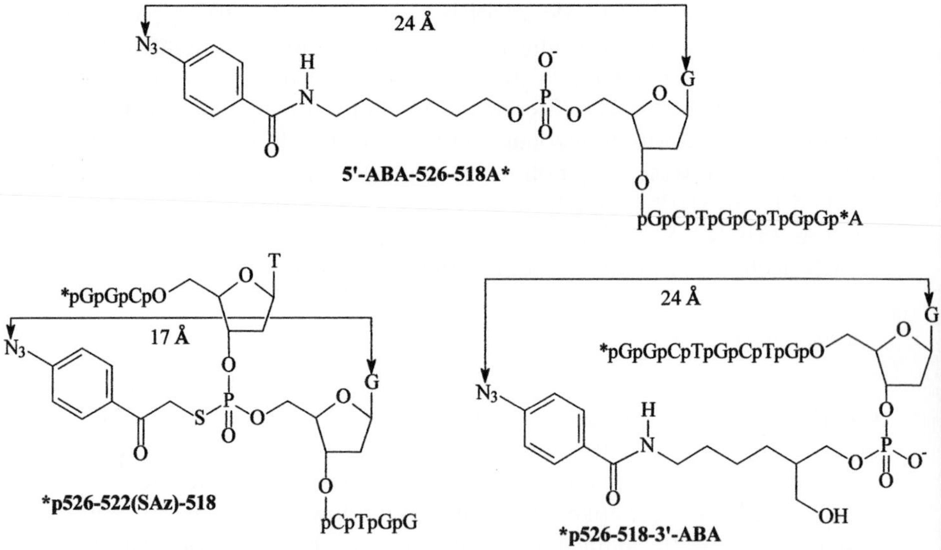

FIG. 2. Typical PHONT structures probing the 530 loop in 16S rRNA. Distances between the photogenerated nitrenes and the bases complementary to 16S nucleotides 526, 518, or 522/3 are indicated. Adapted from R. Wang, R. W. Alexander, M. VanLoock, S. Vladimirov, Y. Bukhtiyarov, S. C. Harvey, and B. S. Cooperman, *J. Mol. Biol.* **286,** 521 (1999).

nmol of amino-derivatized oligonucleotide in 250 $\mu$l of 0.05 $M$ Na$_2$CO$_3$ buffer (pH 10.0) is mixed with 250 $\mu$l of a dimethylformamide (DMF) solution of HSAB (10 mg/ml), and the resulting solution is incubated in the dark for 16 hr at room temperature. Modification of phosphorothioate oligonucleotides with arylazide reagent is achieved by reacting 100 nmol of oligonucleotide with 2 $\mu$mol of 4-azidophenacylbromide in 50 $\mu$l of 100 m$M$ sodium bicarbonate containing 50% dimethyl sulfoxide (DMSO). The reaction is allowed to proceed for 1.5 hr at room temperature, the volume is brought to 160 $\mu$l with DEPC-treated water, and the mixture is extracted twice with 300 $\mu$l of i-BuOH; 1 $M$ triethylammonium acetate, pH 7.6 (20 $\mu$l), is added to the aqueous layer prior to RP-HPLC.

Photolability can also be introduced directly in oligonucleotide synthesis, using the appropriate phosphoramidite synthon, e.g., s$^4$T, also available from Glen Research.

PHONTs resulting from these procedures (some examples are shown in Fig. 2) are purified by RP-HPLC as described earlier, with the PHONT eluting somewhat later than the underivatized oligonucleotide. In the case of phosphorothioates, which are diastereomeric, RP-HPLC resolves the $R_p$

and $S_p$ enantiomers, which can be used in photoincorporation experiments either individually or as a mixture.

Radioactivity may be conveniently introduced into either the 5' end, using [$\gamma$-$^{32}$P]ATP and T4 polynucleotide kinase,[8] or the 3' end, using [$\alpha$-$^{32}$P]ATP and deoxynucleotidyltransferase.[9] Radioactive PHONTs are purified using Sep-Pak $C_{18}$ cartridges[8] (Waters Corp., Milford, MA).

Noncovalent Binding

Overall noncovalent probe:subunit complex formation is measured by filter-binding assays. In a typical assay, 30S or 50S subunits (15 pmol), prepared as described,[10,11] are combined with varying amounts of $^{32}$P-labeled oligonucleotide in the presence of 40 m$M$ Tris–HCl or HEPES (pH 7.6), 60 m$M$ KCl, and 0.45 m$M$ $MgCl_2$ in a total volume of 25 $\mu$l. The mixture is then incubated at 37° for 5 min and on ice for 15 min. The $MgCl_2$ concentration is then raised to 10 m$M$, and the mixture is incubated at 37° for 5 min and on ice for 0.5–2 hr. The samples are diluted with 0.5 ml of cold buffer and applied immediately onto Millipore 0.45-$\mu$m HAWP nitrocellulose membrane filters that have been prewashed with 5 ml of the binding buffer containing 10 m$M$ $MgCl_2$. Filters are washed three times with 1 ml of the buffer, air dried, and counted for radioactivity in EcoLite scintillant (ICN, Irvine, CA).

We find that the preincubation step at lower [$Mg^{2+}$] increases binding stoichiometry, presumably by loosening the subunit structure, which is restored when [$Mg^{2+}$] is raised. These assays typically show two-phase binding, with high-affinity ($K_d \sim 0.1$–5 $\mu M$) sites of stoichiometry 0.1–1.0 per subunit, and an indeterminate number of low-affinity sites (Fig. 1).

Evidence that at least some oligoDNA probe binding occurs from the target site is sought by determining whether RNase H cleavage of the probe:subunit complex generates appropriate length fragments. Typically, RNase H (10 international units, Promega) is added to a sample prepared as described earlier containing 75 pmol of subunits and 150 pmol of complementary oligonucleotide in a total volume of 25 $\mu$l. Incubation proceeds on ice for 10–16 hr and at 4° for 2 hr. Following the addition of 200 $\mu$l of extraction buffer [3 m$M$ EDTA, 0.3 $M$ sodium acetate (pH 5.2), 0.5%

[8] J. Sambrook, T. Maniatis, and E. F. Fritsch, "Molecular Cloning: A Laboratory Manual." Cold Spring Harbor Laboratory Press, Cold Spring Harbor, NY, 1989.

[9] S. I. Yousaf, A. R. Carroll, and B. E. Clarke, *Gene* **27**, 309 (1984).

[10] R. A. Goldman, T. Hasan, C. C. Hall, W. A. Strycharz, and B. S. Cooperman, *Biochemistry* **22**, 559 (1983).

[11] M. A. Buck, T. A. Olah, C. J. Weitzmann, and B. S. Cooperman, *Anal. Biochem.* **182**, 295 (1989).

SDS] the reaction mixture is extracted twice with phenol, once with 1:1 phenol:chloroform, and once with chloroform. The rRNA is precipitated with 3 volumes of 95% ethanol at −20° for 1 hr, pelleted by centrifugation, dissolved in gel loading solution (7 $M$ urea/0.1% xylene cyanol FF/0.1% bromphenol blue), and incubated at 65° for 5 min. It is then analyzed by electrophoresis on a 4% polyacrylamide:bisacrylamide (19:1, w/w) 7 $M$ urea gel made up in 89 m$M$ Tris–borate (pH 8.3) buffer containing 2 m$M$ EDTA. RNA bands are visualized by staining with methylene blue (0.2% in 0.2 $M$ sodium acetate, pH 4.7).

### Photolysis

Samples are prepared in a manner identical to that used for filter-binding assays. Photolysis is performed at 4° in a Rayonet (Branford, CT) RPR 100 reactor equipped with RPR 3000-Å lamps, which have significant intensity from 280 to 320 nm, as described.[12] In this apparatus, aryl azides are completely photolyzed within 2 min, as determined by monitoring of the UV spectrum. In photoincorporation experiments, a photolysis time of 4–5 min is used. Photolysis may be carried out in either quartz or borosilicate tubes with little apparent difference in results. Following photolysis, 2-mercaptoethanol is added immediately to a final concentration of 20 m$M$ to quench any light-independent reaction.[10] Two volumes of 9:1 ethanol: 2-mercaptoethanol is added, the mixture is allowed to stand at −20° for 2 hr, and subunits are collected by centrifugation for 15 min in an Eppendorf 5414 microcentrifuge at 4° (14,000 rpm, 16,000$g$). The pellets are washed with 70% (v/v) ethanol prior to further use.

### Identification of PHONT-Labeled Nucleotides or Limited rRNA Sequences

Localization of photoincorporation sites within 16S or 23S rRNA is accomplished in a two-step process, a partial localization employing RNase H digestion followed by identification of specific labeled nucleotides using pauses or halts in reverse transcriptase primer extension. The first step, which monitors $^{32}$P-labeled PHONT radioactivity, can be carried out even for cross-links formed in very low yield. The second step, which measures pauses or halts in reverse transcription of labeled rRNA, requires significant labeling stoichiometry (≥0.2–0.5%) to be successful.

---

[12] E. N. Jaynes, Jr., P. Grant, R. Wieder, G. Giangrande, and B. S. Cooperman, *Biochemistry* **17,** 561 (1978).

*RNase H Digestion*

Sites are first localized to limited sequences by hybridizing $^{32}$P-labeled PHONT rRNA with pairs of oligoDNAs complementary to two different sequences, digestion with RNase H, and, following PAGE separation of fragments, determination of radioactivity in the resulting RNA fragments by autoradiography.[13] The use of oligoDNAs placed along the entire length of both 16S and 23S rRNA permits rapid localization of labeled rRNA to within 50–200 nucleotides. However, because not all complementary oligonucleotides are equally efficient in generating RNase H sites, it is essential to assess the efficiency of each directly, using urea–PAGE analysis and methylene blue staining to determine the yields of fragment formation. Interestingly, oligoDNA–probe-labeled rRNA is itself a substrate for RNase H. This is because the specific heteroduplex between photolyzed, covalently incorporated oligoDNA probe and rRNA is typically reformed under the conditions used for RNase H digestion and provides a site for RNase H cleavage (see earlier discussion).

Because the RNase H step allows facile quantification of the extent of labeling, it is particularly useful for testing the effects of variables such as light fluence, added complementary oligonucleotide as a competitor, and PHONT concentration. Sample data are presented in Fig. 3. This step is typically carried out at ribosomes in large excess over PHONT in order to maximize the fraction of photoincorporation arising from PHONT bound to the high-affinity target site (see Fig. 1).

A drawback of the use of oligoDNAs to generate RNase H sites is that because RNase H can cut each RNA–DNA heteroduplex at several positions, fragment formation by cutting at two sites generates a family of labeled fragments, typically spreading over 10–20 nucleotides, and limiting localization to 50 nucleotides or so. In a recent improvement, we have generated RNase H cleavage sites with chimeric oligonucleotides containing four consecutive DNA residues, with the remainder of the residues being 2'-*O*-Me-RNAs rather than with oligoDNAs. This results in RNase H cuts at single positions within each of the heteroduplexes,[6] giving rise to a single labeled fragment and making it feasible to localize labeled RNase H fragments to ~10 nucleotides. This is particularly important for cross-links formed in low yields. The only drawback to the general use of chimeric oligonucleotides in generating RNase H sites is that they are substantially more expensive to synthesize.

---

[13] R. Brimacombe, B. Greuer, P. Mitchell, M. Osswald, J. Rinke-Appel, D. Schüler, and K. Stade, "The Ribosome: Structure, Function, and Evolution" (W. E. Hill, A. Dahlberg, R. A. Garrett, P. B. Moore, D. Schlessinger, and J. R. Warner, eds.), p. 93. American Society for Microbiology, Washington, DC, 1990.

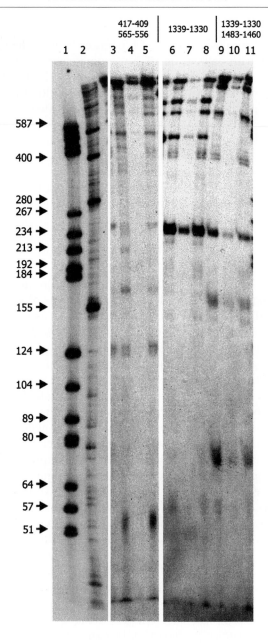

The protocol for oligoDNA-directed RNase H cleavage follows Donis-Keller[14] with the following modifications. RNA from PHONT-labeled ribosomes is extracted as described earlier (noncovalent binding). Typically 10 pmol of extracted $^{32}$P-labeled rRNA is combined with 20 pmol of each oligoDNA in 8 $\mu$l of 25 m$M$ Tris–HCl (pH 7.6), 125 m$M$ NaCl, 1.25 m$M$ DTT. Following incubation at 55° (5 min) and 32° (15 min), MgCl$_2$ (final concentration, 10 m$M$) and RNase H (2 units) are added (total volume 10 $\mu$l) and incubation is continued at 32° for 30 min. The reaction is quenched by the addition of 100% formamide/0.1% xylene cyanol FF/0.1% bromphenol blue (5 $\mu$l), and samples are analyzed by urea/PAGE. Radioactive bands are detected by autoradiographic analysis. For PHONTs that bind noncovalently to the extracted rRNA with very high affinity (e.g., 2'-$O$-Me-oligoRNAs with high G/C content), it is sometimes necessary to add excess complementary unlabeled parent oligonucleotide prior to the 55° heating step in order to reduce background due to noncovalently bound radioactivity.

[14] H. Donis-Keller, *Nucleic Acids Res.* **7**, 179 (1979)

---

FIG. 3. RNase H digestion of 16S rRNA photolabeled with a $^{32}$P-labeled PHONT probing the 530 loop in 16S rRNA. Extracted RNA was incubated with the indicated cDNA probes and digested with RNase H. Photolabeling experiments were carried out in the absence of added cDNA (lanes 3, 6, and 9) or in the presence of a 10-fold (over ribosomes) excess of cDNA complementary to nucleotides 526-518 (lanes 4, 7, and 10) or MM-cDNA 526-518 (lanes 5, 8, and 11). Lanes 1 and 2 are DNA (Boehringer-Mannheim Biochemicals) and RNA (Gibco/BRL) molecular weight markers, respectively, with sizes (in nucleotides) indicated to the left of the gel. The markers were $^{32}$P labeled by treatment with calf intestine alkaline phosphatase (Promega) and T4 polynucleotide kinase (New England Biolabs)-catalyzed phosphorylation with [$\gamma$-$^{32}$P]ATP. Lanes 3–5, with cDNAs 417-409 and 565-556; lanes 6–8, with cDNA 1339-1330; and lanes 9–11, with cDNAs complementary to nucleotides 1339-1330 and 1483-1469. Site-specific photoincorporation into regions 518-526 to 556-565 (stronger) and 409-417 to 518-526 (weaker) is shown by the bands at 50 and 120 nucleotides formed in response to added cDNAs complementary to nucleotides 565-556 and 417-409, respectively (lanes 3–5). The intensity of both bands, each of which reflects incorporation at or near the target site, is reduced severely when photoincorporation is carried out in the presence of cDNA 526-518 (5'-GGCTGCTGG-3'), but is little affected in the presence of MM-cDNA 526-518 (5'-GGCGTCTGG-3'; mismatched bases are underlined), thus demonstrating that labeling is target-site specific. An intense band, at 225 nucleotides, is formed when RNase H digestion is carried out in the presence of cDNA 1339-1330 (lanes 6–8). Further RNase H digestion in the presence of cDNAs complementary to nucleotides 1339-1330 and 1483-1469 (lanes 9–11) produces bands of 150 (weaker) and 70 (stronger), nucleotides. This corresponds to labeling of two regions, 1330-1339 to 1469-1483 and 1469-1483 to the 3' end. Adapted from R. Wang, R. W. Alexander, M. VanLoock, S. Vladimirov, Y. Bukhtiyarov, S. C. Harvey, and B. S. Cooperman, *J. Mol. Biol.* **286**, 521 (1999).

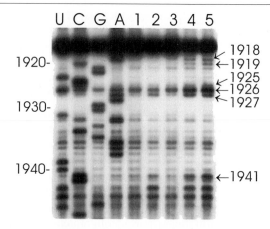

FIG. 4. Reverse transcriptase analyses of 23S rRNA photolabeled with a PHONT probing nucleotides 2604-2612 in 23S rRNA. Photoincorporation was performed using 150 pmol of 50S subunit in a reaction volume of 150 $\mu$l. The cDNA primer used was complementary to nucleotides 1983-1965. Lanes U, C, G, and A are sequencing products generated from control (nonphotolyzed) 23S rRNA in the presence of ddATP, ddGTP, ddCTP, and ddTTP, respectively. Lanes 1–3 are control experiments for RNA isolated from samples with (1) no photolysis and no photoprobe, (2) photolysis and no photoprobe, and (3) no photolysis and 375 pmol photoprobe. Lanes 4 and 5 are for photolyzed samples containing 375 or 750 pmol of photoprobe, respectively. Arrows indicate nucleotides at which pauses or stops induced by photoincorporation are observed. Adapted from S. Vladimirov, Z. Druzina, R. Wang, and B. S. Cooperman, *Biochemistry* **39**, 183 (2000).

## Reverse Transcriptase Primer Extension

In the second step of localization, PHONT-labeled rRNA is hybridized with single-stranded oligoDNA primer complementary to the 3' side of each region of rRNA found to be labeled in the first step. This heteroduplex is then used as a substrate for reverse transcriptase, and the exact position(s) of labeling is identified, using a sequencing gel, as the position(s) following that for which a halt or pause is observed[15] (Fig. 4). These analyses are performed on PHONT-labeled rRNA extracted from subunits with which photolyses have been carried out either with nonradioactive PHONTs or with PHONTs of very low specific radioactivity (<1%) of the levels used in RNase H partial localization experiments. As controls, analyses are also carried out on samples from unmodified subunits, from subunits irradiated in the absence of PHONT, and from samples incubated with PHONT but not irradiated. Simultaneously, rRNA extracted from unmodified subunits

[15] A. Barta, G. Steiner, J. Brosius, H. F. Noller, and E. Kuechler, *Proc. Natl. Acad. Sci. U.S.A.* **81**, 3607 (1984).

is used for sequencing. The procedure employed follows Muralikrishna and Wickstrom.[16] Typically, 5'-$^{32}$P-labeled primer oligoDNA (0.5–1 pmol, 1.5 × 10$^6$ cpm, 3 μl), prepared using T4 polynucleotide kinase as described earlier, is annealed to rRNA (0.6 pmol, 1 μl) by incubation in 50 m$M$ Tris–HCl (pH 8.3), 60 m$M$ NaCl, 10 m$M$ DTT for 5 min at 70°. Samples are frozen in 2-propanol/dry ice bath (1 min) and then kept on ice for at least 30 min, after which time they are microfuged briefly to bring down any liquid from the sides of the tube. To each annealing mix is added extension mix (6 μl), containing 10 units of avian myeloblastosis virus (AMV) reverse transcriptase (Promega) and affording the following final concentrations in 10 μl total volume: each dNTP (0.32 m$M$), 50 m$M$ Tris–HCl (pH 8.3), 50 m$M$ KCl, 10 m$M$ MgCl$_2$, 10 m$M$ DTT, 0.5 m$M$ spermidine. For sequencing samples, one ddNTP is added to a final concentration of 0.16 m$M$. Samples are incubated at 47° for 30 min and then quenched by the addition of 5 μl of 100% formamide/0.1% xylene cyanol FF/0.1% bromphenol blue loading buffer. After 10 min at 70°, samples are loaded on a 6% polyacrylamide urea sequencing gel and electrophoresed at 1400–1500 V for 1–1.5 hr until bromphenol blue reaches the bottom of the gel. The temperature of the gel during the run is maintained at 45–55°.

## Identification of Labeled Proteins

We identify proteins labeled with $^{32}$P-labeled PHONTs by SDS–PAGE, RP-HPLC, and specific immunochemical analyses. Each of these methods has advantages and limitations. Used in conjunction with autoradiographic analysis, one-dimensional SDS–PAGE analysis, in which the labeled protein migrates with an apparent molecular mass equal to the sum of the masses of protein and PHONT, permits rapid analysis of incorporation into proteins. Because many samples can be analyzed simultaneously, it permits facile optimization of experimental conditions as well as an assessment of various control experiments. The major drawback is that because many ribosomal proteins have similar molecular weights, precise identification of the labeled protein is usually not possible except for the heavier proteins (>20 kDa), which are generally fully resolved when analyzed as part of TP30 (total protein from 30S subunit) or TP50 (total protein from 50S subunit). Two-dimensional PAGE analysis is more highly resolving and is useful for identification purposes, but is more time-consuming. Further, changes in migration on protein modification can introduce ambiguity into the identification process.

[16] P. Muralikrishna and E. Wickstrom, *Biochemistry* **28,** 893 (1989).

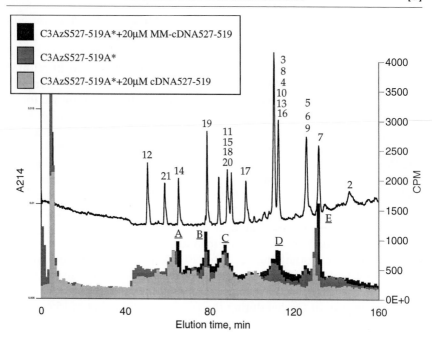

FIG. 5. RP-HPLC analysis of TP30 photolabeled with a $^{32}$P-labeled PHONT probing the 530 loop in 16S rRNA. The photolysis reaction mixture (total volume, 75 $\mu$l) contained 30S subunits (300 pmol) and PHONT (20 pmol) in the absence or presence (3000 pmol) of added cDNA 527-519 (5'-CGGCTGCTG-3') or MM-cDNA 527-519 (5'-CGGTCGCTG-3'; mismatched bases are underlined). For each analysis, proteins were extracted from photolyzed subunits with 67% acetic acid, precipitated with 5 volumes acetone, dissolved in 100 $\mu$l of 0.1% TFA, and applied to a SynChropak RP-P reversed-phase $C_{18}$ column. Following a wash step to remove unbound PHONT (15% acetonitrile/0.1% TFA, 30 min), the protein was eluted with a 120-min 15–45% convex acetonitrile gradient containing 0.1% TFA. The flow rate was 0.7 ml/min. From R. Wang, R. W. Alexander, M. VanLoock, S. Vladimirov, Y. Bukhtiyarov, S. C. Harvey, and B. S. Cooperman, *J. Mol. Biol.* **286,** 521 (1999).

RP-HPLC analysis using multistep acetonitrile–water gradients[17,18] (Fig. 5) complements SDS–PAGE analysis well. We have found it to be generally true that labeled protein coelutes with unmodified protein, or only slightly differently. Thus, labeled proteins that are well resolved can be identified unambiguously by this analysis. However, in more crowded regions of the chromatogram, even small changes in elution volume on protein modification can introduce ambiguity into the identification process. Typically, SDS–

[17] A. R. Kerlavage, T. Hasan, and B. S. Cooperman, *J. Biol. Chem.* **258,** 6313 (1983).
[18] B. S. Cooperman, C. J. Weitzmann, and M. A. Buck, *Methods Enzymol.* **164,** 423 (1988).

PAGE analysis of labeled peaks seen by RP-HPLC permits unambiguous identification of labeled protein.

In some cases, further analysis is needed. At present, agarose antibody affinity chromatography analysis,[19] graciously carried out in the laboratory of Dr. Richard Brimacombe (Max Planck Institute, Berlin), provides the needed additional analysis. To minimize background problems, samples subjected to such analysis are first purified, either by PAGE (using gel elution) or by RP-HPLC. Immunochemical analyses are performed with antibodies to proteins migrating or eluting in the general vicinity of the labeled protein. However, this approach demands a considerable effort for the maintenance of active stocks of some 55 antibodies, and we plan to replace it with MALDI mass spectral analysis of labeled proteins,[20] purified either by RP-HPLC or by one-dimensional SDS–PAGE.

## Controls

The validity of the PHONT approach for generating useful photo-cross-links for model building rests on two assumptions: (1) that the cross-links used as constraints derive from PHONTs bound to their respective target sites, as determined by criteria listed later, and (2) that PHONT binding, with the attendant conversion of a single-stranded loop sequence into a duplex, causes no major disruption of ribosome structure.

### Criteria for Target Site-Specific Labeling

1. Noncovalent binding of the probe to the targeted subunit, with tight binding stoichiometry of 0.1–1.0.
2. RNase H cleavage at the target site on addition of the parent unmodified cDNA to the ribosomal subunit, as demonstrated by the generation of fragments of the correct predicted size. For some targets, the appropriate RNase H site can only be demonstrated in extracted RNA, rather than in the intact subunit, i.e., the target site is accessible to the PHONT, but not to RNase H.
3. PHONT photoincorporation at or adjacent to its target site.
4. A large decrease in yield of each photo-cross-link when photolysis is conducted in the presence of added excess complementary-unlabeled parent oligonucleotide (cDNA or 2'-O-Me-oligoRNA), as contrasted with little effect on yield in the presence of an excess of a similar

---

[19] H. Gulle, E. Hoppe, M. Osswald, B. Greuer, R. Brimacombe, and G. Stoffler, *Nucleic Acids Res.* **16,** 815 (1988).
[20] A. R. Perrault, B. S. Cooperman, L. Montesano-Roditis, and D. G. Glitz, *J. Biol. Chem.* **272,** 8695 (1997).

oligonucleotide containing two to four mismatches to the target at internal positions (denoted MM-oligonucleotides) (Figs. 3 and 5). Such mismatches severely weaken binding to the target site, but not necessarily to secondary sites (e.g., to proteins).

5. Similar to criterion 4, much smaller, if any, decrease in the yield of photo-cross-link for photolysis conducted in the presence of an oligonucleotide complementary to a "fortuitous" site within the subunit rRNA as compared with the oligonucleotide complementary to the targeted site. By "fortuitous," we refer to a site that, by chance, has substantial sequence overlap (e.g., 7–8 nucleotides out of 10) with the targeted site. Fortuitous sites are found by sequence scanning using the freely available program AMPLIFY (authored by Bill Engels. The net address is www.wisc.edu/genestest/CATG/amplify). Should this criterion not be met, it would suggest that labeling is occurring from the PHONT bound to a fortuitous site. This point could be tested directly using a new PHONT targeting such a site.

### Possible Subunit Structure Distortion Induced by Noncovalent PHONT Binding

Although it is to be expected that PHONT binding will in general cause a local distortion of the subunit structure at the current resolution of the ribosomal structure, strictly local distortions are not necessarily problematic. In addition, because intact ribosomes are targeted, PHONTs that are found to bind tightly are unlikely to have caused major structural change, as the energy required to induce such a change would result in overall weak binding. Nevertheless, the effect of complementary oligonucleotide binding on ribosome conformation should be examined directly. To do this we employ a chemical footprinting approach, first introduced to the ribosome field by Moazed and co-workers,[21] consisting of using pauses or halts in reverse transcriptase primer extension to determine which rRNA bases have their chemical reactivities toward dimethyl sulfate or kethoxal altered on complementary oligonucleotide binding. Using a series of oligoDNA primers, it is feasible to scan pertinent sequences of ribosomal rRNA or even the entire 16S and 23S molecules.

The protocol uses samples prepared as described earlier for filter-binding assays. Samples (75 $\mu$l) containing subunits (150 pmol) either in the absence or in the presence (3- to 10-fold molar excess) of complementary oligonucleotide are incubated with either dimethyl sulfate (1 $\mu$l) or kethoxal (8 $\mu$l of 37 mg/ml kethoxal in 20% ethanol) at 4° for 2.5–3 hr with gentle

[21] D. Moazed, S. Stern, and H. F. Noller, *J. Mol. Biol.* **187**, 399 (1986).

shaking. Reactions are stopped by the addition of either 75 $\mu$l of 1 $M$ Tris–acetate (pH 7.5), 1 $M$ 2-mercaptoethanol, 1.5 $M$ sodium acetate (pH 5.2), 0.1 m$M$ EDTA (dimethyl sulfate reaction) or 20 $\mu$l of 0.5 $M$ potassium borate (pH 7.0) (kethoxal reaction). Following incubation on ice for 10 min, subunits are precipitated with 2.5 volumes of ethanol, washed with 70% ethanol, and vacuum dried. RNA extraction and reverse transcriptase primer extension proceeds as described earlier.

The generalization that emerges from these studies is that, as anticipated, oligonucleotide binding does not lead to major alterations in ribosome structure. The reactivity changes that are seen tend to be at or near the targeted site in the primary sequence.[22–24] Less commonly, we sometimes see limited numbers of reactivity changes at positions remote in the primary sequence from the target site. Such changes are evidence of interaction between the site in question and the targeted site. Indeed, in some cases a site of changed reactivity is found very close in the sequence to a site of PHONT cross-linking.

## Construction of Three-Dimensional Models

In collaboration with Steve Harvey, we use the YAMMP protocol[25] for model construction based on the constraints generated from PHONT-derived photo-cross-links and the related work of others. This protocol uses molecular mechanics (including simulated annealing and energy minimization) to convert such constraints into a coherent three-dimensional model. An important attribute of YAMMP is that constraints may be individually weighted, with the result that as the level of resolution of a cross-link is increased (i.e., from a stretch of RNA sequence or ribosomal protein to a single nucleotide or amino acid), its weight in determining the overall structure is also increased. Also important is that the level of detail of the model can be varied within the overall structure, depending on the available information for different parts of the structure, through the use of pseudoatoms of different sizes. Originally, Harvey's group converted electron microscopy (EM) shape data into constraint functions based on spherical harmonics. While that is a suitable approach for very low resolution EM data, it would be computationally prohibitive for surfaces as irregular as those presented by the latest EM and X-ray data. They have

[22] R. W. Alexander, P. Muralikrishna, and B. S. Cooperman, *Biochemistry* **33**, 12109 (1994)

[23] P. Muralikrishna, R. W. Alexander, and B. S. Cooperman, *Nucleic Acids Res.* **25**, 4562 (1997)

[24] S. Vladimirov, Z. Druzina, R. Wang, and B. S. Cooperman, *Biochemistry* **39**, 183 (2000).

[25] A. Malhotra, R. K.-Z. Tan, and S. C. Harvey, *Biophys. J.* **66**, 1777 (1994).

implemented and tested a new real-space refinement procedure that is being used in their current generation models (S. Harvey, personal communication).

We have targeted the 530 loop in ribosomal 30S subunits for study by the PHONT approach. This loop, which includes 16S rRNA nucleotides 518-533, is almost universally conserved and is strongly implicated in both the accuracy control mechanism and tRNA binding by biochemical and genetic evidence. A series of PHONTs were synthesized that together effectively extend helix 18 in the 16S rRNA secondary structure model of Gutell *et al.*[26] (nucleotides 511-516 which are complementary to nucleotides 540-535) by up to 11 bps (517-527) or one full turn of an expected A-form helix. In this extended helix, we placed photolabile groups opposite nucleotides 518, 522/3, 525/6, 526, and 527, with maximal distances between photogenerated nitrene and complemented base in the 530 loop of 19–26 Å (Fig. 6, see color plate). The protein and RNA cross-links identified in this work were added to the existing database for the 30S model, and a new model was generated with the YAMMP protocol (Fig. 7, see color plate). This 30S model uses whole proteins as pseudoatoms, but as both cross-linking data to amino acids or peptides within the proteins and high-resolution structures of the proteins become available (see later), it will be possible to reduce the size of such pseudoatoms to folding domains or even individual amino acids.

## Future Improvements

### PHONT Design

We are currently seeking to expand the versatility of the PHONT approach, as outlined next.

1. Decreasing the distance between the photolabile group and the complemented base will generate cross-links that constrain model building to a greater extent than PHONTs in which this distance is larger, at the probable cost of generating fewer cross-links. One approach will be to use the enzyme terminal deoxynucleotidyltransferase, which accepts both dNTPs and NTPs as substrates, to introduce a photolabile base at the 3′ terminus that is still capable of complementary Watson–Crick base pairing, thus placing photolability within several angstroms of the complemented base. For this purpose, we will employ deoxyribose or ribose triphosphates of 8-azidoadenine, 8-azidoguanine, 5-bromouracil, 5-azidouridine, and 5-azidocytidine.

[26] R. R. Gutell, N. Larsen, and C. R. Woese, *Microbiol. Rev.* **58,** 10 (1994).

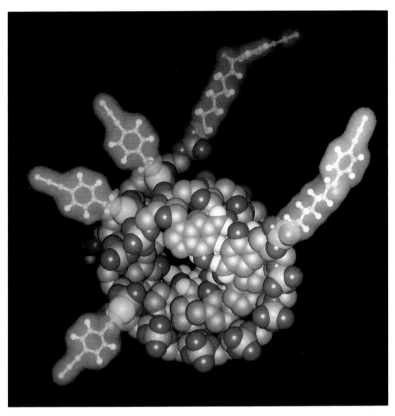

FIG. 6. Space-filling model of helix 18A, with attached photoprobes. Tethered aryl azides are shown as enveloped ball and stick models. Adapted from R. Wang, R. W. Alexander, M. VanLoock, S. Vladimirov, Y. Bukhtiyarov, S. C. Harvey, and B. S. Cooperman, *J. Mol. Biol.* **286,** 521 (1999).

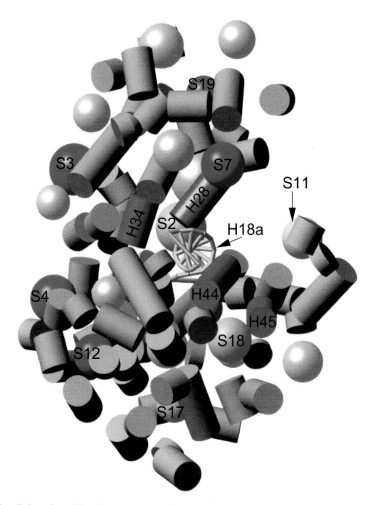

Fig. 7. Interface side of the model for the 30S ribosomal subunit. The 50S subunit binds in the front. RNA helices are represented by cylinders with a diameter of ~18 Å, and those helices involved in a cross-link are colored magenta. Helix 18 is gold. Proteins are shown as spheres of the appropriate diameter. Spheres colored dark green are strongly cross-linked, whereas those colored light green are cross-linked to lesser extents. Helix 18a is shown at the all-phosphate level of detail. Adapted from R. Wang, R. W. Alexander, M. VanLoock, S. Vladimirov, Y. Bukhtiyarov, S. C. Harvey, and B. S. Cooperman, *J. Mol. Biol.* **286,** 521 (1999).

Photolability and radioactivity can be introduced in a single step when photolabile triphosphates are available commercially, as is the case for 8-azido [$\alpha$-$^{32}$P]ATP or 8-azido [$\alpha$-$^{32}$P]GTP (Andotek Life Sciences, Irvine, CA).

2. PHONTs in which photolability is incorporated at internal positions are not subject to the uncertainty that the effective tether length could be larger than that calculated if there is some unraveling of the heteroduplex at terminal positions prior to photoincorporation. We currently introduce photolability at internal positions by derivatizing thiophosphates. Other approaches we are exploring involve the use of two kinds of commercially available purine or pyrimidine phosphoramidite synthons. The first allows direct introduction of a photolabile base and includes s$^4$dT, s$^2$dT, 5-BrdT, and 5-BrdC. The second involves synthons with leaving groups that are displaceable by nucleophiles after oligonucleotide synthesis has been completed and for which Watson–Crick base pairing is preserved. For example, the synthons for 8-BrdA and 8-BrdG can be used to introduce 8-azidoG or 8-azidoA, via displacement of 8-Br by azide ion. Another example is the synthon containing the $O^6$-phenyl derivative of dI. Displacement of the $O^6$-phenyl group with linear diamines of the type H$_2$N(CH$_2$)$_n$NH$_2$, followed by condensation of the resulting primary amine with a photolabile reagent such as HSAB (see earlier discussion), permits the introduction of photolability at a variable distance from an adenosine residue.

3. Low yields of photo-cross-linking (below 0.5%) sometimes prevent us from identifying a labeled nucleotide, largely because of background levels of reverse transcriptase pauses or stops, and will pose difficulties for identifying labeled peptides and amino acids. A strategy we are exploring for overcoming this problem involves the use of biotin-conjugated PHONTs, which will permit us to use magnetic streptavidin beads (Dynal, Inc., Lake Success, NY) to separate biotin-labeled ribosomal components (rRNA and proteins) from those that are not labeled. Here it is important to emphasize that because ribosomes and PHONTs can be prepared readily in large quantities ($\sim$0.2–1.0 $\mu$mol) compared to the sample sizes (typically $\leq$100 pmol) needed for analysis, successful identifications are limited more by the need to reduce background than by low yields of photoincorporation.

## Identification of Labeled Peptides and Amino Acids

In order to allow modeling to be carried out at a higher resolution level with respect to ribosomal protein than we have so far, we will localize labeling to the amino acid, or at least the oligopeptide level. For labeled proteins of known three-dimensional structure (the structures of some 15 isolated bacterial ribosomal proteins are known at this time[27] with the number increasing steadily), such localization makes possible attempts to fit regions of resolved electron density, as determined by both electron microscopy and X-ray crystallography, to specific RNA–protein contacts.[28] To do this we will first purify the labeled protein and then identify the labeled site or sites within the protein.

Several alternatives exist for the purification of the labeled protein. PVDF blotting from two-dimensional electrophoretic gels offers a one-step method. Alternatively, labeled proteins can first be separated from unlabeled proteins before being resolved from one another by one-dimensional SDS–PAGE and/or HPLC. The avidin–biotin methodology described earlier provides one approach for the first step. It should also be possible to exploit the photoincorporated nucleotides as a purification vehicle, using either anion-exchange column chromatography or via an affinity column made by attaching an oligoDNA complementary to the PHONT to CNBr-activated Sepharose.

We use microsequence analysis[29] of labeled peptides to identify sites of photoincorporation at the amino acid level. Following proteolysis of the labeled protein, RP-HPLC and MALDI mass spectrometric analysis will be used to resolve and identify cross-linked peptides. The cross-linked amino acid will be identified by subjecting the labeled peptides to automated Edman analysis. This procedure can be performed on as little as 5–10 pmol of labeled protein.

[27] V. Ramakrishnan and S. W. White, *TIBS* **23,** 208 (1998).
[28] I. Tanaka, A. Nakagawa, H. Hosaka, S. Wakatsuki, F. Mueller, and R. Brimacombe, *RNA* **4,** 542 (1998).
[29] S. F. Best, D. F. Reim, J. Mozdzanowski, and D. W. Speicher, *Techniq. Prot. Chem.* **V,** 565 (1994).

# [10] Photoaffinity Cross-Linking and RNA Structure Analysis

*By* Brian C. Thomas, Alexei V. Kazantsev, Jiunn-Liang Chen, and Norman R. Pace

## Introduction

Photoaffinity cross-linking reagents commonly have been used to demonstrate intermolecular interactions. This article describes the use of photoaffinity agents as measuring sticks to infer *intra*molecular structural relationships and thereby the molecular tertiary structure of the ribozyme ribonuclease P (RNase P). Although the application here is with a specific RNA, the methods for cross-linking and structure analysis are generally applicable to other RNAs or RNA-containing complexes.

RNase P is the ubiquitous RNA processing enzyme that cleaves precursors of tRNAs to form their mature 5' ends. RNase P is unusual because the catalytic subunit of the ribonucleoprotein holoenzyme is composed of RNA, not the usual protein. RNase P RNA, the catalytic RNA, has so far eluded crystallographic structural analysis, but a considerable amount has been learned about its structure and function.[1] The secondary structure was determined by phylogenetic–comparative sequence analysis, and the helical barrels that result from the secondary structure have been ordered in space by site-specific cross-linking experiments. In these studies, photoaffinity cross-linking agents were specifically attached to suspected key nucleotides and, after irradiation, sites of intramolecular insertion elsewhere were used as distance measurements with which to position the RNA helices and thereby infer the tertiary architecture. Photoaffinity cross-linking was also used to study RNase P RNA–tRNA interactions. This article details the photoaffinity agents used in these studies and the methods for their application in the determination of tertiary structure in large RNAs.

## Photoaffinity Cross-Linking Agents

Photoaffinity cross-linking agents are chemical moieties that upon irradiation with light generate highly reactive species capable of forming covalent bonds to atoms within the reach of the reactive group. Aromatic azides are particularly widely used photoaffinity cross-linking agents because they are stable, incorporated easily into biological macromolecules, and upon irradiation generate reactive species with high quantum yields.[2,3] The complex photochemistry of aromatic azides has been studied extensively over the last decades and reviewed.[4] On light excitation a spectrum of reactive species is generated, with reactivities dependent primarily on the nature and distribution of substituents at the aromatic ring.

[1] D. N. Frank and N. R. Pace, *Annu. Rev. Biochem.* **67**, 153 (1998).
[2] H. Bayley, "Photogenerated Reagents in Biochemistry and Molecular Biology." Elsevier, New York.
[3] H. Bayley and J. R. Knowles, *Methods Enzymol.* **46**, 69 (1977).
[4] G. B. Schuster and M. S. Platz, *Adv. Photochem.* **17**, 69 (1992).

Figure 1 shows the photochemical process. Light excitation of unsubstituted phenylazide **Ia** (Fig. 1) results in a loss of molecular nitrogen and transient formation of singlet phenylnitrene **II**[5] that immediately rearranges to form the azacycloheptatetraene species **III**. The latter are electron-deficient species and react predominantly with nucleophiles (Nu), forming azepine adducts **IV**.[4] *p*-Azidobenzoic acid (**Ib**), its dimethylamide (**Ic**), and *p*-azidoacetophenone (**Id**) follow primarily the same pathway.[6] Small amounts of triplet nitrene species **V** are formed in the latter cases through a process of intersystem crossing, catalyzed by a moderate electron-withdrawing effect of substituents. Triplet nitrenes are essentially diradical species that are capable of hydrogen-radical abstraction from the environment. Interaction of radical species with RNA results largely in backbone lesions.[7] Because few chain-breaking events are observed in aryl azide cross-linking experiments, the major cross-linking pathway is thought instead to be an attack of RNA nucleophilic centers (primarily N-7 and N-3 of purines[8]) on ketenimine species **III**.

## Incorporation of Photoaffinity Cross-Linking Agents into Large RNAs

The redundant nature of the RNA chain, composed of only four chemically distinct nucleotides, complicates the labeling of specific nucleotides within the chain. Only at the 5' and 3' ends of RNA molecules do unique chemistries occur and provide specific attachment sites for photoagents. In general, the synthesis of large RNAs is conducted using a phage polymerase, e.g., T7 RNA polymerase. Consequently, the approach toward labeling the RNA is constrained by the properties of the enzymes. We focus here on generally applicable chemical linkers for RNA that can be used for other applications, such as coupling the RNA to support surfaces, as well as to photoagents.[9]

## 5'-End Attachment of Photoagent

5'-end modification relies on the ability of T7 RNA polymerase to initiate transcription with thiol-containing analogs of 5'-GMP. Because only nucleoside triphosphates can support elongation, the addition of guanosine

[5] N. P. Gritsan, T. Yuzawa, and M. S. Platz, *J. Am. Chem. Soc.* **119**, 5059 (1997).
[6] Y.-Z. Li, J. P. Kirby, M. W. George, M. Poliakoff, and G. B. Schuster, *J. Am. Chem Soc.* **110**, 8092 (1988).
[7] D. W. Celander and T. R. Cech, *Biochemistry* **29**, 1355 (1990).
[8] N. K. Kochetkov and E. I. Budovskii (eds.), "Organic Chemistry of Nucleic Acids." Plenum Press, New York, 1971.
[9] D. N. Frank and N. R. Pace, *Proc. Natl. Acad. Sci. U.S.A.* **94**, 14355 (1997).

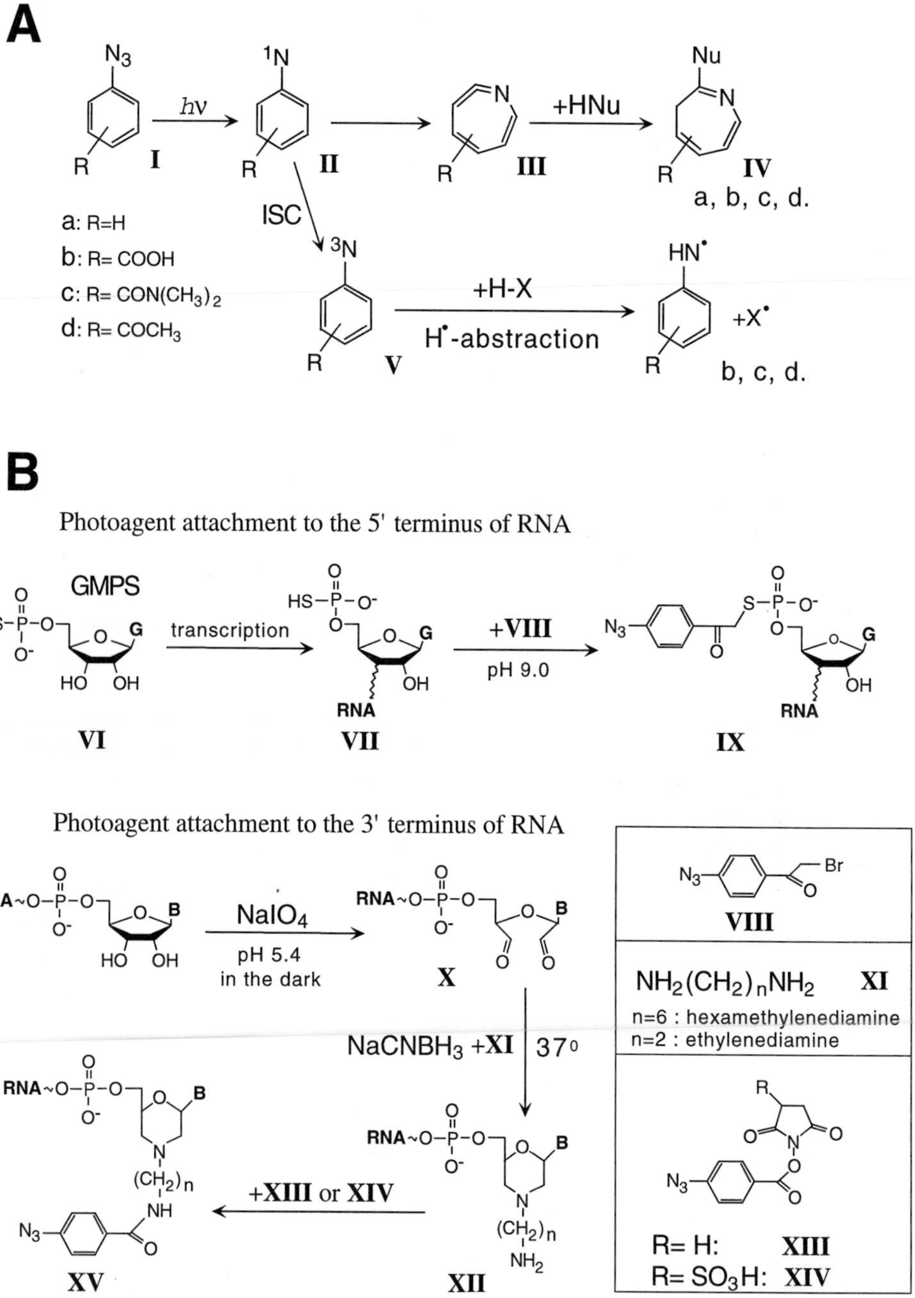

FIG. 1. Photochemistry of phenylazides and site-specific incorporation of a photoaffinity cross-linking agent into RNA. (A) Photochemistry of model phenylazides. (B) Flow chart outline of attachment of photoreactive moieties to the 5' and 3' termini of large RNA.

5'-thiomonophosphate **VI** (GMPS) in concentrations in excess of GTP in transcription results in incorporation of a unique phosphorothioate at the 5'-terminal nucleotide of the transcript.[10] This thiol-containing RNA **VII** is then reacted with an $\alpha$-bromoketone-containing arylazide (e.g., *p*-azido-phenacyl bromide **VIII**), leading to the attachment of the arylazide moiety to the 5'-terminal phosphate of the RNA by the formation of *S*-phenacyl-thiophosphate **IX**.[11,12] The following experimental protocol is outlined.

1. RNA is synthesized *in vitro* in 100 $\mu$l of 40 m$M$ Tris–HCl, pH 8.0, 2 m$M$ spermidine, 5 m$M$ dithiothreitol, 0.05% nonidet P-40 (NP-40), 6 m$M$ magnesium chloride, 1 m$M$ each NTP, 6 to 10 m$M$ GMPS, 5 $\mu$g of linearized template DNA, and 40 U of T7 RNA polymerase for 4–20 hr at 37°. GMPS and other phosphorothioates are available commercially or are synthesized readily.[10] RNA for analytical cross-linking experiments is transcribed on a smaller scale (20–25 $\mu$l reaction) and [$\alpha$-$^{32}$P]NTP is added for internal labeling. Transcribed RNA is then purified by electrophoresis in a denaturing polyacrylamide gel, visualized by UV shadow, the RNA-containing portion of the gel is excised, and RNA is eluted passively and ethanol precipitated, all using standard methods.[13] The RNA pellet is washed twice with 70% ethanol and air dried for 5 min at room temperature.

2. All operations with photoagent-containing materials are carried out in the dark or appropriate low light. The dry, 5'-thiol-labeled RNA pellet is dissolved in 60 $\mu$l of 33 m$M$ sodium bicarbonate, pH 9.0. Forty microliters of fresh 12.5 m$M$ solution of *p*-azidophenacyl bromide in methanol is added and the mixture is incubated for 1 hr at room temperature. Excess *p*-azidophenacyl bromide is then removed by phenol extraction and the photoagent-conjugated RNA is recovered by ethanol precipitation.

3'-End Attachment of Photoagent

3'-end modification takes advantage of the unique chemical properties of the 3' terminus of RNA. The 3'-terminal *cis*-diol group is oxidized easily by periodate[14] (Fig. 1) to give the dialdehyde RNA derivative **X** that is

[10] A. B. Burgin and N. R. Pace, *EMBO J.* **9**, 4111 (1990).
[11] M. M. Hanna and C. F. Meares, *Biochemistry* **22**, 3546 (1983).
[12] S. H. Hixson and S. S. Hixson, *Biochemistry* **14**, 4251 (1975).
[13] J. Sambrook, E. F. Fritsch, and T. Maniatis (eds.), "Molecular Cloning." Cold Spring Harbor Laboratory Press, Cold Spring Harbor, NY, 1989.
[14] S. B. Easterbrook-Smith, J. C. Wallace, and D. B. Keech, *Eur. J. Biochem.* **62**, 125 (1976).

subsequently reduced in the presence of an aliphatic diamine[15] **XI**, resulting in formation of the substituted morpholino derivative of RNA **XII** that bears an aliphatic amino group at the 3' terminus. The latter is then reacted with an *N*-succinimide ester of azidobenzoic acid **XIII**[16,17] (or **XIV**) to conjugate the photoagent to the 3' terminus of RNA as the amide **XV**.
The following experimental protocol is outlined.

1. RNA is obtained by runoff transcription as described earlier. Purified RNA is oxidized in 100 mM sodium acetate, pH 5.4, 3 mM sodium periodate in the dark at room temperature for 1 hr. Oxidized RNA is collected by ethanol precipitation, air dried, and modified in 1 mM alkyldiamine (e.g., 1,6-hexamethylenediamine or ethylenediamine), 20 mM imidazole, pH 8.0, 5 mM NaCNBH$_3$ at 37° for 1 hr. The length of the linker between the photoagent and its attachment site can be manipulated by the molecular length of the alkyldiamine. NaBH$_4$ is then added to 5 mM and the reaction is allowed to proceed for another 5–15 min.

2. Modified RNA is twice ethanol precipitated, air dried and dissolved in 50 mM HEPES/NaOH, pH 9.0. An equal volume of fresh, 20 mM *N*-hydroxysuccinimidyl-4-azidobenzoate **XIII** in DMSO is added, and the mixture is incubated in the dark at room temperature for 1 hr. If a water-soluble sulfo derivative **XIV** is used at this step, DMSO should be substituted by water. The photoagent-conjugated RNA is then recovered by solvent extraction and ethanol precipitation as described earlier.

## Handling and Stability

In general, all operations with arylazide-containing reagents and RNAs prior to cross-linking experiments should be carried out in the dark. Dark vessels such as microfuge tubes should be used. We have found that red Plexiglas boxes are useful for bench-top work. Commercially available bifunctional reagents **VIII**, **XIII**, and **XIV** can be stored desiccated in the cold and dark without detectable loss of activity for a few months. 5'-Thiophosphate RNA derivatives **VII** are slowly oxidized upon workup to form dimers that can be prevented by the addition of reducing compounds (DTT, 2-mercaptoethanol) to the elution (or storage) buffer. However,

[15] R. Rayford, D. D. Anthony, Jr., R. E. O'Neill, Jr., and W. C. Merrick, *J. Biol. Chem.* **260,** 15708 (1985).
[16] B.-K. Oh and N. R. Pace, *Nucleic Acids Res.* **22,** 4087 (1994).
[17] U. C. Krieg, P. Walter, and A. E. Johnson, *Proc. Natl. Acad. Sci. U.S.A.* **83,** 8604 (1986).

reducing agents should be completely removed prior to conjugation with arylazide (usually by ethanol precipitation and washing with 70% ethanol) as they react readily with the $\alpha$-bromoketone group of a photoagent. 3'-end-oxidized RNA **X** should be converted immediately to the amino derivative **XII**, which can be stored under 70% ethanol at $-20°$ for a few weeks. Both 5'-end and 3'-end photoagent-conjugated RNAs can be stored frozen at $-20°$ in the dark either in TE or under 70% ethanol for a few weeks without loss of activity. Even though arylazides are sensitive to heat, the photoagent-conjugated RNA can be annealed prior to cross-linking experiments by heating to $70°$ for a few minutes under a variety of mono- and divalent salt concentrations.

### Circularly Permuted Transcription Templates and RNAs

We have used circular permutation as a method to create ends for a specific modification. The structural integrity of RNase P RNA is maintained globally by secondary and tertiary structures. Consequently, the introduction of nicks internally in the RNA usually has no major impact on its folded structure. Because the 5' and 3' termini of RNase P are paired, they can be linked in the RNA, and the ends relocated to a selected position for specific modification. As outlined in Fig. 2, this is accomplished by the use of polymerase chain reaction (PCR) to fabricate circularly permuted transcription templates.[18] Transcription and modification are carried out as described earlier.

Circular permutation of RNase P RNA usually has little effect on activity, although the global structure seems slightly destabilized in some cases. Destabilization is generally manifest in elevated $K_m$ values (5- to 10-fold) for the affected constructs at ionic strength optimal for the native RNA. Kinetic defects in circularly permuted RNase P RNAs have always been substantially rectified, however, by elevated ionic strength.[19–21] If of interest to do so, ends of circularly permuted RNAs can be joined, to repair the nick, using RNA ligase[22] or deoxyoligonucleotide splint-promoted ligation with DNA ligase.[23]

[18] J. M. Nolan, D. H. Burke, and N. R. Pace, *Science* **261,** 762 (1993).
[19] M. E. Harris, J. M. Nolan, A. Malhotra, J. W. Brown, S. C. Harvey, and N. R. Pace, *EMBO J.* **13,** 3953 (1994).
[20] M. E. Harris, A. V. Kazantsev, J.-L. Chen, and N. R. Pace, *RNA* **3,** 561 (1997).
[21] J.-L. Chen, J. M. Nolan, M. E. Harris, and N. R. Pace, *EMBO J.* **17,** 1515 (1998).
[22] P. J. Romaniuk and O. C. Uhlenbeck, *Methods Enzymol.* **100,** 52 (1983).
[23] M. J. Moore and P. A. Sharp, *Science* **256,** 992 (1992).

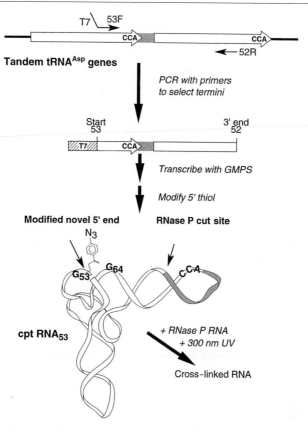

FIG. 2. Engineering and synthesis of circularly permuted RNA. The DNA template for circularly permuted RNA (tRNA in this example) is obtained by PCR from a tandem of tRNA genes with a set of primers, defining new 5' and 3' termini that correspond to adjacent nucleotides within the native sequence. *In vitro* transcription from such a template yields circularly permuted RNA.

## Cross-Linking and Analysis

Irradiation of arylazide-containing RNAs results in intramolecular cross-linking if the photoagent is within the reaction range of some component of the RNA chain. The result is a circle, for end-to-end cross-linking, or a lariat, if the cross-link is internal in the RNA chain. RNA loops and circles are separable from straight-chain RNA by gel electrophoresis. Electrophoretic mobility is retarded by loops; the extent of retardation is dependent on the size of the loop. Figure 3 shows the experimental dependence of electrophoretic mobility on the distance from the cross-linking

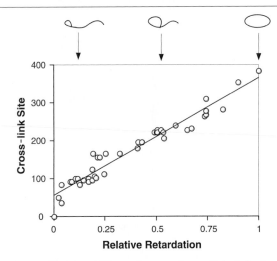

FIG. 3. Dependence of lariat mobility on distance of cross-linked site from the unmodified RNA terminus. Lariat mobility is expressed as retardation relative to circular and linear RNA standards,[20] with retardation of linear RNA set to 0 and retardation of circular RNA set to 1.

end to the point of insertion.[20] The gel mobility information provides some check on cross-link sites determined by primer extension.

Cross-linking reactions are carried out with isotopically labeled (analytical) or nonradioactive (preparative) quantities. Typical cross-linking analyses are as follows.

1. In a 20-$\mu$l reaction volume, approximately 0.02–2 nmol of $^{32}$P-labeled, 5′-APA RNase P RNA (~4 × 10$^5$ cpm) is preincubated at 50° for 1–5 min in RNase P reaction buffer in a 0.5-ml microcentrifuge tube (brown-colored, Fisher Scientific) and then allowed to cool to room temperature (~20°) or placed on ice.

2. Cross-linking is carried out at room temperature or on ice, in the reaction tube, with a hand-held UV light (302 nm, Model UVM-57, UVP, Inc., San Gabriel, CA) fitted with a thin polystyrene sheet (cut from a plastic petri dish) to block UV light <300 nm. The UV lamp is allowed to rest on top of the open reaction vessels, approximately 2 cm above the reaction mixture. UV irradiation is carried out for approximately 30–45 min and then RNAs are precipitated with 3 volumes of ice-cold 100% ethanol. RNAs are collected by centrifugation for 30 min at 4°, washed with 70% ethanol, air dried, and resuspended in 10 $\mu$l of H$_2$O and 10 $\mu$l of an appropriate gel-loading solution containing 10 m$M$ EDTA to bind residual magnesium and 7 $M$ urea.

3. RNAs are resolved by electrophoresis in a 4–6% polyacrylamide gel containing 7 $M$ urea. After electrophoresis, the gel is dried onto 3-mm Whatman (Clifton, NJ) paper for autoradiography in the case of isotopically labeled RNAs. RNA species in preparative gels are visualized by staining with ethidium bromide and eluted from excised slices of the gel. Primer extension analysis to identify the site(s) of insertion is conducted as described.[10] The typical extent of cross-linking in the cases of successfully analyzed examples is 1–2 to 30%. In general, we find that it is not practical to analyze cross-linked conjugates that occur with an extent of much less than 1%.

## Phylogenetic–Comparative Data Sets and Verification of Results

We recommend that any cross-linking analysis be conducted in parallel with the homologous RNAs from at least two different organisms in which the sequences vary. In the case of RNase P RNA, the corresponding RNAs used in structure analysis were from *Escherichia coli* and *Bacillus subtilis*. Cross-linking experiments are fraught with potential artifacts, such as cross-linking in a denatured state. Any artifact is likely to be an idiosyncrasy of the particular RNA, so observation of the corresponding cross-link in two instances of the RNA affords considerable assurance that data reflect the native state. Examples of such corroborative data are shown in Fig. 4. Additionally, the different RNAs often provide different cross-linking efficiencies; for instance, a weak cross-link in one RNA sometimes is manifest as a substantial one in another RNA.

## Structure Modeling with Cross-Linking Data

Cross-linking results and secondary structure are used to develop three-dimensional structure models. In the case of RNase P RNA, the secondary structure was known; 64% of the *E. coli* RNA consists of established, base-paired helices. Presuming that these paired regions adapt the configuration of A-form helices, then >60% of the tertiary structure of the enzyme–substrate complex is known to a reasonable approximation. However, the configurations of "single-strand" regions of the RNA are usually not predictable at this time. Thus, the challenge of building a global tertiary structure model becomes positioning the helical elements relative to one another.

A cross-link is a physical tether between two points in the RNA. Because the length of the APA cross-link is about 9 Å, the two tethering sites in the RNA are expected to be within this distance of one another in the context of the folded RNA. The cross-link is a global distance constraint. As the size of the data set of structural constraints increases, the possible

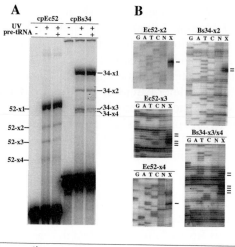

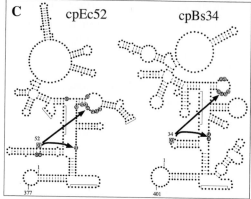

FIG. 4. Phylogenetic–comparative cross-linking data. (A) Photoagents, positioned by circular permutation to the homologous nucleotides in RNase P RNAs from different species (G 52 in *E. coli* and G 34 in *B. subtilis*), form sets of homologous cross-links. Photoagent-containing cpRNAs were irradiated with 302 nm UV in either the presence (+) or the absence (−) of pre-tRNA. (B) Primer extension analysis of cross-linked RNAs. CpEc52-x2, cpEc52-x3, cpEc52-x4, cpBs34-x2, and cpBs34-x3/x4 cross-linked RNAs were purified from preparative gels and analyzed using primer extension with appropriate 5′-³²P-labeled primers. Lanes C, U, A, and G denote sequencing reactions on non-cross-linked RNAs and lane N indicates the same primer extension without adding dideoxynucleotides. The actual cross-link sites are one nucleotide 5′ to primer extension stop sites. (C) Comparison of cross-link sites between cpEc52 and cpBs34 RNA. Square-boxed Gs indicate the photoagent attachment sites, located at the 5′ end of each cpRNA. Highlighted letters represent the corresponding cross-linked sites. Arrowheads indicate the direction of long-range cross-links.

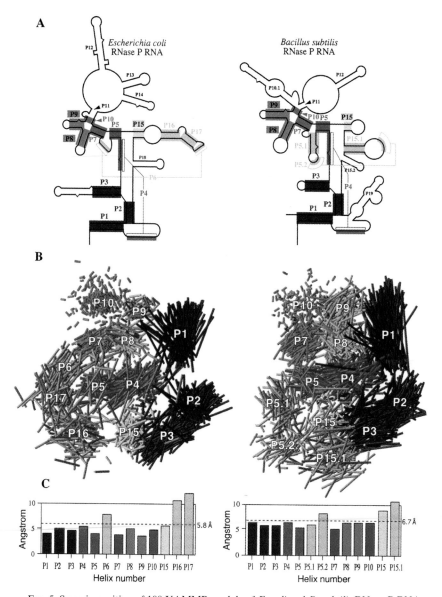

FIG. 5. Superimposition of 100 YAMMP models of *E. coli* and *B. subtilis* RNase P RNAs.
(A) Secondary structures of *E. coli* and *B. subtilis* RNase P RNAs colored according to the
helical domains. (B) Superimpositions of helices of *E. coli* and *B. subtilis* RNase P RNAs.
Helices in each model are represented by a thin cylinder that is 5% the width of a normal
A-form RNA helix. Colored elements correspond to the secondary structure shown in (A).
Each superimposition consists of 100 distinct models derived by molecular mechanics. (C)
Standard deviations of each helix in the superimposed model set. The dashed line represents
the mean of standard deviations of the helices shown. The collection of helical models is
internally consistent, evidenced by their low-average potential energies.

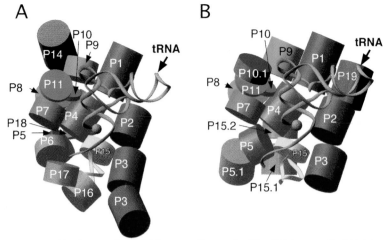

A          B

P10          P10
P14  P9     P9  P1
P8  P11     P8  P10.1
P7  P4      P11
P18  P6     P7  P4
P5          P15.2
P17         P5
P16  P15    P5.1
P3          P15.1
            P15  P3

tRNA        tRNA

P1
P2
P3

P19
P2

E. coli RNase P RNA    B. subtilis RNase P RNA

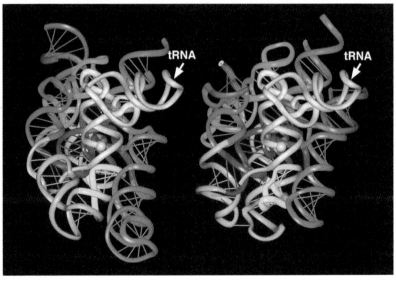

tRNA          tRNA

FIG. 6. Structure models of *E. coli* and *B. subtilis* RNase P RNA–tRNA complexes. (A) Helix barrel models of *E. coli* and *B. subtilis* RNase P RNA–tRNA complexes. Each cylinder represents the position and orientation of an A-form helix of appropriate length. tRNA is displayed as a ribbon in which the 5′-phosphate (the RNase P substrate site) is shown as a gray sphere. CCA-binding elements, J15/16 of *E. coli* RNA and L15 of *B. subtilis* RNA, are shown as yellow ribbons. Elements are colored and numbered according to the secondary structures shown in Fig. 5. (B) Ribbon models of *E. coli* and *B. subtilis* RNase P RNA–tRNA complexes. The backbones of RNA are shown as ribbons and colored as in (A). tRNA is shown as a gray ribbon with the 5′-phosphate indicated as a sphere.

organization that the individual helices can assume is limited, and the model is increasingly "resolved."

Modeling the RNase P tertiary structure was facilitated greatly by the computer program YAMMP-RNA, a suite of molecular modeling programs designed specifically for modeling RNAs.[24] On the basis of the secondary structure and intramolecular cross-linking data, YAMMP-RNA arranges the helices of an RNA structure model and assesses its optimality using standard energy minimization techniques. YAMMP calculations are complex and path dependent. Consequently, no singular global solution is obtained. In order to filter the many solutions, we conducted a bootstrap-like analysis, conducting multiple calculations with the data set and considering the consistency of the helix position in the collection of YAMMP solutions. Figure 5 (see color plate) shows a superimposition of 100 YAMMP solutions and the consistency of the helix positions in the YAMMP analysis for RNase P RNAs from both *E. coli* and *B. subtilis*. The current extrapolated models of the RNase P RNAs from both *E. coli* and *B. subtilis* are shown in Fig. 6 (see color plate). It is encouraging that both RNAs exhibit the core structural elements in the same regions of the structure model. Differences between the two types of RNAs are seen to arise from the peripheral structural elements, hairpin helices. Further details on the structural interpretations of the models have been published.[19,20,21]

[24] R. K. Tan and S. C. Harvey, *J. Comput. Chem.* **14,** 455 (1993).

# [11] Cleavage of RNA with Synthetic Ribonuclease Mimics

*By* Richard Giegé, Brice Felden, Marina A. Zenkova, Vladimir N. Sil'nikov, and Valentin V. Vlassov

## Introduction

Small RNA cleaving molecules are used extensively as probes to investigate the structure of RNAs free in solution or complexed with small molecules or proteins[1,2] (see also elsewhere in this volume). As compared to

[1] N. Kolchanov, V. Vlassov, and I. Vlassova, *Prog. Nucleic. Acids Res. Mol. Biol.* **53,** 131 (1996).
[2] R. Giegé, M. Helm, and C. Florentz, *in* "Comprehensive Natural Product Chemistry" (D. H. R. Barton and K. Nakanishi, eds.), Vol. 6. Pergamon, New York, 1999.

enzymatic probes, small chemical reagents withstand a wider range of conditions and do not perturb the object of study. In the early 1990s, we searched for compounds that (i) would react specifically with single-stranded regions of RNA under physiological conditions, (ii) would enable an easy analysis of natural RNAs, particularly those containing modified bases that are labile under different chemical treatments, and (iii) would be potentially applicable to probe RNA in RNA–protein complexes, viral particles, and directly inside cells. We believed that novel probes fulfilling these requirements can be developed straightforwardly by the synthesis of molecules mimicking essential patterns of functional groups found in catalytic centers of RNases. Feasibility of this goal was suggested by a few studies demonstrating a possibility to catalyze cleavage of RNA in physiological conditions by some metal complexes and organic compounds. Such compounds can be used per se, but alternatively they can be linked to structures having affinity to RNA, as oligonucleotides or other ligands, and produce conjugates capable to cut RNA at specific structural motifs and sequences.[3–10] This article describes molecules mimicking the active center of RNase A and the procedures for mapping single- and double-stranded regions in RNA structures.

## Chemical Basis for Design of Small Ribonuclease Mimics

Figure 1 shows how RNase A cleaves RNAs and emphasizes the role of the two histidine residues in the catalytic mechanism of hydrolysis.[11–13] This mechanism can be achieved by a variety of small molecules. The simplest is imidazole, an histidine mimic. At high concentration and at pH 7.0, the protonated and unprotonated imidazole molecules can play the role of the two histidines in the active site of RNase A, and actually it was

[3] J. K. Bashkin, E. I. Frolova, and U. S. Sampath, *J. Am. Chem. Soc.* **116,** 5981 (1994).

[4] M. Komiyama and T. Inokawa, *J. Biochem.* **116,** 719 (1994).

[5] J. Hall, D. Husken, U. Pieles, H. E. Moser, and R. Haner, *Chem. Biol.* **1,** 185 (1994).

[6] D. Magda, R. A. Miller, J. L. Sessler, and B. L. Iverson, *J. Am. Chem. Soc.* **116,** 7439 (1994).

[7] A. S. Modak, J. K. Gard, M. C. Merriman, K. A. Winkeler, J. K. Bashkin, and M. K. Stern, *J. Am. Chem. Soc.* **113,** 283 (1991).

[8] C.-H. Tung, Z. Wei, M. J. Leibowitz, and S. Stein, *Proc. Natl. Acad. Sci. U.S.A.* **89,** 7114 (1992).

[9] M. A. Podyminogin, V. V. Vlassov, and R. Giegé, *Nucleic Acids Res.* **21,** 5950 (1993).

[10] V. V. Vlassov, G. Zuber, B. Felden, J.-P. Behr, and R. Giegé, *Nucleic Acids Res.* **23,** 3161 (1995).

[11] C. A. Deakyne and L. C. Allen, *J. Am. Chem. Soc.* **101,** 3951 (1979).

[12] H. W. Wyckoff, D. Tsernoglou, A. W. Hanson, J. R. Knox, B. Lee, and F. M. Richards, *J. Biol. Chem.* **245,** 305 (1970).

[13] A. Wlodawer, R. Bott, and L. Sjölin, *J. Biol. Chem.* **257,** 1325 (1982).

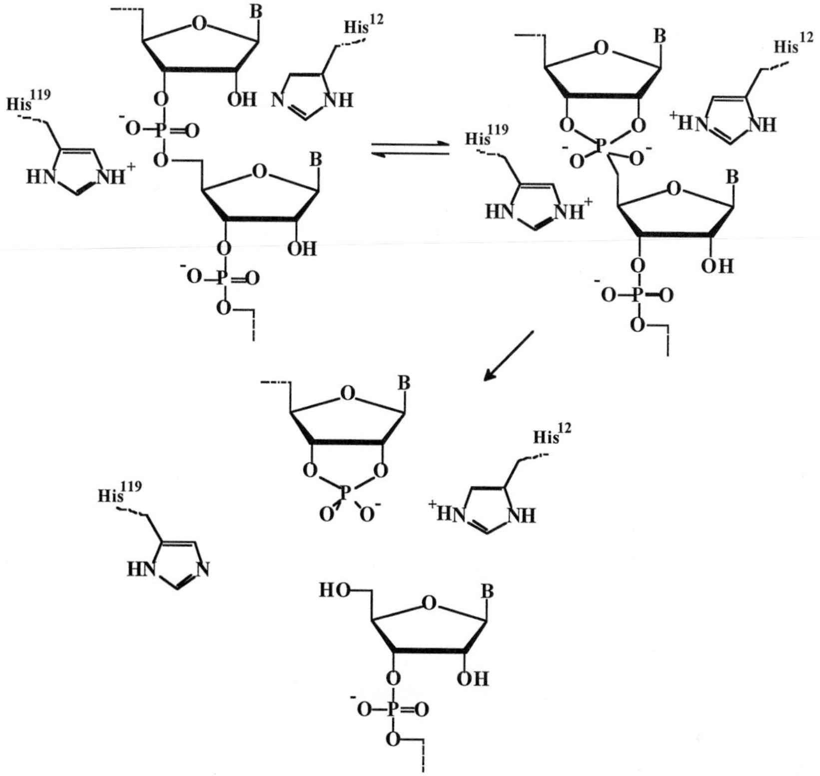

FIG. 1. Scheme of RNA hydrolysis by RNase A on the basis of the crystallographic structure of the enzyme.[11,12] Two imidazoles of histidine residues of the enzyme act as acidic and basic imidazolium and imidazole units, respectively. At the first step of the process, the ester interchange to form a cyclic phosphate occurs, which results in scission of the RNA chain. Then the formed 2′,3′-cyclic phosphate ester is hydrolyzed by the enzyme.

shown that RNA is hydrolyzed in concentrated imidazole (Im) buffer.[14–16] More elaborate compounds contain one or two imidazole residues connected to cationic structures or intercalating dyes by linkers of variable length and flexibility (Fig. 2). Due to the affinity of the cationic or intercalating moieties for RNA, the conjugates bring the imidazole(s) in close proximity to the phosphate–ribose backbone, and thus RNA can be cleaved at low concentrations of the compounds, provided in the case of conjugates

[14] R. Breslow, *Accounts Chem. Res.* **24**, 317 (1991).
[15] R. Breslow and M. Labelle, *Am. Chem. Soc.* **108**, 2655 (1986).
[16] R. Breslow, D.-L. Huang, and E. Anslyn, *Proc. Natl. Acad. Sci. U.S.A.* **86**, 1746 (1989).

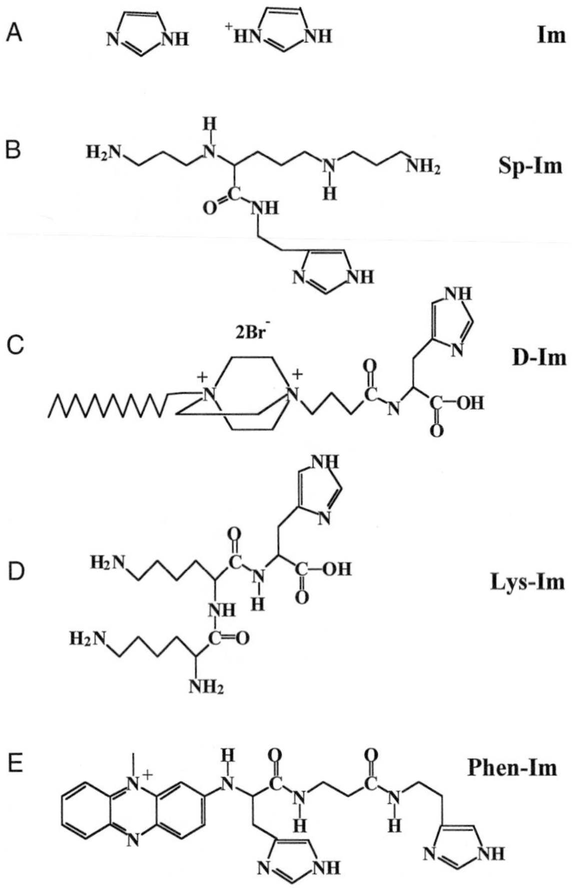

FIG. 2. RNase A mimics. (A) Imidazole (Im) in free and protonated forms and (B–E) conjugates bearing imidazole residues and groups capable of noncovalent interaction with nucleic acids: (B) spermine imidazole (Sp-Im); (C) diazobicyclooctane-based synthetic RNase (D-Im)[30]; (D) oligolysine-based synthetic RNase (Lys-Im)[31]; and (E) phenazine-based synthetic RNase (Phen-Im).[9]

with only one imidazole residue (Sp-Im, D-Im, and Lys-Im) that the second catalytic imidazole is brought by the buffer.

Examples of Utilization of RNase Mimics

*Cleavage of RNAs in Imidazole Buffer*

Figure 3A displays a typical autoradiogram of the cleavage pattern by imidazole of 3'-end-labeled yeast tRNA[Asp]. Cleavage of the RNA is random

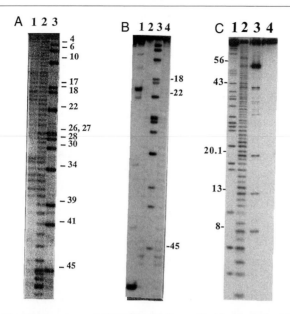

FIG. 3. Probing the structure of tRNA$^{Asp}$ with Im and imidazole-derived RNase mimics, and electrophoretic analysis of cleavage products on 12% denaturing polyacrylamide gels. (A) Cleavage by Im. These experiments were done on 3'-end-labeled native tRNA: lane 1, tRNA incubated at 37° for 12 hr in 2 $M$ Im buffer at pH 7.0 containing 40 m$M$ NaCl, 10 m$M$ MgCl$_2$, and 0.5 m$M$ EDTA; lane 2, tRNA$^{Asp}$ incubated at 90° for 10 min in 2 $M$ Im buffer at pH 7.0 containing 0.5 m$M$ EDTA; and lane 3, partial RNase T1 digest. (B) Cleavage by the histamine conjugate Sp-Im in 50 m$M$ Im buffer at pH 7.0. Experiments were done on 3'-end-labeled native tRNA: lane 1, tRNA incubated with 2.5 m$M$ Sp-Im; lane 2, partial alkaline cleavage of the tRNA; lane 3, partial RNase T1 digest of the tRNA; and lane 4, tRNA incubated for 10 min at 90° in 2 $M$ Im buffer at pH 7.0 containing 0.5 m$M$ EDTA.[10] (C) Cleavage by Phen-Im. Experiments were done on 5'-end-labeled tRNA$^{Asp}$ transcripts: lanes 1 and 2, RNase T1 and formamide ladders, respectively; lanes 3, tRNA incubated in 1 m$M$ Phen-Im and 50 m$M$ Im buffer at pH 7.0 for 12 hr at 37°; and lane 4, control incubation for 12 hr under the reaction conditions without the reagent.

at high temperature when the molecule is unfolded (lane 2, Fig. 3A). When the reaction is performed in conditions stabilizing the RNA structure, i.e., at low temperature and in the presence of magnesium ions, cleavage occurs essentially nonrandomly (lane 1, Fig. 3A). Cuts are located predominantly within the single-stranded regions of the tRNA$^{Asp}$ cloverleaf (Fig. 4). Phosphodiester bonds in double-stranded regions are more resistant to imidazole hydrolysis, and low reactive bonds are at the 5' side of loops. These observations illustrate the potential of Im buffer to cleave RNA in a way reflecting features of its secondary structure. The decreased rate of hydrolysis within helical regions and at the 5' side of the loops, where nucleotides may be

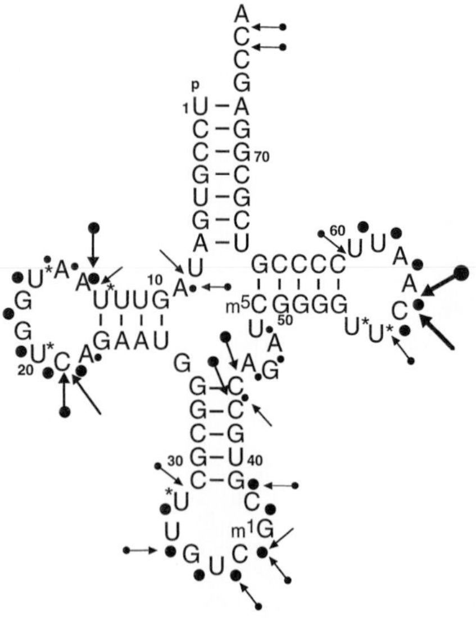

Fig. 4. Cloverleaf structure of yeast tRNA$^{Asp}$ with location of Im-induced cleavages. Phosphodiester linkages displaying enhanced susceptibility to hydrolysis by the Im buffer in conditions stabilizing the RNA structure are indicated by dots whose diameters are proportional to the intensity of the cuts. Phosphodiester linkages attacked by Sp-Im in the presence of Im buffer (→) and by Phen-Im (●——→) are indicated by arrows whose sizes are proportional to the intensity of the cuts.[9,10] Asterisks indicate the presence of modified U residues (Ψ13, 32, 55, D16, 20, and T54) in native yeast tRNA$^{Asp}$; notice that in the transcript the U1-A72 pair is replaced by a G1-C72 pair and that numbering is standardized as for tRNA$^{Phe}$.

stacked,[17] reflects an increased rigidity of the RNA backbone in these regions. This rigidity interferes with the conformational changes needed for the appropriate orientation of phosphodiester bonds and sugars for the transesterification step of the reaction.

Suppression of imidazole-catalyzed cleavages in double-stranded RNA sequences suggested to use this reaction to investigate RNA complexes with oligonucleotides. Probing such complexes is needed when studying binding of antisense agents to their target sequences in RNAs. Figures 5A and 5B compare the cleavage pattern produced by incubation in Im buffer of tRNA$^{Phe}$ (Fig. 5A, lane 3) with those of tRNA$^{Phe}$ complexed to oligonucleotides (Fig. 5B, lanes 1–3). It is seen that binding of an oligonucleotide complementary to the sequence 59–75 suppresses the hydrolysis of phosphodiester linkages in this sequence and promotes new cleavages within the sequence 49–55 (Fig. 5B, lane 3). This is because the cleaved region

[17] E. Westhof, P. Dumas, and D. Moras, J. Biomol. Struc. Dyn. 1, 337 (1983).

becomes single stranded in the complex. Similarly, binding of another oligonucleotide, complementary to the sequence 44–55 in the tRNA (Fig. 5B, lane 2), protects this sequence from cleavage and promotes the hydrolysis of bonds within the sequence 60–66 as a result of unwinding of the T-stem. This binding also promotes a cut at position 26, at the junction of the D and anticodon stems, apparently due to the disruption of stacking interactions in the central part of the molecule. Simultaneous binding of both oligonucleotides to the tRNA results in an efficient protection of both covered sequences and an enhancement of the cleavages at the junction of the anticodon and D arms (Fig. 5B, lane 1). Altogether, these experiments demonstrate that probing with imidazole is a simple way to identify sequences that are involved in oligonucleotide binding and to reveal structure elements perturbed by the complex formation.

## Cleavage of RNAs with Imidazole Conjugates

*Transfer RNAs.* The four conjugates Sp-Im, D-Im, Lys-Im, and Phen-Im (Fig. 2) have affinity to RNA and thus bring the conjugated imidazole residue in contact with the phosphodiester backbone of the RNA. The second imidazole needed for the cleavage reaction would be provided either as a free molecule by the Im buffer or from the second conjugated imidazole moiety bound in the vicinity of the first one.

Cleavage of tRNA by Sp-Im is optimal at concentrations of the Im buffer in the 50–100 m$M$ range. A further increase of the buffer concentration inhibits the reaction, likely because of the displacement of Sp-Im from the tRNA due to the increased ionic strength of the solution. Incubation with Sp-Im for 48 hr under optimal reaction conditions results in the complete hydrolysis of the tRNA. Figure 3B (lane 1) shows the nonrandom hydrolysis of tRNA$^{Asp}$, with two major and several faint bands, generated by Sp-Im. The major cuts occur at two CpA sequences between positions 55–56 and 20–21 in the tRNA; less intense cuts appear after U8, $\Psi$13, C36, and C43. It is noticeable that all cleavages are after pyrimidine residues, and for the most intense they are within 5′-PypA-3′ sequences, in particular CpA (Fig. 4).

The cleavage pattern of a tRNA$^{Asp}$ transcript by the RNase mimic Phen-Im is shown in Fig. 3C (lane 3). Cleavage positions are similar to those observed with Sp-Im, although relative intensities of cuts are not. Investigation of the conditions affecting tRNA hydrolysis by Phen-Im indicates a characteristic bell-shaped relationship between the rate of cleavage and pH.[9] Maximal cleavage occurs at about pH 7.0 when the two imidazole moieties are in equimolar equilibrium between deprotonated imidazole and protonated imidazolium forms. The cleavage yield increases with increasing

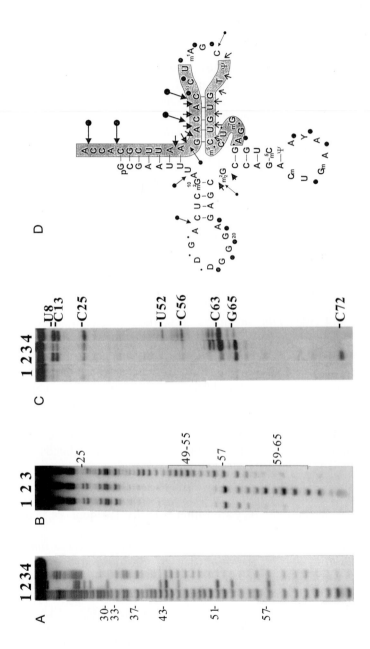

temperature up to 40°. At higher temperatures, the reaction slows down, probably because the RNA structure melts and the helices of tRNA unfold, where the binding of the intercalating groups is supposed to take place.

Fragmentation of yeast tRNA$^{Phe}$ by D-Im, supplemented with diluted Im buffer, is shown in Fig. 5C. The rate of tRNA$^{Phe}$ hydrolysis versus D-Im concentration exhibits a bell-shaped dependence, with a maximum of 0.5 m$M$ D-Im. This likely reflects a requirement of an optimal density of positively charged conjugates bound to RNA for optimal catalysis. The D-Im mimic, similarly to Sp-Im tested on tRNA$^{Asp}$, generates a nonrandom hydrolysis of tRNA$^{Phe}$ with the major cuts within single-stranded regions and at junctions of tRNA domains, i.e., at phosphodiester bonds after U8, C13, G65, C72, and C75 (Fig. 5C, lane 2). Probing tRNA$^{Phe}$ complexed with a 15-mer complementary to the 62–76 sequence (lanes 3 and 4, Fig. 5C) reflects rearrangements of the tRNA$^{Phe}$ structure. Oligonucleotde binding masks from hydrolyzes phosphodiester bonds after G65, C72, and C75 and results in unfolding of the T-stem, leading to new cleavages after C63, C61, C57, U52, and C25.

*Minihelices Derived from 3' Ends of TMV and TYMV RNAs.* Based on functional and structural properties of tRNAs and viral tRNA-like domains, minimalist structures corresponding to the very 3' end of tobacco mosaic virus (TMV) RNA would correspond to histidine-accepting sub-

---

FIG. 5. Probing of free tRNA$^{Phe}$ and of tRNA$^{Phe}$ hybridized to oligonucleotides with Im and the D-Im conjugate, and electrophoretic analysis of cleavage products on 15% denaturing polyacrylamide gels. All tRNAs were 3' end labeled. (A) Cleavage of free tRNA by Im. Lane 1, tRNA incubated at 90° for 10 min in 2 $M$ Im buffer at pH 7.0 containing 0.5 m$M$ EDTA; lane 2, tRNA partially cleaved by RNase T1; lane 3, tRNA incubated at 20° for 20 hr in 2 $M$ Im buffer at pH 7.0 containing 40 m$M$ NaCl and 0.5 m$M$ EDTA; and lane 4, control incubation at 20° for 20 hr of tRNA in 40 m$M$ HEPES at pH 7.0 containing 0.5 m$M$ EDTA. (B) Cleavage of complexed tRNA by Im. Complexes were incubated at 20° for 20 hr in 2 $M$ Im buffer at pH 7.0 containing 40 m$M$ EDTA. Lane 1, tRNA$^{Phe}$ with 5 $\mu M$ 17-mer GGTGCGAATTCTGTGGA (1) and 12-mer GAACACGGACCT (2) each; lane 2, tRNA$^{Phe}$ with 5 $\mu M$ 12-mer (2); and lane 3, tRNA$^{Phe}$ with 5 $\mu M$ 17-mer (1). (C) Cleavage of free tRNA and tRNA hybridized to 15-mer TGGTGCGAATTCTGT (3) by D-Im. Incubations were at 37° for 10 hr in 50 m$M$ Im buffer at pH 7.0 containing 0.2 $M$ KCl and 0.1 m$M$ EDTA. Lane 1, control incubation in the absence of reagent; lane 2, tRNA incubated with 0.5 m$M$ D-Im; and lanes 3 and 4, tRNA bound to the complementary oligonucleotide (5 and 50 $\mu M$, respectively) incubated with 0.5 m$M$ D-Im. (D) Cloverleaf structure of yeast tRNA$^{Phe}$ with location of Im and D-Im-induced cleavages. Phosphodiester bonds, displaying enhanced cleavage rate in the Im buffer, are indicated by dots. The sequences 59–75 or 62–76 and 44–55, complementary to the antisense oligonucleotides 1 or 3 and 2, respectively, are shadowed. Arrows ($\rightarrow$) and ($\rightarrow$) indicate the phosphodiester bonds, where the cleavage by Im is stimulated by the binding of oligonucleotides 1 and 2, respectively. Arrows ($\bullet\!\!\longrightarrow$) indicate cleavages generated by D-Im.

strates of synthetases. Such structures mimicking the amino acid accepting end of a tRNA contain a pseudoknot, and functional assays have indeed verified their ability to be charged by histidine.[18–20] Probing a minihelix derived from the TMV RNA with Im and Sp-Im indicates cleavages in the four single-stranded regions of the 38-nucleotide-long RNA (Fig. 6A, lanes 4 and 5). Interestingly, the two connecting single strands of the 3'-terminal pseudoknot [L1 and L3 in Fig. 6C, (a)] are well cut, except nucleotide A18 in L1, which was proposed in the case of the satellite virus of TMV to mimic the minus one histidine identity nucleotide present in all canonical tRNA[His] species.[19] These data support the proposal that A18 is stacked on the acceptor stem over C19 and that its phosphodiester linkage is less flexible and thus is prevented to be strongly cut by Im. An interesting point concerns the cleavages generated by the Im buffer in the region between nucletodies U28 and U34, which are consistent with a T-loop conformation as in canonical tRNAs. Indeed U28 and especially A29 are weakly cut as compared to the other nucleotides of the loop (Fig. 6A, lane 4), in agreement with a weak flexibility of the phosphodiester bond between C27 and A30, which are constrained geometrically in a T-loop conformation. In such a conformation the A30-U34 reverse Hoogsteen pair is stacked on the last base pair of the acceptor branch and residues U28 and U29 form an internal bulge.

Another example of a pseudoknotted minihelix, mimicking the acceptor branch of the tRNA-like domain of turnip yellow mosaic virus (TYMV) RNA, probed with Lys-Im is given in Fig. 6B (lanes 4 and 5). Cleavages occur exclusively within the single-stranded sequences of this minihelix at phosphodiester bonds after C2, C11, C23, and C37 [Fig. 6C, (b)]. Low reactivity of the C37-A38 bond toward the synthetic RNase reflects the rigidity of the U33-U39 loop, similar to that of the TMV minihelix. The most intensive cuts occur at the CpA and CpG sequences.

*Specific Features of Cleaving Reactions*

Hydrolysis of RNA by imidazole or imidazole conjugates is strongly dependent on architectural features of the nucleic acid. The yield of hydrolysis by free imidazole increases with the first power of the imidazole concentration and significant cuts are generated for concentrations above 1 $M$. This linear dependence can be explained by an increase of the concentration of protonated imidazole in the vicinity of the polyanionic RNA. Because of this, protonated imidazole is always present in excess so that the reaction

[18] B. Felden, C. Florentz, R. Giegé, and E. Westhof, *RNA* **3**, 201 (1996).
[19] J. Rudinger, B. Felden, C. Florentz, and R. Giegé, *Bioorg. Med. Chem.* **5**, 1001 (1997).
[20] B. Felden, C. Florentz, A. McPherson, and R. Giegé, *Nucleic Acids Res.* **22**, 2882 (1994).

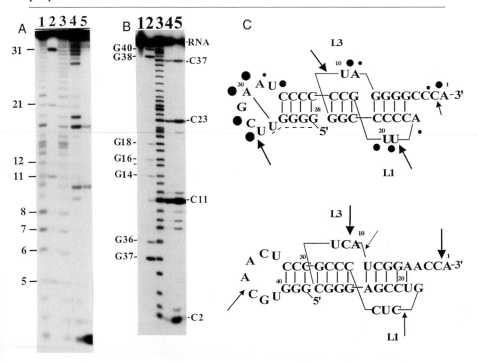

FIG. 6. Probing of pseudoknotted minihelices derived from TMV and TYMV tRNA-like domains with Im buffer, Sp-Im, and Lys-Im. (A) Cleavage of the RNA transcript derived from the TMV RNA by Im buffer. Lane 1, partial alkaline hydrolysis; lane 2, partial RNase T1 hydrolysis; lane 3, incubation for 10 min at 90° in 2 $M$ Im buffer at pH 7.0; lane 4, incubation for 16 hr at 25° in 2 $M$ Im buffer at pH 7.0 containing 40 m$M$ NaCl, 1 m$M$ EDTA, and 10 m$M$ MgCl$_2$; and lane 5, hydrolysis by 2 m$M$ Sp-Im in 50 m$M$ Im buffer at pH 7.0 with an incubation of 7 hr at 37°.[10] (B) Cleavage of the RNA transcript derived from the TYMV RNA by Lys-Im. The minihelix was incubated at 37° for 10 hr in 50 m$M$ Im buffer at pH 7.0 containing 200 m$M$ KCl, 0.1 m$M$ EDTA, and 100 mg/ml of carrier tRNA. Lane 1, control incubation; lanes 2 and 3, RNase T1 and alkaline ladders, respectively; and lanes 4 and 5, incubation with 1 and 0.5 m$M$ Lys-Im, respectively. (C) Secondary structure of the TMV (a) and TYMV (b) minihelices. In the TMV-derived structure, Im-induced cleavage points are indicated by dots. Diameters of the dots are proportional to the intensity of the cuts. Arrows indicate the residues after which phosphodiester bonds are attacked by Sp-Im (TMV) and Lys-Im (TYMV). Loops L1 and L3 emphasize the two single strands of the pseudoknot crossing the deep and the shallow grooves, respectively.[40] The size of the arrows correlates with the intensity of the cleavage. For methodological reasons, the last 5' residues [dashed in (a)] could not be tested; notice that numbering of nucleotides is from 3' to 5'.

should depend linearly on the concentration of the second required species, nonprotonated imidazole. The pH optimum of the reaction is around pH 7.0, consistent with an imidazole-promoted hydrolytic mechanism similar to that of RNase A (Fig. 1) with imidazole moieties in equimolar equilibrium between deprotonated and protonated forms. Increasing temperature results in the enhanced reactivity of phosphodiester bonds otherwise protected from hydrolysis by the RNA structure. By analogy with the enzymatic mechanism demonstrated for RNase A,[11-13] cleavage by imidazole reagents yields two fragments, one terminating with 5'-OH and the other with a 3'-phosphate group.

Hydrolysis in Im buffer at high temperature provides a simple method for producing RNA ladders at neutral pH,[10] which is an advantage for RNAs containing alkali-sensitive bases. The ladders are partially destroyed in traditional hydrolysis conditions, which result in doubling of electrophoretic bands and strongly enhanced bands at the positions of pH-sensitive nucleotides. This is the case for the $m^7G$ containing tRNAs.[21]

Cleavage specificity of the conjugates is drastically different from that observed for RNA hydrolysis by the Im buffer. In Im buffer, all positions in single-stranded regions are hydrolyzed with similar rates, whereas conjugates demonstrate pronounced preference to PypPu sequences, particularly to CpA. An explanation may be that small free imidazole molecules can reach ribose and phosphate moieties in RNA easily from different directions without disturbing the RNA structure. The conjugates are more bulky and their access to phosphodiester bonds in different dinucleotide sequences is affected by the nature of the residues in vicinity of the bond to be cleaved and by conformational features and freedom of the linkage. Apparently, a favorable situation is realized for PypPu sequences. The chemical instability of such sequences is a known phenomenon,[22-26] but its exact nature is still unclear.

The cleavage pattern of tRNA may be affected by the reactivities of individual phosphodiester bonds and the specificity of binding of the cationic molecules at the surface of the tRNA. Probing data indicate that the specificity of the imidazole conjugates is mainly governed by the reactivity

[21] W. Wintermeyer and H. G. Zachau, *FEBS Lett.* **11,** 160 (1970).
[22] C. Florentz, J.-P. Briand, P. Romby, L. Hirth, Ebel, J.-P., and R. Giegé, *EMBO J.* **1,** 269 (1982).
[23] A.-C. Dock-Bregeon and D. Moras, *Cold Spring Harbor Symp. Quant. Biol.* **52,** 113 (1987).
[24] R. Kierzek, *Nucleic Acids Res.* **19,** 5073 (1992).
[25] R. Kierzek, *Nucleic Acids Res.* **19,** 5079 (1992).
[26] H. Hosaka, L. Sakabe, K. Sakamoto, S. Yokoyama, and H. Takaku, *J. Biol. Chem.* **269,** 20090 (1994).

of individual phosphodiester bonds and is modulated by the binding pattern of the cleavers on the RNA structure.

*Applications*

Probing a RNA structure with compounds reacting with phosphodiester linkages is particularly informative because informations about all residues in this structure can be obtained in a single experiment. Some small molecules such as ethylnitrosourea, capable of reacting with RNA, are well known, but present only moderate cleavage efficiency.[1,2] Some metal ions and metal complexes also cleave RNA, and one of them, namely $Pb^{2+}$,[27,28] turned out to be a useful tool in revealing structural features in RNA under mild conditions. Indeed, specific cleavage by this ions requires an appropriate environment of ligands, which is realized in some RNA substrates having the proper tertiary folding.

The RNA cleaving compounds described here are a new family of tools for studying RNA conformation under mild conditions that maintain their native form. Hydrolysis in Im buffer can be used to distinguish easily and readily between single-stranded and double-stranded RNA sequences. Conjugates with imidazole groups are simple mimics of RNase A with the main attributes of an enzyme: they contain structures capable of recognizing RNA and they have functional groups responsible for catalysis. These simple molecules have an advantage as structural probes, as compared to RNase A. The nuclease is known to affect RNA structures on binding and thus can cleave at PypPu sequences in double-stranded regions of its RNA substrates. In contrast to the enzyme, the synthetic mimics cannot unfold RNA. Therefore, these compounds may be used to compare structures of mutant RNAs straightforwardly. The efficiency of RNA cleavage by artificial RNase makes them attractive prototypes for the design of second-generation antisense oligonucleotide derivatives that could find novel applications for inactivation of RNAs in gene-targeted therapeutics. It is hoped that the remarkable properties of the imidazole-based compounds will encourage researchers to apply them to investigate the structure of RNAs and RNA complexes.

Materials

*RNase Mimics*

Nuclease and metal-free imidazole should be used. Prepare a stock solution as follows: dissolve imidazole and EDTA in water, adjust pH to

[27] C. Werner, B. Krebs, G. Keith, and G. Dirheimer, *Biochim. Biophys. Acta* **432,** 161 (1976).
[28] W. J. Krzyzosiak, T. Marciniec, M. Wiewiorowsky, P. Romby, J.-P. Ebel, and R. Giegé, *Biochemistry* **27,** 5771 (1988).

7.0 with HCl, and add water to make the final concentration of imidazole (4 $M$) and EDTA (1 m$M$). To test the solution for the absence of any RNA-cleaving contaminant, incubate end-labeled RNA in 2 $M$ imidazole at 37° for 12 hr. The produced cleavage pattern should not show enhanced intensities of bands at PypPu sequences. RNase-free tested imidazole is available from Euromedex (Strasbourg, France). Synthesis of Sp-Im is according to a published procedure.[29] The compound Phen-Im is synthesized as described.[9] Synthesis of D-Im and Lys-Im is described elsewhere.[30,31] The probe Lys-Im is available from Euromedex (Strasbourg).

*Miscellaneous Chemicals and Enzymes*

Chemicals for electrophoresis, such as Rotiphorese Gel 40 solution of acrylamide and bisacrylamide, can be obtained from Carl Roth GmbH (Karlsruhe, Germany). [$\gamma$-$^{32}$P]ATP (3000 Ci/mmol), [$\alpha$-$^{32}$P]ATP (400 Ci/mmol), and [$\alpha$-$^{32}$P]pCp (3000 Ci/mmol) are from Amersham (Les Ulis, France). All solutions for RNA handling should be prepared using Milli-Q water containing 1 m$M$ EDTA. These solutions should be filtered through membranes with a 0.2-$\mu$m pore size, e.g., Millex GS filters from Millipore (Bedford, MA). We use T4 polynucleotide kinase from Amersham, snake venom phosphodiesterase from Worthington (Freehold, NJ), and bacterial alkaline phosphatase from Appligène (Strasbourg, France). Restriction nucleases are from Gibco-BRL (Bethesda, MD) and RNasin is from Promega (Madison, WI). The T7 RNA polymerase[32] and (ATP:CTP) tRNA nucleotidyltransferase[33] are prepared according to established procedures. Enzymes and chemicals used for end labeling of RNA and electrophoresis are essentially as described.[34] Oligonucleotides are prepared by standard phosphoramidite chemistry and purified by HPLC.

RNAs

Native yeast tRNA[Asp] and tRNA[Phe] are obtained from total brewer's yeast tRNA by established procedures.[34] Wild-type aspartate-specific tRNA transcripts are synthesized enzymatically *in vitro* using T7 RNA

[29] G. Zuber, C. Sirlin, and J.-P. Behr, *J. Am. Chem. Soc.* **115**, 4939 (1993).
[30] D. A. Konewetz, I. E. Beck, N. G. Beloglazova, I. V. Sulimenkov, V. N. Sil'nikov, M. A. Zenkova, G. V. Shishkin, and V. V. Vlassov, *Tetrahedron Lett.* **55**, 503 (1999).
[31] N. S. Zjdan, M. A. Zenkova, A. V. Vlassov, V. N. Sil'nikov, and R. Giegé, *Nucleosides & Nucleotides* **18**, 1491 (1999).
[32] J. R. Wyatt, M. Chastain, and J. D. Puglisi, *BioTechniques* **11**, 764 (1991).
[33] B. Rether, J. Bonnet, and J.-P. Ebel, *Eur. J. Biochem.* **50**, 281 (1974).
[34] P. Romby, D. Moras, M. Bergdoll, P. Dumas, V. V. Vlassov, E. Westhof, J.-P. Ebel, and R. Giegé, *J. Mol. Biol.* **184**, 455 (1985).

polymerase and linearized plasmids essentially as described earlier.[35] A 42-nucleotide-long pseudoknotted minihelix derived from the tRNA-like structure of TYMV RNA is synthesized by established procedures.[36] The 38-nucleotide-long RNA derived from the TMV tRNA-like structure is prepared by similar procedures.[37] Both RNAs recapitulate the acceptor branch of the viral tRNA-like domains.

RNA concentrations are determined spectrophotometrically, assuming 1 $A_{260\ nm}$ unit corresponds to 40 $\mu$g/ml RNA in a 1-cm path-length cell. For unambiguous analysis of cleaving products, RNA samples must be free of background nicks. Care must be taken at all steps of the probing procedures to ensure that RNA is not exposed to metal ions, nucleases, or anything else that might cleave RNA. Bulk yeast tRNA, used as carrier to supplement labeled RNAs, is from Boehringer-Mannheim (Meylan, France).

### General Procedures for RNA Cleavage by RNase Mimics

*End Labeling of RNAs*

RNAs are 5' end labeled by dephosphorylation with alkaline phosphatase followed by phosphorylation with [$\gamma$-$^{32}$P]ATP and polynucleotide kinase.[34,38] To label the 3' end of tRNA$^{Asp}$, its CCA 3' end is removed by partial digestion with phosphodiesterase and restored by (ATP:CTP) tRNA nucleotidyltransferase in the presence of [$\alpha$-$^{32}$P]ATP.[34] Labeling at the 3' end of minihelices is performed by the ligation of [$\alpha$-$^{32}$P]pCp with the T4 RNA ligase.[39] After labeling, RNAs are purified by electrophoresis in 12% denaturing polyacrylamide gels. The labeled RNAs are eluted from gels by 125 m$M$ ammonium acetate at pH 6.0 containing 0.5 m$M$ EDTA and 0.025% SDS. After ethanol precipitation, RNAs are dissolved in water and stored at $-20°$.

*Cleavage Experiments*

Experiments are usually performed on 50,000 to 100,000 Cerenkov cpm of labeled RNA. They comprise a ladder for the assignment of cleavage

[35] V. Perret, A. Garcia, J. D. Puglisi, H. Grosjean, J.-P. Ebel, C. Florentz, and R. Giegé, *Biochimie* **72**, 735 (1990).

[36] J. Rudinger, C. Florentz, T. Dreher, and R. Giegé, *Nucleic Acids Res.* **20**, 1865 (1992).

[37] B. Felden, Thèse de l'Université Louis Pasteur de Strasbourg, 1994.

[38] P. Romby, D. Moras, P. Dumas, J.-P. Ebel, and R. Giegé, *J. Mol. Biol.* **195**, 193 (1987).

[39] T. E. England and O. C. Uhlenbeck, *Biochemistry* **17**, 2069 (1978).

[40] E. Westhof and L. Jaeger, *Curr. Opin. Struct. Biol.* **2**, 327 (1992).

products (here imidazole ladders, and G ladders produced by RNase T1), the actual cleavage assays, and control incubations in the absence of reagents in order to visualize unspecific cuts. The volume of assays is 10 to 20 $\mu$l. All reaction mixtures are supplemented by 1 $\mu$g of carrier RNA to facilitate precipitation steps and to control the stoichiometry between cleavers and RNA. Ladders and cleavage reactions are done according to the following protocols.

### Imidazole Ladders

1. In an Eppendorf tube, add 5 $\mu$l of 4 $M$ Im stock solution at pH 7.0 (pH adjusted with HCl) and 5 $\mu$l solution of end-labeled RNA in water. Cap the tube tightly, vortex to mix, and place in a hot water bath at 90° for 10 min.
2. Transfer the tube to ice, remove the cap, and precipitate RNA by the addition of 200 $\mu$l of 2% lithium perchlorate in acetone. Cap the tube, vortex, and centrifuge for 15 min at 15,000$g$ at 4°.
3. Remove the liquid carefully, wash the precipitate with acetone (200 $\mu$l), remove acetone carefully, dry the pellet in a desiccator, and dissolve it in 6 $\mu$l of loading buffer (7 $M$ urea containing 0.02% bromphenol blue and 0.02% xylene cyanol).

### Structural Probing of RNA with Imidazole

1. In an Eppendorf tube, add 5 $\mu$l Im stock solution, 2.5 $\mu$l end-labeled RNA solution in water, and 2.5 $\mu$l water solution containing 160 m$M$ NaCl, 40 m$M$ MgCl$_2$, and 4 m$M$ EDTA. Cap the tube tightly, vortex to mix, and incubate at 37° for 12 hr.
2. Precipitate the RNA by adding 200 $\mu$l of 2% lithium perchlorate in acetone. Cap the tube, vortex, and centrifuge for 15 min at 15,000$g$ at 4°.
3. Remove the liquid carefully, wash the precipitate with acetone (200 $\mu$l), and continue as described earlier for imidazole ladders.

*Structural Probing of RNA with Imidazole Conjugates.* Procedures with Sp-Im, D-Im, Lys-Im, and Phen-Im are essentially the same as for RNA cleavage by Im, except for the incubation times and the concentration of the probes that have to be adjusted for each particular RNA. The reason is that the susceptibility of RNAs to the imidazole conjugates is affected considerably by the RNA structure. Tight-folded molecules are more resistant to the cleavers than are RNAs with loose structures. For instance, reasonable cleavage of tRNA$^{Phe}$ by D-Im is achieved after a 10-hr incubation in the presence of 0.5 m$M$ D-Im, whereas only a 2-hr incubation in 0.05 m$M$ D-Im is sufficient for the cleavage of influenza virus M2 RNA (data not shown). Only under optimal concentration of the cleaving compound and incubation time can a "one-hit" cleavage pattern be obtained.

Currently, reactions are performed using the following buffer: 50 m$M$ Im at pH 7.0 (pH is adjusted by HCl), 0.2 $M$ KCl, 0.1 m$M$ EDTA, and 100 $\mu$g/ml bulk tRNA. The stock buffer solution is concentrated five fold (250 m$M$ Im at pH 7.0, 1 $M$ KCl, and 0.5 m$M$ EDTA). If needed, 50 m$M$ MgCl$_2$ is added.

### Particular Case of RNA Minihelix Cleavage with Lys-Im

1. Mix 1 $\mu$l of end-labeled RNA solution, 1 $\mu$l of carrier RNA (1 mg/ml) solution in water, 5 $\mu$l water, and 2 $\mu$l of stock Im buffer. Vortex and incubate for 20 min at room temperature.
2. Add 1 $\mu$l of a 10 m$M$ Lys-Im solution. Vortex, cap the tube, and incubate at 37° for 10 hr.
3. Add 200 $\mu$l of 2% lithium perchlorate in acetone. Cap the tube, vortex, and centrifuge at 14,000 rpm for 15 min at 4°.
4. Remove the liquid carefully, wash the precipitate with acetone (200 $\mu$l), and continue as described earlier for imidazole ladders.

Lys-Im Solution. Dissolve 9.6 mg of Lys-Im in 0.2 ml water to obtain a 50 m$M$ stock solution. This solution is kept at room temperature. Before use, mix 2 $\mu$l of the stock solution with 8 $\mu$l water and vortex to prepare the 10 m$M$ solution needed for the reaction.

### Particular Case of tRNA$^{Phe}$ Cleavage with D-Im

1. Mix 1 $\mu$l of labeled tRNA solution, 1 $\mu$l of the carrier RNA solution, 4.5 $\mu$l water, and 2 $\mu$l of stock Im buffer. Incubate for 20 min at 20°.
2. Add 1 $\mu$l of 5 m$M$ D-Im solution. Incubate for 18 hr at 37°.
3. Add 200 $\mu$l of 2% lithium perchlorate in acetone. Cap the tube, vortex, and centrifuge at 14,000 rpm for 15 min at 4°.
4. Remove the liquid carefully, wash the precipitate with acetone (200 $\mu$l), and continue as described earlier for imidazole ladders.

D-Im Solution. Dissolve 26 mg of D-Im in 1 ml DMSO to produce a 33 m$M$ stock solution. Store at ambient temperature in the dark. Before use, mix 1 $\mu$l of the stock with 5.5 $\mu$l water and vortex to get the 5 m$M$ solution used for the reaction.

### Particular Case of in Vitro Transcript tRNA$^{Asp}$ Cleavage with Phen-Im

1. Mix 1 $\mu$l of labeled tRNA solution, 1 $\mu$l of the carrier RNA solution, 4.5 $\mu$l water, and 2 $\mu$l of 5× stock Bis–Tris buffer. Incubate for 20 min at 20°.
2. Add 1 $\mu$l of 10 m$M$ Phen-Im solution. Incubate for 12 hr at 37°.

3. Add 200 $\mu$l of 2% lithium perchlorate in acetone. Cap the tube, vortex, and centrifuge at 14,000 rpm for 15 min at 4°.
4. Remove the liquid carefully, wash the precipitate with acetone (200 $\mu$l), and continue as described earlier for imidazole ladders.

STOCK SOLUTIONS. For Phen-Im, dissolve 5.1 mg of the compound in 1 ml water to produce a 10 m$M$ stock solution. Store at ambient temperature in the dark. The 5× stock of Bis–Tris buffer is 50 m$M$ Bis–Tris–propane–HCl, pH 7.0, 200 m$M$ NaCl, 50 m$M$ MgCl$_2$, and 1 m$M$ EDTA.

*Particular Case of tRNA$^{Asp}$ Cleavage with Sp-Im*

1. Mix 1 $\mu$l of labeled tRNA solution, 1 $\mu$l of the carrier RNA solution (1 mg/ml), 5 $\mu$l water, and 2 $\mu$l of stock Im buffer. Incubate for 20 min at 20°.
2. Add 1 $\mu$l of 25 m$M$ Sp-Im solution. Incubate for 7 hr at 37°.
3. Add 200 $\mu$l of 2% lithium perchlorate in acetone. Cap the tube, vortex, and centrifuge at 14,000 rpm for 15 min at 4°.
4. Remove the liquid carefully, wash the precipitate with acetone (200 $\mu$l), and continue as described earlier for imidazole ladders.

SP-IM SOLUTION. Dissolve 3.4 mg of Sp-Im in 0.4 ml water to produce a 25 m$M$ stock solution. Store at ambient temperature in the dark.

*Analysis of Cleaved RNA and Quantitation of Data*

RNA samples are dissolved in 7 $M$ urea containing 0.02% bromphenol blue and 0.02% xylene cyanol and are subjected to electrophoresis through a denaturing polyacrylamide gel (12% acrylamide, 7 $M$ urea, 30 × 40 × 0.04 cm$^3$). The gel contains 8 $M$ urea in 1× TBE (100 m$M$ Tris, 100 m$M$ borate, and 2.8 m$M$ EDTA). Gels are electrophoresed in 1× TBE running buffer. For calibration of cleavage patterns, end-labeled RNA is cleaved statistically at G residues by digestion with RNase T1 or cleaved by imidazole at 90°, as described earlier. Gels must be run at a temperature of 45–50° to keep RNA denatured. After separation, the gels are transferred onto Whatman (Clifton, NJ) 3MM filter paper, covered with Saran wrap, and dried. Dried gels are exposed to X-ray film (Kodak X-Omat, Rochester, NY) either at room temperature or at −70° with intensifying screens.

Cleavage patterns are quantitated using a Bio-Imaging Analyzer, e.g., the FUJIX BioImaging Analyzer BAS 2000 system. Photostimulatable imaging plates (type BAS-III from Fuji Photo Film Co., Ltd., Japan) are pressed on gels and exposed at room temperature for 30 min. Imaging plates are analyzed by performing volume integrations of specific cleavage sites and reference blocks using the FUJIX BAS 2000 workstation software (version 1.1).

Acknowledgments

We thank J.-P. Behr and G. Zuber who designed and prepared the Sp-Im conjugate, M. Podyminogin who synthesized and first tested the Phen-Im conjugate, and N. Zjdan and D. Konevetz who prepared the Lys-Im and D-Im conjugates. We are indebted to A. Vlassov, I. Kuznetsova, M. Helm, and N. Beloglazova who did the original work with tRNA$^{Asp}$, tRNA$^{Phe}$, and the minihelices; we also thank C. Florentz for helpful discussions and constant interest and A. Théobald-Dietrich for help in tRNA purification. This work was supported by INTAS Grant 96-1418, RFBR Grants 95-03-323691 and 96-15-97732, CNRS, and University Louis Pasteur, Strasbourg.

# [12] Site-Specific Sulfhydryl Groups for Study of RNA Conformation via Disulfide Cross-Linking

*By* SNORRI TH. SIGURDSSON

The hammerhead ribozyme is an RNA motif, which catalyzes the site-specific cleavage of RNA and is utilized by a number of small pathogenic plant viroid and satellite RNAs for processing during rolling cycle replication.[1] The hammerhead consists of three helices (Fig. 1, see color plate), connected by single-stranded regions. The nucleotide sequence in the helical regions has minimal effect on the catalytic activity as long as base pairing is maintained. Substitutions, however, of nucleotides in the single-stranded regions of the catalytic core reduce activity dramatically. Advances in the chemical synthesis of oligoribonucleotides have enabled the incorporation of modified nucleotides to investigate the role of specific functional groups and to probe metal ion-binding sites. However, it has been difficult to correlate functional group data with specific catalytic and structural roles in the absence of a three-dimensional structure of the ribozyme.

Two models of the eagerly awaited tertiary structure of the hammerhead ribozyme were published late in 1994. One model, based on single crystal X-ray crystallography, displayed a wishbone shape with helices II and III almost collinear with helix I and II proximal (Fig. 1A, see color plate).[2] The other model was based on fluorescence resonance energy transfer (FRET) measurements in solution, which yielded the relative orientation of the three helices, allowing the central core to be modeled (Fig. 1B, see color plate).[3] Comparison of the two models revealed striking similarities

[1] R. H. Symons, *Annu. Rev. Biochem.* **61,** 641 (1992).
[2] H. W. Pley, K. M. Flaherty, and D. B. McKay, *Nature* (*London*) **372,** 68 (1994).
[3] T. Tuschl, C. Gohlke, T. M. Jovin, E. Westhof, and F. Eckstein, *Science* **266,** 785 (1994).

regarding the placement of helices in three-dimensional space. However, the relative orientations of helices I and II in the two models were different. At the same point of reference, the minor grooves of helices I and II face each other in the X-ray structure, whereas the major grooves face each other in the FRET model. Thus, helices I and II need to be rotated simultaneously about 180° around their helical axes to convert one structure to the other.

Which one of these two models represented the global shape of the transition state of the hammerhead-catalyzed reaction? Many would argue in favor of the X-ray structure, given the fact that X-ray crystallography is a well-established technique for determining three-dimensional structures of biomolecules, such as proteins. However, biochemical data indicated that the X-ray structure represented an inactive ground state structure.[4] For example, the required 2'-hydroxyl group at the cleavage site was not positioned for an in-line attack on the scissile phosphorus and no metal ion was found close to the cleavage site. In addition, functional group modification experiments have shown that the exocyclic amino group of G5 is essential for activity. This group, however, was exposed to solvent in the X-ray structure. Thus, the X-ray structure was a ground state structure, requiring a conformational change to reach the transition state. Was the required conformational change local or global? Was the FRET model a better representation of the global shape of the active structure?

## Strategy

To determine which one of these structural models represented catalytically active species better, covalent cross-links were used to "lock" each of two ribozymes in the conformation displayed in the two models. Due to the different orientations of helices I and II in the two models, pairs of nucleotides were identified for cross-linking that were proximal in one model and distal in the other. For example, nucleotides L2.4 and 2.6 are separated by about 11 Å in the X-ray structure and by about 32 Å in the FRET model (Fig. 1, see color plate). The catalytic activity of the cross-linked ribozymes would serve as an indicator of cross-link-induced structural perturbations of the active structure. Thus, if the cross-linked ribozyme was as active as a non-cross-linked ribozyme, the model in which the cross-linked nucleotides were proximal would be favored.

Disulfide cross-linking between cysteine residues has been used to study global conformations of proteins.[5] An advantage of using disulfide cross-

[4] D. M. McKay, *RNA* **2**, 395 (1996).
[5] J. J. Falke and D. E. Koshland, Jr., *Science* **237**, 1596 (1987).

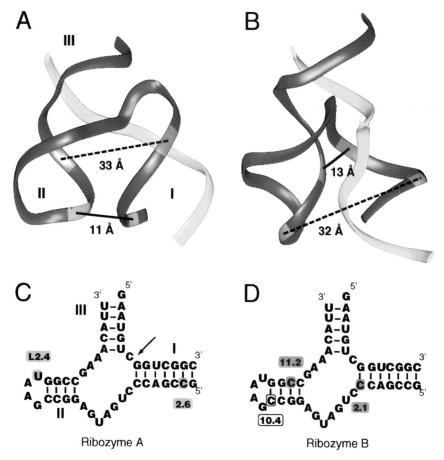

FIG. 1. Ribbon representation of the tertiary structures of the hammerhead ribozyme based on X-ray crystallography (A) and FRET solution measurements (B). The ribozymes and substrates are colored blue and yellow, respectively. Nucleotides that are close in space according to the X-ray structure and the FRET model (green and red, respectively) are connected by solid lines, whereas broken lines represent the corresponding long distances in the other model. The distances between the 2'-hydroxyl groups of connected residues are given in angstroms (Å). Roman numerals indicate the number of the helices. (C and D) RNA constructs chemically synthesized for interhelical cross-linking experiments. The colored residues contain a 2'-amino functionality. An arrow indicates the site of cleavage. Residue 10.4, also utilized for cross-linking to test the FRET model (see text), is boxed. Adapted from S. Th. Sigurdsson, T. Tuschl, and F. Eckstein, *RNA* **1**, 575 (1995), with permission from Cambridge University Press.

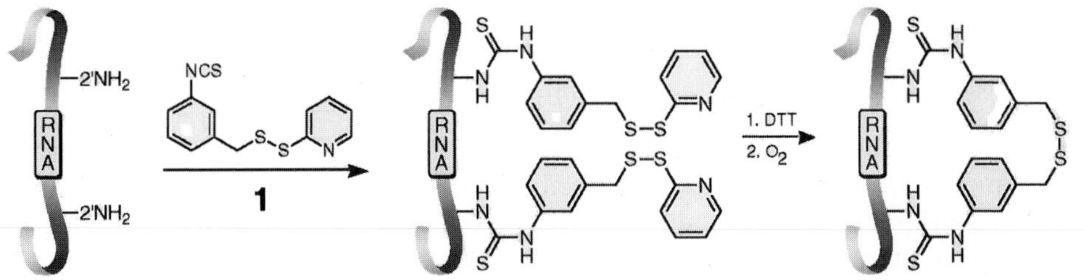

FIG. 2. Formation of an intramolecular disulfide cross-link by chemical modification of a 2'-amino-containing ribozyme. Adapted from S. Th. Sigurdsson *et al., RNA* **1,** 575 (1995), with permission from Cambridge University Press.

links is that they form in high yield between specific functional groups (thiols) and that the cross-linking can be reversed with the addition of thiols. Therefore, a method was sought that enabled the site-specific incorporation of thiols into the hammerhead ribozyme. Thiols can, in principle, be linked to either RNA nucleotide bases[6,7] or the sugar–phosphate backbone. However, to minimize the disruption of base pairing, cross-linking through the sugar–phosphate backbone was preferred. Furthermore, distances between the backbones in helices I and II were shorter than those between bases and allowed the use of a short tether.

The ability to site specifically incorporate 2'-amino nucleotides into RNA and selectively modify the 2'-amino groups with aromatic isothiocyanates was utilized for the incorporation of thiols into the hammerhead ribozyme.[8] Because an aromatic isothiocyanate containing a protected thiol functionality was not available commercially, compound **1** was prepared. Reaction of **1** with the 2'-amino groups in RNA yielded 2'-thiourea-substituted RNA (Fig. 2). Subsequent deprotection and oxidation of the thiols resulted in the formation of disulfide cross-links. It should be noted that although this article only describes the use of aromatic isothiocyanate **1,** reactions of 2'-amino groups with an aliphatic isocyanate containing the same protected thiol attached to a more flexible and less sterically hindered aliphatic tether have also been described.[9–11]

[6] C. R. Allerson and G. L. Verdine, *Chem. Biol.* **2,** 667 (1995).
[7] C. R. Allerson, S. L. Chen, and G. L. Verdine, *J. Am. Chem. Soc.* **119,** 7423 (1997).
[8] S. Th. Sigurdsson, T. Tuschl, and F. Eckstein, *RNA* **1,** 575 (1995).
[9] S. Th. Sigurdsson and F. Eckstein, *Nucleic Acids Res.* **24,** 3129 (1996).
[10] D. J. Earnshaw, B. Masquida, S. Mueller, S. Th. Sigurdsson, F. E. Eckstein, E. Westhof, and M. J. Gait, *J. Mol. Biol.* **274,** 197 (1997).
[11] S. Th. Sigurdsson, METHODS: A Companion to Methods in Enzymology **18,** 71 (1999).

Procedure

*Preparation of Ribozymes and Substrates*

Oligoribonucleotides are prepared by automated chemical synthesis on an Applied Biosystems (Foster City, CA) 380B DNA synthesizer on a 1-$\mu$mol scale using phosphoramidites from MilliGen/Biosearch with the exception of phosphoramidites used for the incorporation of 2'-amino nucleotides. These are now available commercially from Glen Research. The oligoribonucleotides are deprotected by incubation in 3 ml of concentrated aqueous ammonia (32%)/ethanol (3 : 1) for 16 hr at 55° in a screw-top vial. The solvent is removed completely on a Speed-Vac evaporator, and the residue is treated with 0.5 ml of 1.0 $M$ tetrabutylammonium fluoride in tetrahydrofuran (THF) (Aldrich, Milwaukee, WI) at 25° for 16 hr to remove silyl-protecting groups. Next, 0.5 ml of 1.8 $M$ sodium acetate (pH 5.8) is added and the solution is concentrated to a volume of about 0.5 ml and extracted twice with 0.8 ml of ethyl acetate. After precipitation of the RNA by the addition of 1.6 ml of cold absolute ethanol, the mixture is centrifuged and the supernatant discarded. The pellet is dissolved in water and the RNA is purified by 20% denaturing polyacrylamide gel electrophoresis (DPAGE). RNA is detected in the gel by UV shadowing, eluted from the excised gel slices by electroelution in Tris–borate–EDTA buffer, and desalted with a Sep-Pak $C_{18}$ cartridge. The cartridges are washed sequentially with acetonitrile (10 ml) and 100 m$M$ triethylammonium bicarbonate, pH 7.5 (10 ml). The sample is loaded, washed with 4% acetonitrile in 100 m$M$ triethylammonium bicarbonate (20 ml), and eluted with acetonitrile/methanol/water (35 : 35 : 30, v/v). Next, the solvent is removed in a Speed-Vac, and methanol is added (0.2 ml) and evaporated to remove traces of triethylamine. Finally, the RNA is dissolved in water (0.2 ml) and the concentration is determined using the molar extinction coefficient of 6600 $M^{-1}$ cm$^{-1}$ per nucleotide.

*Chemical Modification and Cross-Linking of 2'-Amino
　Ribozymes A and B*

1. The ribozymes are radiolabeled at either the 5' end using T4 polynucleotide kinase and [$\gamma$-$^{32}$P]ATP or the 3' end using T4 RNA ligase and 5'-$^{32}$P-labeled cytidine 3',5'-diphosphate. The buffers are supplied by the manufacturer (Amersham, Piscataway, NJ). After a 15-min reaction time for the 5'-labeling reaction, nonradiolabeled ATP (final concentration of 1.6 m$M$) is added to "chase" the reaction to completion. Most of the experiments are carried out with the 5'-

labeled ribozyme. Analysis of the cross-linked ribozymes by limited alkaline hydrolysis utilizes both 3'- and 5'-labeled ribozymes.

2. Isothiocyanate **1** is reacted with radiolabeled ribozyme for 28 hr at 37° [50 m$M$ **1**; 1 m$M$ ribozyme; 50 m$M$ borate buffer, pH 8.6; 50% dimethylformamide (DMF); final volume 10 $\mu$l].

3. The reaction mixture is diluted and the noncleavable substrate 5'-GAAUGUdCGGUCGGC is added (10 $\mu M$ ribozyme; 20 $\mu M$ substrate; 50 m$M$ sodium cacodylate, pH 7.5; 50 m$M$ NaCl; final volume 1270 $\mu$l). The substrate is annealed to the ribozyme by heating the solution to 90° for 3 min followed by slow cooling to 25° over a period of 2.5 hr.

4. Dithiothreitol (DTT) (1 $M$ aqueous solution, 127 $\mu$l) is added to the solution containing the ribozyme–substrate complex and incubated for 2 hr at 25° to remove thiol-protecting groups.

5. The oligomers are precipitated at −20° after the sequential addition of sodium acetate (3.0 $M$ aqueous solution, pH 5.2; 1140 $\mu$l) and ethanol (9120 $\mu$l). The pellet is collected by centrifugation and redissolved (10 $\mu M$ ribozyme; 50 m$M$ sodium cacodylate, pH 7.5; 50 m$M$ NaCl; 20 m$M$ MgCl$_2$; final volume 1710 $\mu$l).

6. For cross-linking, the solution is divided into six microfuge tubes and an equal volume of DMSO is added to each tube. The solutions are incubated for 30 hr at 25° under an atmosphere of oxygen and the RNA is precipitated as described earlier. It should be noted that the cross-linking conditions are not optimized; DMSO is used for cross-linking because it has been shown to increase the rate of cross-linking but gives only slightly better results than those obtained without DMSO.

## Analysis of Cross-Linking Procedure and Purification of Cross-Linked Ribozymes

The chemical modification of the 5'-radiolabeled, 2'-amino containing ribozymes was monitored by 20% DPAGE (Fig. 3). As can be seen, reactions with isothiocyanate **1** were close to quantitative (Fig. 3, lanes 2), and a major product with different electrophoretic mobility was observed after cross-linking (Fig. 3, lanes 3). The minor products that had strongly retarded electrophoretic mobility were presumably intermolecular cross-links.

It is noteworthy that the electrophoretic mobilities of the putative cross-linked ribozymes were very different, despite the fact that they had the same charge and molecular weight. Cross-linked ribozyme A had electrophoretic mobility similar to that of the non-cross-linked sample. This was of concern because even a minor contamination of non-cross-linked ribozyme could

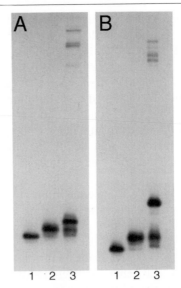

Fig. 3. Analysis of the cross-linking reactions of ribozymes A (**A**) and B (**B**) by 20% DPAGE. Lanes 1, 2′-amino-modified ribozymes; lanes 2, reaction mixtures after treatment with DTT; and lanes 3, crude cross-linking reaction mixtures. Reprinted from S. Th. Sigurdsson *et al., RNA* **1**, 575 (1995), with permission from Cambridge University Press.

affect the measurement of the catalytic activity of the cross-linked ribozyme. This would in turn interfere with the assessment of structural perturbations caused by the cross-link.

It was found that the relative mobility of cross-linked to non-cross-linked material varied greatly with the percentage of acrylamide in the denaturing gels. In a systematic study, ribozymes A and B, as well as the corresponding non-cross-linked ribozymes (obtained by the reduction of the cross-linked samples with dithiothreitol), were subjected to DPAGE containing 8, 12, 16, 20, and 24% acrylamide (0.04 × 20 × 40 cm gel; acrylamide : bisacrylamide, 19 : 1).[12] The distance that the cross-linked and non-cross-linked material migrated in the gels was subsequently measured. To facilitate a comparison of results from all the gel analyses, the mobilities of ribozyme A and B were normalized, assuming that the non-cross-linked samples had migrated 30 cm in all the gels. The results are presented in the simulated gel shown in Fig. 4, which reveals a dramatic difference in the relative mobility of the cross-linked to the non-cross-linked material. This phenomenon has also been observed for "lariat" intermediates in

[12] S. Th. Sigurdsson and F. Eckstein, *Anal. Biochem.* **235**, 241 (1996).

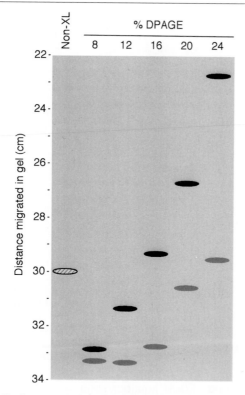

FIG. 4. A normalized representation of the distances migrated by cross-linked ribozymes during DPAGE as a function of percentage acrylamide, assuming that the non-cross-linked (non-XL) material migrated 30 cm under all conditions. The gray and black bands represent ribozymes A and B, respectively.

messenger RNA splicing.[13] It is also striking that in low percentage gels, the cross-linked ribozymes run faster than non-cross-linked ribozymes, whereas the opposite is true in high percentage gels. These results underscore the fact that careful choice of percentage acrylamide in DPAGE gels is important for the optimal separation of RNA containing intramolecular cross-links from non-cross-linked material. Furthermore, these results indicate that DPAGE is a general method for analysis and isolation of nucleic acids containing intramolecular cross-links.

Thus, the major products of the cross-linking reactions of ribozymes A and B were purified by 12 and 20% DPAGE, respectively. The cross-linked material was extracted from the gel slices by a crush and soak procedure

[13] P. J. Grabowski, R. A. Padgett, and P. A. Sharp, *Cell* **37,** 415 (1984).

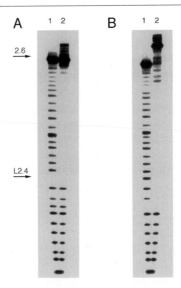

FIG. 5. Limited alkaline hydrolysis of cross-linked ribozymes A (**A**) and B (**B**). Lanes 1, ribozymes A and B, and lanes 2, cross-linked ribozymes A and B. Arrows indicate positions of the 2'-amino-modified residues in ribozymes A and B, where no cleavage occurs. Reprinted from S. Th. Sigurdsson *et al., RNA* **1,** 575 (1995), with permission from Cambridge University Press.

using sodium acetate (1.0 *M* aqueous solution, pH 5.2) from which the products were subsequently precipitated by the addition of ethanol. This resulted in an approximately 15% overall yield of cross-linked material, based on the starting 2'-amino oligomers.

### Analysis of Cross-Linked Ribozymes

*Limited Alkaline Hydrolysis.* To verify that the isolated products did indeed contain an intramolecular cross-link and to verify the location of the cross-link, the putative cross-linked ribozymes were subjected to limited alkaline hydrolysis. $3'-^{32}P$-labeled ribozymes A and B, and the corresponding cross-linked ribozymes, were subjected to limited alkaline hydrolysis (50 m*M* NaHCO$_3$; final volume 20 $\mu$l; 100°; 5 min), followed by analysis by 20% DPAGE (Fig. 5). This procedure was also performed with 5'-radiolabeled ribozymes (data not shown). As expected for ribozymes containing intramolecular cross-links, there was a virtual footprint in the ladder of bands. This was due to the cleavage between cross-linked nucleotides yielding products that have the same molecular weight and charge as the uncleaved oligomer and thus have similar electrophoretic mobility.

*Procedure for Obtaining Kinetic Parameters*

1. Non-cross-linked ribozymes are obtained by reduction of the cross-linked ribozymes with DTT (5 m$M$; 50 m$M$ Tris–HCl, pH 7.5) for 4 hr at 25° prior to the heat shock protocol (step 2).
2. Solutions of ribozyme and substrate RNA (50 m$M$ Tris–HCl, pH 7.5) are preheated separately at 90° for 1 min and cooled to 25° over a 15-min period.
3. MgCl$_2$ is added (final concentration, 10 m$M$; 50 m$M$ Tris–HCl, pH 7.5), and the solutions are incubated for 15 min at 25°.
4. The solutions of ribozymes and substrates are combined in such a way that a trace amount of 5′-$^{32}$P-labeled substrate (<1 n$M$) is incubated in the presence of the ribozymes at varying concentrations (20–300 n$M$) (50 m$M$ Tris–HCl, pH 7.5; final volume 50 $\mu$l; 10 m$M$ MgCl$_2$) at 25°.
5. Aliquots (8 $\mu$l) are withdrawn at appropriate time intervals and added to 16 $\mu$l of urea stop mix (3.5 $M$ urea, 25 m$M$ EDTA, 0.02% bromphenol blue, 0.02% xylene cyanol) and subsequently subjected to 20% DPAGE.
6. The rates of cleavage, at different concentrations of ribozymes (20–300 n$M$), are obtained by plotting the remaining fraction of uncleaved substrate to a single exponential decay as a function of time. The rates reflect time points where less than 30% of the substrate is cleaved because a fraction of the substrate remains uncleaved, even after incubation for a few hours. However, it has been pointed out that the cleavage rate can be determined more accurately by accounting for the uncleaved fraction during curve fitting.[14] The $k'_{cat}$ and $K'_m$ parameters are determined from Eadie–Hofstee plots.

*Analysis of Kinetic Data*

Activities of the cross-linked ribozymes were determined under single turnover conditions (Table I) to monitor the rates of chemical cleavage rather than product release. Non-cross-linked ribozymes yielded activities similar to that of the original 2′-amino-modified ribozymes, demonstrating that the structures were not perturbed by the chemical modifications. However, the activity of the two cross-linked ribozymes differed by about 300-fold: ribozyme A had a cleavage efficiency similar to that of non-cross-linked species, whereas the activity of ribozyme B was reduced dramatically. In fact, the activity of cross-linked ribozyme B was accounted for by a contamination of less than 1% of the non-cross-linked material.

---

[14] T. K. Stage-Zimmermann and O. C. Uhlenbeck, *RNA* **4**, 875 (1998).

TABLE I
KINETIC PARAMETERS FOR CROSS-LINKED AND NON-CROSS-LINKED
RIBOZYMES A AND B[a]

| Ribozyme (Rz) | $k'_{cat}$ (min$^{-1}$) | $K'_m$ (nM) | $k_{cat}'/K_m'$ ($\mu M^{-1}$ min$^{-1}$) |
| --- | --- | --- | --- |
| Rz A | 0.72 | 19 | 37 |
| Rz A XL | 0.60 | 35 | 17 |
| Rz A non-XL | 0.58 | 25 | 23 |
| Rz B | 0.49 | 20 | 24 |
| Rz B XL | 0.002[b] | nd | nd |
| Rz B non-XL | 0.67 | 46 | 15 |

[a] XL, cross-linked; nd, not determined.
[b] Observed rate constant at 200 nM ribozyme.

It might be argued that cross-linked ribozyme B was inactive because of reasons other than distance constraints. For example, there is a possible steric interaction between the cross-linker and nucleotides in the catalytic core. Therefore, a ribozyme containing a cross-link between nucleotides 2.1 and 10.4 was prepared and characterized kinetically (Fig. 1, see color plate). The 2'-amino groups on those nucleotides are separated by 13 Å in the FRET model and would place the cross-link in a different position than in ribozyme B. The results are very similar to those obtained for cross-linked ribozyme 2; the observed rate constant was 0.005 min$^{-1}$ at 500 nM ribozyme concentration and the catalytic efficiency for the non-cross-linked species was 30 $\mu M^{-1}$ min$^{-1}$. The combined results from the kinetic measurements strongly support the X-ray structure as a good representative of the global shape of the transition state structure of the hammerhead ribozyme.

Concluding Remarks

The first step toward understanding the fascinating function of RNAs, such as ribozymes, is to obtain a three-dimensional structure. X-ray crystallography of RNA molecules is still in its early stages, with only a handful of structures having been obtained of complex RNAs. Thus, other physical, biophysical, biochemical, and chemical approaches are being used extensively for obtaining structural information. The cross-linking approach described in this article is one such useful tool to study RNA tertiary structure and has been applied to other ribozymes. For example, it has been used to obtain distance constraints for the generation of a three-dimensional model of the hairpin ribozyme,[10] and an extension of this method has allowed the assessment of the dynamics of helical elements within the *Tetrahymena* group I ribozyme by following the kinetics of cross-linking.[15]

This cross-linking method is advantageous for testing a three-dimensional model: nucleotides that are close spatially can be cross-linked and the activity of the cross-linked species yields structural information. This method will be valuable in conjunction with other techniques, such as FRET, which can be used to generate structural models to be tested by cross-linking experiments. However, this approach becomes increasingly more difficult to apply to three-dimensional structure determination as the complexity of the RNA increases or where there is limited structural data available from other techniques. In addition, the flexibility of some ribozymes may make the interpretation of such cross-linking experiments difficult. For example, the *Tetrahymena* group I ribozyme containing cross-links between nucleotides that were separated by about 50 Å in the current three-dimensional model still retained some catalytic activity.[15] However, this method will be valuable for the study of RNA structure as long as these potential limitations are kept in mind.

### Acknowledgments

I am grateful to Professor F. Eckstein, in whose laboratory this work was carried out. I thank Dr. T. Tuschl and members of my research group for critical reading of the manuscript.

[15] S. B. Cohen and T. R. Cech, *J. Am. Chem. Soc.* **119**, 6259 (1997).

# [13] Directed Hydroxyl Radical Probing Using Iron(II) Tethered to RNA

*By* SIMPSON JOSEPH and HARRY F. NOLLER

### Introduction

Directed hydroxyl radical probing using tethered Fe(II)-EDTA is a powerful tool for studying the tertiary structure of RNA at low resolution.[1,2] In the presence of reducing agents, Fe(II)-EDTA generates neutral, reac-

[1] J. F. Wang and T. R. Cech, *Science* **256**, 526 (1992).
[2] H. Han and P. B. Dervan, *Proc. Natl. Acad. Sci. U.S.A.* **91**, 4955 (1994).

tive species, believed to be hydroxyl radicals, that attack the riboses in the nucleic acid backbone, resulting in strand scission.[3-6] Tethering the Fe(II)-EDTA complex to a fixed position in the RNA causes cleavage to be directed to those regions of the RNA surrounding that position. Fe(II) can be tethered to proteins and RNA using the reagent 1-($p$-bromoacetamido-benzyl)-EDTA (BABE) originally synthesized by Meares and co-workers.[7,8] This reagent can be attached covalently to nucleophilic positions (most coveniently, cysteine sulfhydryls in proteins and phosphorothioates or other thio-substituted nucleotides in RNA) of a macromolecule. Fe(II) is bound firmly to the EDTA moiety of BABE ($K_d \approx 10^{-14}$). Because hydroxyl radicals are small in size, highly reactive, and the reactivity is sequence and structure independent, they can be used to obtain a fairly comprehensive picture of the RNA elements surrounding the probe. Most importantly, the cleavage intensity of the target depends on its distance from the tethered probe. This can be used to obtain information about the structure of the nucleic acid target in three dimensions at low resolution. Two methods have been used to detect the sites of strand scission, depending on the size of the target nucleic acid. For short RNA or DNA chains (less than 200 nucleotides), the sites of cleavage can be detected using [32]P-end-labeled target nucleic acid and separating the cleavage products on denaturing polyacrylamide gels. Alternatively, for longer chains, the sites of strand scission are detected by primer extension using reverse transcriptase.[9] This article describes a simple method for tethering Fe(II)-EDTA to the 5' end of *in vitro*-transcribed RNAs and illustrates the potential of tethered Fe(II)-EDTA for studying the folding of large RNA molecules using ribosomal RNA as an example.

Materials

*Buffers*

    $10\times$ Kinase buffer: 700 m$M$ Tris–Cl (pH 7.6), 100 m$M$ MgCl$_2$, 50 m$M$ dithiothreitol (DTT)

[3] R. P. Hertzberg and P. B. Dervan, *J. Am. Chem. Soc.* **104,** 313 (1982).
[4] C. Bull, G. J. McClune, and J. A. Fee, *J. Am. Chem. Soc.* **105,** 5290 (1983).
[5] R. P. Hertzberg and P. B. Dervan, *Biochemistry* **23,** 3934 (1984).
[6] T. D. Tullius and B. A. Dombroski, *Science* **230,** 679 (1985).
[7] L. H. DeRiemer, C. F. Meares, D. A. Goodwin, and C. J. Diamanti, *J. Label. Compd. Radiopharmaceut.* **18,** 1517 (1981).
[8] T. M. Rana and C. F. Meares, *J. Am. Chem. Soc.* **112,** 2457 (1990).
[9] S. Stern, D. Moazed, and H. F. Noller, *Methods Enzymol.* **164,** 481 (1988).

10× *Bst*NI buffer (pH 7.9 at 25°): 100 m$M$ Tris–Cl, 100 m$M$ MgCl$_2$, 500 m$M$ NaCl, 10 m$M$ DTT

2× gel-loading buffer: 95% formamide, 20 m$M$ EDTA, 0.05% xylene cyanol, 0.05% bromphenol blue

Elution buffer: 0.5 $M$ ammonium acetate (pH 5.2), 0.1% sodium dodecyl sulfate (SDS), 0.1 m$M$ EDTA (pH 8.0)

10× transcription buffer: 400 m$M$ Tris–Cl (pH 7.5), 200 m$M$ MgCl$_2$, 20 m$M$ spermidine, 0.1% Triton X-100

Potassium phosphate buffer: 400 m$M$, pH 8.5

5× C$_{80}$M$_{20}$N$_{150}$ buffer: 400 m$M$ potassium-cacodylate (pH 7.2), 100 m$M$ magnesium acetate, 750 m$M$ ammonium chloride

10× T$_1$ buffer: 100 m$M$ Tris–Cl (pH 7.4), 10 m$M$ EDTA (pH 7.0)

Paper electrophoresis buffer (pH 3.5): 0.5% pyridine, 5% acetic acid (v/v), and 2 m$M$ EDTA

Dye mix: 1% xylene cyanol FF, 1% acid fuschin, and 2% orange G

*Enzymes and Reagents*

T4 polynucleotide kinase, 10 U/$\mu$l (New England Biolabs)

*Bst*NI, 10 U/$\mu$l (New England Biolabs)

T7 RNA polymerase, 25–50 U/$\mu$l (Epicenter or purified in the laboratory)

RQI RNase-free DNase, 1 U/$\mu$l (Promega, Madison, WI)

Avian myeloblastosis virus (AMV) reverse trancriptase, 25 U/$\mu$l (Seikagaku America Inc., Falmouth, MA)

Ribonuclease T$_1$ (Sigma, St. Louis, MO)

Ribonuclease T$_2$ (grade V from Sigma)

Bovine serum albumin (BSA) (Sigma)

Transfer ribonucleic acid from *Escherichia coli* (MRE 600): Phenylalanine Specific (Subriden, Rollingbay, WA or Sigma)

Polyuridylic acid (Sigma)

Tight couple 70S ribosomes[10]

1-(*p*-Bromoacetamidobenzyl)-EDTA [prepared from aminobenzyl-EDTA (Dojindo Labs) as described][7,8]

NTPs, dNTPs, and ddNTPs (Pharmacia, Piscataway, NJ)

5'-Guanosine-$\alpha$-phosphorothioate (United States Biochemical, Cleveland, OH)

Ammonium iron(II) sulfate hexahydrate (Aldrich, Milwaukee, WI)

Diisopropylethylamine (Aldrich)

---

[10] T. Powers and H. F. Noller, *EMBO J.* **10,** 2203 (1991).

Vitamin C/ascorbic acid (Fluka, Ronkonkoma, NY)

30% hydrogen peroxide, Ultrex ultrapure reagent (J. T. Baker, Phillipsburg, NJ)

[γ-$^{32}$P]ATP (4500 Ci/mmol; 10 mCi/ml) (ICN, Costa Mesa, CA)

[α-$^{32}$P]CTP (3000 Ci/mmol; 10 mCi/ml) (ICN)

[α-$^{32}$P]UTP (3000 Ci/mmol; 10 mCi/ml) (ICN)

## DNA Oligonucleotides

The following deoxyoligonucleotides are synthesized on a Millipore (Bedford, MA) automated DNA/RNA synthesizer:

1. 18T7T: 5'-TAATACGACTCACTATAG-3' corresponding to the top strand of T7 RNA polymerase promoter.
2. DNA-T37: 5'-ACGGTGCCTGACTGCGTTA**CTATAGTGAGT-CGTATTA**−3', complementary to nucleotide sequence 65 to 83 in plasmid pBR322. The sequence corresponding to the bottom strand of T7 RNA polymerase promoter is shown in bold. DNA-T37 is used as template to transcribe a 20-mer RNA (RNA-20).
3. 16S-DNA526: 5'-GCTAACTCCGTGCCAGCAGC**CTATAGTG-AGTCGTATTA**−3'. The sequence corresponding to the bottom strand of T7 RNA polymerase promoter is shown in bold. 16S-DNA526 is used to transcribe a 21-mer RNA (16S-cRNA526) complementary to sequence 506 to 526 in 16S rRNA.

## Methods

### A. 5' End Labeling of DNA

End label 10 pmol of DNA-T37 with 10 pmol of [γ-$^{32}$P]ATP using 10 units of T4 polynucleotide kinase and 2.5 μl of 10× kinase buffer in 25 μl final reaction volume. Incubate the reaction at 37° for 30 min. Add 75 μl of water and extract twice with 100 μl phenol, followed by two chloroform extractions. Add 300 μl of 100% ethanol and place the Eppendorf tube in a dry ice–ethanol bath for 25 min. Centrifuge the tubes at 4° in a microfuge at 13,000 rpm for 25 min. Remove the ethanol carefully, wash the DNA pellet with 500 μl of 70% (v/v) ethanol, dry the pellet in a Speed-Vac (Model SC110 from Savant, Farmingdale, NY) for 5 min, and resuspend in 50 μl of gel-loading buffer. Isolate the end-labeled DNA by separating on a 15% denaturing polyacrylamide gel. Identify the band by autoradiography, cut out the gel piece containing the band, and elute passively by soaking overnight at room temperature in 600 μl of elution buffer. Pass the sample through a 0.2-μm syringe filter (MSI Cameo 25NS, Fisher) to remove gel pieces and extract twice with 600 $\mu$l of chloroform. Recover the DNA by ethanol precipitation as described earlier and resuspend in 50 $\mu$l of water. Estimate cpm/$\mu$l by spotting a 1.0-$\mu$l aliquot of the labeled DNA on filter paper and counting using a liquid scintillation counter.

## B. Preparation of E. Coli tRNA$^{Phe}$ Transcription Template

E. Coli tRNA$^{Phe}$ is prepared by in vitro transcription of linearized plasmid p67CF10.[11] Digest 80 $\mu$g of p67CF10 by adding 10 $\mu$l of 10× BstNI buffer, 10 $\mu$l of 1.0 $\mu$g/$\mu$l bovine serum albumin, and 10 $\mu$l of 10 units/$\mu$l BstNI in 100 $\mu$l final reaction volume. Incubate at 60° for 2 hr. Add 30 $\mu$l of 0.5 $M$ Na$_4$EDTA, 30 $\mu$l of 5 $M$ ammonium acetate, pH 5.2, and extract twice sequentially with phenol and twice with chloroform. Recover the DNA by ethanol precipitation as described earlier and resuspend in 50 $\mu$l of water.

## C. In Vitro Transcription Reaction

RNAs are synthesized by in vitro transcription using T7 RNA polymerase. A standard 500-$\mu$l scale transcription reaction contains 30 $\mu$g of linearized plasmid, 50 $\mu$l of 10× transcription buffer, 50 $\mu$l of 50 mM DTT, 80 $\mu$l of 25 mM each NTP mix (final concentration 4 mM each), and 300 units of T7 RNA polymerase. For incorporation of a 5'-phosphorothioate at the 5' terminus of the RNA, perform transcription in the presence of 50 $\mu$l of 50 mM 5'-guanosine $\alpha$-phosphorothioate (GMPS) or guanosine monophosphate (GMP) (5 mM final concentration) and 20 $\mu$l of 25 mM each NTP mix (final concentration 1 mM each). When synthetic oligodeoxynucleotides are used as templates in transcription reaction, anneal 5.0 $\mu$l of 100 pmol/$\mu$l of 18T7T to 5.0 $\mu$l of 100 pmol/$\mu$l of the template strand by heating to 95° for 2 min and cooling to room temperature gradually over 10 min. Incubate the transcription reactions at 37° for 4 hr. Add 10 $\mu$l of RQI RNase-free DNase and incubate for 30 min at 37°. Stop the reaction by adding 50 $\mu$l of 0.5 $M$ Na$_4$EDTA and 50 $\mu$l of 3 $M$ sodium acetate, pH 5.2, and extract sequentially with 600 $\mu$l of phenol and 600 $\mu$l of chloroform. Precipitate the RNA by adding 1.2 ml of 100% ethanol as described earlier. Resuspend the pellet in 100 $\mu$l of gel-loading buffer and isolate on a denaturing polyacrylamide gel. Identify the transcripts by UV shadowing, excise the band, elute passively at 4° (see earlier discussion), and store in water

[11] J. R. Sampson, A. B. DiRenzo, L. S. Behlen, and O. C. Uhlenbeck, Science **243**, 1363 (1989).

at $-20°$. Estimate the concentration of the RNA by measuring absorbance at $A_{260}$.

## D. Efficiency of GMPS (or GMP) Incorporation

For determining the level of incorporation of GMPS (or GMP) at the 5' end of the transcripts, perform transcription in the presence of 2.0 $\mu$l of $[\alpha\text{-}^{32}P]CTP$ (3000 Ci/mmol; 10 mCi/ml) in 500 $\mu$l of the transcription reaction and purify the transcripts (see earlier discussion). Place 5.0 $\mu$l of 10,000 cpm/$\mu$l of $^{32}P$-labeled RNA in an Eppendorf tube and digest to completion by adding 20 $\mu$l of ribonuclease $T_2$ (1 unit/$\mu$l), 4.0 $\mu$l of 0.5 $M$ ammonium acetate (pH 4.7), and 11 $\mu$l of water. Incubate for 24 hr at 37°. Take a 5.0-$\mu$l aliquot of the digested sample and treat with 5.0 $\mu$l of 14.4 $M$ 2-mercaptoethanol for 30 min at 40° to reduce any oxidized GMPS. Spot 5.0 $\mu$l of the sample on Whatman (Clifton, NJ) No. 51 filter paper 10 cm from the cathode end (46 $\times$ 57 cm). Also spot 2.5 $\mu$l of dye mix on the filter paper to serve as markers. Wet the filter paper with paper electrophoresis buffer, pH 3.5, place gently in a high-voltage electrophoresis chamber (Savant), and electrophorese at 2500 V for 30 min.[12] Air dry the filter paper and quantify the amount of 5'-thio-pG[$^{32}P$]p (or pG[$^{32}P$]p) using a Molecular Dynamics PhosphorImager (Sunnyvale, CA). The analysis presented in Fig. 1 shows that at least 80% of the transcripts initiate with GMPS (or GMP).

## E. 5'-Fe(II)-BABE Modification of RNA

We use the reagent 1-(p-bromoacetamidobenzyl)-EDTA (BABE)[7,8] to tether Fe(II) to the 5' terminus of RNA bearing a 5'-phosphorothioate.[13] Mix 20 $\mu$l of 28 m$M$ freshly made ammonium iron(II) sulfate hexahydrate with 25 $\mu$l of 28 m$M$ BABE. Adjust the pH to 3–4 using diisopropylethylamine. Incubate at room temperature for 60 min. Combine 800 pmol of 5'-GMPS-RNA, 5.0 $\mu$l of the Fe(II)-BABE mix, and 1.0 $\mu$l of 400 m$M$ potassium phosphate buffer (pH 8.5) in a final 10-$\mu$l reaction volume. Incubate for 60 min at 37°, add 1.0 $\mu$l of 50 m$M$ EDTA, and incubate for an additional 10 min to chelate excess free Fe(II). Stop the reaction by adding 10 $\mu$l of 5 $M$ ammonium acetate, 80 $\mu$l of water, and extracting twice with water-saturated phenol to remove unreacted Fe(II)-BABE. Recover the 5'-Fe(II)-BABE-RNA by ethanol precipitation as described earlier, resuspend in water at 100 pmol/$\mu$l final concentration, and store at $-20°$.

---

[12] B. G. Barrell, in "Procedures in Nucleic Acid Research" (G. L. Cantoni and D. R. Davies, eds.), p. 924. Harper & Row, New York, 1971.

[13] S. Joseph and H. F. Noller, *EMBO J.* **15** (1996).

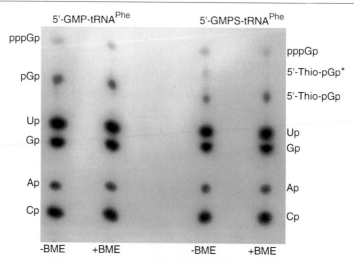

FIG. 1. Efficiency of incorporation of GMPS (or GMP) by T7 RNA polymerase at the 5′ end of tRNA^Phe transcripts. Lanes 5′-GMP-tRNA^Phe and 5′-GMPS-tRNA^Phe correspond to tRNA^Phe transcribed in the presence of GMP and GMPS, respectively. Transcripts were digested to completion using ribonuclease T$_2$ and an aliquot treated with 2-mercaptoethanol (+BME). Mononucleotides were separated by high-voltage paper electrophoresis on Whatman 51 paper at pH 3.5. Cp, Ap, Gp, and Up correspond to the respective mononucleotide-3′-phosphates. pppGp is guanosine 5′,3′-tetraphosphate. pGp is guanosine 5′,3′-diphosphate. 5′-thio-pGp and 5-thio-pGp* correspond to 5′-phosphorothioate–guanosine 3′-monophosphate and an uncharacterized oxidation product of 5′-phosphorothioate–guanosine 3′-monophosphate, respectively.

## Tethered Hydroxyl Radical Probing

### 1. DNA–RNA Duplex

Initially, to optimize tethered hydroxyl radical probing and to calibrate the distance dependence of hydroxyl radical probing, we used a DNA–RNA duplex as a simple model system. A 37-mer target DNA is 5′-end-labeled, hybridized to a probe-modified RNA, cleavage chemistry is activated, and the sites of strand scission in the DNA are detected by separating the cleavage products on a denaturing polyacrylamide gel.

Transcribe a 20-mer RNA (RNA-20) with a 5′-terminal phosphorothioate (either GMPS or 5′-γ-S-GTP) using DNA-T37 as template. Fe(II)-BABE modify RNA-20 as described earlier. Mix 5.0 μl of 5′-end-labeled DNA-T37 ($5 \times 10^5$ cpm/μl) with 20 μl of 50 pmol/μl of unlabeled DNA-T37. Denature DNA-T37 by heating to 95° for 2 min and placing on ice

[14] S. Joseph, B. Weiser, and H. F. Noller, *Science* **278**, 1093 (1997).

FIG. 2. Strand scission of 5'-end-labeled DNA-T37 by hydroxyl radicals generated from Fe(II)-BABE-modified RNA-20 hybridized to DNA-T37. Lane 1, mock reaction performed in the presence of unmodified RNA-20 and solution Fe(II)-EDTA; lane 2, DNA-T37 probed using Fe(II)-BABE attached to RNA-20 via GTPγS; lane 3, DNA-T37 probed using Fe(II)-BABE attached to RNA-20 via GMPS. Positions of strand scission are indicated by the bar.

for 10 min. In an Eppendorf tube, combine 1.0 μl of denatured DNA-T37, 1.0 μl of 40 pmol/μl of Fe(II)-BABE-RNA-20 (or unmodified RNA-20 as mock), 1.25 μl of 500 m$M$ Tris–Cl, pH 7.4, 1.25 μl of 50 m$M$ NaCl, and 4.5 μl of water. Incubate the tube at 25° for 30 min. Add 1.0 μl of 40 m$M$ DTT and incubate at 25° for 60 min.[15] Additionally, to a mock reaction, add 1.0 μl of Fe-EDTA mix [1 m$M$ Fe(II) and 2 m$M$ EDTA, final concentration] to monitor background cleavages. Stop reactions by adding 10 μl of gel-loading buffer, heat to 95° for 2 min, and load on a 20% denaturing polyacrylamide gel. Visualize the cleavage products by autoradiography.

Figure 2 is an example of tethered hydroxyl radical cleavage of the DNA–RNA duplex. Fe(II)-BABE was tethered to the 5' end of RNA-20 via either GTPγS (Fig. 2, lane 2) or GMPS (Fig. 2, lane 3). The autoradio-

[15] G. B. Dreyer and P. B. Dervan, *Proc. Natl. Acad. Sci. U.S.A.* **82,** 968 (1985).

graph shows the sites of strand scission in DNA-T37. The strongest cleavages are at positions 17–22 in DNA-T37, consistent with the location of the Fe(II)-BABE probe close to position 20 in the DNA–RNA duplex. We calculated the distance between the 5′-terminal phosphate on the RNA strand to which Fe(II) is tethered and the 3′-phosphate of the target nucleotide residues cleaved in the DNA by making the assumption that the DNA–RNA duplex has an A-form helical conformation that extends into the unpaired DNA strand. Strong cleavages at nucleotide positions 17, 18, 19, 20, 21, and 22 correspond to distances of 17-9 Å. Medium cleavages at positions 16, 23, and 24 correspond to distances of 19, 18, and 19 Å, respectively, whereas weak cleavages at positions 13, 14, 15, and 25 correspond to distances of 30, 26, 22, and 22 Å, respectively. These distance ranges are consistent with previous measurements.[1,2] Additionally, as noted by Han and Dervan,[2] the cleavage efficiency depends not only on the distance of the target nucleotide from the probe, but also on the accessibility of the ribose, especially in the 11- to 24-Å range. Furthermore, the cleavage efficiency is stronger when Fe(II)-BABE is tethered via GMPS (Fig. 2, lane 3) compared to GTPγS (Fig. 2, lane 2). This may be due to the greater rotational freedom for the Fe(II)-BABE probe when tethered via GTPγS.

## 2. RNA–RNA Duplex

In this example, we used native 16S rRNA as a target for tethered hydroxyl radical probing. A short RNA complementary to 16S rRNA is Fe(II)-BABE modified, hybridized to 16S rRNA, cleavage chemistry activated, and the sites of strand scissions are identified by primer extension reactions.[9]

Modify 16S-cRNA526 (complementary to sequence 506 to 526 in 16S rRNA) with Fe(II)-BABE as described earlier. Combine 5.0 μl of 5 pmol/μl of naked 16S rRNA, 5.0 μl of 5× $C_{80}M_{20}N_{150}$ buffer, and either 2.5 or 12.5 μl of 10 pmol/μl Fe(II)-BABE-16S-cRNA526 (1:1 and 1:5 ratio, respectively). Make the final reaction volume to 24 μl with water. In parallel, set up the following controls: omit 16S-cRNA526 from the reaction (for K and mock lanes); use unmodified 16S-cRNA526 at 1:1 and 1:5 ratios instead of Fe(II)-BABE-16S-cRNA526. Incubate at 42° for 20 min and cool slowly to room temperature. Initiate hydroxyl radical strand scission by adding 0.5 μl of 250 mM ascorbate and 0.5 μl of 2.5% (v/v) hydrogen peroxide to all the tubes, except the K lane reaction. Incubate for 10 min at room temperature. Stop the reaction by adding 300 μl of cold ethanol, 50 μl of 0.3 M sodium acetate, and quick-freezing in a dry ice–ethanol bath for 10 min. Recover RNAs by ethanol precipitation as described earlier

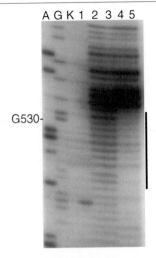

FIG. 3. Strand scission of 16S rRNA by hydroxyl radicals generated from Fe(II)-BABE-16S-cRNA526 hybridized to 16S rRNA. Positions of strand scission were determined by primer extension.[9] A and G, sequencing lanes; K, unmodified rRNA; lane 1, 16S rRNA treated chemically as for the probing reaction, but in the absence of 16S-cRNA526; lane 2, 16S rRNA hybridized with an equimolar amount of Fe(II)-BABE-16S-cRNA526; lane 3, 16S rRNA probed with a fivefold molar excess of Fe(II)-BABE-16S-cRNA526; lane 4, 16S rRNA mock probed with an equimolar amount of unmodified 16S-cRNA526; and lane 5, 16S rRNA mock probed with a fivefold molar excess of unmodified 16S-cRNA526. The main region of strand scission is indicated by the bar.

and resuspend in 100 $\mu$l water. Identify the sites of cleavage in 16S rRNA by primer extension and gel electrophoresis, as described previously.[9]

Strong cleavages are detected at positions G527 and C528, medium-strength cleavages at positions 529 to 532, and weak cleavages at positions 533 to 537 in 16S rRNA (Fig. 3, lanes 2 and 3). Assuming that target nucleotides form part of an A-form RNA helix, the distance between the 5'-terminal phosphate to which Fe(II) is tethered and the 3'-phosphate of the target nucleotide residues can be estimated. Strong cleavages at positions G527 and C528 correspond to distances of 18 and 17.5 Å, respectively; medium-strength cleavages at positions 529 to 532 correspond to distances of 18, 19, 22, and 26 Å, respectively, and weak cleavages at positions 533 to 537 correspond to distances of 30, 34, 37, 40, and 42 Å, respectively. Additionally, strong reverse transcriptase stops are observed at positions 510 to 526 in 16S rRNA in both unmodified (Fig. 3, lanes 4 and 5) and modified (Fig. 3, lanes 2 and 3) reactions. This is due to inhibition of the reverse transcriptase enzyme by 16S-cRNA526 hybridized to position 506 to 526 in 16S rRNA in lanes 2–5 (Fig. 3) during the primer extension reaction.

## 3. Probing rRNA Environment Surrounding 5' End of tRNA$^{Phe}$ in Ribosome

We used 5'-Fe(II)-BABE-modified tRNA bound to the ribosomal A, P, and E sites to map rRNA elements surrounding the 5' end of tRNA.[13] We present a general method for binding tRNAs to the ribosome and carrying out tethered hydroxyl radical probing. Prepare Fe(II)-BABE-modified *E.coli* tRNA$^{Phe}$ as described earlier. Bind unmodified or 5'-Fe(II)-BABE-modified deacylated tRNAs to ribosomes in the absence and presence of poly(U) mRNA by making the following reaction mixtures: (a) Ribosome mixture I, combine 2.0 $\mu$l of 30 pmol/$\mu$l 70S ribosomes, 15 $\mu$l of 5× C$_{80}$M$_{20}$N$_{150}$ buffer, and 58 $\mu$l of water; (b) ribosome mixture II, combine 1.3 $\mu$l of 30 pmol/$\mu$l 70S ribosomes, 4.0 $\mu$l of 10 $\mu$g/$\mu$l poly(U) mRNA, 10 $\mu$l of 5× C$_{80}$M$_{20}$N$_{150}$, and 34.7 $\mu$l of water; (c) control (no tRNA) mixture, combine 5.0 $\mu$l 5× C$_{80}$M$_{20}$N$_{150}$ and 20 $\mu$l of water; (d) 5'-GMP-tRNA$^{Phe}$ mixture, combine 1.0 $\mu$l of 100 pmol/$\mu$l 5'-GMP-tRNA$^{Phe}$, 5.0 $\mu$l 5× C$_{80}$M$_{20}$N$_{150}$, and 19 $\mu$l of water; (e) 5'-GMPS-tRNA$^{Phe}$ mixture, combine 1.0 $\mu$l of 100 pmol/$\mu$l 5'-GMPS-tRNA$^{Phe}$, 5.0 $\mu$l 5× C$_{80}$M$_{20}$N$_{150}$, and 19 $\mu$l of water; and (f) 5'-Fe(II)-BABE-tRNA$^{Phe}$ mixture, combine 1.0 $\mu$l of 100 pmol/$\mu$l 5'-Fe(II)-BABE-tRNA$^{Phe}$, 5.0 $\mu$l 5× C$_{80}$M$_{20}$N$_{150}$, and 19 $\mu$l of water. Incubate all mixtures at 42° for 10 min, cool water bath to 37°, and incubate for an additional 10 min.

*Binding Reaction Mixtures.* Bind tRNAs to ribosomes in the absence of poly(U) mRNA by adding 12.5-$\mu$l aliquots of ribosome mixture I to (i) K lane, 12.5 $\mu$l of control (no tRNA) mixture; (ii) mock lane, 12.5 $\mu$l of control (no tRNA) mixture; (iii) 5'-GMP-tRNA$^{Phe}$ lane, 12.5 $\mu$l of 5'-GMP-tRNA$^{Phe}$ mixture; (iv) 5'-GMPS-tRNA$^{Phe}$ lane, 12.5 $\mu$l of 5'-GMPS-tRNA$^{Phe}$ mixture; and (v) 5'-Fe(II)-BABE-tRNA$^{Phe}$ lane, 12.5 $\mu$l of 5'-Fe(II)-BABE-tRNA$^{Phe}$ mixture.

Bind tRNAs to ribosomes in the presence of poly(U) mRNA by adding 12.5-$\mu$l aliquots of ribosome mixture II to (vi) 5'-GMP-tRNA$^{Phe}$ lane, 12.5 $\mu$l of 5'-GMP-tRNA$^{Phe}$ mixture; (vii) 5'-GMPS-tRNA$^{Phe}$ lane, 12.5 $\mu$l of 5'-GMPS-tRNA$^{Phe}$ mixture; and (viii) 5'-Fe(II)-BABE-tRNA$^{Phe}$ lane, 12.5 $\mu$l of 5'-Fe(II)-BABE-tRNA$^{Phe}$ mixture.

Incubate the binding reaction mixtures at 37° for 30 min on ice for 20 min. Initiate hydroxyl radical strand scission by the addition of ascorbate and hydrogen peroxide to all the reaction mixtures, except the K lane reaction, as described earlier. Incubate the tubes at room temperature for 10 min. Add 300 $\mu$l of 100% ethanol and 50 $\mu$l of 0.3 $M$ sodium acetate and quick freeze in a dry ice–ethanol bath. Extract rRNAs from all samples and identify sites of strand scission by primer extension and gel electrophoresis, all as described previously.[9]

## 4. Probing rRNA in Ribosome Using Fe(II) Tethered to Anticodon Stem–Loop Analogs of tRNA

Transcribe anticodon stem–loop analogs of tRNA using the corresponding oligodeoxynucleotide templates.[14] Isolate ASLs on denaturing polyacrylamide gels and purify as described earlier. Confirm the ASL sequences by complete ribonuclease $T_1$ digestion and analysis by paper electrophoresis. For sequence analysis, transcribe ASLs in the presence of 2.0 $\mu$l of [$\alpha$-$^{32}$P]UTP (3000 Ci/mmol; 10 mCi/ml) in 500-$\mu$l scale transcription reactions. In an Eppendorf tube, combine 5.0 $\mu$l of transcript (10,000 cpm/$\mu$l), 1.0 $\mu$l of 1.0 $\mu$g/$\mu$l of ribonuclease $T_1$, 2.0 $\mu$l of 10 $\mu$g/$\mu$l carrier tRNA, 1.0 $\mu$l of 10$\times$ $T_1$ buffer, and 1.0 $\mu$l of water. Incubate at 37° for 3 hr. Spot 5.0 $\mu$l of the sample on Whatman DE81 paper and electrophorese at 1000 V for 6 hr in 7% formic acid (v/v), pH 1.5. Process the filter paper as described earlier. Verify ASL sequences based on the mobility of the labeled fragments (Fig. 4).

Renature ASLs in 10 m$M$ Tris (pH 7.5), 1 m$M$ EDTA by heating to 90° for 1 min and placing on ice for 10 min. Derivatize the ASLs at their 5′ ends with Fe(II)-BABE and purify essentially as described earlier.

Bind 5′-Fe(II)-BABE-ASLs4–9 to ribosomes in the absence and presence of poly(U) mRNA by following the general method described earlier for binding tRNAs to the ribosome. Initiate hydroxyl radical strand scission by adding ascorbate and hydrogen peroxide to the binding reaction mixtures and process the samples as described earlier and in Ref. 9.

Figure 5 is an example of directed hydroxal radical probing using 5′-Fe(II)-BABE-ASLs4–9 bound to the P site of 70S ribosomes in a poly(U)-dependent manner. The sites of cleavage in 23S rRNA around position C1925 are shown. The cleavages are all mRNA dependent, and the pattern of cleavage is dependent on ASL stem length. This dependence of cleavage intensity on ASL stem length was used to constrain the rRNA targets in three-dimensional space as described later and in Ref. 14.

The cleavage intensities of the target nucleotides were quantified by loading the primer extension samples in every other well on a second high-resolution sequencing gel, run in parallel. Leaving a blank lane between each sample lane facilitates the subsequent quantification of the bands. The gels were quantified by PhosphorImager analysis in the following manner: First, for normalizing any difference in sample loading between lanes, at least two sets of K bands, one above and the other below the band of interest, were also quantified. The cleavage intensity of the target band was then normalized using the K bands as controls for sample loading differences between lanes. Second, to compare the cleavage intensity of a target nucleotide from two or more independent experiments, 10 sequenc-

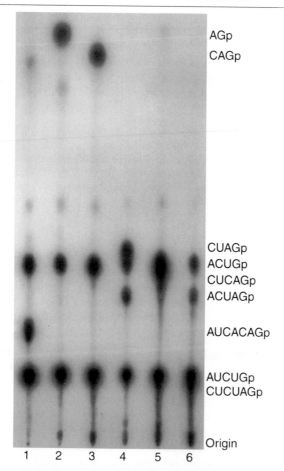

FIG. 4. Sequence characterization of ASL10–15 transcripts. Lanes 1–6 correspond to ASLs10–15. Samples were digested to completion using ribonuclease $T_1$ and separated by high-voltage paper electrophoresis on Whatman DEAE paper in 7% formic acid. Predicted sequences of the fragments are as indicated.

ing bands close to the target nucleotide were quantified and the average counts determined. Then, the cleavage intensities of the target nucleotide from independent experiments were normalized using the average counts of the sequencing bands. Finally, these normalized cleavage intensities were used to calculate the mean cleavage intensity and standard deviation for each ASL probing position.

We used two independent approaches to calibrate target distances. First, we used results from our RNA–RNA duplex study (described earlier) to

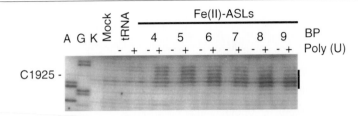

FIG. 5. Strand scission of 23S rRNA by hydroxyl radicals generated from Fe(II)-BABE-modified ASLs4–9 bound to the P site of *E. coli* 70S ribosomes analyzed by primer extension.[9] A and G, sequencing lanes; K, unmodified rRNA; mock, ribosomes treated chemically as for the probing reaction, but in the absence of ASLs; tRNA, ribosome complexes containing unmodified *E. coli* tRNAPhe bound to the P site. Lanes designated Fe(II)–ASLs are ribosomes probed with ASLs of increasing length, as indicated. A minus or plus sign indicates the absence or presence of poly(U) mRNA. Positions of strand scission are indicated by the bar.

estimate the target distances for strong, medium, and weak cleavages. Results indicate that the maximum target distance for strong cleavage is ~18 Å, medium-strength cleavage is ~26 Å, and weak cleavage is ~42 Å. In a second approach, we made use of the internal calibration inherent in ASL cleavage data. Ranges for strong, medium, and weak cleavages can be estimated by determining two widely separated ASL-probing positions that result in similar cleavage strength. Then, the lower limit for the distance range that results in similar strength cleavage is equal to half the distance between the two widely separated ASL-tethering sites plus or minus the length of the tether arm. For example, target 1924 is cleaved strongly by the two widely separated probing positions ASL6 and ASL10, which correspond to a distance of 20 Å. Then, the location of target 1924 can be estimated as half the distance of the two widely separated probing positions (10 Å) plus or minus the length of the tether arm (12 Å). This gives a distance range of 0 to 22 Å between the tethering position and the target position for strong cleavages. In a similar manner, we estimate medium-strength cleavages to indicate a range of 12 to 36 Å and weak cleavages a range of 20 to 44 Å. The results of the aforementioned DNA–RNA and RNA–RNA duplex calibration experiments agree well with these distance ranges and provide independent verification of the distance ranges obtained by internal calibration. Additionally, these values are consistent with those reported in earlier studies.[2,15–17]

[16] H. E. Moser and P. B. Dervan, *Science* **238**, 645 (1987).
[17] T. D. Tullius, *Free Radic. Res. Commun.* **2**, 521 (1991).

Discussion

Tethered Fe(II)-EDTA probing has been exploited by a number of laboratories to study ligand–nucleic acid interaction and has thus been tested extensively under a number of different conditions. Some general precautions should be kept in mind. Use water that has been doubly distilled and deionized and highest purity reagents that are low in heavy metal contaminants; bake glassware to destroy RNases; use sterile, disposable plasticware whenever possible; use only freshly made Fe(II) and ascorbate solutions; combine Fe(II) and BABE before derivatization of GMPS-RNA; after derivatization, add EDTA to chelate any free Fe(II); do not use leftover Fe(II)-BABE mixture for future modifications; and Fe(II)-BABE-modified RNA stored at $-20°$ for more than 2–3 weeks may undergo spontaneous Fenton reaction at a slow rate, destroying the tether and releasing free Fe(II). Additionally, the presence of free-radical scavengers such as ethanol, glycerol, and sucrose in the reaction mixture can quench the cleavage reaction.

We used the DNA–RNA duplex (Fig. 2) as a model system to test different probing conditions. Results indicate that the cleavage efficiency is relatively insensitive to buffer composition, pH, and concentration of monovalent ions. A significantly reduced level of cutting was observed at $0°$ compared to 23 and $37°$.

As described in the Results section, cleavage intensity can be used to estimate target distance, albeit at low resolution. The following two examples further illustrate the potential of tethered Fe(II)-EDTA probing in obtaining useful distance constraints. Wang and Cech[1] probed the tertiary structure of *Tetrahymena* ribozyme using Fe(II) tethered to the substrate GMP. They compared the target cleavage intensity versus distance using a three-dimensional model of the ribozyme.[18] They classified their cleavage intensities as $+++$, $++$ (cleavage intensity is 50% of $+++$), and $+$ (cleavage intensity is 25% of $+++$). They detected $+++$ cleavages at 7.0 Å distance from the 5'-phosphate of GMP. Taking into account the 10-Å length of the tether, the effective distance range for strong cleavages is about 0–17 Å. Similarly, $++$ and $+$ cleavages lie within a distance range of 11 to 37 Å and 14 to 42 Å, respectively. Han and Dervan[2] incorporated Fe(II)-EDTA at position 47 in yeast tRNA[Phe] and correlated the auto-cleavage pattern with the X-ray crystal structure of tRNA. They observed strong cleavages at distances that are less than 10 Å from the probe. Medium-strength cleavages were detected within a distance range of 11 to 24 Å and weak cleavages at distances greater than 40 Å. These distances agree

[18] F. Michel and E. Westhof, *J. Mol. Biol.* **216**, 585 (1990).

well with the distance ranges that we have estimated using the ASL internal calibration method.

Chemical footprinting and photochemical cross-linking are two powerful methods that have been used successfully to study macromolecular interactions; however, they have their limitations. Footprinting cannot distinguish between direct effects due to ligand binding versus indirect effects due to conformational rearrangements caused by ligand binding. The efficiency of cross-linking is often low and depends critically on the reactivity and stereochemistry of the target. Tethered hydroxyl radical probing offers important advantages over solvent-based chemical probing and cross-linking. Tethered Fe(II)-EDTA cleavage sites are restricted to the proximity of the Fe(II) ion, which overcomes the difficulty of distinguishing between direct and indirect effects that is encountered with chemical footprinting. The advantage over cross-linking is that since all accessible riboses are reactive, all accessible targets within range are probed comprehensively and simultaneously in a single experiment. Furthermore, cleavage intensities provide important distance information that can be used to constrain the target. More elaborate strategies, such as the RNA helical ruler approach (Fig. 5), provide three-dimensional information that can be used to triangulate the positions of targets in three-dimensional space.

# Section III

## *In Vitro* Affinity Selection Methodologies

# [14] *In Vitro* Selection of RNA Aptamers

*By* KRISTIN A. MARSHALL and ANDREW D. ELLINGTON

## Introduction

*In vitro* selection has proven to be an extremely useful technique for defining protein-binding sites and for deriving novel binding species (aptamers) from random sequence populations.[1,2] The *in vitro* selection of aptamers is governed by the same principles as natural selection. A population of oligonucleotides with random sequence is screened for fitness, and survivors are culled and amplified (Fig. 1). The primary differences between *in vitro* and natural selection are the degree of heritable variation and the focus of selection on individual molecules rather than on whole organisms. The number of variants that are screened per generation in most natural populations, with the exception of bacteria and viruses, is typically no more than 1000. Furthermore, mutations in a natural population are single base substitutions or deletions. In striking contrast, nearly $10^{15}$ variants can be screened per generation in an artifically generated population of nucleic acids, and sequences can be either partially or completely randomized as desired. Indeed, researchers can exert control over virtually every aspect of the *in vitro* selection process: from pool size, design, and chemistry to selection targets, stringency, and rounds.

## Pool Design

Most random sequence pools will contain a region of random sequence flanked by constant regions required for enzymatic manipulations. The four factors that are most important in pool design are the type of randomization, the length of the random sequence region, the chemistry of the pool, and the utility of the constant regions.

Three types of randomization can be employed: partial, segmental, and complete. Partial randomization involves "doping" mutations into a constant sequence at a fixed rate and can aid in determining which residues contribute to structure and function.[3,4] This method produces arrays of point mutations similar to those seen in nature, albeit at much higher

[1] M. Famulok and A. Jenne, *Curr. Opin. Chem. Biol.* **2,** 320 (1998).
[2] B. E. Eaton, *Curr. Opin. Chem. Biol.* **1,** 10 (1997).
[3] D. P. Bartel, M. L. Zapp, M. R. Green, and J. W. Szostak, *Cell* **67,** 529 (1991).
[4] E. T. Peterson, J. Blank, M. Sprinzl, and O. C. Uhlenbeck, *EMBO J.* **12,** 2959 (1993).

0076-6879/00 $30.00

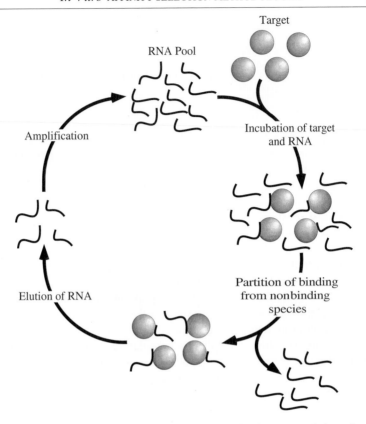

Fig. 1. General selection scheme. This schematic depicts how a population of random sequence RNA molecules is screened for binding. Pool RNA is mixed with target and some members of the pool bind to the target. Target:RNA complexes are then partitioned from unbound RNA molecules. Bound RNA molecules are eluted from the target and amplified. Multiple cycles of selection and amplification result in RNA aptamers that have high affinity and specificity toward the target.

frequencies. For example, constant sequences are frequently randomized at rates of 5 to 30% per position and can contain all single to hextuple sequence substitutions.[5] Conversely, nucleic acid segments of from 30 to 200 residues can be completely randomized, allowing a wide swath of sequence space to be examined for novel nucleic acid structures and functions. The segmental strategy is a compromise between partial and complete randomization and involves the complete randomization of a short segment

[5] R. Green, A. D. Ellington, D. P. Bartel, and J. W. Szostak, *Methods* **2,** 75 (1991).

of an extant functional nucleic acid. This scheme has some of the advantages of both partial and complete randomization in that it can reveal what role wild-type residues play but also identify novel sequence or structural motifs that significantly alter binding or catalytic function.

The length of the region to be randomized depends in part on what function is being explored or exploited.[6] If a target has intrinsic nucleic acid affinity, be it exquisite or nonspecific, then a partial or segmental random sequence pool based on a natural ligand or a short, completely random pool will likely yield aptamers. As a rule, completely random sequence pools used for the selection of aptamers contain from 30 to 60 randomized nucleotides.[7] Aptamers selected from such short pools can be aligned readily, and sequence and structural motifs can be identified easily. However, if a target is not known to bind nucleic acids, then a longer ($>60$ residue), completely random sequence pool may provide better opportunities for identifying aptamers, although short sequence motifs often still predominate following selection. In addition, longer random sequence pools may allow for greater structural or contextual complexity. For example, tRNA molecules are 76 nucleotides in length: it might prove more difficult to select tRNA mimics from a random sequence population containing 30 randomized residues than from a pool spanning 80 randomized residues, and any tRNA mimics selected from the shorter pool might bear little resemblance to a natural tRNA. Irrespective of whether shorter or longer random sequence libraries are used, the ultimate complexity of the population is limited by DNA synthesis chemistry to a total of $10^{13}$–$16^{16}$ different sequences. Therefore, most completely randomized libraries will contain all possible sequences from 22 to 27 residues in length ($4^{22} \approx 10^{13}$; $4^{27} \approx 10^{16}$). A frequent misconception is that longer pools will be less representative than shorter pools: for example, a 22-mer pool might be considered "complete" whereas a 50-mer pool would be decidedly "incomplete." However, longer pools will subsume the shorter pools (i.e., the 50-mer pool also has all possible 22-mers) and will, in addition, contain a diverse and perhaps critical representation of longer sequence motifs.

Selections are frequently carried out with RNA pools, but single-stranded DNA pools can also yield aptamers. In addition, *in vitro* RNA selections need not be restricted to the four naturally occurring ribonucleotides. The T7 RNA polymerase can incorporate many modified ribonucleotides such as 2′-fluoro- and 2′-amino-2′deoxynucleoside 5′-triphosphates[8]

[6] P. C. Sabeti, P. J. Unrau, and D. P. Bartel, *Chem. Biol.* **4,** 767 (1997).
[7] L. Gold, B. Polisky, O. Uhlenbeck, and M. Yarus, *Annu. Rev. Biochem.* **64,** 763 (1995).
[8] H. Aurup, D. M. Williams, and F. Eckstein, *Biochemistry,* **31,** 9636 (1992).

and phosphorothioate nucleotides.[9] Mutant T7 polymerases have also been isolated that will better incorporate modified nucleotides.[10] The chemistry of a nucleic acid pool can strongly bias the course and outcome of a selection experiment. In roughly parallel experiments, aptamers selected to bind the cytokine vascular endothelial growth factor (VegF) had very different sequence and structural motifs depending on whether they were culled from a RNA pool or a pool containing 2' amino pyrimidines.[11,12] Other examples are reviewed in Osborne and Ellington.[13]

Constant sequences should be designed to ensure vigorous amplification of all sequences in the population. The primers should anneal strongly to the template and should not form extensive secondary structures or "primer dimers." Primers at least 20 nucleotides long will have melting points that are convenient for PCR and can be synthesized in high yield. Various programs have been written to help in the design of primers (such as Amplify (University of Wisconsin Genetics Department) and Oligo (Lifescience Software Resource, Long Lake, MN). If a RNA pool will be transcribed from a double-stranded DNA pool, the 5' primer should contain a T7 RNA polymerase (RNAP) promoter. Multiple promoter sequences are recognized by T7 RNAP.[14] 'TAA TAC GAC TCA CTA TA' followed by 'GGG' at positions +1 to +3 allows for near optimal transcription initiation.[14] If the sequence of a natural ligand restricts the incorporation of multiple guanosine residues at the 5' termini, even a single guanosine should support moderate (albeit reduced) *in vitro* transcription.

RNA Pool Synthesis

The original pool is synthesized chemically as a single-stranded DNA oligonucleotide. The oligonucleotide is amplified into a double-stranded DNA library by polymerase chain reaction (PCR). The double-stranded PCR product can then be transcribed into a single-stranded RNA pool or modified RNA pool.

[9] T. Ueda, H. Tohda, N. Chikazumi, F. Eckstein, and K. Watanabe, *Nucleic Acids Res.* **19**, 547 (1991).
[10] K. Raines and P. A. Gottlieb, *RNA* **4**, 340 (1998).
[11] L. S. Green, D. Jellinek, C. Bell, L. A. Beebe, B. D. Feistner, S. C. Gill, F. M. Jucker, and N. Janjic, *Chem. Biol.* **2**, 683 (1995).
[12] D. Jellinek, L. S. Green, C. Bell, and N. Janjic, *Biochemistry* **33**, 10450 (1994).
[13] S. E. Osborne and A. D. Ellington, *Chem. Rev.* **97**, 349 (1997).
[14] J. F. Milligan, D. R. Groebe, G. W. Witherell, and O. C. Uhlenbeck, *Nucleic Acids Res.* **15**, 8783 (1987).

*Synthesis and Purification of Initial DNA Pool*

The chemical synthesis of a random sequence oligonucleotide is an exacting task. Standard procedures for the chemical synthesis of oligonucleotides can be used for the synthesis of most pools. However, the stepwise yield must be high to ensure as complete a representation of sequences in the final population as possible, and care must be taken to avoid compositional biases. While the random region of a pool can be synthesized simply pumping all four individual phosphoramidites simultaneously to the synthesis column, some phosphoramidites (dG and dT) incorporate more readily than others and can skew the composition of the pool. For this reason, it is better to synthesize both partially or completely random regions from a premade mixture of bases with a ratio of dA : dG : dC : dT, 1.5 : 1.15 : 1.25 : 1 (ABI users manual[15]).

Precautions should also be taken to avoid the cross-contamination of a pool and the primers used in the amplification of the pool, lest a small number of sequences be preferentially amplified. We have found that when DNA pools are synthesized immediately prior to their cognate primers, the primers will contain low but significant levels of amplicons. Therefore, pools and primers should be synthesized on different ports on the synthesizer or the machine should be flushed thoroughly with acetonitrile between syntheses. Following synthesis, the single-stranded (ss)DNA pool should be gel purified to remove products from failed syntheses. Gel purification of the primers is not necessary but is highly recommended.

*Assessing Complexity of Initial DNA Pool*

PCR amplification of the ssDNA pool will yield multiple copies of a double-stranded (ds)DNA pool, allowing the same or similar selection experiments to be repeated several times. This step also allows a T7 RNAP promoter to be appended to the pool via the 5′ primer. Because depurination and other chemical lesions that occur during the chemical synthesis will ultimately limit pool size, the complexity of the initial pool should first be determined by following the enzymatic extension of a limited amount of radiolabeled primer. For example, frequently only 10–30% of a longer pool can be extended to full length. The yield can be improved if a reverse transcriptase such as murine Moloney leukemia virus reverse transcriptase (MMLV-RT) is used in the initial extension reaction in place of a standard DNA polymerase.

[15] G. Zon, K. A. Gallo, C. J. Samson, K. L. Shao, M. F. Summers, and R. A. Byrd, *Nucleic Acids Res.* **13**, 8181 (1985).

K30 Pool
Primers are in bold

T7 promotor +1
41.K 5'**GATAATACGACTCACTATA**GGGCCGTTCGAACACGAGCATG 3'
   5'GGGCCGTTCGAACACGAGCATG– N30 –GGACAGTACTCAGGTCATCCTAGG 3'
                          3'**CCTAGGATGACCTGAGTACTGTCC** 5', 24.K

N30 Pool
Primers are in bold

T7 promotor +1
41.30 5'**GATAATACGACTCACTATA**GGGAATGGATCCACATCTACGA 3'
   5'GGGAATGGATCCACATCTACGAATTC– N30 –TTCACTGCAGACTTGACGAAGCTT 3'
                               3'**AAGCTTCGTCAAGTCTGCAGTGAA** 5', 24.30

N71 Pool
Primers are in bold

T7 promotor +1
42.71 5'**GGTAATACGACTCACTATA**GGGAGATACCAGCTTATTCAATT 3'
   5'ATACCAGCTTATTCAATT– N71 –AGATAGTAAGTGCAATCT 3'
                      3'**AGTTGGACACTTACATCT** 5', 18.71

Rex Pool
Bases in bold are doped at a rate of 30%
per position relative to wild type

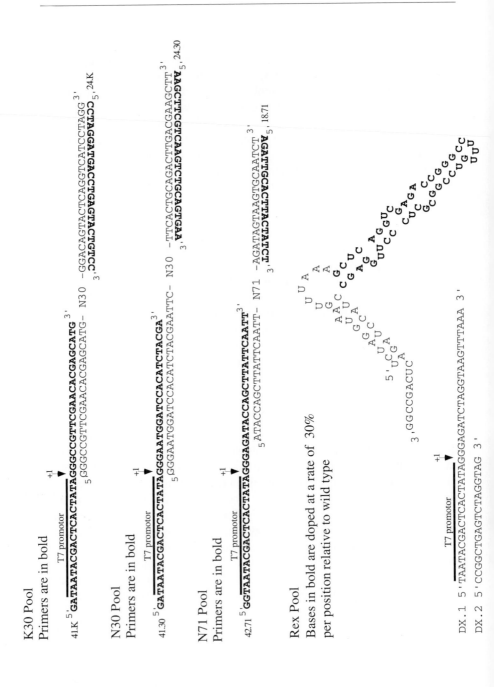

DX.1   5'TAATACGACTCACTATAGGGAGATCTAGGTAAGTTTAAA 3'
DX.2   5'CCGGCTGAGTCTAGGTAG 3'
       T7 promotor +1

*Example: K30 Pool Extension Procedure*

The K30 pool (Fig. 2) is synthesized on an ABI synthesizer (Perkin-Elmer, Foster City, CA) on a 1-$\mu$mol scale. The K30 pool is gel purified on a 6% denaturing polyacrylamide gel.[16] A small amount of the pool is assayed in a trial extension reaction.

*Solutions*

Buffer: 50 m$M$ Tris–HCl (pH 8.3), 75 m$M$ KCl, 3 m$M$ MgCl$_2$, 10 m$M$ dithiothreitol (DTT)
Primer: 2$\mu M$ 24.K (5' CCT AGG ATG ACC TGA GTA CTG TCC), 5' labeled with $^{32}$P
Nucleotides: 200 $\mu M$ each NTP
Template: Gel-purified K30 pool (1 $\mu$g; 4 $\mu M$ final concentration)
Polymerase: MMLV-RT (Gibco-BRL, Rockville, MD; 40 U; 200 U/$\mu$l)

*Procedure*

A 9.8-$\mu$l reaction (minus enzyme) is heated in an MJ Research thermocycler (Watertown, MA) to 70° for 3 min to denature the ssDNA. The primer is annealed to the template by cooling the reaction mixture to 25° at a rate of 5°/min. MMLV-RT (0.2 $\mu$l) is added, and the reaction is incubated at 42° for 1 hr. The reaction is terminated by the addition of stop dye [95% (v/v) formamide, 20 m$M$ EDTA, 0.5% (w/v) xylene cyanol, and 0.5% (w/v) bromphenol blue], heated to 95°, and run on a 10% denaturing polyacrylamide gel. An aliquot of the radiolabeled primer and appropriate radiolabeled size standards are also loaded as controls. The gel is dried, and counts are quantitated using a Molecular Dynamics PhosphorImager (Sunnyvale, CA). The amount of primer that is extended is calculated by dividing the number of counts in or near (+ or − about five residues) the full-length product either by the total counts present in the reaction lane or by the counts present in the unextended, radiolabeled primer in an

---

[16] A. D. Ellington and R. Green, *in* "Current Protocols in Molecular Biology" (F. M. Ausubel, R. Brent, R. E. Kingston, D. D. Moore, J. G. Seidman, J. A. Smith, and K. Struhl, eds.), p. 2.11.12. Wiley, New York, 1989.

---

FIG. 2. Random sequence nucleic acid pools. Pools used in the experiments described in the text are shown along with the primers needed for their amplification. The T7 promoter sequence (TAATACGACTCACTATA) is added with the 5' primer. In all cases the strand corresponding to the RNA transcript is shown. The Rex pool was based on deletion analysis of the Rex-binding element (XBE).[39]

adjacent lane. Approximately 28% of the primer could be extended to full-length or near full-length.

### Large-Scale Amplification of Initial DNA Pool

The chemically synthesized oligonucleotide pool should be amplified enzymatically prior to initiating a selection experiment in order to eliminate damaged templates and provide multiple copies of the original "genome." Optimal amplification conditions can be determined prior to large-scale amplification by carrying out trial 100-$\mu$l reactions. Synthesis of the first (minus) strand should be done under the same conditions as were used during the trial extension reaction (described earlier) to ensure that the final pool has a complexity similar to the calculated complexity. The K30 pool is amplified in multiple, parallel 200-$\mu$l reactions in 96-well microtiter plates. If a high throughput thermocycler is not available, the pool can also be amplified in 5 to 10-ml batches using water baths[17] or even a hotplate or microwave oven.

### Example: K30 Pool Amplification Procedure

The full-scale extension reaction for the K30 pool contains 189 $\mu$g in 943 $\mu$l (200 ng of ssDNA/$\mu$l). The amplification reaction is carried out in a total volume of 83.3 ml.

#### Extension Solutions

Buffer: 50 m$M$ Tris–HCl (pH 8.3), 75 m$M$ KCl, 3 m$M$ MgCl$_2$, 10 m$M$ DTT
Primer: 2 $\mu M$ 24.K and 2 $\mu M$ 41.K (5' GAT ATT ACG ACT CAC TAT AGG GCC GTT CGA ACA CGA GCA TG)
Nucleotides: 200 $\mu M$ each dNTP
Template: Gel-purified K30 pool (189 $\mu$g; 8 $\mu M$ final concentration)
Polymerase: MMLV-RT (Gibco-BRL; 3800 U; 200 U/$\mu$l)

#### Extension Procedure

The 924-$\mu$l reaction (minus enzyme) is heated for 3 min at 70° to denature the ssDNA and then cooled to 25° at a rate of 5° every min. MMLV-RT (19 $\mu$l) is added, and the reaction is incubated at 42° for 1 hr. The reaction is then added to the PCR mixture.

---

[17] R. C. Conrad, L. Giver, Y. Tian, and A. D. Ellington, *Methods Enzymol.* **267,** 336 (1996).

*PCR Mixture*

PCR buffer: 10 m$M$ Tris–HCl, pH 8.3, 50 m$M$ KCl; 1.5 m$M$ MgCl$_2$, 0.1% Triton X-100, 0.005% gelatin
Nucleotides: 200 $\mu M$ each dNTP
Primers: 4 $\mu M$ of 24.K and 41.K (333.2 nmol total)
*Taq* polymerase (PGC, Frederick, MD; 500 units; 6 U/$\mu$l)

*PCR Procedure*

The extension reaction is added to the PCR mixture and aliquoted into each of 384 wells in four 96-well microtiter plates. The reactions are thermocycled 14 times under the following regime: 94° for 30 sec, 50° for 30 sec, and 72° for 1 min. These conditions had previously been shown to yield optimal amplification of a small-scale extension reaction. After the 14th cycle, the reaction is incubated at 72° for 5 min. The reactions are pooled and precipitated in the presence of 0.2 $M$ NaCl and 2.5 volumes of ethanol. The precipitant is collected by centrifugation, washed with 70% ethanol, and resuspended in water (1.5 ml). The use of water for resuspension simplifies the introduction of additional buffers. However, very short double-stranded DNA molecules (<50 residues in length) can potentially be denatured by resuspension in water, potentially making them unsuitable for subsequent procedures, such as *in vitro* transcription. Further, double-stranded DNA molecules suspended should not be heated in the absence of buffer.

*Large-Scale Transcription*

The double-stranded DNA templates are typically suitable for transcription into RNA without further purification. Like the initial pool amplification, the transcription reaction is typically quite large in order to accommodate the large number of DNA templates. To best retain the total complexity of the pool, up to or (optimally) greater than one genome equivalent of the dsDNA should be used to create the RNA pool. For example, for the K30 pool, at least 111 $\mu$g of double-stranded DNA (2 × 198 $\mu$g × 0.28, the fraction of extended templates) should be used. If less complexity is required, a fraction of the pool can be used for transcription. For example, in the following experiments, only ca. $10^{14}$ different sequences were required, and only ca. 1/10 of the total pool complexity was plumbed (111 $\mu$g/62,700 g/mole × 6 × $10^{23}$ molecules/mole = total possible complexity of $1 \times 10^{15}$). It should be noted that the calculated complexity of the initial pool, the calculated complexity of the template population used for transcription, and the calculated complexity of the transcribed RNA pool are all gross approximations that do not take into account the statistical distribution of individual templates. However, these approximations are, for the most part, as useful as more accurate calculations, as other factors may further skew representation in ways that cannot be identified readily (e.g., it is likely that different sequences will amplify to different extents in a PCR reaction, even if they were initially extended by MMLV-RT, and that different templates will be transcribed to different extents).

*Example: K30 Transcription*

The K30 pool (10 $\mu$g; $10^{14}$ variants) is transcribed in a 250-$\mu$l reaction. In order to optimize RNA production, a high-yield, commercial transcription kit was utilized (AmpliScribe from Epicentre, Madison, WI). Based on the amount of nucleic acid recovered, commercial transcription kits are typically more economical than buying individual reagents.

*Solutions*

Buffer: 1× AmpliScribe transcription buffer, 10 m$M$ DTT
Nucleotides: 7.5 m$M$ each NTP
Template: 10 $\mu$g dsDNA ($10^{14}$ molecules) from the large-scale amplification in 50 $\mu$l water
Polymerase: 25 $\mu$l AmpliScribe T7 polymerase solution

*Procedure*

Transcription reactions are set up according to the manufacturer's instructions. The transcription reaction is incubated at 37° for 16 hr. Five units of RNase-free DNase is added and the reaction is returned to 37° for 1 hr. The transcription is then purified on a 10% denaturing polyacrylamide gel to separate full-length RNAs from abortive or incomplete transcripts. The band is excised, and RNA is eluted by incubation in water (9 ml) at 37° for 16 hr. The eluted RNA is ethanol precipitated, washed with 70% ethanol, and resuspended in 150 $\mu$l water.

Selection

*In vitro* selection can be used to target a wide variety of ligands, from small molecules such as inorganic ions[18] or organic cellular metabolites[19–21]

[18] J. Ciesiolka, J. Gorski, and M. Yarus, *RNA* **1,** 538 (1995).
[19] M. Sassanfar and J. W. Szostak, *Nature* **364,** 550 (1993).
[20] A. Geiger, P. Burgstaller, H. von der Eltz, A. Roeder, and M. Famulok, *Nucleic Acids Res.* **24,** 1029 (1996).
[21] P. Burgstaller and M. Famulok, *Angew. Chem. Int. Ed. Engl.* **33,** 1084 (1994).

up through proteins[22–24] of various sizes and even supramolecular complexes such as viruses.[25] The wide range of target sizes and chemistries makes it difficult to advocate fixed rules for how to carry out selection experiments. However, we can suggest some general guidelines. It should be recognized that these guidelines are in fact general and that different targets may require modifications to the example protocols given.

One of the most crucial aspects of a selection experiment is the partition of binding and nonbinding RNA species, which in turn generally requires the immobilization or separation of the target or target:RNA complex. There are a variety of ways in which targets can be selectively immobilized to generate affinity matrices or surfaces. For example, small molecule targets are generally derivatized to activated resins. Protein targets can also be immobilized on resins or deposited nonspecifically on a microtiter plate. In addition, though, protein:RNA complexes can be sieved from free RNA by selective capture on modified cellulose filters,[26] immunoprecipitation,[27] or size separation on native polyacrylamide gels (gel shift[28]). These latter methods are frequently preferred for selection experiments that target proteins over the synthesis of an affinity resin because they require much less sample and, because less target is used, tend to select for higher affinity interactions.

Early rounds of the selection may require less stringent conditions and longer incubation periods to maximize recovery of the relatively few functional molecules present in the initial pool. Stringency can be increased in later rounds as the population winnows to predominantly functional molecules. The stringency of a selection can be affected by numerous variables, including buffer conditions, the time and volume of the binding reaction, and the presence of specific or nonspecific competitors.

Buffer conditions are chosen first and foremost to stabilize or promote the active conformation of a target molecule. Beyond this requirement, high concentrations of monovalent cations tend to quell nonspecific binding

[22] M. Thomas, S. Chedin, C. Carles, M. Riva, M. Famulok, and A. Sentenac, *J. Biol. Chem.* **272,** 27980 (1997).
[23] S. W. Gal, S. Amontov, P. T. Urvil, D. Vishnuvardhan, F. Nishikawa, P. K. Kumar, and S. Nishikawa, *Eur. J. Biochem.* **252,** 553 (1998).
[24] P. K. Kumar, K. Machida, P. T. Urvil, N. Kakiuchi, D. Vishnuvardhan, K. Shimotohno, K. Taira, and S. Nishikawa, *Virology* **237,** 270 (1997).
[25] W. Pan, R. C. Craven, Q. Qiu, C. B. Wilson, J. W. Wills, S. Golovine, and J. F. Wang, *Proc. Natl. Acad. Sci. U.S.A.* **92,** 11509 (1995).
[26] C. Tuerk and L. Gold, *Science* **249,** 505 (1990).
[27] D. E. Tsai, D. S. Harper, and J. D. Keene, *Nucleic Acids Res.* **19,** 4931 (1991).
[28] T. K. Blackwell and H. Weintraub, *Science* **250,** 1104 (1990).

of RNA species to targets.[29] Depending on the target and its ionic requirements, monovalent ion concentrations of from 50 to 1000 m$M$ are typically used. Divalent cations such as $Mg^{2+}$ or $Ca^{2+}$ are important for the formation of nucleic acid secondary and tertiary structures, but can also promote nonspecific binding. Thus, depending on the target and its ionic requirements, divalent ion concentrations from 1 to 10 m$M$ are typically used.

A number of factors besides buffer conditions directly affect how RNA molecules will partition between solution and target. Most of these factors derive directly from attempts to influence the equilibrium of the binding reaction. First, unless a researcher is knowingly attempting to select for fast on rates, the binding reaction should be allowed to come to equilibrium. For almost all aptamers, this will require mere seconds. However, because the association between aptamers and their targets may be dependent on an induced fit,[30] it is wise to allow the binding reaction to equilibrate from 5 to 60 min. While the on rates of aptamers may not generally be a selectable parameter, the off rates are. Aptamers with different off rates can be partitioned most easily by allowing the binding reaction to come to equilibrium and then increasing the available volume prior to separation. Those species that fall off of their targets first will be unlikely to find their way back to another target. This same principle is of course used when aptamers are selected by coimmobilization with a target, and a column or filter is washed with large volumes of solution. The final way to affect the distribution of selected aptamers is to alter target concentration. If a large amount of target is available for binding, then more aptamers with lower association constants will be retained than if a smaller amount of target is available for binding. Thus, increasing the pool:target ratio gradually will also increase the stringency of the selection. We usually start a selection with a low ratio (1:1 to 10:1, RNA:protein) to allow all binding species to be amplified before making the selection more stringent (100:1). Irvine *et al.*[31] have generated detailed formulas for determining the ratios to use in a selection. Although researchers may uniformly attempt to select for extremely high-affinity species, not all aptamers recognize their targets with nanomolar or subnanomolar affinities; for example, even the best anti-ATP aptamers tend to bind their targets with approximately micromolar affinities,[19] whereas antiamino acid aptamers tend to bind their cognate targets with affinities in the millimolar range.[32] Binding and elution condi-

[29] J. Binkley, P. Allen, D. M. Brown, L. Green, C. Tuerk, and L. Gold, *Nucleic Acids Res.* **23,** 3198 (1995).

[30] E. Westhof and D. J. Patel, *Curr. Opin. Struct. Biol.* **7,** 305 (1997).

[31] D. Irvine, C. Tuerk, and L. Gold, *J. Mol. Biol.* **222,** 739 (1991).

[32] I. Majerfeld and M. Yarus, *Nature Struct. Biol.* **1,** 287 (1994).

tions should therefore be adjusted to correspond to rational, not desired, equilibrium constants for target : aptamer complexes.

The addition of competitors to the initial binding reaction or during the elution of bound species can also promote the selection of high-affinity aptamers. During the course of a selection, the amount of competitor added can be increased progressively in order to increase the stringency of the selection. Because many targets or surfaces display nonspecific adherent properties, nucleic acids such as tRNA can be included to prevent the selection of nonspecific binding species. In some cases, a nucleic acid or other ligand that naturally binds a target can be used as a specific competitor.[12,33]

Altering the stringency of a selection not only affects the number and affinities of target-specific aptamers but also the prevalence of matrix-binding species. For example, it is relatively easy to identify nucleic acids that can bind to modified cellulose filters in a target-independent manner.[34] In a selection that employs filter binding as a separation technique, if the stringency of the selection is too high at the start or is increased too quickly in subsequent rounds, then filter-binding species can overrun the population to the detriment of less populous target-binding aptamers. To avoid this problem, a negative, selection step is frequently employed. For example, the RNA pool can be passed over a modified cellulose filter in the absence of a target. This procedure can be carried out more than once to ensure that all matrix-binding species have been removed from the pool. However, if matrix-binding species still arise or predominate in a population, even in the presence of a negative preselection step, then it is best to restart the selection using less stringent selection conditions. In some cases, a population of matrix-binding species may be intermixed with a population of target-specific aptamers; for example, 10% of the population may bind to a modified cellulose filter in the absence of target, whereas 30% of the population may bind to a modified cellulose filter in the presence of target. In this instance, it may be possible to eliminate the filter-binding species from the population by also employing a negative postselection step in which potential aptamers are again passed over a modified cellulose filter prior to amplification. However, because the amount of filtered RNA at the end of a separation is small relative to the amount of RNA pool at the beginning of a separation, this negative postselection step is more technically demanding and can potentially result in the loss of amplifiable product altogether.

[33] L. Giver, D. P. Bartel, M. L. Zapp, M. R. Green, and A. D. Ellington, *Gene* **137,** 19 (1993).

[34] C. Tuerk, S. MacDougal, and L. Gold, *Proc. Natl. Acad. Sci. U.S.A.* **89,** 6988 (1992).

## Selection of RNA Aptamers Utilizing Modified Cellulose Filter Immobilization

We provide two examples of the selection of aptamers using modified cellulose filtration as a selection method. In the first example, novel inhibitors of a bacterial tyrosine phosphatase, *Yersinia enterocolitica* Yop51*, are selected starting from two completely random pools. In the second example, the binding site of a viral regulatory protein, Rex, is defined starting from a partially randomized binding site.

## Example: Selection of RNA Molecules That Bind to and Inhibit Active Site of Tyrosine Phosphatase[35]

### Selection Reagents

Protein target: Full-length Yop51* is from New England Biolabs (Beverly, MA). A deletion variant that lacks the first 162 amino acids at the $NH_2$ terminus, Yop51*$\Delta$162, is from the laboratory of Dr. Jack Dixon at the University of Michigan.

RNA pools: The Yop51*$\Delta$162 selection is initially done with two different pools, one with a random region of 30 nucleotides (N30) and the other with 71 nucleotides (N71). Double-stranded DNA pools (1 $\mu$g; $2 \times 10^{13}$ sequences) serve as templates for *in vitro* transcription. Following transcription, the RNA pools are purified on 10% denaturing polyacrylamide gels, eluted, and quantitated spectrophotometrically. The first rounds of selection start with approximately $9 \times 10^{13}$ N30 RNA molecules and $6.5 \times 10^{13}$ N71 RNA molecules.

Selection buffer: 20 m$M$ Tris–HCl (pH 7.6), 150 m$M$ NaCl, 5 m$M$ $MgCl_2$, 1 m$M$ DTT

Elution buffer: 7 $M$ urea, 100 m$M$ sodium citrate (pH. 5.0), 3 m$M$ EDTA

Reverse-transcription buffer: 50 m$M$ Tris–HCl (pH 8.0), 40 m$M$ KCl, 6 m$M$ $MgCl_2$, 0.8 mM dNTPs

N30 pool PCR buffer: 50 m$M$ KCl, 10 m$M$ Tris–HCl (pH 8.3), 1.5 m$M$ $MgCl_2$, 5% acetamide, 0.1% Nonidet P-40, 0.2 m$M$ dNTPs

N71 PCR buffer: 30 m$M$ Tricine (pH 8.3), 50 m$M$ potassium acetate, 0.5% Triton X-100, 1.5 m$M$ magnesium acetate, 5% acetamide, 0.2 m$M$ dNTPs

---

[35] S. D. Bell, J. M. Denu, J. E. Dixon, and A. D. Ellington, *J. Biol. Chem.* **273**, 14309 (1998).

## Selection Procedure

THERMAL EQUILIBRATION. RNA pools in selection buffer (100 $\mu$l) are thermally equilibrated by heating to 65° for 3 min and cooling to room temperature over a period of 10 min. This step helps ensure that similar distributions of conformers are assessed in each round of selection.

PREFILTRATION. To exclude filter-binding sequences from the selection, the thermally equilibrated pools are passed over a modified cellulose filter. A premoistened HAWP filter (Millipore, Bedford, MA) is first fixed in a pop-top Nuclepore filter holder (Corning Separations, Acton, MA) and mounted on a vacuum manifold (J. T. Baker, Phillipsburg, NJ). A vacuum (5–10 psi) is applied to the manifold, and the pool is passed over the filter and washed once with binding buffer (100 $\mu$l).

BINDING REACTIONS. Following prefiltration, Yop51*$\Delta$162 (0.05 $\mu M$) is added to the pool eluant (200 $\mu$l). The final concentration of the N30 RNA pool is 0.76 $\mu M$, and the concentration of the N71 RNA pool is 0.54 $\mu M$. The binding reactions are incubated at room temperature for 60 min, vacuum filtered over a HAWP filter at 5 psi, and washed once with 2.5 volumes of selection buffer. To increase the stringency of rounds 6–8 of the N30 selection, a nonspecific competitor, tRNA (0.02 $\mu M$), is incubated with Yop51*$\Delta$162 at room temperature for 10 min prior to initiation of the binding reaction.

ELUTION. The filters are eluted twice with elution buffer (200 $\mu$l each elution) for 5 min at 100°. The eluant is precipitated with 2-propanol. The precipitant is collected by centrifugation and washed with 70% (v/v) ethanol.

REVERSE TRANSCRIPTION. The precipitated RNA pools are resuspended in 25 $\mu$l of water. A portion of the pools (10 $\mu$l) is reverse transcribed in reverse transcription buffer (20 $\mu$l total volume) with 2.5 units of avian myeloblastosis virus (AMV) reverse transcriptase (Seikagaku, Ijamsville, MD) for 45 min at 42°. A "no reverse transcriptase control" is carried out in parallel using another 10 $\mu$l of the elution. This control is important to ensure the researcher that molecules for additional rounds of selection are in fact derived from selected species rather than from small amounts of contaminating amplicons (e.g., PCR products from the initial pool amplification). The remaining eluant (5 $\mu$l) is archived in case the next or subsequent rounds of selection fail (e.g., are overrun with filter-binding species)

PCR AMPLIFICATION. A portion of the N30 reverse transcription reaction (10 $\mu$l) or the N71 reverse transcription (RT) reaction (10 $\mu$l) is diluted into 1× N30-optimized PCR buffer (100 $\mu$l final volume) or into 1× N71-optimized PCR buffer (100 $\mu$l final volume), and 2.5 units of *Taq* polymerase are added. Both pools are thermocycled under the following regime: 94°

for 45 sec, 50° for 1 min, 72° for 1.5 min. Again, it is important to carry out "no RT" and "no template" controls to make sure that amplified DNA molecules derive from selected RNA molecules. Small portions (5 μl) of the amplification reactions are analyzed by agarose gel electrophoresis every three to five thermal cycles to determine if amplified, double-stranded DNA is present. By monitoring the amount of amplified DNA closely, the production of catenated or deleted polymerase chain reaction products can be avoided. Amplified, double-stranded DNA is ethanol precipitated, and the precipitant is collected by centrifugation, washed with 70% ethanol, and resuspended in TE (10 μl).

TRANSCRIPTION. One-half of the amplified DNA is used as a template to transcribe the next generation of RNA molecules. Again, AmpliScribe kits are used for transcription reactions. The transcripts are gel purified and eluted into water (500 μl) at 37°. The eluant is ethanol precipitated as described earlier, and the precipitant is resuspended in TE (10 μl). The concentration of the purified RNA sample is quantitated by determining the UV absorbance of appropriate dilutions at 260 nm.

ITERATIVE ROUNDS OF SELECTION. The selection and amplification steps described earlier are repeated for seven additional rounds.

BINDING ASSAYS. After completion of rounds 5 and 8, each pool is assayed for its ability to bind the target, Yop51*Δ162. Binding assays are similar to the selection procedure, except that the RNA is labeled internally with $^{32}$P. RNA pools are radiolabeled by introduced [α-$^{32}$P]UTP (3000 Ci/mmol; 83 nM final concentration) into a 20-μl transcription reaction that also contains 7.5 mM unlabeled UTP. Radioactive transcripts are gel isolated, eluted, and precipitated. The radiolabeled RNA pools (0.76 mM final concentration) are mixed with Yop51*Δ162 (0.5 μM final concentration) in selection buffer (100 μl) for 1 hr at room temperature. The binding reactions are filtered through modified cellulose as described earlier, and the filters are exposed to a PhosphorImager plate (Molecular Dynamics, Sunnyvale, CA). The amount of retained radioactivity on each filter is determined using a PhosphorImager (Molecular Dynamics). The amount of radioactivity retained is compared with the amount of radioactivity originally introduced into the binding reaction to determine the percentage of the population that bound to the protein. In parallel, the radiolabeled RNA is filtered through modified cellulose in the absence of protein to determine if background, matrix-binding species are present in appreciable amounts. The results of the selection are shown in Table I. As can be seen, binding species are recovered easily within five rounds and had been further optimized by eight rounds.

CLONING AND SEQUENCING. PCR products derived from selected pools are ligated into the TA cloning vector (Invitrogen, Carlsbad, CA) and

TABLE I
SUMMARY OF Yop51* SELECTION

|         | N30 selection |          | N71 selection |          |
|---------|---------------|----------|---------------|----------|
| Round   | −protein      | +protein | −protein      | +protein |
| 0       | 0.58%         | 0.49%    | 0.40%         | 0.38%    |
| 5       | 0.51%         | 9.6%     | 0.13%         | 1.7%     |
| 8       | 0.20%         | 23%      | 0.85%         | 53%      |

transformed into *Escherichia coli* InvαF' (Invitrogen). Plasmid DNA is isolated[36] from transformants and sequenced using Sequenase 2.0 (U.S. Biochemical Corp., Cleveland, OH) according to the protocol provided with the enzyme.

APTAMER-BINDING ASSAYS. Individual clones are assayed for binding. Double-stranded templates are amplified from plasmid DNAs and transcribed as described earlier. Transcripts are dephosphorylated with alkaline phosphatase (Boehringer Mannheim, Indianapolis, IN; 15 U; 5 $\mu$l total volume) in a buffer supplied by the manufacturer. RNA aptamers are end labeled with 0.17 n$M$ [$\gamma$-$^{32}$P]ATP (3000 Ci/mmol) using T4 polynucleotide kinase (New England Biolabs Inc., Beverly, MA). Following the labeling reaction, the aptamers are precipitated; approximately 100 ng of radiolabeled product is recovered. Binding curves are determined using radiolabeled RNA (ca. 1 n$M$) incubated in selection buffer (50 $\mu$l) with increasing concentrations of protein. For binding curves with Yop51*$\Delta$162, the reactions are passed over individual nitrocellulose filters as in the selection. For binding curves with Yop51*, the reactions are filtered on a vacuum manifold (Schleicher & Schuell, Keene, NH) containing a piece of pure nitrocellulose (Midwest Scientific, St. Louis, MO) mounted over top of a piece of Hybond nylon membrane (Amersham Corp., Piscataway, NJ) as described in Weeks and Cech.[37] In this method, the nitrocellulose filter captures protein-bound radiolabeled RNA, whereas the Hybond paper captures unbound radiolabeled RNA. For all binding curves, the amount of radioactivity retained at each protein concentration is quantified on a PhosphorImager (Molecular Dynamics). To calculate equilibrium dissociation constants, data are fit to the equation: $f = C*[L]/([L]+K_d)$, where $f$

---

[36] J. Sambrook, E. F. Fritsch, and T. Maniatis, *in* "Molecular Cloning: A Laboratory Manual" (C. Nolan, ed.), 2nd Ed., p. 1.25. Cold Spring Harbor Laboratory Press, Plainview, NY, 1989.

[37] K. M. Weeks and T. R. Cech, *Biochemistry* **34**, 7728 (1995).

is the fraction of the aptamer bound, $C$ is the binding capacity of the aptamer, and [L] is the concentration of protein, using the program Kaleidograph (Abelbeck Software, Reading, PA).

*Example: High-Resolution Mapping of Human T-Cell Leukemia Virus Type 1 Rex-Binding Element by in Vitro Selection[38]*

*Selection Reagents*

Protein target: Rex is from the laboratory of Dr. Maria Zapp at the University of Massachusetts Medical Center.

RNA pool: The extent of the wild-type Rex-binding element (XBE) has been mapped previously by deletion analysis.[39] A 78-residue-long ssDNA oligonucleotide is synthesized in which 42 residues that span the XBE doped at a rate of 70% wild-type and 10% of each non-wild-type residue (Fig. 2); flanking sequences remain constant. The ssDNA pool is amplified using the PCR. In the process, a T7 promoter is appended to the 5' end of the double-stranded DNA, and the resultant pool (0.5 $\mu$g; $8.6 \times 10^{12}$ sequences) serve as a template for *in vitro* transcription.

Selection buffer: 50 m$M$ Tris–HCl (pH 8.0), 50 m$M$ KCl

Elution buffer: 0.2 $M$ Tris–HCl (pH 7.6), 2.5 m$M$ EDTA, 0.3 $M$ NaCl, 2% sodium dodecyl sulfate

Reverse transcription buffer: 50 m$M$ Tris–HCl (pH 8.0), 40 m$M$ KCl, 6 m$M$ MgCl$_2$, 0.4 m$M$ dNTPs, 2.5 $\mu M$ of the 3' primer, 5 m$M$ DTT

PCR buffer: 50 m$M$ KCl, 10 m$M$ Tris–HCl (pH 8.3), 1.5 m$M$ MgCl$_2$, 5% acetamide, 0.1% Nonidet P-40, 0.2 m$M$ dNTPs

Stop dye: 95% formamide, 20 m$M$ EDTA, 0.5% xylene cyanol, 0.5% bromphenol blue

*Selection Procedure*

THERMAL EQUILIBRATION. Following transcription, the RNA pool is purified on 10% denaturing polyacrylamide gel, eluted, precipitated, and quantitated. The RNA pool (1 $\mu M$; Table II) is heated in selection buffer (20 $\mu$l final volume in the first round and 30 $\mu$l in subsequent rounds) to 90° for 2 min and allowed to cool to ambient temperature over 10 min. Rex protein (30 ng) and tRNA (10 $\mu M$) are incubated together in selection

[38] S. Baskerville, M. Zapp, and A. D. Ellington, *J. Virol.* **69**, 7559 (1995).
[39] M. Grone, E. Hoffmann, S. Berchtold, B. R. Cullen, and R. Grassmann, *Virology* **204**, 144 (1994).

TABLE II
SELECTION CONDITIONS AND SUMMARY OF REX SELECTION

| Cycle(s) | [tRNA] ($\mu M$) | [RNA] ($\mu M$) | Rex (ng) | Prefiltration | Binding (%) |
|---|---|---|---|---|---|
| 1 | 10 | 1 | 30 | 3× | 0.6 |
| 2 | 6.6 | 1.2 | 30 | | |
| 3, 4 | 6.6 | 1.2 | 30 | 3× | 5.4 |

buffer (20 $\mu$l final volume in the first round and 30 $\mu$l in subsequent rounds) for 10 min at ambient temperature. The concentration of the pool and tRNA varies in subsequent rounds (Table II).

PREFILTRATION. To exclude filter-binding sequences from the selection, the thermally equilibrated pool is passed over a modified cellulose filter. A premoistened HAWP filter (Millipore) is first fixed in a pop-top Nuclepore filter holder (Corning Separations, Acton, MA) and mounted on a vacuum manifold (J. T. Baker, Phillipsburg, NJ). A vacuum (5 psi) is applied to the manifold, and the pool is passed over the filter and washed once with binding buffer (100 $\mu$l).

BINDING. The REX:tRNA mixture is then added to the eluate from the prefiltration step (final volume of 40 $\mu$l in the first round and 60 $\mu$l in subsequent rounds). The binding reaction is incubated at room temperature for 60 min, vacuum filtered over a HAWP filter at 5 psi, and washed twice with selection buffer (500 $\mu$l each wash).

ELUTION. The filter is eluted twice with elution buffer (400 $\mu$l each elution) for 15 min at 75°. The eluent is ethanol precipitated, and the precipitant is collected by centrifugation, washed with 70% ethanol, and resuspended in 25 $\mu$l of water.

REVERSE TRANSCRIPTION. A portion of the pool (10 $\mu$l) is reverse transcribed in reverse transcription buffer (20 $\mu$l total volume) with 2.5 units of avian myeloblastosis virus reverse transcriptase (Seikagaku, Ijamsville, MD) for 45 min at 42°.

PCR AMPLIFICATION. A portion of the reverse transcription reaction (15 $\mu$l) is combined with PCR buffer and 2.5 units of *Taq* polymerase (85 $\mu$l; 100 $\mu$l final volume). The PCR reaction is thermocycled under the following regime: 94° for 1 min, 45° for 1 min, and 72° for 2 min. A small portion of the amplification reaction is analyzed by agarose gel electrophoresis every three to five PCR cycles to determine if amplified, double-stranded DNA is present. The amplified PCR product is then ethanol precipitated, and the precipitant is collected by centrifugation, washed with 70% ethanol, and resuspended in 25 $\mu$l of water.

TRANSCRIPTION. Two-fifths of the amplified DNA is used as a template to transcribe the next generation of RNA molecules. Transcription reaction components are from the Ampliscribe kit (Epicentre, Madison, WI). The transcription reaction is gel purified on a 10% denaturing polyacrylamide gel, eluted in water, ethanol precipitated, and quantitated.

ITERATIVE ROUNDS OF SELECTION. The selection and amplification steps are repeated for three additional rounds.

CLONING AND SEQUENCING. After the fourth round of selection, PCR products are ligated into the TA cloning vector (Invitrogen, Carlsbad, CA) and transformed into *E. coli* InvαF' (Invitrogen). Plasmid DNA is isolated[36] and sequenced using Sequenase 2.0 (U.S. Biochemical Corp.) according to the protocol provided with the enzyme.

BINDING ASSAYS. Unselected and round 4 pools are assayed for their ability to bind Rex. The binding assays are similar to those used in the selection procedure, except that the RNA is labeled internally with $[\alpha\text{-}^{32}P]UTP$. Thermally equilibrated, radiolabeled RNA pools (0.8 $\mu M$ final concentration) are mixed with Rex (30 ng) and tRNA (8 $\mu M$ final concentration) in selection buffer (50 $\mu l$), and the binding reactions are incubated for 1 hr at room temperature. A small aliquot (5 $\mu l$) is spotted on a filter to determine the total amount of radioactivity present in the reaction. The reactions are then filtered as described earlier. The filters are exposed to a PhosphorImager plate, and the amount of radioactivity retained on the filters is determined using a PhosphorImager (Molecular Dynamics). The number of counts retained on the filter are compared with the total number of counts introduced into the binding reaction to determine percentage binding. As can be seen in Table II, the selection identifies binding species after only four rounds. Although the binding observed in this assay is low compared to that observed for RNA : Yop complexes, values are comparable to those observed for complexes between wild-type XBE and Rex. Because the purpose of these experiments was to use sequence variation to better map and model wild-type XBE, the selection is terminated prior to the fixation of a relatively small number of very high-affinity-binding species.

COMPETITION EXPERIMENTS. Pools and individual clones are assayed for their ability to compete with wild-type XBE for binding to the Rex protein. Radiolabeled RNA (0.6 $\mu M$) is incubated with an equimolar amount of wild-type XBE (0.6 $\mu M$), a 10-fold excess of tRNA (6 $\mu M$), and 30 ng of Rex in selection buffer (60 $\mu l$). The reactions are passed over individual nitrocellulose filters and eluted as in the selection. Eluants are precipitated, resuspended in stop dye, and separated on a 10% denaturing polyacrylamide gel. Counts in each band are quantitated using a PhosphorImager. Binding ratios are calculated using [(counts filtered$_{pool}$ − background)/(counts un-

```
                    N30yc5  CTGCTGGTGACGAGGGCTAGACGACGTACC
N71yc2  CTGGCAATGGGTTATCCCAAGTGCTAAGCTTCAGGGAGCGAGGACCAGACGACGTACCTAACCCCTAAGGTG
```

```
        5'                U •           •              •
           GCUC   •GG         G • •       • • • • • • •     •
           •GAG   •CC        C • •    •G • • • •          •
        3'         •U                  • •            •
```

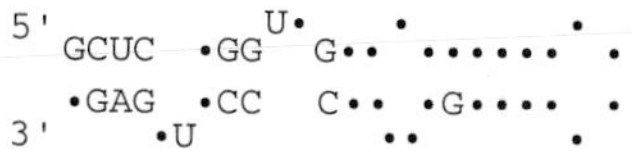

FIG. 3. Aptamer motifs. Sequence motifs or secondary structures of RNA-binding species were mapped from data obtained after multiple rounds of selection. (a) The common binding motif of anti-Yop51* aptamers from N30 and N71 pools is shaded. (b) Anti-Rex aptamers sequenced after four rounds of selection contained significantly conserved residues. Covariations of bases corresponded to regions that were expected to have base pairing interactions, indicating that the proposed secondary structure was correct.

filtered$_{pool}$ − background)]/[(counts filtered$_{XBE}$ − background)/(counts unfiltered$_{XBE}$ − background)]. The binding ratio should represent the dissociation constants for different RNA : Rex complexes relative to the dissociation constant for the XBE : Rex complex.

## Analysis

Selection experiments return either novel or refined RNA-binding species. In addition to the basal binding characterizations cited earlier, the sequences and structures of these RNA-binding species can be mapped. For example, a relatively small number of aptamers or families of related aptamers are frequently culled from a completely random sequence population. Aptamer sequences can be compared to identify binding motifs. For example, anti-Yop aptamers derived from the N30 and N71 pools were different, but contained a common binding motif (Fig. 3). Similarly, by comparing aptamers derived from a doped sequence pool, those residues that are most important for binding function can be identified readily. For example, anti-Rex aptamers obtained after only four rounds of selection already contained a number of uniformly conserved residues, whereas other residues continued to vary randomly (Fig. 3). Davis et al.[40] presented a more thorough analysis of how to identify aptamer sequence motifs and covariations.

[40] J. P. Davis, N. Janjic, B. E. Javornik, and D. A. Zichi, *Methods Enzymol.* **267**, 302 (1996).

Once a functional sequence motif has been identified, a corresponding structure can frequently be predicted using standard folding programs (e.g., Mulfold[41,42]). However, the structural hypotheses yielded by energy minimizing algorithms should be viewed skeptically in the absence of sequence covariations that are consistent with a single structure. Such sequence covariations can frequently be observed in selection experiments that start with heavily doped populations.[3,43] In addition, more conventional techniques for RNA structure mapping, such as nuclease protection, chemical protection, and modification interference analyses, should be used to confirm structural predictions.

[41] J. A. Jaeger, D. H. Turner, and M. Zuker, *Proc Natl Acad Sci U.S.A.* **86,** 7706 (1989).
[42] M. Zuker, *Science* **244,** 48 (1989).
[43] K. B. Jensen, L. Green, S. MacDougal-Waugh, and C. Tuerk, *J. Mol. Biol.* **235,** 237 (1994).

# [15] *In Vitro* Selection and Characterization of RNAs with High Affinity to Antibiotics

*By* Scot T. Wallace and Renée Schroeder

## Introduction

Although antibiotics that bind to RNA have been used therapeutically for decades (streptomycin was discovered in 1944), it was not until recently that RNA was recognized as their targets.[1,2] The mode of action of these antibiotics is still poorly understood. This might be due to the fact that the importance of the role of RNA in the cellular metabolism was underestimated for many years. Antibiotics that interfere with translation most often bind to ribosomal RNA. Additionally, many of these antibiotic translation inhibitors bind and interfere with the function of other catalytic RNAs *in vitro,* giving them a unique feature, namely an exceptional lack of specificity.[3,4] One way to study the mode of action of RNA-binding antibiotics is to determine the structure of RNA–antibiotic complexes.

The natural target sites of RNA-binding antibiotics are often too large

[1] D. Moazed and H. F. Noller, *Nature* **327,** 389 (1987).
[2] U. von Ahsen and R. Schroeder, *Nature* **346,** 801 (1990).
[3] R. Schroeder, U. von Ahsen, F. Eckstein, and D. M. J. Lilley (eds.), "Nucleic Acids and Molecular Biology," Vol. 10, p. 53. Springer-Verlag, Berlin, 1996.
[4] K. Michael and Y. Tor, *Chem. Euro. J.* **4,** 2091 (1998).

for structural characterization. Two methods of reducing the RNA length are (i) dissection of the natural target RNA to that of the antibiotic RNA-binding domain[5] and (ii) isolation, via *in vitro* selection, of small RNAs that bind specifically to antibiotics. RNA aptamers that bind to antibiotics include binders to the aminoglycosides tobramycin,[6] lividomycin and kanamycin A,[7] neomycin B,[8] streptomycin,[9] the peptide antibiotic viomycin,[10] and chloramphenicol.[11]

On characterization of selected antibiotic-binding RNAs, structural motifs that support antibiotic binding to RNA were revealed. Tobramycin binds to the RNA major groove at a stem–loop junction where the major groove and a bulged cytosine sandwich the bound antibiotic.[12,13] A widened major groove in a stem with the loop consensus GNRNA was found as the binding site in the neomycin-binding motif.[8] A pseudoknot structure was found to be necessary for the recognition of viomycin,[10] and a major conformational change in the RNA was observed on RNA binding to the para-positioned guanidino group of streptomycin.[9] The list continues to grow, and it is hoped that our knowledge about the recognition rules governing the interaction between antibiotics and RNA will lead to the rationale design of novel drugs that might escape resistance factors.

For reviews of SELEX (selection of ligands by exponential enrichment) technology, its applications, and specific procedures, see Gold *et al.,*[14] Osborne and Ellington,[15] and Szostak and Ellington.[16]

## In Vitro Selection Procedure

The selection procedure, outlined in Fig. 1, begins with transcription of a beginning DNA pool. The transcription is followed by repeated rounds

[5] P. Purohit and S. Stern, *Nature* **370,** 659 (1994).

[6] Y. Wang and R. R. Rando, *Chem. Biol.* **2,** 281 (1995).

[7] S. M. Lato, A. R. Boles, and A. D. Ellington, *Chem. Biol.* **2,** 291 (1995).

[8] M. G. Wallis, U. von Ahsen, R. Schroeder, and M. Famulok, *Chem. Biol.* **2,** 543 (1995).

[9] S. T. Wallace and R. Schroeder, *RNA* **4,** 112 (1998).

[10] M. G. Wallis, B. Streicher, H. Wank, U. von Ahsen, E. Clodi, S. T. Wallace, M. Famulok, and S. Schroeder, *Chem. Biol.* **4,** 357 (1997).

[11] D. H. Burke, D. C. Hoffman, A. Brown, M. Hansen, A. Pardi, and L. Gold, *Chem. Biol.* **4,** 833 (1997).

[12] L. Jiang, A. K. Suri, R. Fiala, and D. J. Patel, *Chem. Biol.* **4,** 35 (1997).

[13] L. Jiang and D. J. Patel, *Nature Struct. Biol.* **5,** 769 (1998).

[14] L. Gold, B. Polisky, O. Uhlenbeck, and M. Yarus, *Annu. Rev. Biochem.* **64,** 763 (1995).

[15] S. E. Osborne and A. D. Ellington, *Chem. Rev.* **97,** 349 (1997).

[16] J. W. Szostak and A. D. Ellington, *in* "The RNA World" (R. F. Gesteland and J. F. Atkins, eds.), p. 511. Cold Spring Harbor Laboratory Press, Cold Spring Harbor, NY, 1993.

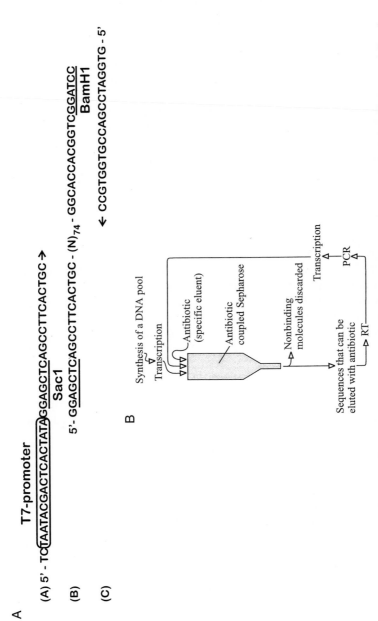

**A**

**T7-promoter**

(A) 5' - TC[TAATACGACTCACTATA]GGAGCTCAGCCTTCACTGC →
                                    Sac1

(B)    5'- GGAGCTCAGCCTTCACTGC - (N)₇₄ - GGCACCACGGTCGGATCC
                                                      **BamH1**

(C)              ← CCGTGGTGCCAGCCTAGGTG - 5'

**B**

Synthesis of a DNA pool
Transcription
Antibiotic (specific eluent)
Antibiotic coupled Sepharose
Nonbinding molecules discarded
Transcription
PCR
RT
Sequences that can be eluted with antibiotic

Fig. 1. (A) Example of a template DNA pool used to isolate antibiotic-binding RNAs. The template DNA consists of 5' and 3' invariable domains as primer binding sites and of a variable domain of 74 random nucleotides. The sense primer contains the T7 RNA polymerase promoter in addition to the invariable domain. Restriction sites are essential for cloning the amplified DNA after selection. (B) A schematic representation of the *in vitro* selection procedure.

of selection and enrichment of selected RNA binders. Each round consists of affinity chromatography, reverse transcription, polymerase chain reaction (PCR), and transcription. The number of rounds necessary to find winning RNA binders depends on the enrichment after the affinity chromatography step. The enrichment of antibiotic-binding RNAs is detected when the fractions of RNA specifically eluted from the column become greater than the column-loading and equilibration wash fractions. An enrichment plateau is reached when the percentage of RNA in the enrichment fractions no longer increases relative to the previous selection round. We first noted enrichment in the fourth round of selection and found a plateau in the eighth round of *in vitro* selection. On detection of an enrichment plateau, the selection cycles can be stopped, the winning RNAs retrotranscribed into DNA, cloned, sequenced, and analyzed for consensus sequences and common structural motifs.

## Selection of Starting DNA Pool

The starting DNA pool (see Fig. 1A) should be carefully designed according to the experiences made during previous selections. The template DNA usually consists of a variable sequence domain of variable size flanked by invariable 5' and 3' sequences required for primer hybridization, amplification, and T7 RNA polymerase transcription.[17] The lengths of the variable regions in the DNA pools used for the isolation of antibiotic binders were 60,[6] 30,[7] 74,[8,9] and 70 or 80 nucleotides.[11] Although for the aminoglycoside binders the binding motifs were well imbedded within the variable region, both for the viomycin and the streptomycin binders, the constant primers were part of the binding motif. For the viomycin binders the 3' primer sequence was part of the pseudoknot, and for the streptomycin binders the 3' primer sequence formed a significant part of the conserved binding motif. The aminoglycoside-binding RNA motifs are stem–loops, whereas for the peptide viomycin and the aminocyclitol, more complex motifs, which include three-dimensional structural elements, were isolated. This raises the question whether and how much the constant regions bias the outcome of the selection.

Another possibility for template design is not to start from a totally random region, but to degenerate ("dope") a natural antibiotic-binding site. This was done for the decoding site of the 16S ribosomal RNA.[18] Interestingly, this selection for neomycin binding did not yield the wild-type sequence, but the same motif isolated from a random pool.[8] The usage

---

[17] S. J. Klug and M. Famulok, *Mol. Biol. Rep.* **20,** 97 (1994).
[18] M. Famulok and A. Huttenhofer, *Biochemistry* **35,** 4267 (1996).

of genomic DNA pools[19,20] to isolate RNAs that bind antibiotics should reveal whether there are additional antibiotic-binding sites in the cell, which might explain the high toxicity of the aminoglycosides. For reviews on the detailed construction of DNA templates see Refs. 21–23.

### Coupling of Antibiotic to Stationary Phase

In the affinity chromatography selection procedure, antibiotic immobilized to a matrix serves as the stationary phase. It is important to choose a matrix substance that, without coupled antibiotic, does not have a high affinity to nucleic acids. Affi-Gel (Bio-Rad, Hercules, CA) and Toyopearl (Tosohaas, Montgomeryville, PA) were both found to bind large amounts of randomized pool RNA nonspecifically.[7] Epoxy-activated agarose (Pierce, Rockford, IL) or Sepharose (Pharmacia, Piscataway, NJ) under 250–500 m$M$ NaCl conditions have been shown to bind less than 1% of randomized pool RNAs. Although the exact binding positions between an epoxy-activated matrix and its ligand cannot be predicted or determined easily, the use of affinity elution and the amplification of only those RNAs that bind the antibiotic in solution reduce the risk of column ligand heterogeneity biasing the final selected RNAs.

For the selection procedures in our laboratory, epoxy-activated Sepharose 6B (Pharmacia) was used. The coupling of antibiotic to this matrix is described in this article. Via free 1,4-bis(2,3-epoxypropoxy)butane (oxirane) groups, epoxy-activated Sepharose forms stable ether bonds with hydroxyl-containing molecules, alkylamine linkages with ligands containing amino groups, and thioether linkages with ligands containing thiol groups.

One milliliter of antibiotic-coupled Sepharose is used per round of *in vitro* selection. The Sepharose must be swelled (1 g freeze-dried Sepharose powder yields 2.5 ml swelled material) and then coupled with the antibiotic. For swelling of the Sepharose, 1 liter of doubly distilled $H_2O$ is added to 5 g of freeze-dried Sepharose powder and mixed by swirling. The mixture is left at room temperature for 1 hr before being loaded onto a sintered filter under suction. As the water is sucked through the filter, the Sepharose will appear dry and cracked. The Sepharose is immediately washed six times with 50 ml doubly distilled $H_2O$, allowing the Sepharose to be sucked dry between each wash.

[19] B. S. Singer, T. Shatland, D. Brown, and L. Gold, *Nucleic Acids Res.* **25,** 781 (1997).
[20] L. Gold, D. Brown, Y. He, T. Shatland, B. S. Singer, and Y. Wu, *Proc. Natl. Acad. Sci. U.S.A.* **94,** 59 (1997).
[21] D. Irvine, C. Tuerk, and L. Gold, *J. Mol. Biol.* **222,** 739 (1991).
[22] R. Green, A. D. Ellington, D. P. Bartel, and J. W. Szostak, *Methods* **2,** 75 (1991).
[23] T. Fitzwater and B. Polisky, *Methods Enzymol.* **267,** 275 (1996).

A final concentration of 1 m$M$ antibiotic is desired on the coupled Sepharose. As the binding efficiency of antibiotic to Sepharose is less than 50%, 31.5 $\mu$mol (based on 25 ml swelled Sepharose) of antibiotic is added to 25 ml coupling buffer (10 m$M$ NaOH, pH 10) in a clean, sterile 50-ml plastic tube. The washed Sepharose is added directly from the sintered filter to the antibiotic solution, sealed shut, and gently shaken for 16 hr at 37°. In order to wash away excess, uncoupled antibiotic, the solution is loaded onto a sintered filter under suction and washed four times with 50 ml coupling buffer. In order to deactivate any uncoupled epoxy groups on the Sepharose, the coupled Sepharose, still loaded on the filter, is washed with three cycles of alternating pH solutions: one cycle consists of 50 ml, pH 4, wash buffer (0.1 $M$ sodium acetate, 0.5 $M$ NaCl, pH 4) followed by 50 ml, pH 8, wash buffer (0.1 $M$ NaHCO$_3$, 0.5 $M$ NaCl, pH 8). In order to neutralize and wash away any buffer, the Sepharose is washed four times with 50 ml doubly distilled H$_2$O. The Sepharose is added directly from the sintered filter to 25 ml of doubly distilled H$_2$O in a clean, sterile 50-ml plastic tube. EDTA is added to a final concentration of 1 m$M$, and the coupled Sepharose is stored at 4°.

### Beginning Pool Transcription

Combine 0.8 nmol of PCR-amplified pool DNA, 40 m$M$ Tris–HCl (pH 7.9), 3 m$M$ spermidine, 26 m$M$ MgCl$_2$, 5 m$M$ ATP, 5 m$M$ CTP, 5 m$M$ GTP, 5 m$M$ UTP, 10 m$M$ dithiothreitol (DTT), 25 $\mu$Ci [$\alpha$-$^{32}$P]GTP, 7500 units T7 RNA polymerase in a final volume of 5 ml at 37° for 16 hr. In order to rid the samples of DNA, 350 U of DNase I (RNase-free) is added and incubated 37° for 1 hr. Pyrophosphates are brought into solution by the addition of 2 ml 500 m$M$ EDTA (pH 8) followed by a phenol extraction. Samples are precipitated by bringing the reaction to 2 $M$ ammonium acetate, adding 3$\times$ volumes of 96% ethanol, and incubating at 20° for 30 min. The RNA is purified by gel filtration over a Sephadex G-50 column at 20°.

### Affinity Chromatography

*Equilibration/Selection Buffer.* In order to hinder unspecific RNA binding to the column matrix, as well as to promote RNA folding, it is important to perform the selection in an appropriate buffer. High concentrations of monovalent salts (250–500 m$M$) decrease unspecific RNA binding to the matrix, thus increasing stringency toward the immobilized antibiotic. Furthermore, RNA folding is highly dependent on divalent salts, especially Mg$^{2+}$. RNAs, which are the natural target for many antibiotics, depend on Mg$^{2+}$ for their folding and therefore for their antibiotic affinity. We include 250 NaCl and 5 m$M$ MgCl$_2$ in our selection buffer.

*Counter Selection.* A very effective way to increase selection stringency toward an antibiotic or small molecule is to perform a counterselection against a counterligand, which is highly homologous but significantly different from the ligand. By preeluting RNA off a stationary phase using an analog of the desired ligand, only those RNAs that recognize the analog–ligand difference survive the selection. RNA aptamers that have undergone a counterselection have shown up to five orders of magnitude selection preference for the desired ligand over the analog that was used as the counterselector. Counterselection was used successfully in the selection of RNAs, which recognize theophylline and discriminate against caffeine.[24] A counterselection was also used to force an RNA aptamer to recognize a specific chemical group on streptomycin. By using bluensomycin, a streptomycin analog that lacks streptomycin's active guanidino group as the counterselector, the aptamer showed five orders of magnitude selection preference toward the desired streptomycin target.[9]

*Procedure.* The stationary phase is prepared by adding 1 ml of resuspended coupled Sepharose into a Poly-Prep column (Bio-Rad, Hercules, CA) and equilibrating it with 10 ml of selection buffer [5 m$M$ MgCl$_2$, 50 m$M$ Tris–HCl (pH 7.6), 250 m$M$ NaCl]. Molecules of random sequence RNA ($5 \times 10^{15}$) in doubly distilled H$_2$O are incubated at 90° for 3 min and left to cool at 20° for 5 min. The solution is brought to 1× selection buffer conditions in a final volume of 500 $\mu$l and incubated at 20° for 5 min.

From this point on, 1-ml fractions are collected. The RNA solution is loaded onto the selection column. In order to wash nonspecific binding RNAs off the column, selection buffer and possible counterselection solutions can be loaded onto the column. In order to specifically elute RNAs, three column volumes of 5 m$M$ antibiotic solutions in selection buffer are loaded onto the column. Specific elutions are pooled and precipitated using glycogen as a carrier. The pellet is resuspended in 50 $\mu$l doubly distilled H$_2$O. During ongoing selection, the concentration of eluant antibiotic can be decreased (e.g., 1–100 $\mu M$) to increase stringency.

### Reverse Transcription, PCR, Transcription

*Reverse Transcription.* In this step, 0.5 nmol of Primer DNA (5′ primer) is added to 50 $\mu$l RNA from affinity chromatography, incubated at 95° for 3 min, and cooled at 20° for 10 min. The solution is brought to the following conditions: 50 m$M$ Tris–HCl (pH 8.3), 75 m$M$ KCl, 3 m$M$ MgCl$_2$, 10 m$M$ DTT, 0.3% Tween 20, 0.4 m$M$ dATP, 0.4 m$M$ dCTP, 0.4 m$M$ dGTP, 0.4 m$M$ dTTP, 0.5 units reverse transcriptase, 100 $\mu$l total volume, incubated

---

[24] R. D. Jenison, S. C. Gill, A. Pardi, and B. Polisky, *Science* **263,** 1425 (1994).

at 42° for 45 min. Some RNA-binding antibiotics, such as neomycin B and viomycin, inhibit the reverse transcriptase. Therefore, the antibiotic has to be removed before reverse transcription, preferentially with phenol extraction and ethanol precipitation.

*PCR.* The reverse transcription solution is aliquoted into five tubes. Each tube is brought to the following conditions: 10 m$M$ Tris–HCl (pH 8.3), 50 m$M$ KCl, 1.5 m$M$ MgCl$_2$, 0.001% (w/v) gelatin, 0.3% Tween 20, 0.2 m$M$ dATP, 0.2 m$M$ dCTP, 0.2 m$M$ dGTP, 0.2 m$M$ dTTP, 1 $\mu M$ 5′ PCR primer, 1 $\mu M$ 3′ PCR primer, 4 units *Taq* DNA polymerase, 100 $\mu$l total volume, 10 rounds, denatured at 94° for 50 sec, annealed at 55° for 2 min, and elongated at 72° for 2 min. Samples are pooled, phenol extracted, precipitated, and resuspended in 100 $\mu$l TE buffer (10 m$M$ Tris–HCl, pH 7.6, 1 m$M$ EDTA).

*Transcription during Selection Rounds.* One-third of the DNA solution from PCR is brought to the following conditions: 40 m$M$ Tris–HCl, pH 7.9, 26 m$M$ MgCl$_2$, 3 m$M$ spermidine, 5 m$M$ ATP, 5 m$M$ CTP, 5 m$M$ GTP, 5 m$M$ UTP, 10 m$M$ DTT, 10 $\mu$Ci [$\alpha$-$^{32}$P]GTP, 200 units T7 RNA polymerase, 100 $\mu$l reaction, incubated at 37° for 16 hr. In order to rid the RNA from DNA, 5 units of DNase (RNase free) is added and incubated at 37° for 1 hr. Pyrophosphates are brought into solution by adding 40 $\mu$l of 500 m$M$ EDTA (pH 8). Samples are precipitated by bringing the reaction to 2 $M$ ammonium acetate, adding 3× volumes of 96% (v/v) ethanol, and incubating at 20° for 30 min. The RNA is purified using an 8% denaturing PAGE. Samples are precipitated and resuspended in 100 $\mu$l doubly distilled H$_2$O. Fifty microliters is used for the next round of selection.

## Characterization of Antibiotic-Binding RNAs

One of the greatest challenges after the selection of antibiotic-binding RNAs is the characterization of the RNA–antibiotic interaction. Upon obtaining sequences of single RNAs, the sequences can be analyzed by a computer in an attempt to find primary, conserved sequences necessary for antibiotic binding. Davis *et al.*[25] have identified consensus patters and secondary structure in multiple sequences using computer analyses.

Once individual sequences have been identified and tested for antibiotic-binding capability, the following questions arise: How strong is the binding between RNA and antibiotic? What is the minimal RNA length needed to bind to the antibiotic? Does the RNA undergo a conformational change on antibiotic binding? Does the antibiotic recognize a specific secondary

[25] J. P. Davis, N. Janjic, B. E. Javornik, and D. A. Zichi, *Methods Enzymol.* **267**, 302 (1996).

sequential structure? What segment of the RNA is directly involved in antibiotic recognition?

*Binding Constants*

Dissociation constant ($K_d$) characterizes the binding strength between selected RNA and the antibiotic. To date there have been five methods used to calculate RNA aptamer–antibiotic $K_d$ values. (1) Lividomycin: The RNA pool populations adhered to the column with such vehemence that well over 30 column washes of selection buffer were unable to elute the RNAs.[7] The $K_d$ values were determined using a modification of the method by Ellington and Szostak.[26] (2) Viomycin: The availability of radioactively labeled viomycin led to $K_d$ determination via equilibrium gel filtration.[27] (3) Neomycin and streptomycin: Chemical modification experiments give a clear signal on the interaction between antibiotic and RNA.[8,9] Antibiotics were therefore titrated into the chemical modification reaction and the $K_d$ was determined. (4) Chloramphenicol[11]: [14]C-labeled chloramphenicol was available and used in a microconcentrator spin assay described in Jenison *et al.*[24] (5) Tobramycin: Pyrine derivatives of tobramycin were synthesized and used for $K_d$ measurements.[6] Once bound to RNA, these derivatives display a marked change in fluorescence. This method is described in Kierzek *et al.*[28] and Bevilacqua *et al.*[29]

*Boundary Mapping*

To narrow down the sequence length required to bind an antibiotic, it is necessary to perform boundary mapping. This consists of radioactively labeling RNA both 5' and 3', performing partial alkaline hydrolysis, selecting the RNA over an affinity column, and analyzing it via PAGE. Figure 2 shows a typical gel obtained from boundary mapping.

*Partial Hydrolysis.* Ten microliters of radioactively labeled (5'- or 3'-labeled) RNA in doubly distilled $H_2O$ (0.5 pmol/$\mu$l) is combined with 1.5 $\mu$l (1 $\mu$g/$\mu$l) tRNA and 10 $\mu$l (50 m$M$) NaHCO$_3$. The reaction tube is sealed and incubated at 90° for 10 min. Affinity chromatography is essentially the same as described with the exception that the column is washed with 2 column volumes of selection buffer before 2 column volumes of affinity

---

[26] A. Ellington and J. W. Szostak, *Nature* **346,** 818 (1990).

[27] M. Misumi and N. Tanaka, *Biochem. Biophys. Res. Commun.* **82,** 971 (1978).

[28] R. Kierzek, Y. Li, D. H. Turner, and P. C. Bevilacqua, *J. Am. Chem. Soc.* **115,** 4985 (1993).

[29] P. C. Bevilacqua, Y. Li, and D. H. Turner, *Biochemistry* **33,** 11340 (1994).

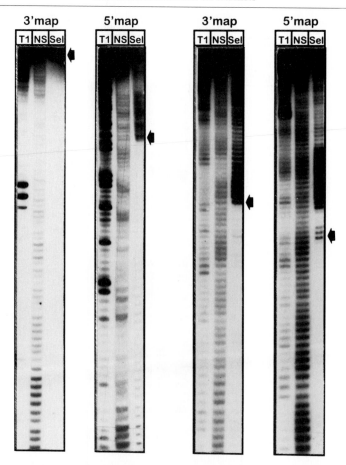

FIG. 2. 5' and 3' boundary mapping using partial alkaline hydrolysis and affinity chromatography of a streptomycin-binding aptamer.[9] T1, partial RNase T1 digestion showing the position in the RNA, which are single-stranded guanosines. NS, partially hydrolyzed, non-selected RNA. Sel, partially hydrolyzed, specifically eluted RNA with arrow indicating the extreme ends of the sequence required for binding.

elution. The affinity elutions are pooled, precipitated, resuspended in loading buffer, and applied to an 8% denaturing polyacrylamide gel.

## $Pb^{2+}$-Induced Cleavage

$Pb^{2+}$ induces cleavage of the RNA phosphate backbone preferentially in single-stranded regions and in regions of high flexibility and is thus a

A

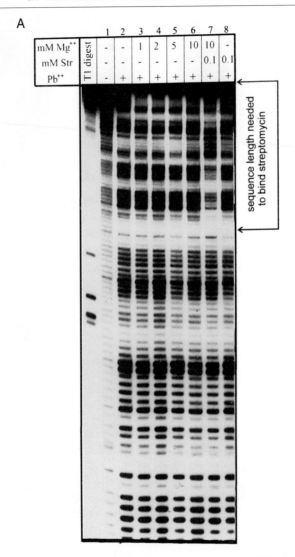

| mM Mg⁺⁺ | T1 digest | 1 | 2 | 3 | 4 | 5 | 6 | 7 | 8 |
|---|---|---|---|---|---|---|---|---|---|
| mM Mg⁺⁺ |  | - | - | 1 | 2 | 5 | 10 | 10 | - |
| mM Str |  | - | - | - | - | - | - | 0.1 | 0.1 |
| Pb⁺⁺ |  | - | + | + | + | + | + | + | + |

sequence length needed to bind streptomycin

FIG. 3. (A) Autoradiogram of gels after $Pb^{2+}$-induced cleavage on RNA from the full-size clone containing a streptomycin-binding domain. Different lanes represent $Pb^{2+}$ cleavage performed under the indicated conditions. Changes in the cleavage pattern, on the addition of streptomycin and $Mg^{2+}$, are indicated. (B) A phosphoimager-generated graph of $Pb^{2+}$-induced cleavage from (A): an individual band $(3' \rightarrow 5')$ from the gel is shown as a function of band intensity. For clarification, the graph and gel are separated in regions that correspond to the proposed secondary structure. P1–P3, paired regions with subscripts 5' or 3' indicating location in RNA; J1/2, J2/3, and J3/2, unpaired regions; B2/1, bulged guanosine.

B

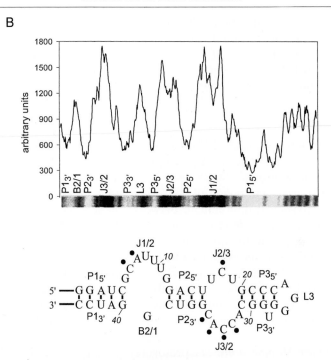

FIG. 3. (*continued*)

good method to probe RNA secondary structure.[30,31] Figure 3B shows a single lane from a gel obtained after lead cleavage and its quantification. Comparing the intensities of single bands, both paired and unpaired regions become apparent and are shown at the bottom of Fig. 3. Furthermore, this procedure is a relatively quick and easy method for detecting changes in RNA conformation under different buffer and ligand conditions. Figure 3A shows a gel where a conformational change on the addition of both streptomycin and magnesium ligands becomes apparent (lane 7).

*Pb²⁺ Cleavage.* 5′-radioactively end-labeled RNA (~0.5 pmol in doubly distilled $H_2O$) is incubated at 90° for 3 min and allowed to cool to 20° for 5 min. Samples are brought to selection conditions by the addition of concentrated selection buffer containing antibiotic and incubated at 20° for 5 min. Controls can be obtained by varying selection buffer conditions

[30] P. Gornicki, F. Baudin, P. Romby, M. Wiewiorowski, W. Kryzosiak, J. P. Ebel, C. Ehresmann, and B. Ehresmann, *J. Biomol. Struct. Dynam.* **6**, 971 (1989).
[31] J. Ciesiolka, D. Michalowski, J. Wrzesinski, J. Krajewski, and W. J. Krzyzosiak, *J. Mol. Biol.* **275** (1998).

([$Mg^{2+}$], [$Na^+$], [antibiotic]). Cleavage is initiated by the addition of lead acetate (final concentration 0.5 m$M$ in reaction volume of 50 $\mu$l) and is incubated for 30 min at 20°. The reaction is stopped by the addition of 1 $\mu$l (1 $\mu$g/$\mu$l glycogen and 250 m$M$ EDTA). Samples are precipitated and analyzed on an 8% denaturing polyacrylamide gel.

### Chemical Modification

A very sensitive method of determining base-specific effects due to the binding of an antibiotic to RNA employs chemical modification interference. Chemical probing can be used to (1) confirm a secondary structure model, (2) locate the region of RNA–antibiotic interaction, (3) ascertain whether RNAs can discriminate between the selected antibiotic and an antibiotic analog, (3) determine an approximate dissociation constant between the RNA and the drug, and (5) verify the buffer conditions dependence of the RNA and antibiotic interaction. Reverse transcription analysis of DMS-modified RNA reveals non-base-paired adenosines and cytosines, whereas kethoxal modification reveals non-base-paired guanosines.[32] Figure 4 shows a typical gel obtained after chemical modification with DMS. A clear protection of adenosines and cytosines are detected and shown in Fig. 4.

*Chemical Modifications.* The protocol is split into five sections: (1) renaturation, (2) modification, (3) hybridization, (4) extension and (5) PAGE.

RENATURATION. With minor changes, the renaturation protocol used in the selection should be followed: 10 pmol RNA in doubly distilled $H_2O$ is incubated at 90° for 3 min, followed by incubation at room temperature for 5 min. Replacing the Tris buffer used in the selection by 50 m$M$ potassium cacodylate buffer (same pH as used in the selection), the solution is brought to 1× selection buffer conditions, including the desired salt and antibiotic concentration in a final volume of 50 $\mu$l. The solution is incubated at room temperature for 15 min and transferred to ice.

MODIFICATION. The modification reaction is started by the addition of 1 $\mu$l of 1 : 10 (in ethanol) diluted DMS or 1 $\mu$l of 1 : 5 (in doubly distilled $H_2O$) diluted kethoxal stock solution (37 mg/ml) to the buffered RNA. On addition, the reaction is mixed with a pipette tip and incubated at 20° for 10 min. The DMS reaction is stopped by the addition of 1 $\mu$l of 1 : 50 (in doubly distilled $H_2O$) diluted 2-mercaptoethanol, and the kethoxal reaction is stopped by the addition of 1 $\mu$l 0.5 $M$ potassium borate, pH 7. Reactions are vortexed and centrifuged for 20 sec at 16,000 rpm. Samples are precipitated by the addition of 1 $\mu$l 10 $\mu$g/$\mu$l glycogen, 5.2 $\mu$l 3 $M$ sodium

---

[32] S. Stern, D. Moazed, and H. F. Noller, *Methods Enzymol.* **164,** 481 (1988).

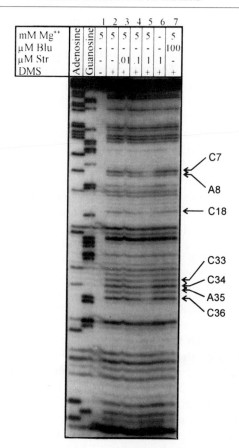

| mM Mg$^{++}$ | Adenosine | Guanosine | 1 | 2 | 3 | 4 | 5 | 6 | 7 |
|---|---|---|---|---|---|---|---|---|---|
| mM Mg$^{++}$ | | | 5 | 5 | 5 | 5 | 5 | - | 5 |
| μM Blu | | | - | - | - | - | - | - | 100 |
| μM Str | | | - | - | .01 | .1 | 1 | 1 | - |
| DMS | | | - | + | + | + | + | + | + |

C7
A8
C18
C33
C34
A35
C36

FIG. 4. Autoradiogram of a gel of chemical protection experiments using DMS performed on the streptomycin-binding aptamer.[9] Adenosine and guanosine, reverse transcription with ddTTP or ddCTP to indicate position of adenosines or guanosines. Different lanes represent DMS modification performed under the indicated conditions. Where chemical protection is detected, nucleotides are indicated by black dots in Fig. 3B. Numbering of the nucleotides is indicated.

acetate, and 171 μl 96% ethanol incubated at −80° for 20 min. The samples are centrifuged at 4° for 15 min at 16,000 rpm at room temperature, and the supernatant is carefully lifted and disposed of. Pellets are dried at room temperature for 5 min. DMS-modified RNA is resuspended thoroughly in 8 μl doubly distilled $H_2O$ and kethoxal-modified RNA in 8 μl 25 m$M$ potassium borate, pH 7. To ensure resuspension, we shake for 5 min at 20° in a high-speed shaker.

HYBRIDIZATION. A hybridization mix is prepared: 4.5× hybridization buffer (1× = 50 m$M$ K–HEPES, pH 7, 100 m$M$ KCl) and 5′-labeled DNA primer (0.1 pmol/$\mu$l) mixed in a proportion of 1:1. Two microliters of the hybridization mix is pipetted carefully onto the side wall of a reaction tube. Similarly, 2 $\mu$l of the modified RNA solution is also pipetted onto the side of the same reaction tube. The two drops are mixed by centrifugation for 10 sec at 16,000 rpm at room temperature. The RNA/DNA is denatured by incubation at 90° for 1 min in a water bath and renatured by letting cool in the same water bath until 47° is obtained. In order to retrieve condensation on lid of the reaction tube, the sample is centrifuged for 10 sec at 16,000g. Compared to the DNA primer used for extension, note that the RNA is in excess and, in order to make quantitative conclusions, great care should be taken that the pipetted DNA volumes are exact.

EXTENSION. Two microliters of extension mixture (recipe below) is added directly to the renatured RNA/DNA solution and mixed thoroughly by pipetting the mixture in and out of the pipette tip 15–20 times.

Extension mixture for 15 samples: 10 $\mu$l 10× extension buffer (1× = 130 m$M$ Tris–HCl, pH 8.5, 10 m$M$ MgCl$_2$, 10 m$M$ DTT), 5 $\mu$l dNTP mix (containing 1 m$M$ dATP, 1 m$M$ dCTP, 1 m$M$ dGTP, 1 m$M$ dTTP), 16 $\mu$l doubly distilled H$_2$O, 1.5 $\mu$l reverse transcriptase (4 U/$\mu$l). The samples are centrifuged for 10 sec at 16,000 rpm at room temperature and incubated at 41° for 30 min. Based on our experience, DNA bands detected by PAGE are sharper when the RNA in the reactions is completely hydrolyzed. The RNA is hydrolyzed by the addition of 1.5 $\mu$l 1 $M$ NaOH and incubation at 41° for 20 min. In order to precipitate, the solution is neutralized by the addition of 1.5 $\mu$l 1 $M$ HCl and precipitated by the addition of 1 $\mu$l 10 mg/ml glycogen, 1 $\mu$l 3 $M$ sodium acetate, and 30 $\mu$l 90% ethanol. The solution is incubated at 20° for 10 min and centrifuged at 20° for 10 min at 10,000g. The supernatant is removed carefully and dried for 5 min at 60°. The DNA pellet is then resuspended thoroughly in 10 $\mu$l PAA loading buffer (89 m$M$ Tris–base, 89 m$M$ boric acid, 2 m$M$ EDTA, 7 $M$ urea, 0.25% (w/v) bromphenol blue, 0.25% (w/v) xylene cyanol). Fully resuspending the DNA pellet is essential. We resuspend using a high-speed shaker for 10 min at 20°.

PAGE. The samples are heated at 90° for 3 min and immediately transferred to ice and incubated for 10 min Retrieve condensation from the lid by centrifuging for 10 sec at 16,000g and immediately replacing the samples on ice. Three microliters of the resuspended and renatured DNA is loaded onto an 8% denaturing PAA gel. Again, very careful pipetting is required in order to obtain quantifiable results.

## Future Perspectives

After an aptamer is well characterized, i.e., the minimal recognition domain is defined, the binding site determined via chemical modification, and the $K_d$ measured, then the aptamer is ready for high-resolution structure determination via nuclear magnetic resonance (NMR) or X-ray crystallography. The structural characterization of several aptamers via NMR revealed surprising features of RNA structure and ligand binding.[33–35] So far, the solution structure of two tobramycin aptamers has been determined via NMR. To recognize general features and define canonical elements of antibiotic binding sites, several aptamers need to be analyzed and compared. To date, all selections for antibiotic binders have been undertaken *in vitro.* We have observed that the aminoglycosides neomycin B and gentamicin, which are both potent inhibitors of group I intron splicing *in vitro,* inhibit splicing *in vivo* only indirectly. They do not interfere directly with splicing *in vivo.*[36] This observation has a big impact on future *in vitro* selections if the desired aptamers are to be used intracellularly.

[33] D. Patel, *Curr. Opin. Chem. Biol.* **1**, 32 (1997).
[34] D. J. Patel, A. K. Suri, F. Jiang, L. Jiang, P. Fan, R. A. Kumar, and S. Nonin, *J. Mol. Biol.* **272**, 645 (1997).
[35] G. R. Zimmermann, R. D. Jenison, C. L. Wick, J.-P. Simorre, and A. Pardi, *Nature Struct. Biol.* **4**, 644 (1997).
[36] C. Waldsich, K. Semrad, and R. Schroeder, *RNA* **4**, 1653 (1998).

## [16] Selection for RNA: Peptide Recognition through Sulfur Alkylation Chemistry

*By* MATT WECKER and DREW SMITH

We describe a method for the *in vitro* selection of 5'-phosphorothioate-modified RNA (GMPS–RNA) having enhanced sulfur alkylation reactivity with the substrate *N*-bromoacetylbradykinin (BrBK). This study shows that selection for catalytic activity actually coselected for binding specificity to a small peptide normally having little ground state structure in an aqueous environment. It also enlarges our understanding of what chemistries can be carried out by RNA. This article outlines our methods for selection, partitioning, and analysis and reviews some of our previously published results.[1]

[1] M. Wecker, D. Smith, and L. Gold, *RNA* **2**, 982 (1996).

METHODS IN ENZYMOLOGY, VOL. 318                                    0076-6879/00 $30.00

Materials and Reagents

*Synthesis of N-Bromoacetylbradykinin*

N-Bromoacetylbradykinin is synthesized by reacting 50 $\mu$l of 5 m$M$ bradykinin (BK; Sigma, St. Louis, MO) with three successive 250-$\mu$l portions of 42 m$M$ bromoacetic acid N-hydroxysuccinimide ester (Sigma) at 12-min intervals at room temperature. Excess bromoacetic acid N-hydroxy-succinimide ester is removed by filtration over 5 ml of AE-P2 aminoethyl-acrylamide (Bio-Rad, Hercules, CA) after 5 min of reaction at room temperature. This is followed by separation of BrBK over GS-10 Sepharose (Pharmacia, Piscataway, NJ). This material is assayed by reactivity with GMPS–RNA, is found to be essentially pure, and is used in the selections. However, after selections are completed, high-performance liquid chromatography (HPLC) analysis shows that 45% of the material is unreacted BK. For all kinetic analysis, BrBK is purified using HPLC. In this case, BrBK is synthesized by mixing 250 $\mu$l of 1 m$M$ BrBK with three consecutive additions of 3.5 mg bromoacetic acid N-hydroxysuccinimide at room temperature with 12-min reaction times. The reaction is injected in 80-$\mu$l volumes on a $C_8$, 100-Å 5-$\mu$m column (Rainin, Woburn, MA) at a flow rate of 1 ml/min using a 20–45% gradient of acetonitrile in water and 0.1% trifluoroacetic acid over 25 min. After overnight lyophilization, high-performance liquid chromatography (HPLC) analysis indicates a homogeneous product. BrBK concentrations are assigned using an absorption coefficient of 39,000 cm$^{-1}$ $M^{-1}$.

*Synthesis of Guanosine 5'-Monophosphorothioate*

GMPS is synthesized essentially as described.[2] The product is purified on a 2.5 × 15-cm DEAE column (Whatman, Clifton, NJ) with a linear gradient of from 0 to 0.5 $M$ ammonium acetate. Fractions are collected, assayed by absorbance at 260 nm, and peak fractions are run on PEI cellulose (Universal Scientific, Mesa, AZ) with 0.75 $M$ $KH_2PO_4$. Those fractions showing the purest migrations are pooled and lyophilized three times, resuspending in distilled water between lyophilizations. The finished product is aliquoted and stored frozen at $-70°$.

*Transcription with Guanosine Monophosphorothioate*

The influence of GMPS concentration on both overall yield of RNA and yield of GMPS–RNA from transcription is given in Table I. Reactions

---

[2] A. B. Burgin and N. R. Pace, *EMBO J.* **9,** 4111 (1990).

TABLE I
OPTIMIZATION OF GMPS–RNA TRANSCRIPTION

| [GTP] (mM) | [GMPS] (mM) | pmol GMPS–RNA produced | % RNA as GMPS–RNA |
|---|---|---|---|
| 0.2 | 0 | 0 | 0 |
| 0.2 | 1 | 15 | 88 |
| 0.2 | 2 | 14 | 93 |
| 0.2 | 8 | 12 | 95 |
| 0.2 | 20 | 8 | 95 |
| 2.0 | 0 | 0 | 0 |
| 2.0 | 1 | 81 | 64 |
| 2.0 | 2 | 116 | 80 |
| 2.0 | 8 | 112 | 89 |
| 2.0 | 20 | 11 | 69 |

are carried out in 10-$\mu$l volumes at 0, 1, 2, 8, or 20 mM GMPS with 0.2 or 2.0 mM GTP and 1.0 mM ATP, UTP, and CTP. Either [$\alpha$-$^{32}$P]GTP or [$\gamma$-$^{32}$P]GTP is added at 66 nM. For those trials at 8 and 20 mM GMPS, an equivalent concentration of $MgCl_2$ is added to compensate for the titration of free magnesium by the nucleotide. Reactions are carried out for 2 hr at 37°. RNA produced is measured through acid precipitation and factored by the specific activity of the label. The percentage of total RNA that has GMPS incorporated into the first position is followed by the decrease in incorporation of labeled [$\gamma$-$^{32}$P]GTP (where only those RNAs initiated with [$\gamma$-$^{32}$P]GTP are radioactive). The total amount of RNA produced from transcription at a given level of GMPS is determined by the amount of [$\alpha$-$^{32}$P]GTP incorporated and precipitated in cold, 1 M HCl on glass fiber filters (Whatman). This amount is then adjusted by a factor that accounts for the loss of radiolabel due to partial substitution of GMPS in the first position:

$$RNA_{total} = RNA_{obs} \times \{1 - p + p[n/(n - 1)]\}$$

where $p$ is the fraction of RNA that is GMPS–RNA and $n$ is the total number of G residues in the molecule.

*Synthesis of [($\beta$-Acryloylamino)phenyl]mercuric chloride (APM)*

APM is synthesized by the method of Igloi[3] and is stored at −70° in 40% acrylamide at a concentration of 10 M. Storage of 1 M APM in formamide for 3 months at room temperature, however, does not reduce

---

[3] G. L. Igloi, *Biochemistry* **27,** 3842 (1988).

its ability to retard GMPS–RNA or lead to less clean-running gels. On storage, any precipitate of APM is resolublized by gentle heating. All waste resulting from the preparation and use of APM should be handled and disposed of as mercuric waste.

## Purification of GMPS–RNA by APM–Polyacrylamide Gel Electrophoresis (APM–PAGE)

Transcriptions of GMPS–RNA are first purified over Microcon-30 filters (Amicon, Danvers, MA) to remove unincorporated GMPS. The reactions are then combined with 1 volume of loading buffer (7 $M$ urea, 0.025% (w/v) bromphenol blue, 0.025% (w/v) xylene cylenol, 1 m$M$ EDTA, 10 m$M$ Tris, pH 7.0) and heated at 70° for 5 min. This is then loaded on 7 $M$ urea, 8% polyacrylamide 25 $\mu M$ APM gels. At 25 $\mu M$ APM, a 2-cm well on a 1.5-mm-thick gel is sufficient for a 50-$\mu$l transcription reaction; higher loading of GMPS–RNA leads to smearing of the retarded band. The concentration of APM in the gel, however, can be increased to well over 250 $\mu M$ without anomalous running behavior. After running 20 min at 50 W, the retarded band containing GMPS–RNA is excised and eluted by crushing the band in 2 volumes of a buffer containing 100 m$M$ dithiothreitol (DTT) (to disrupt the bond between the APM and the thioate) and 1 $M$ sodium acetate, pH 5.2. Elution is followed by precipitation in 70% ethanol and washing in 70% ethanol.

## Purification of GMPS–RNA Using Thiopropyl-Sepharose 6B

The just-described process for the purification of GMPS–RNA is time-intensive, but gives a product of established size. A much quicker, potentially automated method for purification utilizes thiopropyl-Sepharose to covalently separate the thioate group of GMPS–RNA from RNA, followed by release of the GMPS–RNA by cleavage from the column with DTT.

1. Remove free GMPS from the transcription reaction by separation on a Microcon-30 column. Centrifuge at 12,000$g$ for 10 min at room temperature, followed by two consecutive 200-$\mu$l washes with TE (10 m$M$ Tris, pH 7.5, 1 m$M$ EDTA).
2. Bring this RNA to a final concentration of 10 m$M$ EDTA, 20 m$M$ HEPES, pH 7.0, and 500 m$M$ NaCl (this salt concentration is found to lessen the background binding of non-GMPS–RNA to thiopropyl-Sepharose).
3. We then use a "spin column" consisting of a 500-$\mu$l, thin-wall Eppendorf tube pricked on the bottom with a No. 1 lettering pen tip (Keuffel and Esser). This spin column retains the Sepharose while eluant is collected in a 1.5-ml Eppendorf tube. Prewash the thiopro-

pyl-Sepharose with 3 volumes of column buffer at room temperature (500 m$M$ NaCl, 20 m$M$ HEPES, pH 7.0) prior to use, and spin dry for 10 sec at 12,000$g$ (room temperature). Fifty microliters of Sepharose (wetted) is sufficient for an initial 25 $\mu$l of transcription reaction.

4. React the transcription mix with the Sepharose for 5 min at 70°. Spin at 12,000$g$ at room temperature, and spin wash with four 100-$\mu$l aliquots of 90% formamide, 50 m$M$ MES, pH 5.0, at 70°, then spin wash with four room temperature, 100-$\mu$l aliquots of 500 m$M$ NaCl in 50 m$M$ MES, pH 5.0, and finally spin elute with four room temperature, 100-$\mu$l aliquots of 100 m$M$ DTT in 50 m$M$ MES, pH 5.0.

6. Separate the DTT from the GMPS–RNA over a Microcon-30 spin filter at 12,000$g$ for 10 min, followed by two 200-$\mu$l washes with TE, inversion of the filter, and spinning at 3000$g$ for 30 sec, all at room temperature.

### In Vitro Selection

The initial library of GMPS–RNA molecules is transcribed from a polymerase chain reaction (PCR) library[4] and consists of approximately $5 \times 10^{13}$ GMPS–RNA molecules 76 nucleotides in length with a central randomized region 30 nucleotides in length. All reactions (see Table II) are carried out at 650 $\mu M$ BrBK and either 40 $\mu M$ (rounds 1–6) or 20 $\mu M$ (rounds 7–12) GMPS–RNA. Before reaction, 4.88 $\mu$l of GMPS–RNA is added to 1.63 $\mu$l heating buffer (4 m$M$ EDTA, 40 m$M$ MES, pH 5.0) and heat denatured for 3 min at 70°. The mix is then cooled at room temperature for 5 min and put on ice before adding 2.5 $\mu$l of ice-cold 4× reaction buffer (600 m$M$ NaCl, 22.6 m$M$ MgCl$_2$, 50 m$M$ HEPES, pH 8.0). The mix is then preincubated for 3 min at the reaction temperature prior to the addition of 1 $\mu$l of BrBK. Final reaction conditions are thus 150 m$M$ NaCl, 5.65 m$M$ MgCl$_2$, 0.65 m$M$ EDTA, 6.5 m$M$ MES, and 12.5 m$M$ HEPES, with a final pH of 7.0. Reactions are quenched with a final concentration of either 235 m$M$ sodium thiophosphate (rounds 1–8) or 235 m$M$ sodium thiosulfate (rounds 9–12). Sodium thiophosphate is found to not affect the capacity of APM–polyacrylamide appreciably, but gives slightly higher backgrounds than sodium thiosulfate during thiopropyl-Sepharose partitioning.

Because the reaction proceeds very quickly while being heated in loading buffer, those separations using APM–PAGE are loaded and run in 1 volume loading buffer without heating During the selection, each round is performed under several different reaction conditions, including the conditions used in the previous round as well as a more stringent condition

---

[4] D. Schneider, L. Gold, and T. Platt, *FASEB J.* **7**, 201 (1993).

TABLE II

BrBK SELEX PROFILE FOR RNA REACTANTS TO BRADYKININ

| Round | Temperature (°C) | Time (sec) | % RNA reacted | Ratio to background | Separation |
|-------|------------------|------------|---------------|---------------------|------------|
| 1 | 37 | 60 | 1.3 | 3.2 | Gel |
| 2 | 30 | 60 | 0.7 | 3.2 | Gel |
| 3 | 30 | 30 | 0.8 | 3.4 | Gel |
| 4 | 24 | 60 | 0.7 | 1.7 | Gel |
| 5 | 24 | 30 | 1.9 | 2.5 | Gel |
| 6 | 20 | 30 | 2.0 | 4.2 | Column |
| 7 | 0 | 30 | 1.2 | 3.1 | Column |
| 8 | 0 | 30 | 3.4 | 4.0 | Column |
| 9 | 0 | 60 | 4.5 | 4.9 | Column |
| 10 | 0 | 60 | 5.0 | 10.1 | Column |
| 11 | 0 | 30 | 2.5 | 9.0 | Column |
| 12 | 0 | 60 | 2.8 | 4.5 | Column |

[a] Percentage RNA reacted indicates the percentage of the total GMPS–RNA present as product (BK-S-RNA) from acrylamide gel partitioning or freely eluting BK-S-RNA from thiopropyl-Sepharose partitioning. Background was subtracted from the recovered RNA in both of these cases. Background was determined from the amount of RNA recovered in a control reaction quenched prior to the addition of BrBK. Ratio to background is the ratio of reacted RNA to that present as background.

(shorter reaction time, lowered temperature). For each condition a control reaction is run in which the reaction is quenched prior to the addition of BrBK. The ratio is then assessed between product formation in the experimental reaction versus control background reaction. Of those reactions that have this ratio falling within the range of from 2 to 10, the one carried out under the most stringent condition is used for the next round of SELEX after reverse transcription, PCR amplification, and transcription.[4]

*Kinetic Analysis*

All kinetic analyses are carried out in 10-$\mu$l reaction volumes as described for the selection. The reaction temperature is 0° and the GMPS–RNA concentration is set to 2 $\mu M$ unless otherwise indicated. Reactions are quenched with 1.6 $\mu$l of 1.7 $M$ sodium thiophosphate. Loading buffer (11.6 $\mu$l) is added and the tube is transferred to a −70° dry ice bath. Reactions are thawed immediately prior to loading. Ten microliters of the reaction in loading buffer is loaded onto 8-mm-wide lanes on 0.75–mm-thick gels prerun at 50 W for 30 min. Gels are run until the xylene cylenol is within 2 cm of the bottom.

For 30N1 data, reactions are analyzed at 1-, 3-, and 6-hr time points at 74, 148, 295, 590, 1200, 4700, and 9400 $\mu M$ BrBK. For reactant 12.1 data, reactions are analyzed at 10-sec time points between 0 and 100 sec, with additional time points of 3.5, 10, 20, 100, 200, and 400 min. For the $N$-bromoacetamide trials, reactions are analyzed at 40, 120, 360, 1080, 3240, and 9720 $\mu M$ $N$-bromoacetamide at 1, 3.3, and 10 hr and at 0.23, 0.7, 2.1, 6.3, 18.9, and 56.7 m$M$ $N$-bromoacetamide for 0.2, 0.5, 1.5, and 4.5 hr. For those data sets that follow monophasic behavior, kinetic data are fit to

$$S_t/S_0 = e^{-k_{obs}t}$$

where $S_t/S_0$ is the fraction of unreacted GMPS–RNA remaining in the reaction at time $t$. When biphasic, reactions are fit to

$$S_t/S_0 = Fe^{(-k_1t)} + (1 - F)e^{(-k_2t)}$$

where $F$ is the fraction of fast-reacting GMPS–RNA in the reaction and $k_1$ and $k_2$ are the faster and slower rate constants, respectively. Inhibition kinetics are carried out at 0, 40, 120, 360, 1080, and 3240 $\mu M$ inhibitor for time points of 1, 3, and 9 hr for 30N1 GMPS–RNA and 24, 60, and 120

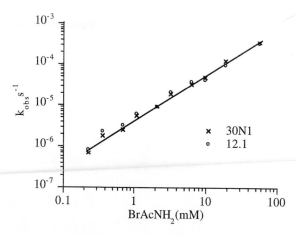

FIG. 1. The use of APM–PAGE to compare the activity of reactant 12.1 and the 30N1 pool RNA for BrBK. The second phase of a two-phase gel is shown. The first phase (not shown) consisted of 10 cm of a 7 $M$ urea 8% denaturing polyacrylamide gel and was used to discriminate between any artifacts left in the well and Hg-retarded GMPS–RNA. The second phase of the gel was identical to the first but included 25 $\mu M$ APM. As is typical with APM gels, approximately 3% of the GMPS–RNA (purified previously as GMPS–RNA using APM–PAGE) is not retarded by APM in the second purification. The cause of this is not known.

TABLE III
KINETIC DATA FOR REACTANT 12.1 AND ROUND ZERO 30N1 GMPS–RNA

| GMPS–RNA | $k_{obs}/[S]$ $(M^{-1} \, sec^{-1})$ | Inhibitor | $K_i$ (m$M$) |
|---|---|---|---|
| 30N1 | 0.062 | — | — |
| 30N1 | — | BK | >10 |
| Reactant 12.1 | 150 | — | — |
| Reactant 12.1 | — | BK | 0.23 ± 0.08 |
| Reactant 12.1 | — | de-Arg$^9$-BK | 1.8 ± 0.4 |
| Reactant 12.1 | — | de-Arg$^1$-BK | >10 |
| Reactant 12.1 | — | Arginine | >10 |

sec for reactant 12.1. $K_i$ values (concentration of inhibitor giving half-maximal inhibition) are determined using

$$k_{obs} = k_0/(1 + [I]/K_i)$$

where $k_0$ is the rate of the uninhibited reaction and I is the inhibitor concentration.

## Results

### *Optimizing Transcription with Guanosine Monophosphorothioate*

GMPS is more efficient than GTP in priming the synthesis of RNA by T7 DNA-dependent RNA polymerase (see Table I). To get the greatest amount of GMPS-initiated RNA from a transcription, GTP and GMPS

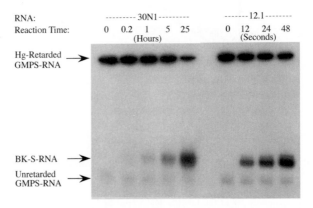

FIG. 2. A comparison of $k_{obs}$ of reactant 12.1 and the 30N1 GMPS–RNA pool with *N*-bromoacetamide.

should be used at a level of 2.0 m$M$, resulting in approximately 80% of the total RNA being GMPS–RNA. For those situations where a greater percentage of the transcripts must be initiated with GMPS, the GMPS concentration should be raised to 8 m$M$ and the concentration of GTP should be lowered to 0.2 m$M$.

## Selection for GMPS–RNA Catalysts

The selection was essentially complete by round 10, as seen by comparison of the values for percentage RNA reacted in Table II. In comparison of $k_{obs}$ values for the pools at rounds 0, 2, 4, 6, 8, 10, and 12, there was an approximate 360-fold increase in the $k_{obs}$ value from the original starting pool to the pool at round 10.[1] Most of the increase was seen between rounds 4 and 6 with no increase seen between rounds 10 and 12. Fifty-six clones were isolated from the round 10 and round 12 pools, leading to 29 distinct sequences (not shown).[1]

One reactant, 12.1, was analyzed for reactivity to BrBK (Fig. 1, Table III), as well as for reaction kinetics with various inhibitors. This reactant had a nearly 2500-fold increase in reaction rate as compared to the starting pool. The selected reactant had a $K_i$ of 230 $\mu M$ for bradykinin, whereas the starting pool had a $K_i$ greater than 10 m$M$. Recognition of bradykinin ($NH_2$-Arg-Pro-Pro-Gly-Phe-Ser-Pro-Phe-Arg-COOH) by reactant 12.1 required the arginines on both the amino and the carboxyl termini for full inhibition of the reaction. Reactant 12.1 was compared to the 30N1 GMPS–RNA pool in reaction with the minimal substrate $N$-bromoacetamide (Fig. 2). At all $N$-bromoacetamide concentrations tested, the reaction rates were identical, indicating that the increased rate of reaction of reactant 12.1 with BrBK was not primarily due to increased availability or reactivity of the 5′-thioate.

We have used the SELEX process to isolate a catalytic RNA with the ability to carry out sulfur alkylation chemistry on a specific substrate. Our belief is that in the original pool of GMPS–RNA molecules there are individual molecules that would have the inherent ability to specifically bind and react with a discreet subset of molecules from even a very large library of halide–acetyl compounds. Coupled with the potential for substitution of modified nucleotides in the original RNA library,[5,6] we should be able to generate some very interesting chemistries and potentially a wide variety of compounds for pharmaceutical and agricultural uses.

[5] T. M. Tarasow, S. L. Tarasow, and B. Eaton, *Nature* **389,** 54 (1997).
[6] T. W. Wiegand, R. C. Janssen, and B. E. Eaton, *Chem. Biol.* **4,** 675 (1997).

# [17] *In Vitro* Selection of RNA Substrates for Ribonuclease P and Its Catalytic RNA

By FENYONG LIU, JUN WANG, and PHONG TRANG

## Introduction

*In vitro* selection procedures have been widely used to generate new RNA catalysts and evolve novel ribozyme variants from those found in nature.[1-6] Moreover, these procedures have been used to select nucleic acid-based substrates for different enzymes and to generate ligands that bind to different molecules such as proteins and small compounds.[7-13] The following *in vitro* selection protocol has been used successfully to generate novel substrates for ribonuclease P (RNase P) holoenzyme and its catalytic RNA subunit.[9] RNase P is a ribonucleoprotein complex responsible for the 5' maturation of tRNAs.[14,15] This essential enzymatic activity has been found in all organisms examined. It catalyzes a hydrolysis reaction to remove a 5' leader sequence from tRNA precursors (pre-tRNA) and several small RNAs. In *Escherichia coli*, RNase P consists of a catalytic RNA subunit (M1 RNA) and a protein subunit (C5 protein).[14,15] M1 RNA is 377 nucleo-

[1] B. Cuenoud and J. W. Szostak, *Nature* **375**, 611 (1995).
[2] D. P. Bartel and J. W. Szostak, *Science* **261**, 1411 (1993).
[3] R. R. Breaker and G. F. Joyce, *Chemistry and Biology* **1**, 223 (1994).
[4] L. Gold, P. Allen, J. Binkley, D., B., D. Schneider, S. R. Eddy, C. Tuerk, L. Green, S. MacDougal, and D. Tasset, *in* The RNA world (R. R. Gesteland and J. F. Atkins, eds.), pp. 497–509. Cold Spring Harbor Laboratory: Cold Spring Harbor, New York, 1993.
[5] G. F. Joyce, *Sci. Am.* **267**, 90 (1992).
[6] J. W. Szostak and A. D. Ellington, *in* The RNA world (R. R. Gesteland and J. F. Atkins, eds.), pp. 511–533. Cold Spring Harbor Laboratory Press: Cold Spring Harbor, New York, 1993.
[7] L. C. Bock, L. C. Griffin, J. A. Latham, E. H. Vermaas, and J. J. Toole, *Nature* **355**, 564 (1992).
[8] A. D. Ellington and R. Conrad, *Biotechnol. Annu. Rev.* **1**, 185 (1995).
[9] F. Liu and S. Altman, *Cell* **77**, 1083 (1994).
[10] T. Pan, *Biochemistry* **34**, 8458 (1995).
[11] E. T. Peterson, J. Blank, M. Sprinzl, and O. C. Uhlenbeck, *EMBO J.* **12**, 2959 (1993).
[12] C. Tuerk and L. Gold, *Science* **249**, 505 (1990).
[13] Y. Yuan and S. Altman, *Science* **263**, 1269 (1994).
[14] S. Altman, L. Kirsebom, and S. J. Talbot, *FASEB J.* **7**, 7 (1993).
[15] N. R. Pace and J. W. Brown, *J. Bact.* **177**, 1919 (1995).

tides long and the C5 protein consists of 119 amino acids. In the presence of a high concentration of salt, such as 100 m$M$ Mg$^{2+}$, M1 RNA acts as a catalyst and cleaves pre-tRNAs *in vitro* by itself.[16] The addition of C5 protein increases the rate of cleavage by M1 RNA *in vitro* dramatically and is required for RNase P activity and cell viability *in vivo*.[17] It has been proposed that C5 protein, an extremely basic protein, functions to stabilize the conformation of M1 RNA by providing electrostatic shielding.[17,18] RNase P and M1 RNA recognize a common high-order structure of their natural tRNA substrates but not the primary sequence because there is little sequence homology among these substrates.[14,15] Any RNA can be hydrolyzed by M1 ribozyme and RNase P if this RNA can be folded into a tRNA-like structure.[19] Moreover, an mRNA can be hydrolyzed by these enzymes if a custom-designed sequence can be constructed to hybridize to the target mRNA to form a tRNA-like structure (Fig. 1B).[19] This custom-designed sequence is called an external guide sequence (EGS) because it hybridizes with the target mRNA and directs M1 ribozyme and RNase P to recognize the structure of the mRNA–EGS complex and cleave at the site where the EGS hybridizes to the target mRNA (Fig. 1B). Subsequent studies have shown that EGSs efficiently target both viral and host mRNAs for cleavage by RNase P from HeLa cells and *E.coli* and effectively inhibit expression of these mRNAs in cell culture.[20–22] These studies demonstrate the feasibility to use the EGS technology as a unique gene-targeting approach in both basic research and clinical applications. The EGS may consist of two sequence elements: a sequence complementary to the targeted sequence and a sequence that resembles a portion of the T-loop and stem, and the variable region of a tRNA (Fig. 1B). The following protocol is for selecting novel substrates for studies of substrate recognition by RNase P and its catalytic RNA and for generating highly active EGSs that can be used for gene-targeting applications.

[16] C. Guerrier-Takada, K. Gardiner, T. Marsh, N. Pace, and S. Altman, *Cell* **35,** 849 (1983).

[17] V. Gopalan, S. J. Talbot, and S. Altman, *in* "RNA-protein interactions" (K. Nagai and I. W. Mattaj, eds.), pp. 103–126. Oxford University Press: Oxford, 1995.

[18] C. Reich, G. J. Olsen, B. Pace, and N. R. Pace, *Science* **239,** 178 (1988).

[19] A. C. Forster and S. Altman, *Science* **249,** 783 (1990).

[20] D. Plehn-Dujowich and S. Altman, *Proc. Natl. Acad. Sci. U.S.A.* **95,** 7327 (1998).

[21] Y. Yuan, E. Hwang, and S. Altman, *Proc. Natl. Acad. Sci. U.S.A.* **89,** 8006 (1992).

[22] Y. Li, C. Guerrier-Takada, and S. Altman, *Proc. Natl. Acad. Sci. U.S.A.* **89,** 3185 (1992).

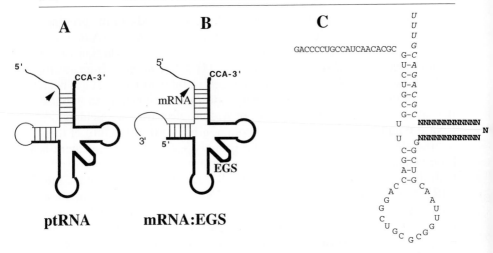

FIG. 1. Schematic representation of substrates for *E. coli* RNase P holoenzyme and its catalytic RNA subunit (M1 RNA). (A) A natural substrate (ptRNA) for RNase P and M1 ribozyme. (B) A hybridized complex of a target RNA (e.g., mRNA) and an EGS that resembles the structure of a tRNA and can be cleaved by RNase P and M1 ribozyme. The anticodon domain of the EGS is dispensable for EGS targeting activity.[13] (C) The substrate used for selection. The 5' portion of the sequence shown (position 1 to 55), including the large loop, represents a part of oligonucleotide TK21 and consists of part of the sequence of the mRNA that encodes thymidine kinase from herpes simplex virus 1. In addition to the sequence shown, TK21 also contains the sequence coding for the promoter for T7 RNA polymerase at its 5' region and is used as the 5' primer for PCR to amplify the DNA molecules coding for the RNA substrates during each cycle of selection. The rest of the sequence of the substrate molecule was designed to hybridize to the thymidine kinase-encoding sequence or was randomized. The 25 nucleotides that were randomized are each indicated by N. They are shown in boldface type. Note that the 3' invariant CCA sequence in all mature tRNAs has been replaced by the sequence UUU.[9]

## Reagents

### Oligonucleotide Primers

5' polymerase chain reaction (PCR) primer
TK21 (GGAATTCTAATACGACTCACTATAGACCCCTGCCA-
TCAACACGCGTCTGCGTTCGACCAGGCTGCGCGGTTAA-
CGTCG)
3' PCR primer
TK23(AAACGTCTGCG)
3' primer for the construction of randomized substrate sequences

TK22 (5'-AAACGTCTGCGNNNNNNNNNNNNNNNNNNNNNNNNN-
NNNCCGACGTTAA-3')
5' primer for the construction of EGSs
TK24 (5'-GGAATTCTAATACGACTCACTATAGGTTAACGTC
-3')

All oligonucleotides are synthesized chemically with a DNA synthesizer.
Equal molar amounts of A, C, G, and T are incorporated in the positions
represented as N during the chemical synthesis.

### PCR Reagents

10× PCR buffer (500 m$M$ KCl, 100 m$M$ Tris–HCl, pH 8.3)
25 m$M$ MgCl$_2$
10 m$M$ dATP
10 m$M$ dGTP
10 m$M$ dTTP
10 m$M$ dCTP
*Taq* DNA polymerase (Perkin-Elmer, Norwalk, CT)
Each dNTP is diluted from stock solution of 100 m$M$. The stock solution
is dissolved from solid powder and adjusted to pH 7 by the addition of
1 $M$ NaOH.

### Reverse Transcription

5× Moloney murine leukemia virus (MMLV) RT buffer [250 m$M$
Tris–HCl, pH 8.3, 375 m$M$ KCl, 15 m$M$ MgCl$_2$, 25 m$M$ dithiothreitol (DTT)]
MMLV reverse transcriptase (GIBCO-BRL, Grand Island, NY)
5× Avian myeloblastosis (AMV) RT buffer (500 m$M$ Tris–HCl, pH
8.3, 50 m$M$ KCl, 30 m$M$ MgCl$_2$, 50 m$M$ DTT)
AMV reverse transcriptase (Boehringer Mannheim, Indianapolis, IN).

### In Vitro Transcription

5× buffer for *in vitro* transcription (200 m$M$ Tris–HCl, pH 7.9, 30 m$M$
MgCl$_2$, 50 m$M$ DTT, 10 m$M$ spermidine)
100 m$M$ DTT
10 m$M$ ATP
10 m$M$ GTP
10 m$M$ CTP
10 m$M$ UTP
RNasin RNase inhibitor (Promega Inc., Madison, WI)
T7 RNA polymerase (Promega Inc.)

Each NTP is diluted from a stock solution of 100 m$M$. The stock solution is dissolved from solid powder and adjusted to pH 7 by the addition of 1 $M$ NaOH.

### Radioisotope and Gel Electrophoresis

[$\alpha$-$^{32}$P]GTP
5% nondenaturing polyacrylamide gels
8% denaturing gels that contain 7 $M$ urea

### Reagents for Selection Procedure

10× buffer A (500 m$M$ Tris–HCl, pH 7.5, 1000 m$M$ NH$_4$Cl, 1000 m$M$ MgCl$_2$) (for digestion by M1 RNA)
10× buffer B (500 m$M$ Tris–HCl, pH 7.5, 1000 m$M$ NH$_4$Cl, 100 m$M$ MgCl$_2$) (for digestion by RNase P holoenzyme)
RNase P holoenzyme (synthesized as described previously[9,23])
RNase P catalytic RNA (M1 RNA) (synthesized *in vitro* as described previously[9,23])

### Design and Construction of DNA Library That Codes for RNA Substrate Pool

RNA molecules that contain only a stem structure and a randomized sequence of 25 nucleotides are used as substrates. The stem structure contains only seven base pairs and resembles the acceptor stem of a tRNA (Fig. 1C). None of the sequences in the substrate are derived from a tRNA in order to exclude the possibility that a tRNA sequence might specifically interact with RNase P and M1 RNA. The 5′ portion of the sequence shown (position 1 to 55), including the large loop, consists of part of the sequence of the mRNA that encodes thymidine kinase from herpes simplex virus 1 (Fig. 1C). The rest of the sequence is designed to hybridize to the thymidine kinase-encoding sequence or is randomized. Procedures to construct the DNA pools that serve as the templates for synthesis of RNA substrates are as follows.

### Annealing

Equal amounts (5 nmol) of oligonucleotides TK21 and TK22 are resuspended into 500 $\mu$l of annealing buffer (10 m$M$ Tris, pH 7.5, 10 m$M$ KCl).

---

[23] S. J. Talbot and S. Altman, *Biochemistry* **33,** 1399 (1994).

The reaction mixture is heated at 95° for 2 min and then allowed to anneal by dropping the temperature gradually to 37°.

## Synthesis of Double-Stranded DNA Molecules from Single-Stranded Oligonucleotides by Reverse Transcription

Different methods (including using *E. coli* DNA polymerase I Klenow fragment and PCR with *Taq* DNA polymerase) are used to generate the double-stranded DNA templates from single-stranded oligonucleotides. Our results indicate that reverse transcription by MMLV RT yields the most efficient synthesis (about 60%) from double-stranded DNA templates. For a 40-$\mu$l reaction, assemble the reaction mixture as follows.

16 $\mu$l of annealing mixture

8 $\mu$l of 5× MMLV reverse transcription buffer (500 m$M$ Tris–HCl, pH 8.3, 50 m$M$ KCl, 30 m$M$ MgCl$_2$, 30 m$M$ DTT)

4 $\mu$l of dATP (10 m$M$)

4 $\mu$l of dCTP (10 m$M$)

4 $\mu$l of dGTP (10 m$M$)

4 $\mu$l of dTTP (10 m$M$)

2 $\mu$l MMTV reverse transcriptase

Incubate at 37 °C for 3–4 hr

The synthesized double-stranded DNA templates are loaded on 5% nondenaturing polyacrylamide gels, and the full-length DNA products are excised from the gel, eluted from the gel pieces, and finally purified by phenol/chloroform extraction followed by ethanol precipitation. This step serves two purposes. First, gel purification eliminates the non-full-length DNA sequences that might be generated due to the premature termination of reverse transcription. Second, the primers (TK21 and TK22) are removed. We found that a combination of these primers with the 3' PCR primer TK23 (see later) interferes with the subsequent PCR reaction.

## Amplification of Double-Stranded DNA Templates

It is important to preamplify the DNA oligonucleotides prior to *in vitro* transcription of the RNA library. The main purpose of this step is to increase the copy numbers of each individual sequence. There is only a single copy of each specific sequence in our RT-generated DNA pool. Some of these sequences might be lost during the selection procedures in which complete recovery of nucleic acid molecules might not be possible (e.g., gel purification, ethanol precipitation). This amplification step would increase the probability that a particular sequence, which was present in the original pool and might be efficiently cleaved by the enzyme, would not be lost during selection. In our experience, an amplification of 5- to 10-fold is usually adequate. The following is the protocol (see Fig. 2).

For a 100-$\mu$l reaction, add the following reagents.

10 $\mu$l of 10× PCR buffer
6 $\mu$l of 25 m$M$ MgCl$_2$
2 $\mu$l of 10 m$M$ dATP
2 $\mu$l of 10 m$M$ dGTP
2 $\mu$l of 10 m$M$ dCTP
2 $\mu$l of 10 m$M$ dTTP
20 pmol of RT-generated DNA templates
Water filled to 100 $\mu$l
400 pmol of 5′ primer TK21
400 pmol of 3′ primer TK23
*Taq* DNA polymerase 2.5 unit

The PCR reaction mixture is overlaid with mineral oil and cycled at 94° for 1 min, at 47° for 1 min, and at 72° for 1 min. To avoid overamplification and the accumulation of artifacts, samples (3–5 $\mu$l) are checked by agarose gel electrophoresis every two cycles. Amplification reactions are stopped in the PCR cycle when the intensity of the amplified products on the gel is not increased in the next cycle. The PCR products are loaded on 5% nondenaturing polyacrylamide gel, and the full-length DNA products are excised from the gel, eluted from the gel pieces, and finally purified by phenol/chloroform extraction followed by ethanol precipitation. We usually obtain 100–150 pmol of PCR products after gel purification.

## Production of RNA Library by *in Vitro* Transcription

For a 100-$\mu$l *in vitro* transcription reaction, add the following reagents sequentially.

**TK21**
5′-GGAATTCTAATACGACTCACTATAGACCCCTGCCATCAACACGCGTCTGCGTTCGACCAGGCTGCGCGGTTAACGTCG-3′
                                                                    ||||||||||
                                    5′-AATTGCAGCCNNNNNNNNNNNNNNNNNNNNNNNNNNNNNNNGCGTCTGCAAA-3′
**TK22**

                                                    5′-GCGTCTGCAAA-3′
**TK23**

FIG. 2. Schematic representation of the hybrid formed between oligonucleotides TK21 and TK22. Reverse transcription of the hybrids gave rise to double-stranded DNA molecules. These DNA molecules were subsequently amplified by PCR to generate the DNA pool coding for the RNA substrates used for the selection (see text and Fig. 3).[9] The boxed sequence of TK22 represents that of oligonucleotide TK23, which was used as the 3′ primer for the PCR and RT-PCR to amplify the DNA pool and the cleavage products generated in each cycle of the selection.

20 $\mu$l of 5× buffer for *in vitro* transcription
10 $\mu$l of 100 m$M$ DTT
5 $\mu$l of 10 m$M$ ATP
5 $\mu$l of 10 m$M$ GTP
5 $\mu$l of 10 m$M$ CTP
5 $\mu$l of 10 m$M$ UTP
Water filled to 90 $\mu$l
7 $\mu$l of PCR-generated DNA templates (about 10 $\mu$g DNA)
2–3 $\mu$l (20–30 $\mu$Ci) [$\alpha$-$^{32}$P]GTP
1.5 $\mu$l of RNasin RNase inhibitor
1.5 $\mu$l of T7 RNA polymerase

The reaction mixture is incubated at 35–37° for at least 16 hr and is stopped by adding 100 $\mu$l of denaturing dye [1 m$M$ Tris–HCl, pH 7.0; 8 $M$ urea, 0.01% (w/w) xylene cyanol, and 0.01% (w/w) bromphenol blue]. The reaction mixture is loaded on 8% denaturing gels that contain 7 $M$ urea. The full-length RNA products are visualized either by autoradiography or by UV shadowing, excised from the gel, and purified by extraction followed by ethanol precipitation. The yield of the *in vitro* RNA synthesis is determined by measuring either the concentration of the RNA using a spectrophotometer or quantitation of the radioactivity using a phosphorimager STORM840 (Molecular Dynamics, Sunnyvale, CA). For a typical 100-$\mu$l reaction, we routinely obtain 50–100 pmol of RNA products.

Selection Protocols

The selection procedures are shown in Fig. 3 and can be summarized as follows.[9]

*Cleavage Reaction*

In the first (initial) selection cycle, 20 nmol of the pool of RNA substrates that contain the randomized sequence are digested either by M1 RNA or RNase P holoenzyme in a volume of 2 ml. The buffers used for digestion with M1 RNA and RNase P holoenzyme are buffer A (50 m$M$ Tris–HCl, pH 7.5, 100 m$M$ NH$_4$Cl, 100 m$M$ MgCl$_2$) and buffer B (50 m$M$ Tris–HCl, pH 7.5, 100 m$M$ NH$_4$Cl, 10 m$M$ MgCl$_2$), respectively. One hundred picomoles of M1 RNA is used in the selection experiments with the catalytic RNA subunit alone, whereas 10 pmol of M1 RNA and 300 pmol of C5 protein are used in those with the RNase P holoenzyme. The reaction mixture is incubated at 37° for 16 hr in the presence of the RNasin RNase inhibitor. The digestion is stopped by adding 1.5 ml denaturing dye.

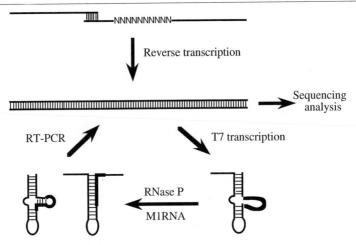

FIG. 3. Schematic representation of the *in vitro* selection procedure (see text).

In subsequent cycles of selection, 5–10 pmol of radiolabeled substrates is used and digested with the appropriate enzyme in a volume of 20 $\mu$l. Between the second and the fourth cycles of selection, substrates are incubated for 120 min at 37° with either 20 pmol of M1 RNA or 2 pmol of RNase P holoenzyme. During the final four cycles of selection, the incubation time is shortened to 10 min. The amounts of M1 RNA and of RNase P holoenzyme used are reduced by 100- and 200-fold, respectively. This strategy allows enrichment of all sequences that exhibit similar susceptibility to cleavage by either enzyme.

## Gel Purification

The reaction mixture is loaded on 8% denaturing gels. A radiolabeled RNA sample, FL113, which serves as a size marker for the cleavage product, is also loaded on the gels. This RNA is a truncated substrate and contains the sequence of the substrate from position 21–91 (Fig. 1C). The gel is subjected to autoradiography. The gel section containing the 3′ proximal cleavage products (i.e., the sequence from position 21–91) that comigrate with RNA FL113 is excised from the gel. RNA cleavage products are extracted by soaking the gel piece in diethyl pyrocarbonate (DEPC)-treated water and finally purified by ethanol precipitation.

During the first two cycles of selection, very little amount of cleavage products is generated and little radioactivity of the products is detected

(Fig. 4A). Therefore, it is important to include the size marker RNA FL113. We usually use the same amount of substrate for each cycle of selection (e.g., 5–10 pmol). The selection stringency is adjusted by changing the length of time for digestion and the amount of the enzymes used. In order to achieve optimal enrichment and selection, the amount of cleavage products usually accounts for not more than 0.5% of input substrates for cycles 2 and 5 and not more than 20% for cycles 4 and 8.

*Amplification by Reverse Transcription-PCR*

Seven microliters of purified RNA cleavage products is mixed with 1 $\mu$l (about 50 pmol) reverse transcription primer (TK23). The mixture is heated at 95° for 2 min and then put on ice immediately. The mixture is added with 2 $\mu$l each of 10 m$M$ dATP, dGTP, dCTP, and dTTP, 4 $\mu$l of 5× RT buffer for AMV reverse transcriptase (500 m$M$ Tris–HCl, pH 8.3, 50 m$M$ KCl, 30 m$M$ MgCl$_2$, 50 m$M$ DTT), 0.5 $\mu$l of RNasin RNase inhibitor, and 1 $\mu$l of AMV reverse transcriptase. The reaction is carried out at 42° for 2 hr. Then the PCR reaction mixture is assembled in a volume of 100 $\mu$l that contains 10 m$M$ Tris–HCl, pH 8.3, 15 m$M$ MgCl$_2$, 0.01% gelatin, 50 m$M$ KCl, 0.2 m$M$ dNTPs, 2.5 unit *Taq* DNA polymerase (Perkin-Elmer), and 200 pmol each of oligodeoxynucleotides TK21 and TK23. The reaction mixture is overlaid with mineral oil and cycled at 94° for 1 min, at 47° for 1 min, and at 72° for 1 min. To avoid overamplification and the accumulation of artifacts, samples (5–10 $\mu$l) are checked by agarose gel electrophoresis every three cycles, and the optimal number of cycles is determined. In our experience, a PCR of 18–24 cycles is adequate for amplification. The PCR products are loaded on 5% nondenaturing polyacrylamide gel, and the full-length DNA products are excised from the gel, eluted from the gel pieces, and finally purified by phenol/chloroform extraction followed by ethanol precipitation.

*In Vitro Transcription*

The transcription reactions are carried out as described in the previous section for the generation of a RNA substrate library from the DNA pool. The full-length RNA products are separated on a 8% denaturing gel, visualized either by autoradiography or by UV shadowing, excised from the gel, and finally purified by extraction followed by ethanol precipitation. These RNA molecules are used as the substrates for cleavage in the next cycle of selection (Fig. 3).

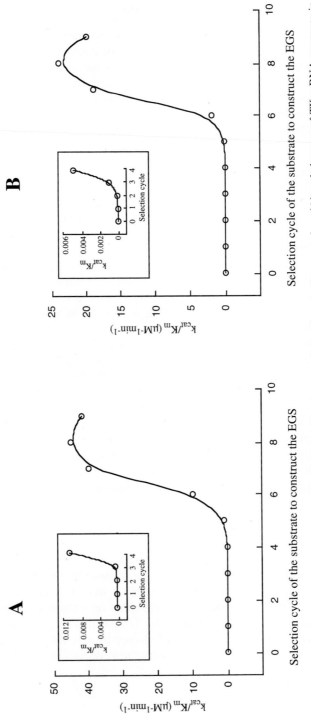

Fig. 4. Cleavage of substrates in populations of randomized substrates during the selection procedure (A) and cleavage of TK mRNA sequence in the presence of the EGSs generated from each of these substrate populations (B). Only reactions catalyzed by RNase P holoenzyme are shown. Kinetic analyses were carried out under single-turnover conditions, and the values of $k_{cat}/K_m$ were determined as described previously.[9,13] No significant increase in the rate of cleavage of substrates was observed after seven cycles.

Assessing Progress of Selection

The susceptibility of the substrate population to cleavage by RNase P holoenzyme and M1 RNA was determined after each cycle of selection in order to assess the progress of the selection. The assays for cleavage of substrates by RNase P holoenzyme and M1 RNA have been described previously.[9,16]

For a 10-$\mu$l M1 RNA reaction, add

1 $\mu$l of 10× buffer A (500 m$M$, pH 7.5, 1000 m$M$ NH$_4$Cl, 1000 m$M$ MgCl$_2$)

1000 cpm of $^{32}$P-radiolabeled substrate population (about 0.1 pmol)

water filled to 10 $\mu$l

0.5 $\mu$l of RNasin RNase inhibitor

1 $\mu$l M1 RNA of different concentrations

For a 10-$\mu$l of RNase P holoenzyme reaction, add

1 $\mu$l of 10× buffer B (500 m$M$, pH 7.5, 1000 m$M$ NH$_4$Cl, 100 m$M$ MgCl$_2$)

1000 cpm of $^{32}$P-radiolabeled substrate population (about 0.1 pmol)

water filled to 10 $\mu$l

0.5 $\mu$l of RNasin RNase inhibitor

1 $\mu$l RNase P holoenzyme of different concentrations

The cleavage reactions are incubated at 37°. Aliquots are withdrawn from reaction mixtures at regular intervals, and the reactions are stopped by adding 10 $\mu$l of denaturing dye.[9] The mixtures are loaded on a 8% polyacrylamide gel that contains 7 $M$ urea. The cleavage products are quantitated by a STORM840 phosphorimager (Molecular Dynamics, Sunnyvale, CA), and the rates of formation of cleavage products (e.g., $k_{cat}/K_m$) are determined (Fig. 4).

Potential Problems and Applications of Selection Procedure

This procedure has been used to select substrates for *E. coli* RNase P holoenzyme and M1 RNA.[9] Similar procedures have also been used for the selection of substrates for RNase P holoenzyme from HeLa cells.[13] Moreover, the selected substrates can be used for construction of efficient EGSs for gene targeting applications.[13] We have also determined whether the substrate population selected can be used to generate EGSs that efficiently target TK mRNA sequence (represented in TK21 sequence) for cleavage by the enzymes (e.g., *E. coli* RNase P holoenzyme) (Fig. 4B) (F. Liu, J. Wang, and P. Trang, unpublished data). DNAs coding for the EGSs were generated by PCR using the substrate populations as the templates and oligonucleotides TK24EGS (5'-GGAATTCTAATACGACTCAC-TATAGGTTAACGTC-3') and TK23 as the 5' and 3' primers, respectively.

EGS RNAs were synthesized from these DNA templates by T7 RNA polymerase. EGSs and *E. coli* RNase P holoenzyme were incubated with substrate tk46, which contained the sequence between position 1 and 46 in the substrate shown in Fig. 1C. The overall cleavage efficiency, indicated as the values of $K_{cat}/K_m$, showed that the targeting activity of EGSs improved as the susceptibility of substrate population for cleavage by RNase P increased (Fig. 4B). These results are consistent with the previous observations that EGSs generated from the substrates selected for cleavage by RNase P from HeLa cells can target mRNAs for cleavage by human RNase P.[13] Moreover, these data demonstrate the feasibility to use the selection approach to generate efficient EGSs for gene targeting applications.

Experiments have also been carried out under low salt conditions (i.e., buffer B) to select substrates for M1 RNA alone in the absence of C5 protein. No detectable substrate species were found after 8 cycles of selection. While an RNA population was selected under these conditions after 16 cycles, these RNA molecules turned out to be self-cleaving RNAs and are not substrates for M1 RNA. Similar results were also found in our *in vitro* selection experiments with H1 RNA, the RNA subunit of human RNase P (P. Trang, S. Jo, J. Kim, and F. Liu, unpublished data). These results are consistent with the notions that high ionic strength might be required for substrate binding and folding of the active conformation of the ribozymes, possibly by neutralizing the negative charges of these molecules.

### Acknowledgments

The selection procedures were initially developed in the laboratory of Dr. Sidney Altman at Yale University while one of us (F.L.) was a Parke-Davis postdoctoral fellow of Life Sciences Research Foundation. We are grateful to Dr. Yan Yuan for invaluable advice during the course of the study. J.W. is a visiting scientist from Jinan University of China. F.L. is currently a Pew Scholar in Biomedical Sciences and a recipient of Regent's Junior Faculty Fellowship of University of California and a Basil O'Connor Starter Scholar Research Award from March of Dimes Birth Defects Foundation. Research in this laboratory has been supported by grants from University of California Cancer Research Coordinating Committee, Hellman Family Fund, University of California AIDS Research Program (R98-B-146), and U.S. Public Health Services (AI41927 and GM54875).

## [18] RNA–Protein Interactions in Ribosomes: *In Vitro* Selection from Randomly Fragmented rRNA

*By* Ulrich Stelzl, Christian M. T. Spahn, and Knud H. Nierhaus

### Introduction

The ribosome is the largest ribonucleoprotein particle (RNP) in the cell. The *Escherichia coli* ribosome consists of 54 different proteins and three RNAs: 5S RNA (120 nucleotides), 16S RNA (1542 nucleotides), and 23S RNA (2904 nucleotides). A detailed picture of the ribosome has been achieved by cryoelectron microscopy (cryo-EM),[1,2] reaching about 15 Å resolution with a programmed ribosome that carries a formylmethionine (fMet)-tRNA at the P site.[3] Cryo-EM models were initially used to solve the phase problem of X-ray diffraction patterns from crystals of ribosomal subunits. Thus, a 9-Å resolution map was derived from crystals of the large ribosomal subunit of *Haloarcula marismortui*.[4] Modeling of biochemical data in combination with cryoelectron microscopy led to a comprehensive model for 16S rRNA folding within the subunit.[5-7] In addition, the tertiary structures of 15 isolated ribosomal proteins are known.[8,9]

Assembly and functions of the ribosome are dependent on the highly cooperative interaction of the different components. Out of 33 large subunit proteins (L-proteins), 21 are known to interact with 23S or 5S RNA, and at least 12 out of the 21 proteins of the small subunit (S-proteins) bind directly to 16S RNA in the 30S subunit.[10] In hydroxyl-radical footprinting experiments, each S-protein induces protections of specific nucleotides in

[1] J. Frank, J. Zhu, P. Penczek, Y. H. Li, S. Srivastava, A. Verschoor, M. Radermacher, R. Grassucci, R. K. Lata, and R. K. Agrawal, *Nature* **376,** 441 (1995).

[2] H. Stark, F. Mueller, E. V. Orlova, M. Schatz, P. Dube, T. Erdemir, F. Zemlin, R. Brimacombe, and M. van Heel, *Structure* **3,** 815 (1995).

[3] A. Malhotra, P. Penczek, R. K. Agrawal, I. S. Gabashvili, R. A. Grassucci, R. Jünemann, N. Burkhardt, K. H. Nierhaus, and J. Frank, *J. Mol. Biol.* **280,** 103 (1998).

[4] N. Ban, B. Freeborn, P. Nissen, P. Penczek, R. Grassucci, R. Sweet, J. Frank, P. B. Moore, and T. A. Steitz, *Cell* **93,** 1105 (1998).

[5] F. Mueller and R. Brimacombe, *J. Mol. Biol.* **271,** 524 (1997).

[6] F. Mueller and R. Brimacombe, *J. Mol. Biol.* **271,** 545 (1997).

[7] F. Mueller, H. Stark, M. van Heel, J. Rinke-Appel, and R. Brimacombe, *J. Mol. Biol.* **271,** 566 (1997).

[8] S. V. Nikonov, N. A. Nevskaya, R. V. Fedorov, A. R. Khairullina, S. V. Tishchenko, A. D. Nikulin, and M. B. Garber, *Biol. Chem.* **379,** 795 (1998).

[9] V. Ramakrishnan and S. W. White, *Trends Biochem. Sci.* **23,** 208 (1998).

[10] K. H. Nierhaus, *Biochimie* **73,** 739 (1991).

16S RNA on assembly, with the single exception of S5.[11] Only a very few of these RNA–protein interactions are characterized.[12]

We are interested in the interaction of ribosomal proteins with RNA pieces for the following reasons: (i) RNA–protein interactions are crucial for the assembly, stability, and function of the ribosome. Dissection of the ribosome into pieces results in smaller particles suitable for the extensive analysis of RNA structure, and even short isolated rRNA fragments are useful model systems.[13] A prime example of this is the so-called A-site region of 16S RNA. The structure of a 3′ fragment from the 16S rRNA in complex with the error inducing antibiotic paromomycin was obtained by nuclear magnetic resonance (NMR) and revived the discussion about the mechanism of the decoding step in elongation.[14,15] (ii) Many ribosomal proteins act as translational regulators and, in the case of L4, as a transcriptional regulator via binding to mRNAs.[16] (iii) A defined interaction between a ribosomal protein and an rRNA oligomer can serve as a paradigm for the study of RNA–protein interactions (cf. ribosomal protein S8[17,18] and S15[19,20]). RNA–protein interactions of ribosomal components are probably prototype interactions, at least in those cases where proteins present in the ribosomes of all kingdoms are involved, thus indicating their presence very early in evolution.

We developed an *in vitro* selection variant to detect specific interactions between ribosomal proteins and rRNA fragments. The method follows in principle the SELEX (systematic evolution of ligands by expontential enrichment) technology introduced by Turek and Gold[21] and by Ellington and Szostak[22] except that we use a pool of randomly fragmented rRNA generated by a random cutting of rDNA followed by a transcription step (SERF, *in vitro* selection from random RNA fragments). This has the

[11] T. Powers and H. F. Noller, *RNA* **1**, 194 (1995).
[12] D. E. Draper, *in* "Ribosomal RNA: Structure, Evolution, Processing and Function in Protein Biosynthesis" (R. A. Zimmermann and A. E. Dahlberg, eds.), p. 171. CRC Press, Boca Raton, FL, 1996.
[13] R. Schroeder, *Nature* **370**, 597 (1994).
[14] P. Purohit and S. Stern, *Nature* **370**, 659 (1994).
[15] D. Fourmy, M. I. Recht, S. C. Blanchard, and J. D. Puglisi, *Science* **274**, 1367 (1996).
[16] J. M. Zengel and L. Lindahl, *Prog. Nucleic Acid Res. Mol. Biol.* **47**, 331 (1994).
[17] H. Moine, C. Cachia, E. Westhof, B. Ehresmann, and C. Ehresmmann, *RNA* **3**, 255 (1997).
[18] K. Kalurachchi, K. Uma, R. A. Zimmermann, and E. P. Nikonowicz, *Proc. Natl. Acad. Sci. U.S.A.* **94**, 2139 (1997).
[19] R. T. Batey and J. R. Williamson, *J. Mol. Biol.* **261**, 536 (1996).
[20] A. A. Serganov, B. Masquida, E. Westhof, C. Cachia, C. Portier, M. Garber, B. Ehresmann, and C. Ehresmann, *RNA* **2**, 1124 (1996).
[21] C. Turek and L. Gold, *Science* **249**, 505 1990.
[22] A. D. Ellington and J. W. Szostak, *Nature* **346**, 818 (1990).

following advantages: (i) the selection aims just at the native rRNA–protein interaction; (ii) the size of the winner rRNA can be chosen, overlapping sequences of selected fragments reveal the minimal binding site; and (iii) the selection should be fast and efficient because the variability of the pool is low compared to the starting pool used in classical SELEX experiments.

Expression libraries consisting of rRNA fragments were used for *in vivo* selection experiments by two groups. Prescott and colleagues[23,24] found a 16S rRNA fragment (helix 34) that could sequester the antibiotic spectino-mycin, thereby conferring resistance to it. Tenson *et al.*[25,26] selected cells that are resistant against the antibiotic erythromycin. The clones revealed a 23S rRNA fragment containing an open reading frame for a pentapeptide. Not the rRNA fragment itself but rather expression of the corresponding MRMLT peptide renders cells resistant to erythromycin.

Figure 1 outlines the experimental strategy for the selection of randomly fragmented rRNA. rDNA is digested with DNase I, producing random rDNA fragments with a defined length distribution (100 to 300 nucleotides). After cloning into a vector and creating an rDNA fragment pool via poly-merase chain reaction (PCR), a pool of randomly fragmented rRNA is prepared by T7 transcription. The rRNA–protein complexes are collected on nitrocellulose filters. Bound rRNA is recovered from the filter, reverse transcribed to cDNA, and amplified in a PCR reaction. The PCR products are the template for further T7 *in vitro* transcription to produce rRNA fragments for the next round of selection. This article demonstrates the power of this method with ribosomal protein L11, the rRNA-binding site of which is one of the best characterized.[27,28]

### Example: L11–23S rRNA Interaction

L11 binds to an rRNA fragment (rRNA$^{L11}$) of domain II in 23S rRNA.[29] The 58-nucleotide-long minimal binding site comprises the residues 1051–1108 of 23S RNA (Fig. 2). The pentameric complex L8 [L10(L7/L12)$_4$] binds adjacent to the L11 binding site via L10. L8 and L11 bind in a mutually

[23] B.-A. Howard, G. Thom, I. Jeffery, D. Colthurst, D. Knowels, and C. D. Prescott, *Biochem. Cell Biol.* **73**, 1161 (1995).

[24] G. Thom and C. D. Prescott, *Bioorgan. Med. Chem.* **5**, 1081 (1997).

[25] T. Tenson and A. Mankin, *Biochem. Cell Biol.* **73**, 1061 (1995).

[26] T. Tenson, A. DeBlasio, and A. S. Mankin, *Proc. Natl. Acad. Sci. U.S.A* **93**, 5641 (1996).

[27] J. Thompson and E. Cundliffe, *Biochimie* **73**, 1131 (1991).

[28] D. E. Draper, *in* "The Many Faces of RNA" (D. S. Eggleston, C. D. Prescott, and N. D. Pearson, eds.), p. 113. Academic Press, London, 1998.

[29] F. J. Schmidt, J. Thompson, K. Lee, J. Dijk, and E. Cundliffe, *J. Biol. Chem.* **256**, 12301 (1981).

## *In vitro* selection from randomly fragmented rRNA (SERF)

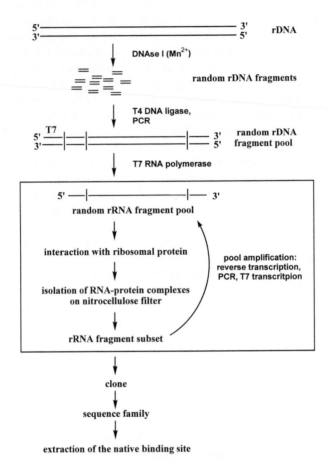

FIG. 1. Outline of the random rRNA fragment SELEX procedure.

cooperative manner to an rRNA sequence extending from nucleotide 1040 to 1120.[30] This region, in particular A1067, is protected against base-modifying reagents by the elongation factor EF-G[31] and an mutation of A1067 impaired binding of EF-Tu.[32] EF-G binds to an 84-mer that includes

[30] A. A. D. Beauclerk, E. Cundliffe, and J. Dijk, *J. Biol. Chem.* **259,** 6559 (1984).
[31] D. Moazed, J. M. Robertson, and H. F. Noller, *Nature* **334,** 362 (1988).
[32] U. Saarma, J. Remme, M. Ehrenberg, and N. Bilgin, *J. Mol. Biol.* **272,** 327 (1997).

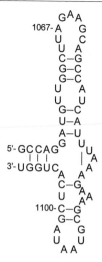

Fig. 2. Secondary structure of the minimal L11-binding region (rRNA$^{L11}$) in domain II of *E. coli* 23S ribosomal RNA (1051–1108).

this region (nucleotides 1036–1119).[33] The antibiotics thiostrepton and micrococcin interfere with EF-G function and interact with this rRNA region as well.[34]

L11 binds with a binding constant of $1 \times 10^7\ M^{-1}$,[35] thiostrepton with $3 \times 10^6\ M^{-1}$,[27] and both bind together with a cooperativity factor of 200 to the RNA fragment.[36] Because the L11-binding site on 23S RNA is an interaction site with both the elongation factors and thiostrepton, this rRNA region is sometimes erroneously termed the "ribosomal GTPase center." Erroneously, because the *E. coli* ribosome cannot bind GTP and does not contain a GTPase center, but it triggers the cleavage of GTP by the GTPase centers of the factors in a way not yet understood.

The L11–rRNA interaction depends on the correct folding of the RNA tertiary structure, a three helix junction, which is stabilized by $Mg^{2+}$ and $NH_4^+$ as well as by the protein. The C-terminal part of the protein L11 comprising the last 76 aminoacyl residues (L11-C76) binds to the L11-binding region of 23S rRNA with the same efficiency as the whole protein. The structure of L11-C76 in complex with RNA and free in solution was determined by NMR spectroscopy (helix III at the C-terminal end is likely

[33] A. Munishkin and I. G. Wool, *Proc. Natl. Acad. Sci. U.S.A.* **94,** 12280 (1997).
[34] J. Egebjerg, S. Douthwaite, and R. A. Garrett, *EMBO J.* **8,** 607 (1989).
[35] P. C. Ryan and D. E. Draper, *Biochemistry* **28,** 9949 (1989).
[36] Y. Xing and D. E. Draper, *Biochemistry* **35,** 1581 (1996).

the main part of the RNA-binding site). The protein is structurally similar to the homeodomain class of DNA-binding proteins.[28]

The L11–rRNA interaction is highly conserved throughout all three kingdoms. The functions of the ribosomes are maintained if the rRNA part or the protein L11 is interchanged between organisms from different kingdoms.[37–40] We tested the binding of *E. coli* L11 and of *Thermus thermophilus* L11 to the *E. coli* RNA$^{L11}$ in nitrocellulose filter binding and gelshift assays (described later): Binding constants for both the homologous and the heterologous complex are in the same order of magnitude as determined from half-saturation of binding curves ($K_B \cong 10^7 M^{-1}$). L11 from *T. thermophilus* shows a higher retention efficiency (87%; i.e., amount of RNA retained on the filter at an infinitely high protein concentration, see Ref. 41) than *E. coli* L11 (65%) in filter-binding assays. Binding of L11 to the minimal rRNA fragment is stimulated by thiostrepton and also by ribosomal protein L7/L12. In competition experiments, a random rRNA fragment pool does not reduce binding of L11 to its rRNA fragment, indicating the specificity of the L11–RNA$^{L11}$ interaction.

Purification of Ribosomal Proteins

*Escherichia coli* ribosomal proteins are isolated from total ribosomal proteins according to Diedrich *et al.*[42] The corresponding ribosomes are derived from the strain MRE 600. Ribosomal protein L11 is purified by ion-exchange chromatography on Mono Q columns (Pharmacia, Uppsala) in 20 m$M$ Tris–HCl, 6 $M$ urea at pH 9.5 (4°) using a linear gradient of 0–500 m$M$ KCl with 1m$M$ KCl/min. The protein is dialyzed against water and lyophilized. After resuspending it in Rec-4 buffer containing 6 $M$ urea, the protein is dialyzed against Rec-4 buffer containing 6 $M$ urea and Rec-4 buffer (transient urea dialysis for activation of L11) followed by an extensive dialysis against Rec-4 buffer. *E. coli* protein L11 is essentially pure as determined by two-dimensional gel electrophoresis (Coomassie staining).

Ribosomal protein L11 from *T. thermophilus* is a kind gift from Dr. François Franceschi. The expression using an *E. coli* pET expression system

[37] A. A. D. Beauclerk, H. Hummel, D. J. Holmes, A. Böck, and E. Cundliffe, *Eur. J. Biochem.* **151,** 245 (1985).

[38] T. T. A. L. El-Baradi, V. H. C. F. D. Regt, S. W. C. Einerhand, J. Teixido, R. J. Planta, J. P. G. Ballesta, and H. A. Raue, *J. Mol. Biol.* **195,** 909 (1987).

[39] W. Musters, P. M. Goncalves, K. Boon, H. A. Raue, H. Vanheerikhuizen, and R. J. Planta, *Proc. Natl. Acad. Sci. U.S.A.* **88,** 1469 (1991).

[40] J. Thompson, W. Musters, E. Cundliffe, and A. E. Dahlberg, *EMBO J.* **12,** 1499 (1993).

[41] D. E. Draper, I. C. Deckman, and J. V. Vartikar, *Methods Enzymol.* **164,** 203 (1988).

[42] G. Diedrich, N. Burkhardt, and K. H. Nierhaus, *Prot. Express. Purif.* **10,** 42 (1997).

and further purification are described elsewhere. *Thermus thermophilus* L11 contains less than 5% of T7 lysozyme (EC 3.5.1.28).

## Pool of Randomly Fragmented rRNA

An rRNA pool is constructed that consists of a statistical distribution of rRNA fragments of certain lengths. This pool contains the rRNA sites to which ribosomal proteins bind within the ribosome, provided that the length of the rRNA fragments covers the corresponding binding site. Affinity selection with this pool should pick the binding site.

The pool construction starts from the plasmid ptac-1, which is 10.2 kB in size and carries the 6.2-kB part of the *rrnB* operon that codes for 16S, tRNA$^{Glu2}$, 23S, and 5S rRNA and is flanked by *Bam*HI restriction sites.[43] The plasmid ptac-1 is digested with DNase I. DNase I cuts dsDNA in the presence of $Mn^{2+}$ via a "double hit" mechanism[44] and produces blunt-ended DNA or overhangs with one to two nucleotides. T4 DNA polymerase is used to make the fragments completely blunt ended. The fragments are ligated into the *Sma*I site of the vector pGem-3Zf(−) (Promega, Madison, WI), thereby introducing the 5′ and 3′ constant regions for PCR amplification.

DNase I digestion is performed in 50 m$M$ Tris–HCl (pH 8.0 at 25°), 0.01 m$M$ MnCl$_2$ (freshly prepared) at 16° for 15 to 50 min (Fig. 3). The concentration of DNA is 35 ng/$\mu$l and that of DNase I (RNase free) is 0.3 $\mu$g per $\mu$g DNA. The reaction is stopped by the addition of EDTA. The reaction mixture is phenol extracted twice and the DNA fragments are concentrated via precipitation with 3 volumes of ethanol in the presence of 0.3 $M$ sodium acetate (pH 6). It is difficult to remove DNase I from the mixture completely. Because the rate of DNase I digestion decreases when the fragments reach a size around 100–250 (Fig. 3, lanes 1–7), the remaining activity does not interfere with further procedures, e.g., blunt ending with T4 DNA polymerase.

Blunt ending of the fragments is done with T4 DNA polymerase in 10 m$M$ Tris–HCl (pH 7.9 at 25°), 10 m$M$ MgCl$_2$, 50 m$M$ NaCl, 1 m$M$ dithiothreitol (DTT), 0.1 mg/ml bovine serum albumin (BSA), and 100 $\mu M$ of each dNTPs. The mixture is incubated at 25° for 15 min, followed by the addition of Klenow fragment, and a further incubation at 25° for 15 min: The mixture contains 0.05 $\mu$g DNA per $\mu$l, 2.5 U T4 DNA polymerase per $\mu$g DNA, and 1 U Klenow fragment per $\mu$g DNA.

---

[43] B. T. U. Lewicki, T. Margus, J. Remme, and K. H. Nierhaus, *J. Mol. Biol.* **231**, 581 (1993).

[44] E. Melgar and D. A. Goldthwait, *J. Biol. Chem.* **243**, 4409 (1968).

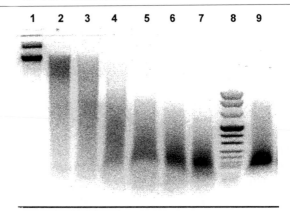

FIG. 3. Construction of a random rRNA fragment pool (2% agarose gel). Lanes 1–7, time course of DNase I digestion of ptac-1 (0.01 m$M$ Mn$^{2+}$, 16°). Lane 1, 0 min (two bands: circular and supercoiled form of the plasmid, respectively); lane 2, 10 min; lane 3, 15 min; lane 4, 20 min; lane 5, 35 min; lane 6, 50 min; lane 7, 70 min; lane 8, marker (1114, 900, 692, 501, 404, 320, 242, 190, 147, 124); and lane 9, PCR template for T7 transcription of the initial pool.

The vector pGem-3Zf($-$) is digested with the restriction enzyme *Sma*I in 20 m$M$ Tris–acetic acid (pH 7.9 at 25°), 10 m$M$ magnesium acetate, 50 m$M$ sodium acetate, and 1 m$M$ DTT at 37° for 4 hr (1 $\mu$g DNA per $\mu$l, 2 U enzyme per $\mu$g DNA).

Dephosphorylation of the vector with calf intestine alkaline phosphatase (CIP): 50 m$M$ Tris–HCl (pH 8.3 at 25°), 1 m$M$ MgCl$_2$, and 1 m$M$ ZnCl$_2$ at 37° for 15 min (0.05 $\mu$g DNA per $\mu$l, 2 U CIP per $\mu$g DNA). A second aliquot of CIP is added and the incubation is continued at 55° for 1 hr. The vector is purified after removal of CIP by phenol extraction on a 0.8% "low melting" agarose gel.

Ligation of the fragments into the digested and dephosphorylated vector with T4 DNA ligase is performed in 50 m$M$ Tris–HCl (pH 7.5 at 25°), 10 m$M$ MgCl$_2$, 10 m$M$ DTT, 10 m$M$ ATP, and 25 mg/ml BSA at 16° for 5 to 7 hr: 0.01–1 $\mu$g DNA (vector + fragment) per $\mu$l and 0.2 U T4 DNA ligase per $\mu$l. Best results are obtained with an excess of vector over rDNA fragments in the ligation reaction. The cut as well as the religated vector alone does not yield any PCR product. The multiple cloning site on the vector, flanking the insert, is used as constant regions for PCR amplification. The 5'($+$) primers introduce a T7 promoter.

The sequence of an rRNA fragment in the pool is

GGGCGAATTCGAGCTCGGTACCC-ptac-1 fragment-
                    GGGGATCCTCTAGAGTCGAC

and those of the corresponding primers are:

T7-5'(+): 5'-TAATACGACTCACTATAGGGCGAAT

TCGAGCTCG-3'

3'(−): 5'-GTCGACTCTAGAGGATCC-3'

The PCR reaction produces a template with a length of 100 to 300 nucleotides (Fig. 3, lane 9) for T7 transcription. The size distribution is restricted further in the following RNA purification step.

Diversity of Random rRNA Fragment Pool

A circular DNA (such as a plasmid) of the size $n$ can yield $n^2$ different fragments. The blunt-end DNA fragments can be inserted in either direction and thus the transcription produces both sense and antisense RNA fragments. Therefore, the theoretical maximum of RNA fragments is $2n^2$. The ptac-1 plasmid (10,231 bp) can thus yield $2 \times 10^8$ fragments with lengths ($l$) varying from 1 to 10,231. We use fragments ranging in length from 70 to 130 nucleotides, which reduces the maximum number of different fragments by two orders of magnitude (to $1.2 \times 10^6$) in the starting pool. The diversity is up to 10 orders of magnitude lower than a classical SELEX pool with a randomized 26-mer containing $4.5 \times 10^{15}$ different sequences.

Assume that the winner sequence has the length $x$ and that the pool consists solely of fragments of one size $l_i$. Then the number of winners would be $w = 1 + (l_i - x)$, if $x \leq l_i$. If the pool contains fragments of various lengths, with $l_{min}$, the length of the shortest and $l_{max}$, the length of the largest fragments, the number $w$ of winners is

$$w = \sum_{l_{min}}^{l_{max}} 1 + (l - x)$$

With the assumption that the minimal binding site consists of 50 nucleotides for a pool with the fragment size distribution of 70 to 130, the fraction of putative winning sequences in the pool will be 0.25%. Furthermore, two or more short fragments could be ligated to a larger one contributing to the pool diversity. It is worth mentioning that the 16S fragment, which conferred streptomycin resistance in selection experiments *in vivo*,[23,24] consisted of two fragments joined together. With the starting fraction size of 0.25% within the initial pool, the binding region can be enriched very fast. Assuming that one round of selection leads to an enrichment of about 10-fold,[45] one can expect that the binding region can be detected after three

[45] D. Irvine, C. Tuerk, and L. Gold, *J. Mol. Biol.* **222**, 739 (1991).

**1    2    3**

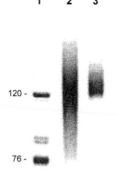

120 -

76 -

FIG. 4. Random rRNA fragment pools before and after four rounds of selection (7% acrylamide/urea gel). Lane 1, marker [5S RNA (120 nucleotides)] and tRNA[bulk] (76 nucleotides; the two bands above the 76 nucleotide band and tRNAs with a long extra arm); lane 2, RNA-0 (starting pool, binding to L11: 1.5%); and lane 3, RNA-4 (RNA after four rounds of selection, binding to L11: 12.1%)

to four rounds of selection, whereas a classical SELEX experiment usually requires more rounds.

Cycles of Selection

SELEX is a widely used and well-established method[46] and detailed descriptions of the experimental procedure are available.[47] Nevertheless, the selective step of each SELEX experiment needs to be optimized in order to avoid artifacts and to be efficient. The specialty of the variant SERF presented here is that the RNA pool consists of sequences of different sizes derived from an rRNA operon (see Fig. 4). During the various cycles of selection, a pressure favoring shorter fragments exists mainly due to two reasons: (i) possible processivity problems of the reverse transcriptase and (ii) a too strong amplification of the cDNA during PCR. Therefore, it is important to be specific in the reverse transcription reaction [see later; reaction time not longer than 30 min; stoichiometry of $3'(-)$ primer:template $\leq 2:1$] and not to perform too many PCR cycles. Furthermore, a control without RNA during the whole procedure is required.

The PCR reactions (150 $\mu$l) are carried out in 10 m$M$ Tris–HCl (pH

[46] L. Gold, B. Polisky, O. Uhlenbeck, and M. Yarus, *Annu. Rev. Biochem.* **64,** 763 (1995).

[47] *Methods Enzymol.* **267** (1996).

8.3 at 25°), 1.5 m$M$ MgCl$_2$, 50 m$M$ KCl, 0.5 m$M$ of each of the T7-5'(+) and 3'(−) primers, 2 m$M$ of each of the dNTPs, and 20 U *Taq* polymerase/ ml. In the case of the initial pool, up to 10 vol% of the ligation reaction is used as a template, and during the subsequent selection rounds, as well up to 10 vol% of a reverse transcription reaction mixture is used. After 2 min at 92°, 10–20 cycles are run in a Bio-Med thermocycler 60 machine (Theres, Germany): 92° for 30 sec, 56° for 30 sec, and 70° for 45 sec. The product is purified with the QIAquick PCR purification kit (Qiagen, Hilden, Germany) according to the user's manual, thereby reducing the volume by one-half. Seven microliters of the sample is used for analysis on a 2% agarose gel and 30–50 $\mu$l for *in vitro* transcription.

T7 *in vitro* transcription is done in 100 $\mu$l according to Triana-Alonso *et al.*[48] The reaction contains 30–50 $\mu$l template (50–150 pmol/ml), 40 m$M$ Tris–HCl (pH 8.0 at 25°), 22 m$M$ MgCl$_2$, 5 m$M$ DTT, 1 m$M$ spermidine, 100 $\mu$g/ml BSA, 1 U/$\mu$l RNasin, 5 U/ml inorganic pyrophosphatase, 3.75 m$M$ of each NTPs, and 40 $\mu$g/ml of purified T7 RNA polymerase and is incubated at 37° for 5 hr or more. For $^{32}$P labeling of RNA, the reaction contains 2–5 $\mu$Ci [$\alpha$-$^{32}$P]UTP (Amersham, Braunschweig). After the incubation, DNase I (RNase free) is added (10 $\mu$g/ml) and kept at 37° for a further 10 min. The mixture is diluted to double the volume, extracted with phenol, and precipitated with ethanol. The rRNA fragments are seen as a smear on a 7% denaturing acrylamide gel (Fig. 4, lane 2). RNA of the desired size (110 to 170 nucleotides) is cut out, and the gel is smashed into pieces by pressing through a syringe and extracting in 10 m$M$ Tris–HCl (pH 7.8 at 25°), 100 m$M$ NaCl, 1% SDS, 1 m$M$ dithioerythritol, and 25% phenol overnight. Phenol and gel pieces are removed by centrifugation. The aqueous phase is extracted with chloroform, and the RNA is precipitated with ethanol twice to remove traces of phenol. A 100-$\mu$l T7 transcription reaction purified this way yields about 2000 pmol of rRNA fragments, based on the assumption that 1 $A_{260}$ of a fragment about 140 nucleotides long corresponds to about 800 pmol.

## Collecting Complexes on Nitrocellulose Filters

A critical step in the SELEX procedure is the separation of bound from nonbound fragments. In our experiment, we separate the RNA–protein complexes from free RNA by filtration through a nitrocellulose filter. The level of RNA background binding (3–5% of the input RNA) to nitrocellu- lose and nonspecific binding of RNA by the protein are factors that interfere

---

[48] F. J. Triana-Alonso, M. Dabrowski, J. Wadzack, and K. H. Nierhaus, *J. Biol. Chem.* **270**, 6298 (1995).

with a specific selection. Furthermore, the actual concentrations of rRNA fragments and protein and their ratio (see Results and Discussion) are crucial for the efficient selection of specific interactions. In general, one may follow Irvine et al.,[45] who developed an equation for a near optimal target concentration. The selection with ribosomal proteins has to be optimized for each individual protein, as the nonspecific binding of the initial pool RNA to the ribosomal proteins varies considerably.

## Formation of RNA–Protein Complexes

rRNA fragments are incubated for 3 min at 70° and slowly cooled to 37° in 20 mM HEPES–KOH (pH 7.5 at 4°), 4 mM $MgCl_2$, 400 mM $NH_4Cl$, and 6 mM 2-mercaptoethanol (Rec-4 buffer). Proteins are incubated for 15 min at 37° in the same buffer before mixing with the rRNA fragments. After incubation for 10 min at 37° and 10 min at 0° (ice bath), the mixture is applied to a 0.45-μm nitrocellulose filter (prewetted and degassed in an evacuator for 30 min in Rec-4 buffer). The filter is washed with 500 μl of ice-cold Rec-4 buffer, cut into pieces, and extracted with 400 μl of phenol saturated with Tris–HCl (pH 7.8 at 25 °; Roth, Karlsruhe) and 200 μl of 7 M urea (freshly prepared) for 60 min at room temperature. Two hundred microliters of chloroform/isoamyl alcohol (24 : 1) is added to facilitate phase separation by centrifugation. The aqueous phase is treated with chloroform/isoamyl alcohol, and the RNA is recovered from the aqueous phase by ethanol precipitation in the presence of 1 μg/ml RNase-free glycogen.

For reverse transcription the selected RNA is dissolved in 10 μl MilliQ–water containing 1.5 pmol/μl 3′(−) primer. The RNA is heated at 70° for 5 min and cooled to room temperature for 5 min. The reverse transcription reaction with avian mycloblastosis virus (AMV) reverse transcriptase is carried out in 30 μl containing 50 mM Tris–HCl (pH 8.5 at 25°), 8 mM $MgCl_2$, 30 mM KCl, 1 mM DTT, and 500 μM of each dNTPs for 30 min at 42°. About 10 μl of the reaction is used directly in a 150-μl PCR reaction.

## Monitoring Progress of Selection by Binding Assays

Progress in the selection procedure can be detected by measuring the binding of the newly synthesized RNA after each round of selection. This is done most easily with a filter-binding assay, where the protein is incubated with radiolabeled RNA and filtered through a nitrocellulose filter. Another widely used method is a band shift assay, by which the protein–RNA complex is separated from free RNA on a native acrylamide gel. Because the random rRNA fragment pool contains RNA of different lengths and might therefore overlap with the gel position of the complex, only the filter-binding assay is used for measuring the progress of selection. However, the binding of cloned fragments (that have a defined length) can be measured in a band shift assay as well.

### Nitrocellulose Filter-Binding Assay[41,49]

rRNA fragments are incubated for 3 min at 70° in Rec-4 buffer and then cooled to 37°. After incubation for 15 min at 37° in Rec-4 buffer, proteins are mixed with rRNA fragments, incubated for 10 min at 37°, and incubated further for 10 min on ice. The mixture is put directly onto a 0.45-$\mu$m nitrocellulose filter (Sartorius, Göttingen) under suction. The filter has been prewetted and degassed for 30 min in Rec-4 buffer previously. Radioactivity retained on the filter is determined by liquid scintillation counting.

Nitrocellulose filter-binding assays are carried out in 300 $\mu$l. Protein concentrations vary from 33 to 240 n$M$ while keeping RNA concentrations at about 70 n$M$. Nonspecific binding of rRNA fragments to a protein of the initial pool should not exceed 1 to 8% of the input RNA after correcting for background binding. Background binding, i.e., the amount of RNA retained on the nitrocellulose filter without protein, usually amounts to 6–10% and can be reduced by washing once with 500 $\mu$l buffer to 3–5%. Starting with these binding parameters, the progress of the selection can be followed easily. Ribosomal protein L11 shows a low nonspecific binding of the initial pool (RNA-0) of about 1.5%.

### Band-Shift Assay in Native Gels

For band-shift assays the Mini-PROTEAN II Bio-Rad (Richmond, CA) casting system is used (gel size: 90 × 60 × 0.7 mm). A 10-$\mu$l sample of an RNA–protein complex is prepared as described earlier containing 60 pmol RNA and, e.g., 60 pmol protein in Rec-4 buffer. After the addition of 1 $\mu$l glycerol (78 vol% stock solution) the complex is loaded onto a Rec-4-buffered 7% acrylamide/bisacrylamide (19/1) gel. It is electrophoresed at 80 V (about 220 mA) for 2.5 hr at 4°. The pH of the electrophoresis buffer (Rec-4) is kept constant at 7.5 with a circulation pump. On a separate lane, a bromphenol blue solution (0.6% in Rec-4 buffer) is applied, and the electrophoresis is stopped when the stain has reached the last third of the gel. RNA is visualized by ethidium bromide staining, i.e., by shaking the gel in water containing 5 $\mu$g/ml ethidium bromid for 10 min (Fig. 5).

For cloning of sequences that bind to the protein, the corresponding

[49] J. Carey, V. Cameron, P. L. deHaseth, and O. C. Uhlenbeck, *Biochemistry* **22**, 2601 (1983).

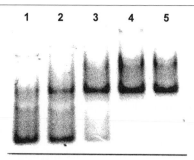

Fig. 5. Band-shift assay showing binding of one selected fragment rRNA$^{L11}$ [rRNA sequence 1042 to 1120 plus flanking sequences (115 nucleotides)] with the ribosomal protein L11 from *T. thermophilus* (7% AA/Rec-4 gel). Lane 1, rRNA$^{L11}$, no protein; lane 2, molar ratio of rRNA$^{L11}$:L11 = 1:0.5; lane 3, rRNA$^{L11}$:L11 = 1:1; lane 4, rRNA$^{L11}$:L11 = 1:1.5; and lane 5, rRNA$^{L11}$:L11 = 1:3.

PCR product is digested with *Bam*HI and *Eco*RI and ligated into a vector for sequencing.

Results and Discussion

A general selection strategy (SERF) is described to identify native occurring RNA sequences that interact specifically with a given protein. The strategy was tested with the ribosomal protein L11. The starting material was a pool of RNA fragments derived from a plasmid with 10,200 bp containing the complete *rrnB* operon. The known binding site was found with a high preponderance in this pool of rRNA fragments after the fourth round of selection. Rec-4 buffer is used in the selection experiments because this condition is essential for the first step of the reconstitution of 50S subunits.[50] High monovalent ion concentrations (400 m$M$ NH$_4^+$) reduce nonspecific binding. The RNA:protein ratio was increased sequentially: 0.5 $\mu M$ rRNA fragments and 0.75 $\mu M$ protein L11 in round one; 0.75 $\mu M$ RNA and 0.5 $\mu M$ L11 in round two; 0.7 $\mu M$ RNA and 0.2 $\mu M$ protein in round three; and 0.8 $\mu M$ RNA and 0.13 $\mu M$ protein L11 in round four. As shown in Fig. 6, binding of the fragments of the initial pool to L11 was about 1.5% (RNA-0) under the conditions of the filter-binding assay; a significant increase in binding was detected already after four rounds of selection (RNA-4). Attempts to increase the binding further succeeded over two more rounds of selection (round 5 was done with 0.5 $\mu M$ RNA and 0.02 $\mu M$ protein, round 6 under the same conditions).

[50] K. H. Nierhaus, *in* "Ribosomes and Protein Synthesis: A Practical Approach" (G. Spedding, ed.), p. 161. IRL Press, Oxford, 1990.

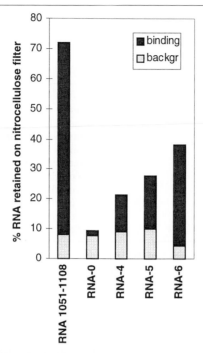

FIG. 6. Progress of random rRNA fragment selection with ribosomal protein L11.

One might expect that distinct bands appear on the denaturing RNA gel during several rounds of selection, but this is not the case; no clear bands appeared during selection (Fig. 4, lane 3). For sequence determination, rRNA fragments after six rounds of selection (RNA-6) were cloned. Eleven out of 42 sequences contained the L11-binding site (Table I). The sequences ranged from nucleotide 1015 to 1125 of 23S rRNA (Fig. 7). The smallest fragment consisted of nucleotides 1052 to 1112 (71 nucleotides). Because the size of the selected fragment is determined by the size distribution of the initial RNA pool, no smaller fragment was found. Nevertheless, the overlapping part of the rRNA sequences (common to all selected rRNA fragments, see Fig. 7) revealed precisely the minimal rRNA fragment shown to bind protein L11 previously: 1052–1108 (note that G1051 described as the first nucleotide of the minimal binding site by others is included due to the usage of T7 *in vitro* transcription). Stoichiometric binding of one selected rRNA fragment to L11 is demonstrated in Fig. 5.

Two other random rRNA fragment pools (pools 2 and 3) were used in independent SERF experiments. rDNA fragments for control pool 2 were obtained by the combined use of a number of restriction enzymes and

TABLE I
SEQUENCES FOUND IN RNA-6

| Selected fragments | Number out of 42 | Comment[b] |
|---|---|---|
| L11-binding region | 11 | Most prominent (see Fig. 7) |
| 23S RNA (DIII) | 6 | Overlap 23S: 1478–1521 (44 nucleotides) |
| *rrnB* 16S–23S interregion | 4 | ptac-1: 2124–2198 (75 nucleotides) |
| *rrnB*, in front of 23S rRNA | 3 | Overlap ptac-1: 2207–2281 (75 nucleotides) |
| 16S RNA | 2 | Overlap 16S: 1282–1317 (36 nucleotides) |
| 23S RNA (DI) | 2 | Overlap 23S: 61–140 (78 nucleotides) |
| 23S RNA | 3 | No overlapping regions |
| 16S RNA | 2 | No overlapping regions |
| Antisense sequences | 5 | No overlapping regions |
| ptac-1 sense | 2 | No overlapping regions |
| Hybrid sequences[a] | 2 | No overlapping regions |

[a] Hybrid sequences composed of two fragments.

[b] Numbering 16S and 23S rRNA according to *E. coli*, and numbering of the ptac-1 according to B. T. U. Lewicki, T. Margus, J. Remme, and K. H. Nierhaus, *J. Mol. Biol.* **231**, 581 (1993).

DNase I. The fragments were cloned into another vector (pBlue KSII+; Stratagene, Heidelberg, Germany), thus yielding different flanking sequences. Pool 3 originated from the sequence of the *rrnB* operon rather than from the whole plasmid. Both control pools 2 and 3 showed increased binding affinity to L11 after four rounds of selection (26 and 14%, respectively). In the case of pool 2, no further increase in binding was obtained after rounds five and six. Further selection rounds with the control pool 3 favored rRNA fragments that bound nitrocellulose over those binding to ribosomal protein L11. Sequences of rRNA fragments after the fourth round of selection were determined: 37 of the 97 cloned rRNA fragments (26% binding) derived from pool 2 and 4 out of 17 (14% binding) from pool 3 contained the L11-binding site.

In summary, in all three pools the L11-binding site was the most abundant (overall one-third of the cloned sequences). Other rRNA overlapping fragments of interest (i.e., rRNA sense strands) appeared more than once, but each only in one of the three pools: nucleotides 1478–1521 of 23S RNA (14% of the clones derived from pool 1, see Table I), 1922–1984 of 23S rRNA (8% of control pool 2), 1282–1317 of 16S RNA (5% of pool 1, see Table I), 61–140 of 23S RNA (5% of pool 1, see Table I), and 1644–1726 of 23S rRNA (3% of control pool 2). Binding of these isolated fragments to protein L11 could not be detected in nitrocellulose filter binding or in a band-shift assay.

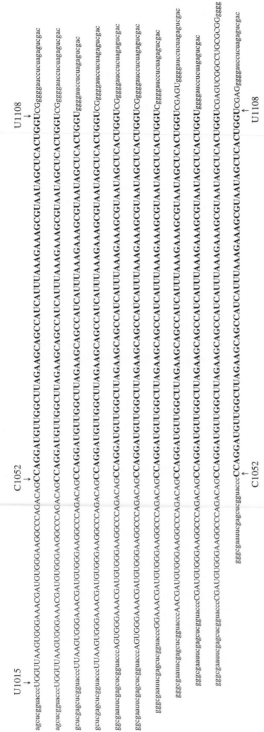

Fig. 7. Selected rRNA fragments (rRNA$^{L11}$) reveal the minimal binding site for protein L11 (bold): capital letters show rRNA sequences, whereas small letters indicate constant primer regions.

Conclusions

*In vitro* selection of random rRNA fragments (SERF) is a simple and straightforward method to find the native rRNA-binding site for ribosomal proteins. As demonstrated, it has the potential to find a minimal rRNA fragment, which may be suitable for structural investigations by means of NMR spectroscopy or X-ray crystallography.

The atomic structure of 15 ribosomal proteins has been solved, and one of the surprising findings was that many of the proteins have two potential RNA-binding sites. It might be difficult to find both interactions in a single SERF experiment. However, in some cases the proposed binding sites are located on different domains, examples include the ribosomal proteins S4, S5, S8, L1, L6, and L9. One possible strategy is to separate the domains of those proteins by genetic means and to perform the selection with each domain separately.

Acknowledgments

We thank Dr. François Franceschi for kindly providing purified L11 from *Thermus thermophilus*, Sean Connell for help and discussions, and Detlev Kamp for expert assistance in the purification of *E. coli* ribosomal proteins.

# [19] Optimized Synthesis of RNA–Protein Fusions for *in Vitro* Protein Selection

*By* Rihe Liu, Jeffrey E. Barrick, Jack W. Szostak, and Richard W. Roberts

Introduction

The extension of *in vitro* selection technology to the *in vitro* selection of peptides and proteins represents an area of great interest, in large part because the chemistry of molecular recognition and catalysis in living systems is largely mediated by polypeptides. The difficulty in designing schemes for *in vitro* protein selection is that while *in vitro* selection involves an iterated cycle of selection and amplification, there is no simple way to amplify protein molecules that have been selected for function. In order to isolate peptides or proteins with a desired function, the genetic information must be kept topologically linked to the protein in the form of a coding sequence such as RNA or DNA, a set of chemical tags, or a physical address. A major advantage of fully *in vitro* approaches is that they allow

0076-6879/00 $30.00

the isolation of proteins with desired properties even when no *in vivo* selection strategy exists or can be designed.

*In vitro* selection experiments begin with the generation of a population or pool containing many different sequences. This pool is then sieved to identify those individuals that have the desired functional properties. The number of different molecules that can be examined (the pool size or complexity) is one of the most important variables in a combinatorial experiment. For example, RNA molecules that bind ATP occur with a frequency of $1/10^{11}$ in random sequence RNA libraries.[1,2] Thus, an RNA selection designed to isolate ATP-binding aptamers would be unlikely to succeed if the starting library contained *only* 1 billion sequences.

Until recently, *in vivo* and *in vitro* protein selection experiments (e.g., the yeast two-hybrid system[3] and phage display[4–7]) were limited to complexities of about 1 million to 1 billion molecules, respectively. The main limitation on library size results from transfecting the starting cDNA library into the organism of choice. In contrast, *in vitro* RNA and DNA selection experiments routinely involve generating and screening libraries containing more than $10^{15}$ independent sequences. Two approaches have been developed that provide for totally *in vitro* selection of peptides and proteins: ribosome display and mRNA–protein fusions. Ribosome display, first examined in 1961,[8] has been the subject of previous reports and will not be covered in detail here.[9–11]

This article describes improvements in the development and use of mRNA–protein fusions for *in vitro* protein selection. An mRNA–protein fusion consists of a protein sequence covalently linked via its C terminus to the 3' end of its own messenger RNA (Fig. 1).[12] The fusions are generated by *in vitro* translation of appropriate mRNA templates containing puromycin at their 3' end. Because the coding and polypeptide sequences are

[1] M. Sassanfar and J. Szostak, *Nature* **364,** 550 (1993).

[2] D. H. Burke and L. Gold, *Nucleic Acids Res.* **25,** 2020 (1997).

[3] S. Fields and O.-K. Song, *Nature* **340,** 245 (1989).

[4] G. P. Smith, *Science* **228,** 1315 (1985).

[5] J. K. Scott and G. P. Smith, *Science* **249,** 386 (1990).

[6] J. J. Devlin, L. C. Panganiban, and P. E. Devlin, *Science* **249,** 404 (1990).

[7] S. E. Cwirla, E. A. Peters, R. W. Barrett, and W. J. Dower, *Proc. Natl. Acad. Sci. U.S.A.* **87,** 6378 (1990).

[8] D. B. Cowie, S. Spiegelman, R. B. Roberts, and J. D. Duerksen, *Proc. Natl. Acad. Sci. U.S.A.* **47,** 114 (1961).

[9] L. C. Mattheakis, R. R. Bhatt, and W. J. Dower, *Proc. Natl. Acad. Sci. U.S.A.* **91,** 9022 (1994).

[10] L. C. Mattheakis, J. M. Dias, and W. J. Dower, *Methods Enzymol.* **267,** 195 (1996).

[11] J. Hanes and A. Pluckthun, *Proc. Natl. Acad. Sci. U.S.A.* **94,** 4937 (1997).

[12] R. W. Roberts and J. W. Szostak, *Proc. Natl. Acad. Sci. U.S.A.* **94,** 12297 (1997).

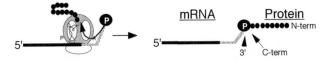

FIG. 1. Formation of mRNA–protein fusions on the ribosome. The ribosome pauses at an RNA/DNA junction, allowing puromycin to thread into the ribosome, entering the A site and forming fusion by accepting the nascent peptide from the peptidyl-tRNA in the P site.

united in a single molecule, RNA–protein fusions provide a means for reading and amplifying a protein sequence after it has been purified based on its function, effectively reverse translating the protein. The maximum library size that may be generated is limited by the size and efficiency of the translation reaction and by the efficiency of fusion formation on the ribosome. At the present time, libraries containing more than $10^{13}$ different sequences can be readily generated. Further, peptides and proteins synthesized as fusions commonly retain the binding properties of the unfused polypeptides. Finally, proteins ranging in size from 1 to at least 30 kDa can be synthesized as fusions, thus opening a great diversity of systems to examination.

Selection Scheme

The basic scheme in an RNA–protein fusion selection experiment is highlighted in Fig. 2, divided into 10 discrete steps: (1) generation of the

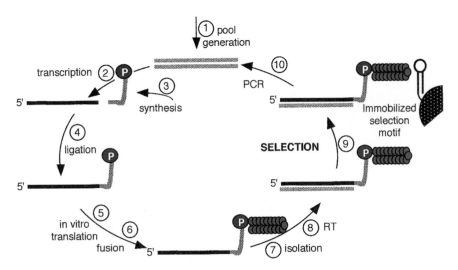

FIG. 2. Generalized selection scheme using mRNA–protein fusions. See text for description and optimization of individual steps.

initial double-stranded DNA sequence or pool, (2) transcription of the DNA into mRNA, (3) synthesis of the 3'-puromycin oligonucleotide, (4) ligation of the puromycin oligonucleotide to the mRNA, (5) *in vitro* translation of mRNA–puromycin templates, (6) generation of the mRNA–protein fusion, (7) isolation of the fusion, (8) reverse transcription to generate the cDNA/mRNA–protein fusion, (9) isolation of functional fusions with an immobilized selection motif, and (10) polymerase chain reaction (PCR) to generate an enriched dsDNA pool.

## 1. dsDNA Library

The starting library is constructed as a mixture of double-stranded DNA sequences. The DNA sequence contains several important design features. A T7 promoter is present at the 5' end to allow large-scale synthesis of mRNA *in vitro* using T7 polymerase.[13] The transcript begins with 3 G nucleotides to aid transcription initiation. The remainder of the 5'-untranslated region (5'UTR) should be chosen according to the *in vitro* translation system to be used for fusion generation. For translation in reticulocyte lysate, we commonly use a deletion mutant of the tobacco mosaic virus 5' UTR (ΔTMV) that provides efficient translation initiation.[12,14] For translation in *Escherichia coli*, a Shine–Dalgarno sequence appropriately spaced with respect to the start codon should be chosen.[15,16]

In contrast with bacteria, the 5' UTR in eukaryotes does not contain a ribosome-binding site or an extensive 5' consensus sequence. Although reports have been made of sequences that greatly enhance translation,[17] in reticulocyte lysate many sequences function efficiently as the 5' UTR.[18] In general, eukaryotic translation systems use the first AUG codon in the mRNA to initiate protein synthesis. The precise sequence context surrounding this codon influences the efficiency of translation.[18,19] The sequence 5'RNN<u>AUG</u>R provides a good start context for most sequences, with a preference for A as the first purine (−3) and G as the second (+4).[18,20]

The open reading frame (ORF) can be constructed from either a defined sequence(s) or a random sequence library. The most important feature of the ORF and adjacent 3' constant region is that neither contain stop codons.

[13] J. F. Milligan and O. C. Uhlenbeck, *Methods Enzymol.* **180,** 51 (1989).
[14] D. R. Gallie, D. E. Sleat, J. W. Watts, P. C. Turner, and T. M. A. Wilson, *Nucleic Acid Res.* **16,** 883 (1988).
[15] J. A. Steitz and K. Jakes, *Proc. Natl. Acad. Sci. U.S.A.* **72,** 4734 (1975).
[16] G. D. Stormo, T. D. Schneider, and L. M. Gold, *Nucleic Acids Res.* **10,** 2971 (1982).
[17] S. A. Jobling and L. Gehrke, *Nature* **325,** 622 (1987).
[18] M. Kozak, *Microbiol. Rev.* **47,** 1 (1983).
[19] M. Kozak, *J. Biol. Chem.* **266,** 19867 (1991).
[20] M. Kozak, *J. Mol. Biol.* **196,** 947 (1987).

TABLE I
STOP FREQUENCIES OF RANDOM CODONS

| Codons | Stop frequency (%) | Length with 50% chance of one stop codon[a] |
|---|---|---|
| NNN | 4.7 | 15 aa |
| NNG/C | 3.1 | 22 aa |
| $N_1N_2N_3$ | 1.0 | 69 aa |
| Insertion/deletion | 1.8 | 39 aa |

[a] aa, amino acid.

The presence of stop codons would allow premature termination of the protein synthesis, preventing fusion formation. A defined sequence can be used to generate a library of mutants via error-prone PCR,[21,22] cassette mutagenesis,[23] or by assembly with synthetic sequences containing a defined mutation frequency at each position.[24] Further complexity can be introduced using *in vitro* recombination via DNA shuffling.[25] Finally, a set of two or more homologous genes can be recombined *in vitro* to generate a starting library.[26]

Open reading frames can be constructed from random sequences in a variety of ways depending on the codons chosen. Stop codons in the ORF are a major concern. Totally random sequence libraries may be used (NNN coding) but contain a proportion of stop codons (3/64 = 4.7% per codon) that is unacceptably high for all but the shortest libraries. Such libraries also contain rarely used codons, resulting in poor translation. NNG/C codons provide a slightly reduced stop frequency (1/32 = 3.1% per codon) while providing access to the best codon for all 20 amino acids for mammalian translation systems. NNG/C codons are less optimal when applied in bacterial translation systems where the best codon ends in A or T in seven cases (AEGKRTV). Several solutions exist that provide for very low stop codon frequency (~1.0%), with amino acid content similar to globular proteins using three different nucleotide mixtures, $N_1N_2N_3$ codons (Table I)[27] (and references therein). Finally, an almost infinite variety of semirational design strategies may be employed to pattern libraries according to amino acid

[21] R. C. Cadwell and G. F. Joyce, *PCR Methods Appl.* **2**, 28 (1992).
[22] J. Tsang and G. F. Joyce, *Methods Enzymol.* **267**, 410 (1996).
[23] J. F. Reidhaar-Olsen *et al., Methods Enzymol.* **208**, 564 (1991).
[24] E. H. Ekland and D. P. Bartel, *Nucleic Acids Res.* **23**, 3231 (1995).
[25] W. P. C. Stemmer, *Nature* **370**, 389 (1994).
[26] A. Crameri, S. A. Raillard, E. Bermudez, and W. P. C. Stemmer, *Nature* **391**, 288 (1998).
[27] T. H. LaBean and S. A. Kauffman, *Prot. Sci.* **2**, 1249 (1993).

type. For example, hydrophobic (h) or polar (p) amino acids can be chosen using NTN or NAN codons, respectively.[28] These can be patterned to give preference to $\alpha$ helix (phpphhpp. . .) or $\beta$ sheet (phphph. . .) formation.

Open reading frames constructed from synthetic sequences may also contain stop codons resulting from insertions or deletions in the synthetic DNA. These defects have serious consequences due to alteration of the translation reading frame. Examination of a number of pools and synthetic genes constructed from synthetic oligonucleotides indicates that insertions and deletions occur with a frequency of ~0.6% per position, or 1.8% per codon (B. Seed, D. Wilson, J. W. Szostak, and R. W. Roberts, unpublished observations). The precise frequency of these occurrences is variable and is thought to depend on the source and length of the synthetic DNA. In particular, longer sequences show a higher frequency of insertions and deletions.[29] A simple solution to reducing frame shifts within the ORF is to work with relatively short segments of synthetic DNA (80 nucleotides or less) that can be purified to homogeneity. Longer ORFs can then be generated by restriction and ligation of several shorter sequences.

The 3' end of the ORF must contain a constant region approximately 21 nucleotides long. This sequence is needed for PCR amplification of the library and for splinted ligation of the puromycin oligonucleotide to the mRNA template. This region must be examined carefully in the design process to assure that it does not contain stop codons or secondary structures that could prohibit binding of either splint or primer oligonucleotides. Finally, the amino acid sequence in this region can be chosen to encode an epitope tag or a flexible linker sequence between the puromycin template and the protein (Fig. 3).

## 2. Synthesis of mRNA Pool

Once generated, the dsDNA library is used to generate large quantities (0.1–10 mg) of RNA enzymatically using T7 polymerase.[13] Transcription is generally performed in the same volume as the PCR reaction (PCR DNA derived from a 100-$\mu$l reaction is used for 100 $\mu$l of transcription). This RNA can be generated with a 5' cap if desired using a large molar excess of m$^7$GpppG to GTP in the transcription reaction.[30] Full-length RNA samples are then purified from transcription reactions as described previously using urea PAGE followed by desalting on NAP-25 (Pharmacia, Piscataway, NJ).[12]

---

[28] J. R. Beasley and M. H. Hecht, *J. Biol. Chem.* **272,** 2031 (1997).
[29] J. Haas, E. C. Park, and B. Seed, *Curr. Biol.* **6,** 315 (1996).
[30] N. K. Gray and M. W. Hentze, *EMBO J.* **13,** 3882 (1994).

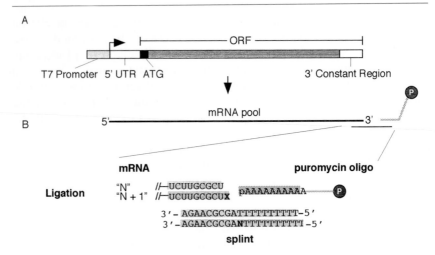

FIG. 3. Design and synthesis of template. (A) DNA template used to generate mRNA. Several features are indicated, including a promoter for *in vitro* transcription, a 5'-untranslated region (UTR) appropriate for the translation system used, a start codon, an open reading frame (ORF), and a 3' constant region. (B) Enzymatic joining of puromycin oligonucleotides to the mRNA. T4 DNA ligase and a splint oligonucleotide are used to join 5'-phosphorylated oligonucleotides.[33] A splint containing an additional random base at the ligation junction improves the efficiency of the reaction by allowing N + 1 transcription products to ligate.

## 3. Synthesis of 3'-Puromycin Oligonucleotides

In order to synthesize oligonucleotides containing puromycin at their 3' end, we first prepare a synthesis resin containing protected puromycin coupled to controlled pore glass (CPG–puromycin). The synthesis of CPG–puromycin follows the general path used for deoxynucleosides, as outlined previously.[31,32] Major departures include the choice of an amino protecting group, the linkage to the solid support through the puromycin 2'-OH, and the concentration of the coupling reaction with the support (see Fig. 4). In particular, the coupling reaction is done at very low concentrations of activated nucleotide, as this material is significantly more precious than the solid support. Despite the low concentration, the solid support can be consistently derivatized at a concentration of 5–20 $\mu$mol/g, about half the activity of commercial supports containing standard nucleotides. The following protocol represents preparation of one batch of puromycin–CPG.

[31] R. A. Jones, *in* "Oligonucleotide Synthesis: A Practical Approach" (M. J. Gait, ed.), p. 23. IRL Press, Oxford, 1984.
[32] T. Atkinson and M. Smith, *in* "Oligonucleotide Synthesis: A Practical Approach" (M. J. Gait, ed.), p. 35. IRL Press, Oxford, 1984.

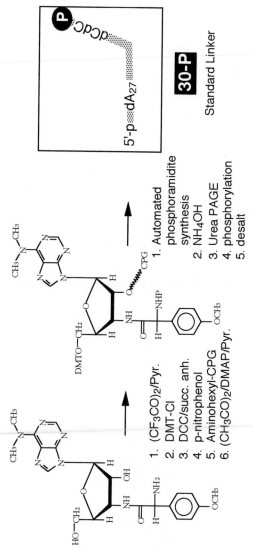

FIG. 4. Synthesis of CPG–puromycin and 3′-puromycin oligonucleotides. See text for details.

Infrared, nuclear magnetic resonance, and mass spectrometer measurements have been made to confirm the formation of N-trifluoroacetyl-puromycin, N-trifluoroacetyl-5'-dimethoxytrityl (DMT)-puromycin, and N-trifluoroacetyl-5'-DMT-2'-succinylpuromycin, as well as to confirm the final puromycin product after removal from the column and deprotection (R. W. Roberts and J. Xu, unpublished observations).

*N-Trifluoroacetylpuromycin.* Two hundred sixty-seven milligrams (0.490 mmol) puromycin hydrochloride (Sigma, St. Louis, MO) is converted to the free base form by dissolving in water, adding pH 11 carbonate buffer, and extracting (3×) into chloroform. The organic phase is evaporated to dryness and weighed (242 mg, 0.513 mmol). The free base is dissolved in 11 ml dry pyridine (Fluka, Ronkonkoma, NY), and 11 ml dry acetonitrile containing 139 μl (2.0 mmol) TEA (Fluka) and 139 μl (1.0 mmol) of trifluoroacetic anhydride (TFAA) (Fluka) are added with stirring. Additional TFAA is then added to the turbid solution in 20-μl aliquots until none of the starting material remains by thin-layer chromatography (93:7, chloroform/methanol) (a total of 280 μl). The reaction is allowed to proceed for 1 hr. At this point, two bands are seen by TLC, both of higher mobility than the starting material representing the mono- and di-TFA adducts. Workup of the reaction with $NH_4OH$ and water reduces the product to a single band, the N-protected monoadduct. Silica chromatography (93:7 chloroform/methanol) gives 293 mg (0.515 mmol) of product, indicating quantitative reaction.

*N-trifluoroacetyl-5'-DMT-puromycin.* The previous product is coevaporated 2× with dry pyridine in order to remove water. Two hundred sixty-two millgrams (0.467 mmol) N-TFA-Pur is dissolved in 2.4 ml pyridine followed by the addition of 1.4 equivalents of TEA, 0.05 equivalent of DMAP (Sigma), and 1.2 equivalent of dimethoxytrityl chloride (Sigma). After some time an additional 50 mg (0.3 eq) DMT-Cl is added and the reaction is allowed to proceed for an additional 20 min. The reaction is stopped with the addition of 3 ml of $H_2O$ and coevaporated 3× with $CH_3CN$. The reaction is purified by 95:5 chloroform/methanol on a 100-ml silica (dry) 2-cm-diameter column. Due to incomplete purification, a second identical column is run with 97.5:2.5 chloroform/methanol. Total yield is 325 mg or 0.373 mmol, equal to 72%.

*N-Trifluoroacetyl-5'-DMT-2'-succinylpuromycin.* Two hundred ninety-two milligrams (0.336 mmol) of the previous product is combined with 1.2 equivalent DMAP in 3 ml of dry pyridine. To this is added 403 μl of 1 *M* succinic anhydride (Fluka) in dry $CH_3CN$, and the mixture is allowed to stir overnight. TLC again showed little of the starting material remaining. This reaction is combined with a smaller scale reaction (32 mg), and an additional 0.2 equivalent of DMAP and succinate is added. The product is coevaporated with toluene 1× and dried to yellow foam in high vacuum. $CH_2Cl_2$ is added (20 ml), and this solution is extracted with 15 ml of 10% ice-cold citric acid 2× and then 2× of pure $H_2O$. The product is dried and redissolved in 2 ml of $CH_2Cl_2$. The product is precipitated by the addition of 50 ml of hexane with stirring. The product is vortexed and then centrifuged at 600 rpm for 10 min in a clinical centrifuge. The majority of the eluent is drawn off, and the rest of the product is dried, first at low vacuum, then at high vacuum in a desiccator. The yield of this reaction is approximately 0.260 mmol for a stepwise yield of ~70%.

*N-Trifluoroacetyl-5′-DMT-2′-succinylpuromycin–CPG.* The tube containing the product from the previous step is dissolved with 1 ml of dioxane (Fluka) followed by 0.2 ml dioxane/0.2 ml pyridine. To this solution, 40 mg of *p*-nitrophenol (Fluka) and 140 mg of DCC (Sigma) are added, and the reaction is allowed to proceed for 2 hr. The insoluble cyclohexylurea produced by the reaction is removed by centrifugation, and the product solution is added to 5 g of long-chain alkylamine CPG (Sigma) suspended in 22 ml of dry DMF and stirred overnight. The resin is then washed with DMF, methanol, and ether and dried. The resulting resin is assayed as containing 22.6 $\mu$mol of trityl per gram, by absorbance at 498 nm in 2% dichloroacetic acid/$CH_2Cl_2$[32] (note: subsequent syntheses using this protocol have produced supports containing 5–20 $\mu$mol of puromycin per gram). The resulting support is then capped by incubation with 15 ml of pyridine, 1 ml of acetic anhydride, and 60 mg of DMAP for 30 min. The resulting column material produces a negative (no color) ninhydrin test, whereas prior to blocking the test is positive (dark blue). Puromycin–CPG can now be obtained commercially (Glen Research, Sterling, VA).

*30-P and Other Puromycin Oligonucleotides.* CPG puromycin is packed into empty synthesis columns (Millipore, Bedford, MA or Glen Research, Sterling, Virginia) and used for automated synthesis with standard protocols (Millipore). Two linker sequences are used extensively: the "standard" (S) 30-P ($dA_{27}dCdCP$) and a "flexible" sequence (F) analogous to 30-P ($dA_{21}[C9]_3dAdCdCP$) (C9 is triethylene glycol phosphate or Spacer 9, Glen Research). Prior to synthesis, the CPG support is sedimented repeatedly in dry acetonitrile to remove small particles and reduce back pressure during automated synthesis.

Several features are critical in the production of high-quality puromycin oligonucleotides. Efficient capping of the solid support is essential. For resins derivatized at low concentration (~5 $\mu$mol/gram), sequences lacking the 3′-puromycin have the potential to contaminate the syntheses. To test for the presence of 3′-hydroxyl groups, the puromycin oligonucleotide is radiolabeled at the 5′ end using T4 polynucleotide kinase and then used as a primer for extension with terminal deoxynucleotidyltransferase: no extension product is observed. The presence of the primary amine in the puromycin can be assayed by reaction with amine-derivatizing reagents such as NHS-LC-biotin (Pierce, Rockford, IL). 30-P shows a detectable mobility shift by denaturing PAGE on reaction, indicating quantitative reaction with the reagent. Oligonucleotides lacking puromycin do not react with NHS-LC-biotin and show no change in mobility.

*Phosphorylated Puromycin Oligonucleotides.* Puromycin oligonucleotides are either phosphorylated using T4 polynucleotide kinase (New England Biolabs) in ligase buffer containing 1 m$M$ ATP and desalted on NAP-25 columns or phosphorylated chemically during automated synthesis (Glen Research).

### 4. Ligation of mRNA and 3'-Puromycin Oligonucleotides

Translation templates are generated using a splinted ligation[33] of mRNA with 5'-phosphorylated puromycin oligonucleotides. Ligation reactions are conducted with mRNA, DNA splint, and puromycin oligonucleotide in a molar ratio of 1.0–1.2 : 1.5–2.0 : 1.0 or 1.0 : 1.2 : 1.4 and 1.6–2.5 units of T4 DNA ligase per picomole of puromycin oligonucleotide or mRNA, respectively. The mixture is first heated in $H_2O$ to 94° for 1 min and is then cooled on ice for 15 min. Ligation reactions are performed for 1 hr at room temperature in 50 m$M$ Tris–HCl (pH 7.5), 10 m$M$ $MgCl_2$, 10 m$M$ DTT, 1 m$M$ ATP, 25 $\mu$g/ml BSA, 15 $\mu M$ puromycin oligonucleotide, 15 $\mu M$ mRNA, 22.5–30 $\mu M$ DNA splint, and RNasin inhibitor (Promega, Madison, WI) at 1 U/$\mu$l and are usually complete after 40 min. After the incubation, EDTA is added to a final concentration of 30 m$M$ and the reaction mixture is extracted with phenol/chloroform. Full-length mRNA–puromycin templates are purified on denaturing PAGE, isolated by electroelution, and desalted.

Several features are important for efficient ligation. First, the splint should overlap both the 3' end of the mRNA and the 5' end of the puromycin oligonucleotide by ~10 bases, bringing the ends into perfect register. One difficulty with this strategy is that runoff transcription products synthesized with T7, T3, or SP6 RNA polymerase are usually heterogeneous at their 3' end.[34] Because T4 DNA ligase prefers precise base pairing near the ligation junction, only those RNAs containing the correct 3'-terminal nucleotide can be ligated efficiently. When a standard DNA splint is used, approximately 40% of runoff transcription products are ligated to the puromycin oligonucleotide. The amount of ligation product is increased when

[33] M. J. Moore and P. A. Sharp, *Science* **256**, 992 (1992).
[34] J. F. Milligan, D. R. Groebe, G. W. Witherell, and O. C. Uhlenbeck, *Nucleic Acids Res.* **15**, 8783 (1987).

an excess of mRNA is used, but not with an excess of puromycin oligonucleotide.

Apparently, the limiting factor for ligation is the amount of RNA that is fully complementary to the corresponding region of the DNA splint. To allow ligation of those transcripts ending with an extra nontemplated nucleotide at the 3′ terminus (N + 1 products), the standard DNA splint (5′-TTTTTTTTTTAGCGCAAGA) is mixed with a new DNA splint containing an additional random base at the ligation junction (5′-TTTTTTTTTTNAGCGCAAGA). Enzymatic ligation of the myc RNA (RNA124; 5′-GGGACAAUUACUAUUUACAAUUACAAUGGACG-AAGAACAGAAACUGAUCUCUGAAGAAGACCUGCUGCGUA-AACGUCGUGAACAGCUGAAACACAAACUGGAACAGCUGCG-UAACUCUUGCGCU-3′) with the standard puromycin linker and the splint mixture results in a yield of more than 70% ligation product, the mRNA–puromycin template.

Two other factors can affect a ligation reaction adversely. First, the mRNA template to be ligated should be devoid of secondary structure at its 3′ end as this can completely abolish ligation. Second, it is important to purify and desalt the template, the splint, and the puromycin oligonucleotide as high concentrations of salt can interfere with ligation.

## 5. In Vitro Translation

In principle, many *in vitro* translation systems could be adapted to produce mRNA–protein fusions (see Fig. 5). We examined three of the most commonly used systems and found the reticulocyte lysate system[35] to be superior to the wheat germ[36] and bacterial systems[37] (and references therein).

Our first experiments were performed using a bacterial extract derived from the RNase⁻ strain MRE600 commonly used for ribosome preparations. Templates containing a single codon in the open reading frame were found to incorporate [³⁵S]Met at a level similar to tRNA in both 30S extracts[37] and ribosome-enriched fractions.[38] Templates containing 12 codons showed no fusion formation and are degraded quickly. Indeed, *in vitro* selection studies using stalled bacterial ribosomes indicate that very little mRNA template is present after 10 min without the addition of RNase inhibitors.[11]

[35] R. J. Jackson and T. Hunt, *Methods Enzymol.* **96**, 50 (1983).
[36] A. H. Erickson and G. Blobel, *Methods Enzymol.* **96**, 38 (1982).
[37] J. Ellman, D. Mendel, S. Anthony-Cahill, C. J. Noren, and P. G. Schultz, *Methods Enzymol.* **202**, 301 (1991).
[38] W. Kudlicki, G. Kramer, and B. Hardesty, *Anal. Biochem.* **206**, 389 (1992).

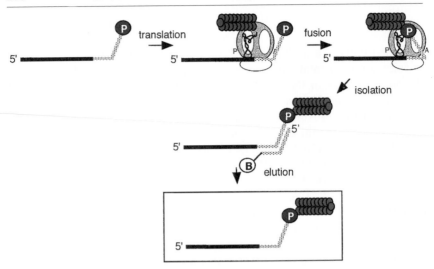

FIG. 5. Translation, fusion formation, and isolation of mRNA–puromycin conjugates.

We next explored the wheat germ and reticulocyte lysate translation systems. Although we could detect fusion with the myc template in the wheat germ system, the resulting mRNA template showed significant degradation after the normal incubation period. The wheat germ system also has the disadvantage of being relatively dependent on the presence of the 5′ m⁷GpppG cap structure for initiation. Initial experiments with the reticulocyte system showed little degradation. Indeed, incubation of 3′-puromycin templates coding for 12 and 33 amino acid myc peptides showed fusion formation and little template degradation as judged by [³⁵S]Met incorporation.[12] Further, reticulocyte lysates are known to support relatively cap-independent translation and can be prepared easily from reticulocyte-rich rabbit blood (Pel-Freez, Rogers, AR)[35] or obtained from a commercial source.

*Reticulocyte Translation Reactions.* Translation reactions were performed in reticulocyte lysate from different commercial sources (Novagen, Amersham, Boehringer, Ambion, or Promega). A typical reaction mixture (25 μl final volume) consisted of 20 m$M$ HEPES, pH 7.6, 2 m$M$ DTT, 8 m$M$ creatine phosphate, 100 m$M$ KCl, 0.75 m$M$ magnesium acetate, 1 m$M$ ATP, 0.2 m$M$ GTP, 25 μ$M$ of each amino acid (0.7 μ$M$ of methionine if [³⁵S]Met is used), RNasin at 1 U/μl, and 60% (v/v) of lysate. The final concentration of uncapped template was in the range of 50 to 800 n$M$. For each incubation, all the components except lysate were mixed carefully on ice, and the frozen lysate is thawed immediately before use. After the addition of lysate, the reaction was mixed thoroughly by gentle pipetting and is incubated at 30° to start translation. The optimal concentrations of $Mg^{2+}$ and $K^+$ vary with different mRNAs and were determined in preliminary experiments for each template. For poorly translated mRNA, it is also worth optimizing the concentrations of hemin, creatine phosphate, tRNA, and amino acids. KCl usually gives better translation and fusion formation than potassium acetate, although occasionally a mixture of the two leads to the best results.

## 6. Formation of RNA–Protein Fusion in Vitro

By the end of a normal translation reaction, only a small amount of the in vitro-synthesized protein has been converted to its mRNA–protein fusion.[12] However, early experiments pointed to certain cases where as much as 50% of the translated protein could be converted to fusion. Subsequent experiments clearly demonstrate that three factors can routinely provide for more than 40% of the input RNA and 50–80% of in vitro-translated protein to be synthesized as fusions: (1) posttranslational addition of $Mg^{2+}$ and $K^+$, (2) use of a flexible linker of the correct length, and (3) incubation after completion of translation at room temperature if $K^+$ and $Mg^+$ are added or long incubations (12–48 hr) at low temperature ($-20°$) if they are not.

Fusion formation (i.e., transfer of the nascent peptide chain from the peptidyl-tRNA to the tethered puromycin moiety) is a slow reaction that follows the initial, relatively rapid translation of the open reading frame. This can be seen in two ways. First, very little of the peptide attached as fusion results from premature entry of the puromycin into the ribosome as judged by the homogeneity and properties of the peptide product.[12] Second, incubation of translation reactions at low temperatures under conditions favoring intact 80S ribosomes results in a gradual increase in the formation of fusion product. These results, taken together with previous data indicating that termination is a relatively slow step in translation,[39] are consistent with a model where 3'-puromycin templates are sequestered on the ribosome following translation. These complexes slowly give rise to the fusion product as the puromycin finds its way to the peptidyl transferase site.

Posttranslational Addition of $Mg^{2+}$ and $K^+$. On discovery that low temperature incubation ($-20°$) increases fusion formation, we examined the effect of various ions present in the translation extract ($Mg^{2+}$ and $K^+$) that might be concentrated on partial freezing. Figure 6A indicates the effect of added $Mg^{2+}$ on fusion formation. The addition of 50–100 m$M$ of $Mg^{2+}$

---

[39] S. L. Wolin and P. Walter, EMBO J. **7**, 3559 (1988).

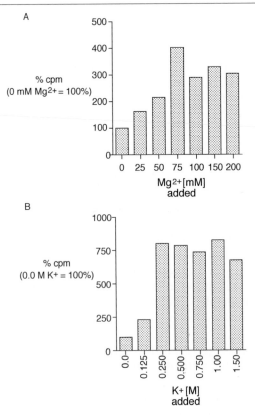

FIG. 6. Effect of posttranslational addition of $Mg^{2+}$ and $K^+$ on fusion formation. (A) Effect of $Mg^{2+}$ addition on fusion formation. After translation, extra magnesium acetate was added corresponding to 0–200 mM final concentration. The translation mixture was then incubated at $-20°$ for 16 hr and assayed by SDS–PAGE. Data are expressed as percentage radioactivity of the 0 mM $Mg^{2+}$ added control. Addition of 75 mM $Mg^{2+}$ produces a fourfold fusion increase relative to the control. (B) Effect of $K^+$ addition on fusion formation. After translation, extra KCl was added corresponding to a concentration of 0–1.5 M, followed by incubation at room temperature for 1 hr. Data are expressed as percentage radioactivity of the 0 mM $K^+$ added control. Addition of 250 mM or higher $K^+$ improves fusion formation by more than sevenfold relative to the control. All translation reactions were performed at 30° for 1 hr using 400 nM myc RNA, the standard linker (dA$_{27}$dCdCP), and [$^{35}$S]Met as label.

increases the amount of fusion formed by two- to fourfold relative to the no $Mg^{2+}$ control. Similarly, the addition of 250–500 mM $K^+$ (Fig. 6B) increases fusion formation by greater than sevenfold relative to the no added $K^+$ control. The posttranslational addition of $NH_4Cl$ also increases fusion formation. The choice of acetate ion vs $Cl^-$ as the anion does not have a profound effect on fusion formation.

Optimally, after translation at 30° for 30 to 90 min, the reaction is cooled on ice for 40 min and $Mg^{2+}$ and/or $K^+$ is added. The resulting mixture is then incubated at $-20°$ for 16 to 48 hr if $Mg^{2+}$ alone is used or at room temperature for 1 hr if $K^+$ or $Mg^{2+}/K^+$ is added. The final concentration of each added cation should be optimized for different mRNA templates and is usually in the range of 50 to 100 m$M$ for $Mg^{2+}$ (50 m$M$ may be used for pools of mixed templates) and 0.30–0.60 $M$ $K^+$ (0.50 $M$ for pools). After incubation, 2 $\mu$l of reaction mixture is mixed with 4 $\mu$l loading buffer and heated at 75° for 3 min. The resulting mixture is then loaded onto a 6% glycine SDS–PAGE (for $^{32}$P-labeled RNA templates) or an 8% tricine SDS–PAGE (for $^{35}$S-labeled peptides) for analysis. The fusion products can also be isolated using $dT_{25}$ streptavidin agarose or thiopropyl-Sepharose[12] (described in detail later).

*Quantitation of Fusion Efficiency.* Fusion efficiency may be expressed as either the fraction of translated peptide or the fraction of input template converted to fusion product. The fraction of translated peptide converted is determined using [$^{35}$S]Met as the label and counting the fusion vs the peptide band intensity by phosphorimager (Molecular Dynamics). After a 1-hr translation reaction under standard conditions, about 3.5% of the *in vitro*-synthesized peptide is converted to fusion using either the $dA_{27}dCdCP$ or the $dA_{27}rCrCP$ linker. This value increases to 12% after overnight incubation at $-20°$ and to more than 50% when $Mg^{2+}$ is added to the posttranslational incubation. Templates containing the flexible linker show sevenfold higher fusion formation than the standard linker after the initial translation reaction (Fig. 7). Posttranslational incubation with optimal $Mg^{2+}$ and $K^+$ concentrations results in a further increase in fusion formation of between 1.2- and 1.6-fold relative to the standard linker (data not shown). Although the standard linker gives adequate results after posttranslational incubation, the flexible linker is superior overall, particularly if the addition of high concentrations of $Mg^{2+}$ is not desired.

The fraction of input template converted is determined using $^{32}$P-labeled template with a flexible linker followed by PAGE analysis and phosphorimager quantitation. The mobility difference between mRNA and mRNA–peptide fusions on SDS–PAGE may be very small if the mRNA template is long. For analysis, the puromycin oligonucleotide is 5′ phosphorylated using [$\gamma$-$^{32}$P]ATP and T4 polynucleotide kinase prior to ligation of the mRNA–puromycin conjugate. After translation and incubation, the long RNA portion is then digested with RNase H in the presence of a complementary DNA splint. Because RNase H leaves a 5′ phosphate, this technique allows quantitation of the modified and unmodified linker after denaturing PAGE. This protocol avoids a second kinase step necessary if RNase A is used, which leaves a 5′-OH. For RNase H treatment, EDTA is added

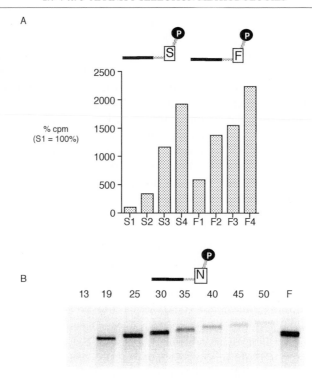

FIG. 7. Optimization of the flexibility and length of the 3′-puromycin oligonucleotide linker. (A) Fusion formation for myc RNA templates containing the standard linker, [S] (dA$_{27}$dCdCP), or the flexible linker, [F] (dA$_{21}$[C9]$_3$dAdCdCP). Posttranslation incubations were carried out under four sets of conditions for each linker (S1–4 and F1–4, respectively): (1) no incubation, (2) incubation at −20° for 16 hr, (3) 50 m$M$ magnesium acetate added and incubation at −20° for 16 hr, and (4) 50 m$M$ magnesium acetate and 500 m$M$ KCl added and incubation at room temperature for 1 hr. Translations were performed with 400 n$M$ template for 1 hr at 30° with [$^{35}$S]Met as the label. Data are expressed as percentage [$^{35}$S]Met incorporation relative to the control without −20° incubation. (B) Effect of linker length on fusion formation. myc templates containing linkers [N] = 13, 19, 25, 30, 35, 40, 45, and 50 nucleotides long (dA$_{10-47}$dCdCP) were assayed for fusion formation by SDS–PAGE. The flexible linker F is also shown. Translations were performed with 600 n$M$ template at 30° for 90 min, followed by the addition of 50 m$M$ Mg$^{2+}$ and incubation at −20° for 2 days.

after posttranslational incubation to disrupt ribosomes, and the reaction mixture is desalted using a Microcon-10 (or Microcon-30, Amicon, Danvers, MA) column. Two microliters of the resulting mixture is combined with 18 μl of RNase H buffer [30 m$M$ Tris–HCl, pH 7.8, 30 m$M$ (NH$_4$)$_2$SO$_4$, 8 m$M$ MgCl$_2$, 1.5 m$M$ 2-mercaptoethanol, and an excess of complementary DNA splint] and incubated at 4° for 45 min. RNase H is then added and digestion is performed at 37° for 20 min.

When the flexible linker is used with the myc template for translation reactions, between 20 and 40% of the input template is converted to fusion after posttranslational incubation without added $Mg^{2+}$ (see Table II). Results indicate that between 2 and $9 \times 10^{13}$ molecules of fusion can be synthesized per milliliter of translation. Similar results are obtained when the posttranslational incubation is performed in the presence of 50 m$M$ $Mg^{2+}$.

*Optimizing Linker Length and Sequence.* The dependence of fusion formation on the length and sequence of the linker was also investigated. In the range between 19 and 30 nucleotides (d(A)$_n$dCdCP, $n$ = 16, 22, and 27), relatively little change is seen in the efficiency of the fusion reaction (Fig. 7B), with the highest yield occurring with $N$ = 25. Even the optimal sequence did not form fusion as efficiently as the flexible linker (Fig. 7). Linkers longer than 40 nucleotides in length and shorter than 16 nucleotides show a greatly reduced efficiency of fusion formation, suggesting that the linker may take an internal pathway through the ribosome to the peptidyl transferase center.

A second factor that could effect the efficiency of fusion formation is the backbone composition and sequence surrounding the puromycin. The standard linker contains a deoxyribose backbone near the 3′-terminal puromycin (5′-dAdCdCP) as compared with the ribose sequence present at the 3′ end of all aminoacyl-tRNA molecules (5′-rCrCrA-amino acid). However, the addition of rCrC to the 3′ end of the standard linker (dA$_{27}$rCrCP) gives little improvement in fusion formation, indicating that the deoxy backbone by itself does not hinder the reaction. However, the importance of the sequence adjacent to the puromycin is highlighted by the observation that rUrUP does not form fusion efficiently (data not shown).

*cis vs trans Fusion Formation.* One essential feature of the fusion formation reaction is that it occurs *in cis,* i.e., within a single ribosome–template–peptide complex. Fusion formation that occurs *in trans,* either by entry of

TABLE II

PERCENTAGE, MOLES, AND MOLECULES OF myc TEMPLATE
CONVERTED TO FUSION

| Concentration of myc template (n$M$) | Template modified (%) | Moles of fusion (pmol) | Molecules of fusion/ml of translation |
|---|---|---|---|
| 800 | 20 | 160 | $9 \times 10^{13}$ |
| 400 | 40 | 160 | $9 \times 10^{13}$ |
| 200 | 40 | 80 | $5 \times 10^{13}$ |
| 100 | 35 | 35 | $2 \times 10^{13}$ |

puromycin from a free template or a template complexed with a different ribosome, would scramble the protein and mRNA sequences, destroying our ability to perform selections. The fact that the bulk of fusion formation occurs during the posttranslational incubation raised the concern that at least some of the fusion formed resulted from the *trans* reaction.

In order to test the extent of the *trans* reaction, we coincubated templates and puromycin oligonucleotides whose fusion products and cross-products (templates fused to the wrong protein) could be separated by electrophoresis. In all the template and linker combinations examined to date, no cross-product formation has been observed. Fusion cross-products could form via two different *trans* mechanisms: (1) reaction of free templates or linkers with the peptide in a peptide–mRNA–ribosome complex or (2) reaction of the template of one complex with the peptide in another. An example testing the latter is shown in Fig. 8. There, the λ protein phosphatase (λPPase) template, which synthesizes a protein 221 amino acids long, was coincubated with the myc template, which generates a 33 amino acid peptide. By themselves, both templates demonstrate fusion formation after posttranslation incubation. When mixed together, only the individual fusion products are seen. No cross-products resulting from fusion of the

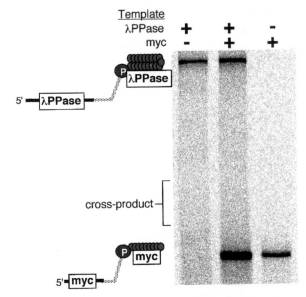

Fig. 8. Cotranslation of myc and λPPase mRNA: 200 n*M* of λPPase RNA (RNA716) and/ or 50 n*M* myc RNA (RNA152) containing the flexible linker (F) was translated with [$^{35}$S]Met. Mg$^{2+}$ (75 m*M*) was added, followed by incubation at $-20°$. No bands are seen from cross-products (myc templates fused to λPPase protein).

λPPase protein with the myc template are seen. Similar experiments show no cross-product formation with several different combinations: the myc template + the single codon template, a 20:1 ratio of the standard linker + the myc template, and the flexible linker + the myc template. These experiments argue strongly against both possible mechanisms of *trans* fusion formation.

The effect of linker length on fusion formation is also consistent with an *in cis* mechanism. Reduction of the linker length from 19 to 13 nucleotides results in the abrupt decrease in the amount of fusion product we would expect if the chain could no longer reach the peptidyltransferase center from the decoding site (Fig. 7). However, this effect could also be due to occlusion of the puromycin within the ribosome if the *trans* mechanism dominated (e.g., if ribosome-bound templates formed fusion via a *trans* mechanism). The decrease in fusion formation with longer linkers again argues against this type of reaction, as no decrease should be seen for the *trans* reaction once the puromycin is free of the ribosome.

*Model for Fusion Formation.* Taken together, our data currently support a *cis* model where all fusion formation (both during translation and afterward in the incubation) occurs between the nascent peptide and the bound mRNA on the same ribosome. In this model, the 3′ end of the message threads through the ribosome, enters the peptidyl transferase site, and fuses with the nascent peptide. The notion that the mRNA must thread into the ribosome is supported by the effect of linker length, flexibility, and sequence on fusion formation and by the 25-Å structure of the 70S ribosome.[40] Observations that the optimal linker is 25 nucleotides long and that only very short linkers (≤13 nucleotides) are able to abolish fusion formation are consistent with the linker sequence directly spanning the distance between the decoding site and the peptidyl transferase site (approximately 75 Å) rather than looping around the ribosome. Longer linker sequences also support a threading model, as the sharp decrease in fusion formation for linkers longer than 30 nucleotides would not be expected if looping occurred.

## 7. Isolation of Fusions

Once synthesized, the fusions are isolated for the next step in the selection cycle. Normally, all the templates in the translation extract are isolated using an affinity tag complementary to the linker sequence. If the yield of fusions is low relative to the input template, a second purification step can be used to isolate only the fusion products.[12]

---

[40] J. Frank *et al.*, *Nature* **376**, 441 (1995).

An alternative is to perform the selective step directly using the fusion/ lysate mixture. The main advantage of this approach is that selection and purification occur in a single step. However, there are several disadvantages. For example, the RNA structure may interfere with the binding of the selection motif. Because the RNA in the ORF is random, RNA aptamers that bind the selection target may be isolated rather than peptides or proteins. An intriguing possibility is that the RNA and protein can collaborate to provide the desired function. Thus, selection prior to cDNA synthesis provides a possible means of isolating functional RNPs.

*Generation of $dT_{25}$ Streptavidin Agarose.* 3'-Biotin–$dT_{25}$ oligonucleotides are generated by automated synthesis using BiotinTEG CPG (Glen Research) and desalted on NAP-25 columns (Pharmacia). The biotin $dT_{25}$ is added to a 50% (v/v) slurry of ImmunoPure streptavidin agarose (Pierce) and 1× TE 8.2 (10 m$M$ Tris–HCl, pH 8.2, 1 m$M$ EDTA) to a final concentration of 10 or 20 $\mu M$ and incubated for 1 hr at room temperature with shaking. The binding capacity of the agarose is then estimated optically by the disappearance of biotin–$dT_{25}$ from solution and/or by titration of the resin with known amounts of complementary oligonucleotide

*Isolation of Fusion with $dT_{25}$–Streptavidin Agarose or Oligo (dT) Cellulose.* After posttranslational incubation, the translation reaction is diluted approximately 150-fold into isolation buffer (1.0 $M$ NaCl, 0.1 $M$ Tris–HCl, pH 8.2, 10 m$M$ EDTA, 0.2% Triton X-100) containing $dT_{25}$–streptavidin agarose or oligo(dT) cellulose (Pharmacia) and incubated with agitation at 4° for 1 hr. The agarose is then removed from the mixture by either filtration or centrifugation and washed with cold isolation buffer two to four times. The template is then liberated from the $dT_{25}$–streptavidin agarose by repeated washing with either pure water at room temperature or 15 m$M$ NaOH, 1 m$M$ EDTA at 4°. The eluent is neutralized immediately and either ethanol precipitated or used directly for the next step in the purification. This protocol generally allows recovery of more than 70% of the input template.

*Isolation of Fusion with Thiopropyl-Sepharose.* Fusions containing cysteine can be purified via disulfide bond chromatography using thiopropyl (TP)-Sepharose 6B (Pharmacia). TP-Sepharose can be used on the translation mixture directly, but generally results in the undesired copurification of many proteins from the extract. The technique is best applied after a prior purification step, such as $dT_{25}$ affinity isolation, and affords a means to specifically purify fusion products away from unmodified templates.[12] For direct isolation of myc peptide fusions attached to the standard linker, the translation reaction ([$^{35}$S]Met-labeled) is mixed with a 10-fold excess of a 50/50 (v/v) slurry of washed TP-Sepharose 6B in 1× TE 8.2 containing DNase-free RNase A (Boehringer Mannheim) and rotated for 1 hr at 4°.

Excess liquid is removed, and the mRNA–peptide is eluted with two washes of isolation buffer containing 20 m$M$ DTT and analyzed by SDS–Tricine–PAGE.

For isolation of intact fusion products, templates are purified first by dT$_{25}$ affinity chromatography. For selections involving the myc template, 25 $\mu$l of the translation reaction is mixed with 7.5 ml of isolation buffer and 125 $\mu$l of 20 $\mu M$ dT$_{25}$ streptavidin agarose and incubated at 4° for 1 hr with rotation. The tubes are centrifuged, the eluent is removed, and the agarose is washed 4× with 1-ml aliquots of ice-cold isolation buffer. The eluent is then removed totally using a Millipore MC filter unit, and templates are isolated by elution with 2 volumes of 100 $\mu$l of 0.1 m$M$ DTT (25°) and 2 volumes of 15 m$M$ NaOH, 1 m$M$ EDTA (4°) followed by neutralization. The purified template/fusion mixture is combined with 40 $\mu$l of a 50/50 (v/v) slurry of washed TP Sepharose in 1× TE 8.2 and rotated for 1 hr at 4°. The Sepharose is washed three times with 1 ml 1× TE, pH 8.2, and the supernatant is removed. One milliliter of 1 $M$ DTT is added to the solid (20–30 ml total volume), incubated for several hours, and washed 4× with 20 $\mu$l H$_2$O four times to remove liberated fusion products. The eluent contains fusion products as well as 2.5 m$M$ thiopyridone liberated from the Sepharose by DTT treatment. Myc fusions are either ethanol precipitated by adding 3 $M$ sodium acetate, pH 5.2, 10 m$M$ spermine, 1 $\mu$l glycogen (10 mg/ml, Boehringer Mannheim), and 3 volumes of 100% ethanol or used directly for the subsequent reverse transcriptase step.

## 8. Generation of cDNA/mRNA–Protein Fusion

For the majority of the selection experiments using the fusion system, it is desirable to remove the secondary and tertiary structures from the RNA so that these do not interfere or compete for binding. To do this, the cDNA/mRNA hybrid fusion product is generated using reverse transcriptase (RT) prior to the selective step.

The reverse transcription step is generally performed after purification of the template from the fusion reaction by dT$_{25}$-agarose or oligo(dT)-cellulose affinity chromatography. Purification of the template is essential for efficient extension with RT as reactions run in the presence of the lysate often give no or very poor cDNA synthesis. For reverse transcription of myc sequences, the templates are first purified with dT$_{25}$-agarose followed by disulfide bond chromatography.[12] Superscript II (GIBCO-BRL) is used for all RT reactions according to the manufacturer's specifications. For ethanol-precipitated samples, 30 $\mu$l of resuspended template, 200 pmol of primer, and water (final volume, 48 $\mu$l) are mixed and heated to 70° for 5 min and then cooled on ice. Sixteen microliters of first strand buffer, 8

$\mu$l 100 m$M$ DTT, and 4 $\mu$l 10 m$M$ dNTP are added and equilibrated at 42° prior to the addition of 4 $\mu$l Superscript II reverse transcriptase. When the TP-Sepharose eluent is used directly for cDNA synthesis, water is added (final volume, 35 $\mu$l) and reactions are performed as described previously. For selections, aliquots both before and after the RT reactions are reserved for subsequent PCR analysis.

### 9. Isolation of Functional Fusions with Immobilized Selection Motif

The selection target (protein, nucleic acid, or small molecule) is generally immobilized on a solid support. The pool of fusion products is then incubated with this support affording enrichment of sequences with the desired binding properties away from the nonfunctional majority.

*Immunoprecipitation of myc Peptide Fusions.* The contents of the RT reaction are mixed with 5 volumes of dilution buffer [10 m$M$ Tris–Cl, pH 8.2, 140 m$M$ NaCl, 1% (v/v) Triton X-100] and 20 $\mu$l of protein G/A conjugate (Calbiochem, La Jolla, CA) and precleared by incubation with rotation for 1 hr at 4°. The eluent is removed, and 20 $\mu$l G/A conjugate and 20 $\mu$l of anti-myc monoclonal antibody 9E10 (Calbiochem) are added and incubated for hr at 4°. The conjugate is precipitated by microcentrifuge at 2500 rpm for 5 min, the eluent is removed, and the conjugate is washed 3× with 1-ml aliquots of ice-cold dilution buffer. The sample is then washed with 1 ml ice-cold 10 m$M$ Tris–Cl, pH 8.2, 100 m$M$ NaCl. The bound fragments are removed using 3 volumes of 4% acetic acid, which is frozen and evaporated to dryness for use in PCR reactions.

*Purification of ARM Motif Peptides and Fusions with Immobilized RNA.* RNA-binding sites for the $\lambda$-boxBR,[41] BIV-TAR,[42] and HIV-RRE[43] are synthesized containing a 3'-biotin moiety using standard phosphoramidite chemistry (Fig. 9A). The synthetic RNA samples are deprotected, desalted, and gel purified as described.[12] The 3'-biotinyl-RNA sites are then immobilized by mixing a concentrated stock of the RNA with a 50% (v/v) slurry of ImmunoPure streptavidin agarose (Pierce) in 1× TE 8.2 at a final RNA concentration of 5 $\mu$M for 1 hr (25°) with shaking. Two translation reactions are performed containing (1) the template coding for the $\lambda$N peptide fragment (Fig. 9B) or (2) globin mRNA (Novagen) as a control. Aliquots [50 $\mu$l of a 50% slurry (v/v)] of each immobilized RNA are washed and resuspended in 500 $\mu$l of binding buffer [100 m$M$ KCl, 1 m$M$ MgCl$_2$, 10 m$M$ HEPES–KOH, pH 7.5, 0.5 m$M$ EDTA, 0.01% Nonidet P-40 (NP-40), 1 m$M$ DTT, 50 $\mu$g/ml yeast tRNA]. Binding reactions are performed by

[41] C. D. Cilley and J. R. Williamson, *RNA* **3**, 57 (1997).
[42] J. D. Puglisi, L. Chen, S. Blanchard, and A. D. Frankel, *Science* **270**, 1200 (1995).
[43] J. L. Battiste *et al.*, *Science* **273**, 1547 (1997).

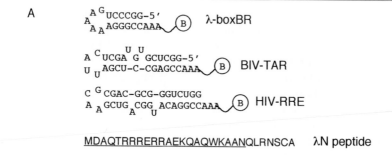

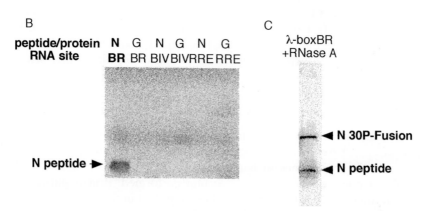

FIG. 9. Binding of the λN arginine-rich motif (ARM) peptide and fusion to immobilized RNA sites. (A) RNA-binding sites for the λN peptide (λ-boxBR),[41] BIV-Tat peptide (BIV-TAR),[42] and HIV-Rev peptide (HIV-RRE)[43] and the λN peptide used. The underlined peptide sequence corresponds to N$^\lambda$(1–22).[41] (B) Isolation of *in vitro*-translated λN peptide (N) or β-globin (G) using immobilized RNA sites. Only the λN/λ-boxBR combination allows isolation. (C) Isolation of λN fusions using immobilized λ-boxBR RNA by RNase A digestion. N peptide fusions containing the standard linker were incubated with immobilized λ-boxBR RNA.[41] Peptides and fusions were released by RNase A treatment of the immobilized RNA and assayed by SDS–Tricine–PAGE.[44] Both λN peptide and λN-30-P fusion are isolated.

adding 15 μl of the translation reaction containing either the N peptide or globin templates to tubes containing one of the three immobilized binding sites followed by incubation at room temperature for 1 hr. The beads are precipitated by centrifugation and are washed 2× with 100 μl of binding buffer. RNase A (DNase free, 1 μl, 1 mg/ml) (Boehringer Mannheim) is added and incubated for 1 hr at 37° to liberate bound molecules. The supernatant is removed and mixed with 30 μl of SDS loading buffer[44] and

[44] H. Schagger and G. V. Jagow, *Anal. Biochem.* **166,** 368 (1987).

analyzed by SDS–Tricine–PAGE (Fig. 9B). The same protocol is used for isolation of N peptide fusions (Fig. 9C), with the exception that 35 m$M$ MgCl$_2$ is added after the translation reaction followed by incubation at room temperature for 1 hr to promote fusion formation.

Figures 9B and 9C demonstrate that the N peptide retains its normal binding specificity both when synthesized *in vitro* and when generated as an RNA–peptide fusion with its own mRNA. This result is of critical importance. The attachment of a long nucleic acid sequence to the C terminus of a peptide or protein (e.g., fusion formation) has the potential to disrupt the polypeptide function relative to the unfused sequence. Arginine-rich motif (ARM) peptides represent a stringent functional test of the fusion system due to their relatively high nonspecific nucleic acid-binding properties. The fact that the N peptide–mRNA fusion (prior to cDNA synthesis) retains the function of the free peptide indicates that specificity is maintained even when there is a likelihood of forming either self- or nonspecific complexes.

## 10. Polymerase Chain Reaction to Generate Enriched dsDNA Pool

Polymerase chain reaction protocols used to regenerate the enriched pool after each round of selection have been covered in detail elsewhere.[45,46] We note the importance of performing PCR controls to assure that the amplified pool results from the selection performed. Primer purity is of central importance. The pairs should be amplified in the absence of input template, as contamination with pool sequences or control constructs can occur. New primers should be synthesized if contamination is found. The isolated fusions should also be subjected to PCR prior to the RT step to assure that they are not contaminated with cDNA. Finally, the number of cycles needed for PCR reactions before and after selection should be compared. Large numbers of cycles needed to amplify a given sequence (>25–30 rounds of PCR) may indicate failure of the RT reaction or problems with primer pairs.

## Conclusions

Through a combination of a flexible linker and appropriate posttranslational incubation, the efficiency of fusion formation has been increased from less than 1% to approximately 40% of the input template. Thus, the mRNA fusion system allows libraries containing as many as 10$^{14}$ members

[45] T. Fitzwater and B. Polisky, *Methods Enzymol.* **267,** 275 (1996).
[46] R. C. Conrad, L. Giver, Y. Tian, and A. D. Ellington, *Methods Enzymol.* **267,** 336 (1996).

to be generated, more than any other protein selection technique. Further, our experiments indicate that sequences synthesized as fusions commonly retain the function of their unfused counterparts. Fusion-based selections promise to be a powerful approach to understanding and discovering protein interactions with nucleic acid, protein, and small molecule targets. If natural sources of genomic or cDNA are used for the library construction, mRNA–protein fusions provide the potential to elucidate protein-binding partners for organisms where genetic analysis is difficult, such as humans.

# Section IV

## Genetic Methodologies for Detecting RNA–Protein Interactions

A. Bacterial Systems
*Articles 20 through 22*

B. Eukaryotic Systems
*Articles 23 through 27*

# [20] Screening RNA-Binding Libraries by Transcriptional Antitermination in Bacteria

*By* Hadas Peled-Zehavi, Colin A. Smith, Kazuo Harada, and Alan D. Frankel

## Introduction

A number of systems have been developed to screen libraries for RNA-binding proteins or peptides. In one of the first examples, phage display was used to identify RNP domain variants with altered specificities.[1] Subsequently, systems have been reported in which RNA-binding domains are used to activate transcription in a yeast three-hybrid assay[2,2a] or in a mammalian Tat-fusion system (see Tan *et al.*[3] and Landt *et al.*,[3a] this volume) or to alter translation[4-7] (see articles 21, 22, 24, and 25 in this volume). This article describes a system based on transcription antitermination in bacteria that has a relatively high screening capacity and has been used to examine libraries of RNA-binding peptides.[8,9]

## Description of Antitermination System

The bacteriophage λ N protein is important for the expression of late phage genes and functions by allowing *Escherichia coli* RNA polymerase to read through transcription termination sites. N promotes the assembly of a multicomponent antitermination complex[10] by binding to an RNA hairpin, box B, present in the nut (N utilization) site on the nascent tran-

[1] I. A. Laird-Offringa and J. G. Belasco, *Proc. Natl. Acad. Sci. U.S.A.* **92**, 11859 (1995).
[2] D. J. SenGupta, B. Zhang, B. Kraemer, P. Pochart, S. Fields, and M. Wickens, *Proc. Natl. Acad. Sci. U.S.A.* **93**, 8496 (1996).
[2a] *Methods Enzymol.* **318**, [27], (2000) (this volume).
[3] R. Tan and A. D. Frankel, *Proc. Natl. Acad. Sci. U.S.A.* **95**, 4247 (1998).
[3a] S. G. Landt, R. Tan, and A. D. Frankel, *Methods Enzymol.* **318**, [23], (2000) (this volume).
[4] C. Jain and J. G. Belasco, *Cell* **87**, 115 (1996).
[5] H. Kollmus, M. W. Hentze, and H. Hauser, *RNA* **2**, 316 (1996).
[6] S. Wang, H. L. True, E. M. Seitz, K. A. Bennett, D. E. Fouts, J. F. Gardner, and D. W. Celander, *Nucleic Acids Res.* **25**, 1649 (1997).
[7] E. Paraskeva, A. Atzberger, and M. W. Hentze, *Proc. Natl. Acad. Sci. U.S.A.* **95**, 951 (1998).
[8] K. Harada, S. S. Martin, and A. D. Frankel, *Nature* **380**, 175 (1996).
[9] K. Harada, S. S. Martin, R. Tan, and A. D. Frankel, *Proc. Natl. Acad. Sci. U.S.A.* **94**, 11887 (1997).
[10] J. Greenblatt, J. R. Nodwell, and S. W. Mason, *Nature* **364**, 401 (1993).

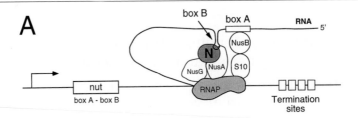

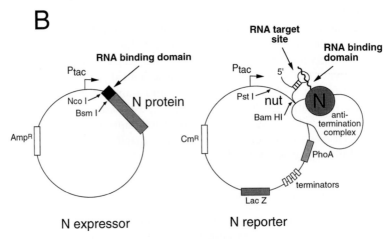

Fig. 1. (A) Schematic model of the λ N antitermination complex. *E. coli* proteins known to participate in antitermination and the approximate arrangement of the complex are shown.[10] (B) The two-plasmid system for detecting transcription antitermination by heterologous RNA-binding polypeptides fused to the N protein.

script (Fig. 1A). A variety of host proteins, including Nus A, Nus G, Nus B, and S 10, and another RNA element in nut (box A) also are important, although antitermination can occur *in vitro* in the absence of most of these components.[11] Franklin[12] has devised a bacterial two-plasmid reporter assay that accurately monitors N-mediated antitermination, and this system has been modified to study heterologous RNA–protein interactions.[8,13] Our use of the system has focused largely on RNA recognition by short arginine-rich peptides[8,9]; however, experiments with the MS2 coat protein,[13] zinc

[11] W. A. Rees, S. E. Weitzel, T. D. Yager, A. Das, and P. H. von Hippel, *Proc. Natl. Acad. Sci. U.S.A.* **93**, 342 (1996).
[12] N. C. Franklin, *J. Mol. Biol.* **231**, 343 (1993).
[13] J. E. Wilhelm and R. D. Vale, *Genes Cells* **1**, 317 (1996).

finger domains,[13a] and the RNP domain of U1A (C. Escùde and A. D. Frankel, unpublished results) suggest that the system can be adapted to work with several different types of RNA-binding domains.

In the two-plasmid system, the N protein is expressed under the control of a *tac* promoter on a pBR322-derived plasmid, and a reporter gene, LacZ, is expressed from a *tac* promoter on a compatible pACYC184-derived plasmid and is located downstream of four tandem termination sites (Fig. 1B). By placing the nut site near the beginning of the transcript, N binds to box B, assembles an antitermination complex, and transcription of the β-galactosidase gene ensues. It is possible to replace box B with other RNA sites of interest and to replace the N-terminal arginine-rich RNA-binding domain of N with other RNA-binding domains, leading to antitermination and β-galactosidase expression via a heterologous RNA–protein interaction. The PhoA reporter located upstream of the termination sites (Fig. 1B) can be used as an internal control for the level of transcription initiation[12,14]; however, we have observed that deletion of the PhoA region increases antitermination levels in the heterologous system, perhaps by reducing the distance between the nut site and termination sites.[10]

RNA-binding domains are fused to N by cloning into the unique NcoI and BsmI sites of the N-expressor plasmid (Figs. 1B and 2A), thereby replacing the N-terminal 18 amino acids of N. Reporter plasmids are constructed using restriction sites in the polylinker (Fig. 2B) to introduce box A followed by the RNA target site. In our experience, the spacing between box A and target site can affect the level of antitermination markedly and it is advisable to approximate the spacing of the wild-type box A–box B configuration as closely as possible (see Fig. 2B). The spacing will be somewhat arbitrary in that it depends on the size and shape of the inserted element, and we do not yet know the precise restrictions on RNA size or structure. There may be less restriction on the types of RNA-binding domains accommodated by the N-fusion proteins, as peptides with different secondary structures, the MS2 coat protein, zinc finger domains, and the U1A RNP domain all create functional fusions. The interactions studied to date have binding constants for specific sites in the nanomolar to subnanomolar range. It is not yet known what affinity and specificity limits will be tolerated by the system, but an excellent correlation between RNA-binding affinity and antitermination has been observed. The caveats notwithstanding, the system has been useful for screening large peptide libraries and is likely to be applicable to a wider variety of interactions.

[13a] D. J. McCall, C. D. Honchell, and A. D. Frankel, *Proc. Natl. Acad. Sci. U.S.A.* **96,** 9521 (1999).

[14] J. H. Doelling and N. C. Franklin, *Nucleic Acids Res.* **17,** 5565 (1989).

A

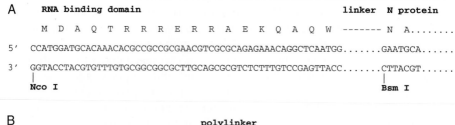

B

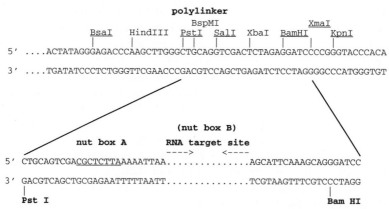

FIG. 2. (A) Sequence of the N-terminal region of the N protein. *Nco*I and *Bsm*I sites in the N-expressor plasmid are used to clone heterologous RNA-binding domains or libraries, generating fusions to amino acid 19 of N. The N-terminal arginine-rich RNA-binding domain of N forms a bent α helix and recognizes the box B hairpin of the nut site.[24,25] (B) Polylinker region of the pACYC N-reporter plasmid. An oligonucleotide containing box A of the nut site followed by the RNA target site is cloned into the polylinker, effectively replacing box B with the RNA site of interest. Unique restriction sites are underlined, and we typically use *Pst*I and *Bam*HI sites for cloning.

## Procedure: Scoring RNA-Binding Interactions by Colony Color

For initial assessment of antitermination activities or for screening libraries, a β-galactosidase colony color assay is used. N-expressor and N-reporter plasmids are transformed into *E. coli* N567 host cells[14] and blue colony color is scored on a relative scale of zero to five pluses, with the wild-type N reporter cells in the absence of N scoring zero and the N reporter in the presence of wild-type N scoring + + + +. Although both expressor and reporter plasmids may be cotransformed, it often is difficult to obtain a uniform number of colonies from plate to plate. Because accurate scoring of colony color requires uniform colony sizes and numbers, it is suggested that competent cells containing either the reporter or the expressor plasmid be used. For most screening experiments, standard heat-shock competent cells prepared by $CaCl_2$ treatment are sufficient. The specificity of an interaction is tested by measuring antitermination activities using cells containing a variety of reporters or expressors. In a typical experiment, a panel of different RNA targets and/or mutant RNA sites is used to assess the binding specificity of a particular RNA-binding domain, and more quantitative assessment of $\beta$-galactosidase activity is carried out using a solution assay.

*Reagents*

5-Bromo-4-chloro-3-indolyl-$\beta$-D-galactopyranoside (X-Gal, 40 mg/ml in dimethyl formamide)

Isopropyl-1-$\beta$-D-galactoside (IPTG, 100 m$M$)

Tryptone broth

Tryptone plates containing 0.05 mg/ml ampicillin, 0.015 mg/ml chloramphenicol, 0.08 mg/ml X-Gal, and 0.05 m$M$ IPTG

Competent N567 cells containing N-expressor (Amp$^R$) or N-reporter plasmids (Cm$^R$)

*Method*

1. Transform the N-expressor or reporter plasmid into competent N567 cells containing the cognate plasmid. Typically 10 ng is used to transform 100 $\mu$l competent cells.
2. Add 1 ml tryptone broth and shake at 37° for 1 hr. Note: Do not use LB or other types of broth that contain yeast extract for growing N567 cells.
3. Plate an appropriate amount of transformation mixture onto tryptone plates containing antibiotics, X-Gal, and IPTG (to induce the *tac* promoters) and incubate for up to 48 hr at 34°. Plate relatively densely to help ensure uniform colony size. Note: Do not use LB or other types of agar plates that contain yeast extract. X-Gal plates are best prepared 1 day before use and stored at room temperature for no more than 3–4 days, or X-Gal may be spread onto plates just prior to use.
4. Score colony color at various times by comparing to appropriate positive and negative controls.

Procedure: Quantitating Antitermination Activities by $\beta$-Galactosidase Solution Assays

*Reagents*

*o*-Nitrophenyl-$\beta$-D-galactopyranoside (ONPG; 4 mg/ml in A medium and stored at 4°)

A medium [10.5 g $K_2HPO_4$, 4.5 g $KH_2PO_4$, 1.0 g $(NH_4)_2SO_4$, 0.5 g sodium citrate dihydrate, $H_2O$ to 1 liter, autoclaved]

Z buffer (60 m$M$ Na$_2$HPO$_4$, 40 m$M$ NaH$_2$PO$_4$, 10 m$M$ KCl, 1 m$M$ MgSO$_4$, 5 m$M$ 2-mercaptoethanol)

*Method*

1. Grow overnight cultures from single representative blue colonies at 37° in tryptone broth containing 0.05 mg/ml ampicillin and 0.015 mg/ml chloramphenicol.
2. Dilute cultures 50-fold into 3–5 ml A medium containing 22.4 m$M$ glucose, 1 mg/ml vitamin B1, 1 m$M$ Mg$_2$SO$_4$, and both antibiotics. Grow at 37° to an OD$_{600}$ ~ 0.2 (3 to 6 hr). Add IPTG (to 0.5 m$M$) and grow an additional hour to OD$_{600}$ ~ 0.4–0.5. Record OD$_{600}$.
3. Mix 0.1 ml cell culture, 0.9 ml Z buffer, 25 $\mu$l chloroform, and 12 $\mu$l 0.1% SDS and vortex vigorously for 10 sec. The solution will be cloudy and the cells permeabilized.
4. Add 0.2 ml ONPG solution at 10- to 15-sec intervals and incubate at 28°. When yellow color develops, add 0.4 ml 1 $M$ sodium carbonate to stop the reaction. Record the times when ONPG was added and when the reactions were stopped.
5. Microfuge to pellet cell debris and measure OD$_{420}$ and OD$_{550}$.
6. Calculate $\beta$-galactosidase activity as follows: Units activity = 1000 × (OD$_{420}$ − 1.6 × OD$_{550}$)/($t$ × 0.1 × OD$_{600}$) where $t$ is reaction time in minutes.

Considerations for Constructing N-Fusion Proteins

The antitermination system was tested initially by replacing the 18 amino acid N-terminal RNA-binding domain of N with the arginine-rich domains of human immunodeficiency virus (HIV-1) Rev and bovine immunodeficiency virus (BIV) Tat proteins and by replacing the box B hairpin of nut with the Rev response element (RRE) and BIV transactivation response element (TAR) (see Smith *et al.,*[14a] this volume). Antitermination was observed only with the cognate interactions, as determined by the $\beta$-galactosidase colony and solution assays[8] (Fig. 3). In general, colony color observed on plates correlates well with $\beta$-galactosidase activities measured in solution and with RNA-binding affinities measured *in vitro*. Particularly at low activities, the colony color assay is more sensitive than the solution assay, presumably because colony color reflects $\beta$-galactosidase accumulated during many hours of growth on plates whereas solution assays reflect enzyme accumulated after 1 hr of IPTG induction. Substitution of the $\lambda$ N–nut interaction with these heterologous interactions usually results in

[14a] C. A. Smith, L. Chen, and A. D. Frankel, *Methods Enzymol.* **318,** [28], (2000) (this volume).

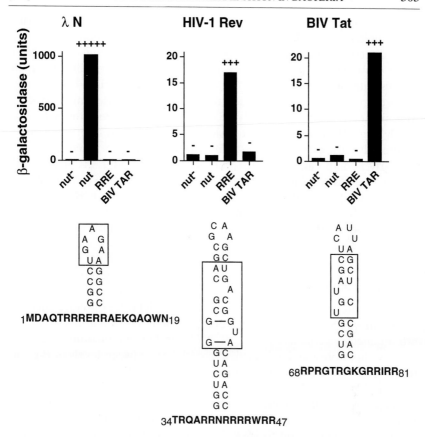

FIG. 3. Antitermination by N-fusion proteins containing the RNA-binding domains of HIV-1 Rev and BIV Tat. Activities of wild-type λ N and the two fusion proteins were measured using reporter plasmids containing the binding sites indicated. The nut⁻ vector contains the polylinker with no box A or box B sites. Values show β-galactosidase activities measured with the solution assay, and pluses and minuses indicate activities scored by colony color. Box B, RRE, and BIV TAR RNA hairpins and the sequences of the corresponding RNA-binding domains are shown (numbering indicates positions in the wild-type protein).

antitermination levels considerably lower than the wild-type case, although they are well above background. We presume that the low activities result from the absence of a Nus A–box B interaction[15] and possibly other interactions that are thought to contribute to the overall stability of the complex.[11,16]

[15] S. Chattopadhyay, J. Garcia-Mena, J. DeVito, K. Wolska, and A. Das, *Proc. Natl. Acad. Sci. U.S.A.* **92,** 4061 (1995).

[16] J. Mogridge, T.-F. Mah, and J. Greenblatt, *Genes Dev.* **9,** 2831 (1995).

When constructing fusions to N, it is important to consider the nature of the linkage to the RNA-binding domain. While some domains may be fused directly to amino acid 19 of N (see Fig. 2A), a short linker is inserted more commonly. The importance of the linker is clearly illustrated by fusions with the arginine-rich RNA-binding domain of HIV-1 Rev, which requires an $\alpha$-helical conformation for RRE binding. Inserting a linker of four alanines, which presumably helps stabilize or nucleate the helix (see Smith et al.[14a]), creates an active fusion protein, whereas inserting a linker of three glycines, expected to destabilize a helix, creates an inactive fusion.[17] Similarly, adding alanines to the C-terminal helix of the U1A RNP domain[18] is important for optimal activity (C. Escude and A. D. Frankel, unpublished results). In contrast, fusions with the arginine-rich RNA-binding domain of BIV Tat, which adopts a $\beta$-hairpin conformation on BIV TAR binding (see Smith et al.[14a]), are slightly more active with a glycine rather than an alanine linker.[17]

## Library Screening Strategies

The antitermination system is suitable for detailed structure/function studies in which the RNA-binding domain, the RNA site, or both partners are mutated systematically. For all RNA–protein interactions studied to date, we have observed an excellent correlation between levels of $\beta$-galactosidase expression in the antitermination assay and RNA-binding affinities measured in vitro. Because the antitermination system is amenable to screening $\sim 10^6$–$10^7$ clones, it is possible to generate a large number of RNA-binding domain or RNA variants by randomizing several positions within the sequence and to identify those with desired ranges of binding affinity. In one relatively simple example, one position in the RNA-binding domain of BIV Tat was randomized and effects on BIV TAR binding were assessed. In the BIV TAR–Tat complex, isoleucine at position 79 makes a hydrophobic contact to a bulge nucleotide in TAR,[19,20] and mutations that reduce hydrophobicity show reduced RNA-binding affinities in vitro.[21] A BIV Tat library randomized at codon position 79 ($4^3 = 64$ sequences) was screened on a BIV TAR reporter, plasmids from 75 colonies displaying varying blue intensities were sequenced, and an excellent correlation was

[17] K. Harada and A. D. Frankel, in "RNA:Protein Interactions: A Practical Approach" (C. W. J. Smith, ed.), p. 217. Oxford Univ. Press, Oxford, 1998.
[18] J. M. Avis, F. H. Allain, P. W. Howe, G. Varani, K. Nagai, and D. Neuhaus, J. Mol. Biol. 257, 398 (1996).
[19] J. D. Puglisi, L. Chen, S. Blanchard, and A. D. Frankel, Science 270, 1200 (1995).
[20] X. Ye, R. A. Kumar, and D. J. Patel, Chem. Biol. 2, 827 (1995).
[21] L. Chen and A. D. Frankel, Proc. Natl. Acad. Sci. U.S.A. 92, 5077 (1995).

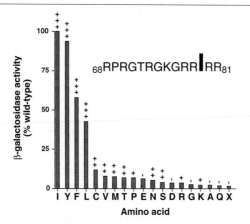

FIG. 4. Antitermination activities of a set of N-fusion proteins containing variants of the BIV Tat RNA-binding domain at position 79. Values are expressed as a percentage of the activity of the wild-type BIV Tat domain having isoleucine at this position, and pluses and minuses indicate activities scored by colony color.

found to the known RNA-binding affinities.[8] All 15 dark blue colonies (+++) encoded large hydrophobic residues (I, Y, F, and L), 29 medium color colonies (++) encoded smaller hydrophobic or uncharged residues, and 31 light blue (+) or background (−) colonies generally encoded amino acids with charged or small side chains (Fig. 4). In another experiment with BIV Tat, lysine-75 and glycine-76, which make up the $\beta$ turn, were randomized simultaneously ($4^6 = 4096$ sequences) and the best binders were found to contain the wild-type residues, with an absolute requirement for the glycine (T. Nagasaki and A. D. Frankel, unpublished results). Thus, the system is well suited to such randomization experiments and, with the current screening capacity, four positions can be fully randomized using 32 codons to encode all 20 amino acids ($32^4 \approx 1 \times 10^6$ sequences).[17]

Procedure: Preparation of Randomized Libraries

In the protocols described here, we focus on generating libraries of RNA-binding domain variants, although similar strategies can be applied to randomization of the RNA target site. In general, library oligonucleotides are prepared by synthesizing a template strand containing appropriately randomized codons, and the second strand is synthesized enzymatically. The library oligonucleotide must possess a fixed 3′ sequence to allow primer binding for second-strand synthesis and restriction sites on the 5′ and 3′ ends for ligation into the NcoI and BsmI sites of the N-expressor plasmid.

Because restriction sites used for cloning may arise in some of the randomized portions, it is advisable to minimize their frequency by choosing alternative nucleotides at the third codon positions of such sequences. It also is advisable to avoid polymerase chain reaction amplification of the cassette, which might further bias the library. Following synthesis and cloning of the library cassette, at least 10–20 random clones should be sequenced to confirm the expected codon and amino acid frequencies. Detailed considerations for generating randomized oligonucleotides may be found elsewhere.[17,22,23]

1. Anneal primer oligonucleotide (400 pmol) to the library template oligonucleotide (300 pmol) by heating to 65° and slow cooling in 1× Sequenase buffer (USB). Synthesize the second strand by primer extension in a reaction mixture (0.5 ml) containing the annealed oligonucleotide, 0.3 m$M$ of each of the deoxynucleotide triphosphates, 10 m$M$ dithiothreitol, 1× Sequenase buffer, and 100 units Sequenase 2.0 for 20 min at 37°. Quench the reaction with 4 $\mu$l 0.5 $M$ EDTA. Run aliquots of the reaction before and after addition of Sequenase on a 4% agarose gel to confirm efficient primer extension. Single-stranded and double-stranded oligonucleotides are distinguished readily.

2. Extract the oligonucleotide twice with 0.5 ml phenol:chloroform [phenol saturated with 1 $M$ Tris–HCl (pH 7.5):chloroform:isoamyl alcohol (25:24:1)] and precipitate with 0.1 volume 3 $M$ sodium acetate (pH 5.2) and 2.5 volumes ethanol. Digest with appropriate restriction enzymes (*Nco*I and *Bsm*I for an N-expressor library), extract twice with phenol:chloroform, and ethanol precipitate. Purify oligonucleotide on a native polyacrylamide gel using UV shadowing to visualize the band. Elute into 400 $\mu$l 300 m$M$ sodium acetate at room temperature, ethanol precipitate, resuspend in 100 $\mu$l H$_2$O, and estimate DNA yield by gel electrophoresis and ethidium bromide staining.

3. Perform a small-scale ligation to determine transformation efficiency (using the appropriate N567 competent cell strain) and estimate the amount required to generate ~2000–3000 colonies per 100-mm plate. This density gives appropriately sized colonies for assessing blue color. Scale the ligation reaction to achieve the desired number of

[22] S. M. Glaser, D. E. Yelton, and W. D. Huse, *J. Immunol.* **149**, 3903 (1992).
[23] B. P. Cormack and K. Struhl, *Science* **262**, 244 (1993).
[24] P. Legault, J. Li, J. Mogridge, L. E. Kay, and J. Greenblatt, *Cell* **93**, 289 (1998).
[25] M. A. Weiss, *Nature Struct. Biol.* **5**, 329 (1998).

clones, depending on the library complexity and how exhaustively sequence space is to be searched.

## Procedure: Selection of RNA Binders from Libraries

Several difficulties may be encountered when screening for rare sequences from large libraries that can be addressed through the series of sequential screens described later. First, the two-plasmid system generates a relatively high rate of false positives (~0.1–0.5%) that appear to result from spontaneous mutation of the reporter plasmid. This high background is observed in the absence of the N-expressor plasmid and may be due to rearrangements of the four tandem terminator elements. Second, nonspecific positive clones are found that antiterminate through a variety of different RNA targets. Third, mutations may arise in the N-expressor plasmid outside the RNA-binding domain that enhance antitermination, although we have not yet observed this problem.

1. Primary screen. Transform the ligated library into competent N567 cells containing the appropriate reporter plasmid (or expressor plasmid when screening RNA libraries). Pick colonies having the desired blue color (typically all colonies showing any blue color are chosen in initial screens for RNA binders) and suspend in individual wells of 96-well plates containing tryptone broth and antibiotics. Grow bacteria overnight at 34° so that each clone is equally represented, pool the cultures, and isolate plasmid DNA by alkaline lysis. Purify the library plasmid (pBR N-expressor or pACYC N-reporter) on agarose gels to eliminate the second plasmid.

2. Eliminate reporter-related false positives. Retransform the plasmid pool into fresh competent cells, plate, and isolate plasmid DNA from individual blue clones. For RNA libraries generated using the pACYC plasmid, reporter-related rearrangements may remain a significant source of false positives; however, most of these can be eliminated by gel purification as we have observed that the rearranged plasmids migrate more slowly on agarose gels and have higher copy numbers than the parent reporter.

3. Eliminate nonspecific positives. Transform plasmids from each clone into competent N567 cells containing the specific cognate plasmid and several nonspecific plasmids and select clones with the desired specificities. The frequency of nonspecific positives varies with the nature of the library, and we have observed frequencies of 50 and 80% (of the clones isolated in step 2) using two highly randomized libraries.[8]

4. Eliminate positives due to changes outside of the randomized region. Sequence the library portion of positive clones remaining from step 3. Synthesize and reclone the corresponding oligonucleotides and retest antitermination activities in appropriate competent cells.

## Types of Libraries and Future Considerations

Clearly there are many types of libraries that may be useful for screening, and much of our work has focused on creating arginine-rich peptide libraries that may contain novel RNA-binding molecules.[3,8,9] For these experiments, we typically generate libraries with a restricted set of amino acids in order to reduce the library complexity and have been able to exhaustively screen libraries with nine randomized positions.[8] It also is possible to generate "doped" libraries in which a prototype sequence is mutated at many positions with a chosen degree of randomness at each codon. Using this strategy, it has been possible to evolve tighter and more specific RRE-binding peptides beginning with a modest binder.[9] Strategies for designing doped libraries and for codon-based mutagenesis schemes that allow efficient, nonbiased searches of sequence space have been considered in detail.[17,22,23]

The antitermination system appears to accommodate relatively diverse RNA-binding domains in that peptides with different secondary structures, zinc finger domains, the MS2 coat protein, and the U1A RNP domain all can be fused to N. Although it is not yet clear what restrictions may exist to the geometry of the antitermination complex, it seems reasonable that the system may be adapted to screen cDNA libraries as well. Given our experience with different sizes and types of RNA sites and with different configurations of box A relative to the inserted RNA, we suspect that the system will be most limited by the nature of the RNA target. We have been testing several modifications to the reporter system and have observed that the spacing between the inserted RNA site and the termination elements may alter the efficiency of antitermination significantly (H. Peled-Zehavi et al., in preparation). One modification that has extended the screening capacity has been the addition of a kanamycin resistance reporter gene that allows direct selection of RNA binders (H. Paled-Zahavi et al., in preparation). Given the large potential signal to noise of the antitermination system (1000-fold is observed with wild-type N), further improvements may be anticipated as restrictions to the system are better characterized.

## Acknowledgments

This work was supported by a postdoctoral fellowship from the Human Frontiers Science Foundation to H.P.-Z., by a postdoctoral fellowship from the American Cancer Society to C.A.S., and by grants from the National Institutes of Health.

# [21] Rapid Genetic Analysis of RNA–Protein Interactions by Translational Repression in *Escherichia coli*

By CHAITANYA JAIN and JOEL G. BELASCO

## Introduction

RNA-binding proteins (RNA-BPs) are involved in a wide variety of cellular and viral processes, including transcription, RNA processing, RNA localization, translation, and mRNA degradation, and defects in these proteins have been implicated in congenital health disorders. Many RNA-BPs can be categorized into families that share certain sequence motifs common to each member of the family.[1,2] Among the major classes of RNA-binding domains are those that contain a conserved RNA recognition motif (RRM, also called RBD), a 10–20 amino acid arginine-rich motif (ARM), or a K homology (KH) motif.

Most RNA-BPs bind to their RNA targets with high affinity and specificity. Dissociation constants of $<10$ n$M$ are common, and single nucleotide substitutions in an RNA target can reduce the binding affinity of an RNA-BP by a factor of 100–1000. Despite the importance of target recognition for RNA-BP function, relatively little is yet understood about the structural basis for the binding specificity and affinity of these proteins. One useful strategy for addressing this question is to identify and characterize RNA-BP variants that are defective in target binding or that bind to new targets.

To facilitate such studies, we have developed a broadly applicable *Escherichia coli*-based genetic method for rapidly identifying RNA-BP variants with altered RNA-binding properties. In this method, the RNA target of a heterologous RNA-BP is inserted immediately upstream of the translation initiation signals of a reporter mRNA (e.g., *lacZ*). The RNA-BP is expressed in *E. coli*, and on binding to its target, it sterically hinders ribosome binding, thereby reducing the translation of *lacZ*. RNA-BP variants that enhance or diminish binding are identified by a change in $\beta$-galactosidase synthesis. This genetic method was first used to examine RNA recognition by the HIV-1 Rev protein. The altered specificity Rev mutants identified in this manner enabled us to construct a structural model for Rev bound to its RNA target.[3] Subsequently, the method has been used to study other

[1] C. G. Burd and G. Dreyfuss, *Science* **265,** 615 (1994).
[2] D. Draper, *Annu. Rev. Biochem.* **64,** 593 (1995).
[3] C. Jain and J. G. Belasco, *Cell* **87,** 115 (1996).

RNA-BPs, such as nucleolin.[4] This article describes the genetic method in detail and provides suggestions and strategies for applying this method to various investigative objectives. We expect that the application of these strategies will greatly expedite future studies of a variety of RNA-BPs.

## Background

In *E. coli,* translation of a typical mRNA is dependent on sequences surrounding the initiation codon. Various studies have indicated that sequences extending from approximately 20 nucleotides upstream to about 13 nucleotides downstream of the initiation codon are nonrandom and can affect translation efficiency.[5,6] In this region, the initiation codon is the single most critical determinant of translation. A second region of great importance, usually located a few nucleotides upstream of the initiation codon, is called the Shine–Dalgarno (S/D) region. The S/D region is complementary to the 3' end of 16S ribosomal RNA, and the degree of complementarity between ribosomal RNA and the S/D region partially determines the efficiency of mRNA translation.[5]

To develop a genetic method for studying RNA-BPs, we took note of the observation that translation is inhibited when ribosomal access to the translation initiation region is blocked by the formation of a secondary structure that occludes this region.[7] In addition, there are natural systems in which proteins autoregulate their synthesis by binding the translation initiation region of their own mRNA.[2] We reasoned that binding by a heterologous RNA-BP close to the initiation codon might cause a similar translation defect. To allow such binding to occur, we decided to insert the RNA target of a heterologous RNA-BP immediately upstream of the S/D region of *lacZ* mRNA, which is a convenient reporter of gene expression. Expression of the heterologous RNA-BP and the binding of this protein to its RNA target within the modified transcript could then impair *lacZ* translation by sterically hindering ribosome binding, thereby causing a reduction in cellular β-galactosidase activity (Fig. 1).

This genetic method requires that the RNA-BP bind close to the translation initiation codon. If the actual target for high-affinity RNA-BP binding is located too far upstream of the initiation codon, binding of the protein to the target may not sterically hinder ribosome binding. Therefore, if an RNA target has not been defined in detail, incorporating a lengthy sequence

[4] P. Bouvet, C. Jain, J. G. Belasco, F. Amalric, and M. Erard, *EMBO J.* **16,** 5235 (1997).
[5] L. Gold, *Annu. Rev. Biochem.* **57,** 199 (1988).
[6] M. Dreyfus, *J. Mol. Biol.* **204,** 79 (1988).
[7] M. H. de Smit and J. van Duin, *Proc. Natl. Acad. Sci. U.S.A.* **87,** 7668 (1990).

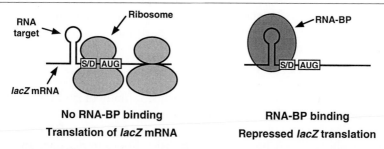

No RNA-BP binding                          RNA-BP binding
Translation of *lacZ* mRNA                 Repressed *lacZ* translation

FIG. 1. Diagrammatic representation of translational repression by a heterologous RNA-BP. The *lacZ* reporter transcript contains an RNA-BP binding target (shown here as a stem–loop structure) upstream of the initiation codon and S/D region (boxed). (Left) In strains lacking the heterologous RNA-BP, translation occurs at normal levels. (Right) On expression of the RNA-BP (shaded oval), it binds to the RNA target and sterically hinders the access of ribosomes to the *lacZ* translation initiation site, thereby repressing *lacZ* expression.

that includes a high-affinity RNA target somewhere within it may not lead to translational repression. In general, we believe that RNA targets comprising ≤50 nucleotides are best suited for this method. Small, high-affinity RNA targets can often be defined by iterative *in vitro* selection from a pool of randomized RNA sequences (SELEX analysis)[8a] or by deletion analysis of a larger RNA to delineate the 5′ and 3′ boundaries of the minimal element that binds the protein with high affinity.

Quantitative Model for Translation Repression by Heterologous RNA-BPs

A mathematical model for translational repression by an RNA-BP can be developed by assuming that when the protein is expressed, some of the reporter mRNA molecules bind the RNA-BP and become completely translationally inactivated, whereas others remain unbound. The fraction of reporter molecules that remain translationally active will then depend on the cellular concentration of the RNA-BP and its *in vivo* dissociation constant $(K_d)$, provided that the concentration of the RNA-BP exceeds the concentration of the reporter mRNA. This condition is achieved by expressing the reporter mRNA at low levels and the RNA-BP at high levels.

The degree of translational repression caused by binding of the RNA-BP to the target transcript can be quantified by introducing a term, the repression ratio $(R)$, which is the ratio of measured $\beta$-galactosidase activity stemming from the fully derepressed *lacZ* transcript in cells that do not

[8a] C. Tuerk and L. Gold, *Science* **249**, 505 (1990).

produce the RNA-BP to the $\beta$-galactosidase activity derived from free *lacZ* transcripts (those not bound by the RNA-BP) in cells that express the RNA-BP. It can be shown that

$$R - 1 = [\text{RNA-BP}]/\alpha K_d \qquad (1)$$

where $\alpha$ is a constant that reflects the intrinsic affinity of ribosomes for the translation initiation region of the reporter mRNA.[8b]

Equation (1) provides a theoretical basis for identifying RNA-BP variants with altered RNA-binding characteristics. For example, once an RNA-BP has been shown to repress translation of a reporter mRNA, a plasmid library encoding randomly mutated variants of the RNA-BP can be transformed into cells that express the reporter mRNA, and the transformants plated on X-Gal indicator plates. RNA-BP variants that bind to the target with altered affinity compared to the wild-type RNA-BP should repress $\beta$-galactosidase expression to different extents [Eq. (1)]. By comparing the color phenotype of individual transformants with similar colonies that express the wild-type RNA-BP, transformants producing RNA-BP variants with the desired characteristics (increased or decreased $K_d$) can be identified.

Equation (1) also can be rearranged to provide estimates of relative $K_d$ values from repression ratios determined using quantitative $\beta$-galactosidase assays. In cells that express a mutant RNA-BP or a mutant RNA target,

$$\frac{K_{d(\text{mutant})}}{K_{d(\text{wild type})}} = \frac{R_{\text{wild type}} - 1}{R_{\text{mutant}} - 1} \qquad (2)$$

## Methods

### Strains

Host strains used for expressing the modified *lacZ* reporter mRNAs have a *lacZ⁻* genotype, i.e., they make no endogenous $\beta$-galactosidase. Although many commonly available strains are *lacZ⁻*, we have mainly used the strain WM1 (*recA56 arg⁻ lac-proXIII nalʳ rifʳ*), which is highly transformable (see later). WM1 also encodes a mutant *recA* gene, ensuring the increased stability of transformed plasmids and greater reproducibility of $\beta$-galactosidase assays. A second strain, WM1/F′, is a WM1 derivative that contains a *lacIq* gene on an F′ plasmid. In this strain, the expression of the heterologous RNA-BP (see later) is controlled by a high cellular

---

[8b] D. E. Draper, T. C. Gluick, and P. J. Schlax, *in* "RNA Structure and Function" (R. W. Simons, and M. Grunberg-Manago, eds.). Cold Spring Harbor Laboratory Press, Cold Spring Harbor, NY, 1998.

level of the *lac* repressor (the *lacI* gene product), which ensures tight control of RNA-BP synthesis in cases where excessive expression of the RNA-BP could be toxic for *E. coli* cells.

## Plasmids

The genetic system comprises two plasmid components. One plasmid expresses the *lacZ* reporter mRNA bearing an RNA target sequence just upstream of the S/D region, and the other expresses the RNA-BP that binds this RNA target. The plasmids have compatible origins of replication, ensuring that residence of either plasmid will not interfere with the replication of the other when both plasmids are present in the same cell. The plasmids used for our studies of Rev encode either the Rev protein or *lacZ* mRNA containing the Rev target, RRE stem–loop IIB.[3] Protocols for making new plasmids that encode a different RNA-BP and its corresponding RNA target are described.

*RNA-BP Plasmid.* The RNA-BP expression plasmid pREV1 confers resistance to chloramphenicol and contains a Rev gene expressed from a strong A1 promoter ($P_{A1}$) that is repressible by the *lac* repressor (Fig. 2). Transcription of the Rev gene initiates at bp 2560 and proceeds in a clockwise direction. The Rev coding region starts at bp 2627 and ends at bp 2974. Important restriction sites are unique *Nde*I and *Cla*I sites close to the start of the Rev coding region, as well as a *Sal*I site (bp 2984) and a unique *Bsu*36I site (bp 2989) just downstream of the Rev coding region (a second *Sal*I site is present within the Rev coding region). Translation begins at an initiation codon (ATG) 15 bp upstream of the *Nde*I site and adds a short peptide sequence (MRGSIH) to the amino terminus of Rev.

To express a different RNA-BP from pREV1, the Rev gene can be excised as a restriction fragment flanked upstream by an *Nde*I or *Cla*I end and downstream by a *Bsu*36I or *Sal*I end, and replaced by a compatible fragment encoding another RNA-BP in the proper reading frame for translation. Because the *Cla*I site of pREV1 is methylated in most *E. coli* strains by dam methylase, digestion with *Cla*I requires prior propagation of the plasmid in *dam⁻* cells. DNA inserts encoding other RNA-BPs are prepared most readily by polymerase chain reaction (PCR), using an upstream primer that creates an *Nde*I-compatible site (*Nde*I, *Ase*I) or a *Cla*I-compatible site (e.g., *Cla*I, *Nar*I, *Psp*1406I, *Bst*BI, certain *Acc*I sites) at or near the RNA-BP start codon and a downstream primer that creates a *Bsu*36I site (CCTCAGG) or a *Sal*I-compatible site (e.g., *Sal*I, *Xho*I) downstream of the RNA-BP stop codon. An advantage of amplifying the insert fragment by PCR is that a short epitope tag (e.g., a FLAG tag) can be incorporated simultaneously at the amino or carboxy terminus of the RNA-BP, which could prove useful in analyzing RNA-BPs for which antibodies are unavailable.

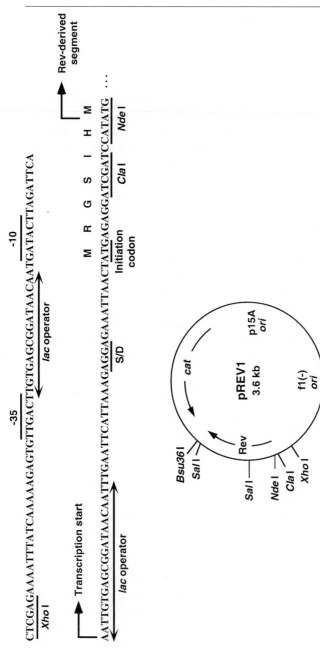

CTCGAGAAAATTTATCAAAAAGAGTGTTGACTTGTGAGCGCGATAACAATGATACTTAGATTCA
<u>Xho</u>I

-35                                                          -10

← Transcription start

→ Rev-derived segment

*lac* operator                                              *lac* operator

AATTGTGAGCGGATAACAATTTGAATTCATTAAAGAGGAGAAATTAACTATGAGAGGATCGATCCATATG · · ·
                                         S/D        Initiation        *Cla*I        *Nde*I
                                                    codon

                                    M   R   G   S   I   H   M

Fig. 2. Plasmid for RNA-BP expression in *E. coli*. The plasmid pREV1 contains a P15A origin of replication derived from pACYC184[13] and a *cat* gene that confers resistance to chloramphenicol. An f1 phage replication origin also present is useful for producing single-stranded plasmid DNA *in vivo*. Important restriction sites are indicated; each is unique except for the *Sal*I site. The sequence of the plasmid region controlling transcription and translation of the RNA-BP gene is shown above the plasmid map. This region, downstream of the *Xho*I site, includes the −35 and −10 regions of the $P_{A1}$ promoter and the *lac* operator sites that make the promoter repressible by the *lac* repressor. Downstream of the transcription initiation site are the S/D region and the RNA-BP initiation codon. The Rev coding region starts at the ATG codon within the *Nde*I site and extends for 116 codons, ending with two tandem termination codons. Downstream of the termination codons are *Sal*I and *Bsu*36I sites. The amino acid sequence of a short peptide preceding the native Rev sequence (MRGSIH) is shown, and the unique *Nde*I and *Cla*I restriction sites that map to this region are indicated. The reconstructed DNA sequence of pREV1 is available at http://saturn.med.nyu.edu/groups/BelascoLab/tech.html.

Expression of a heterologous RNA-BP may be toxic to *E. coli* cells. Until the possible toxic effects of an RNA-BP have been investigated, expression of the RNA-BP should be minimized to avoid inadvertently selecting for mutations that may affect the RNA-binding properties or expression of the protein in *E. coli*. When constructing the multicopy plasmid for expression of the RNA-BP, this can best be accomplished by transforming the ligation mixture into WM1/F' or an equivalent strain containing a *lacI*$^q$ gene. Overproduction of the *lac* repressor (LacI) will ensure that transcription from the RNA-BP promoter is repressed. Selection for transformants containing the RNA-BP expression plasmid is achieved by adding chloramphenicol (35 $\mu$g/ml) to the growth medium.

*Reporter Plasmid.* The plasmid pLACZ-IIB is a pUC19 derivative conferring ampicillin resistance (Fig. 3). It encodes an IS*10*-*lacZ* gene fusion, which is transcribed clockwise starting at bp 81. The weak IS*10* translation initiation signals ensure that *lacZ* expression is kept at a low level,[9] which is a sensitive range for screening changes in $\beta$-galactosidase activity. Upstream of the initiation codon is a weak S/D sequence preceded by stem–loop IIB, the HIV-1 RNA target recognized by Rev. Further upstream are other important elements, such as an IS*10* promoter and a T7 RNA polymerase promoter useful for synthesizing transcripts *in vitro*. These are preceded by signals for blocking transcriptional and translational read through.

Replacement of stem–loop IIB with a different RNA target can be conveniently achieved by oligonucleotide-directed mutagenesis. The plasmid pLACZ-IIB has an f1 replication origin, which allows the production of single-stranded DNA needed for some mutagenesis protocols.[10,11] The mutagenic oligonucleotide should be complementary to the sense strand of the reporter mRNA, and the 3' boundary of the RNA target sequence should be placed immediately adjacent to the S/D sequence to increase the chances of achieving high-level translational repression.

Another way to introduce a new RNA target is through PCR cloning, taking advantage of the *Kpn*I site immediately preceding the pLACZ-IIB sequence corresponding to stem–loop IIB. Although pLACZ-IIB has two *Kpn*I sites, a derivative of this plasmid (pLACZ-U1hpII)[3] contains a unique *Kpn*I site. Thus, PCR amplification can be performed on pLACZ-IIB or pLACZ-U1hpII using a forward primer that has the sequence 5' NNNGGTACC-minimal RNA target sequence-AGACAACAAGAT-GTGCGAACTCG 3' and a reverse primer (5' CGACGGGATCG-

[9] C. Jain and N. Kleckner, *Mol. Microbiol.* **9**, 233 (1993).
[10] T. A. Kunkel, K. Bebenek, and J. McClary, *Methods Enzymol.* **204**, 125 (1991).
[11] D. B. Olsen, J. R. Sayers, and F. Eckstein, *Methods Enzymol.* **217**, 189 (1993).

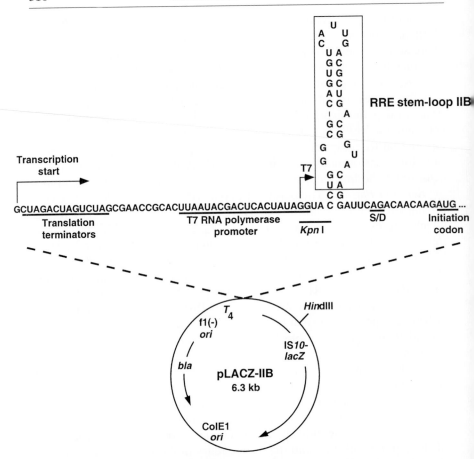

FIG. 3. Reporter plasmid for *lacZ* expression. The plasmid pLACZ-IIB contains an IS*10*-*lacZ* gene fusion consisting of the first 63 codons of the IS*10* transposase gene fused at the *Hin*dIII site to the ninth codon of *lacZ*.[9] The plasmid also contains a ColE1 replication origin, a *bla* gene conferring resistance to ampicillin, and an f1 replication origin useful for the production of single-stranded plasmid DNA. A set of four tandem transcriptional terminators derived from the *E. coli rrnB* operon precede the sequences important for expression of *lacZ*. The RNA sequence of the 5'-untranslated region of the IS*10-lacZ* reporter transcript is shown above the plasmid map. Within this RNA sequence are translation termination codons (UAG) in all three reading frames, a T7 RNA polymerase promoter, RRE stem–loop IIB, and the IS*10* S/D element and initiation codon. Stem–loop IIB is boxed; formation of this stem–loop has been reinforced by the addition of two extra base pairs. The *Kpn*I site preceding stem–loop IIB is useful when PCR is employed to introduce sequences that encode new RNA targets. Although this *Kpn*I site is not unique in pLACZ-IIB, a related plasmid, pLACZ-U1hpII, contains a unique *Kpn*I site at this position and can be used for cloning PCR products encoding new RNA targets. The reconstructed DNA sequence of pLACZ-IIB is available at http://saturn.med.nyu.edu/groups/BelascoLab/tech.html.

ATCCCCCC 3') that is complementary to a sequence immediately down-stream of the unique *Hin*dIII site of the template. The ~0.25-kb PCR product is then digested with *Kpn*I and *Hin*dIII and subcloned between the corresponding sites of pLACZ-U1hpII. Note that pLACZ-U1hpII does not carry a binding site for Rev, and translation of its *lacZ* transcript is not repressed by pREV1.

The ligation mixture is transformed into WM1 and plated on LB medium containing ampicillin (100 μg/ml) and X-Gal (60 μg/ml). Plasmid DNA from blue or light blue colonies, indicative of *lacZ* expression, can then be tested for the incorporation of the desired RNA targets.

## Preparation and Use of Competent Cells

Strains WM1 and WM1/F' can be made very competent for plasmid transformation using a modification of the calcium/manganese-based (CCMB) procedure.[12] Cells prepared by this procedure are highly trans-formable by isolated plasmids and ligation products.

## Modified Procedure for Preparing Competent Cells

1. Grow an overnight culture of the desired strain in LB medium. If the strain contains a plasmid, appropriate antibiotics should be added to maintain plasmid selection.

2. On the next day, add 0.1 ml of the saturated overnight culture to 10 ml LB medium (or LB plus antibiotics, if relevant). Grow at 37° until the culture reaches midlog phase ($OD_{600} \sim 0.5$).

3. Transfer the cells to tubes prechilled on ice. Spin at 5000 rpm in a cold (4°) centrifuge for 5 min. If the cells do not pellet, increase the speed by 2000 rpm or more, as needed.

4. Discard the supernatant. Resuspend the cell pellet in 3 ml cold CCMB solution [80 m$M$ calcium chloride, 20 m$M$ manganese chloride, 10 m$M$ magnesium chloride, 10 m$M$ potassium acetate, 10% glycerol (v/v), pH 6.4].[12]

5. Incubate on ice for 20 min.

6. Pellet the cells as before, discard the supernatant, and resuspend in 0.8 ml cold CCMB. Transfer to microfuge tubes and store at −80°.

Competent cells work best after they have been frozen once followed by thawing on ice. Competent cells may be stored for several months and can endure a few freeze–thaw cycles without a great loss of transformation efficiency. However, for optimum transformation efficiency, smaller vol-umes of cells should be aliquoted into microfuge tubes, frozen, and used

---

[12] D. Hanahan, *Methods Enzymol.* **204,** 63 (1991).

once. Using this procedure to prepare competent cells, one can reasonably expect to obtain ~$10^8$ ampicillin-resistant WM1 transformants per microgram pBR322 DNA using the following transformation procedure.

### Transformation

1. Prechill 1.5-ml microfuge tubes on ice.
2. Add plasmid DNA or the ligation mixture (up to 5 $\mu$l in volume).
3. Thaw the competent cells on ice and resuspend them gently. Add 20 $\mu$l to the DNA.
4. Place in a 37° heat block for 45 sec.
5. Return to ice. Add 200 $\mu$l LB medium, mix, and transfer to a 37° heat block for 1 hr to allow expression of antibiotic resistance.
6. Plate on selective antibiotic-containing medium.

### Testing for Translational Repression

Once the new reporter and RNA-BP expression plasmids have been constructed, the ability of the RNA-BP to repress translation of the modified reporter can be tested. This is done by cotransforming the two plasmids into a *lacZ⁻ E. coli* host strain and growing the transformants on X-Gal indicator plates. Cleavage of X-Gal by the $\beta$-galactosidase encoded by the reporter plasmid yields a product that imparts a blue color to colonies of cells that express this enzyme. The intensity of this color is directly related to the amount of $\beta$-galactosidase produced by the cells and allows the detection of differences in reporter gene expression as small as twofold. To determine whether expression of the *lacZ* reporter is reduced in the presence of RNA-BP, the colony color of cells containing the reporter plasmid and the RNA-BP plasmid is compared to that of otherwise identical cells containing the reporter plasmid and a compatible control plasmid (pACYC184)[13] in place of the RNA-BP plasmid. A significant reduction in the intensity of the blue colony color in the presence of the RNA-BP is indicative of translational repression.

In practice, optimizing translational repression may require identifying the best conditions for RNA-BP expression, as excessive production of some RNA-BPs may be toxic for *E. coli*. Although the elements in pREV1 that control transcription and translation of the cloned RNA-BP are relatively efficient, the presence of *lac* operators in the promoter region allows transcription to be reduced substantially in strains containing an elevated concentration of the *lac* repressor. In such strains (e.g., WM1/F′), intermedi-

---

[13] A. C. Y. Chang and S. N. Cohen, *J. Bacteriol.* **134,** 1141 (1978).

ate levels of RNA-BP production can be achieved by adding the *lac* inducer isopropylthiogalactoside (IPTG) to the growth medium at micromolar concentrations.

### Visualizing Translational Repression

1. Cotransform competent WM1/F' cells with the reporter plasmid and either the RNA-BP expressing plasmid or a control plasmid (pACYC184).

2. Plate serial dilutions (100 $\mu$l of a $10^{-1}$, $10^{-2}$, $10^{-3}$, and $10^{-4}$ dilution) of each transformation mix on LB-agar plates containing ampicillin and chloramphenicol (no X-Gal or IPTG). Freeze the remainder of each transformation mix at $-80°$ after adding glycerol to a final concentration of 15–20% (v/v). Incubate the plates at 37° overnight. Determine the volume of each transformation mix that yields ~500 colonies per plate.

3. Thaw the transformation mixes on ice. Plate ~500 colony-forming units of each in 100 $\mu$l of LB on LB-agar plates containing ampicillin, chloramphenicol, X-Gal, and 0, 1, 2, 5, 10, 20, 50, or 1000 $\mu M$ IPTG. Incubate the plates at 37° overnight. At each IPTG concentration, compare the color and size of isolated colonies containing the RNA-BP expression plasmid versus the control plasmid. If the RNA-BP represses translation of the reporter mRNA, colonies expressing the RNA-BP should be less blue than control colonies lacking this protein. Relative to the control transformants containing pACYC184, determine the highest IPTG concentration at which RNA-BP expression results in a marked reduction in colony color intensity without causing growth inhibition (significantly decreased colony size or colony count) or genetic instability (individual colonies on the same plate should be uniform in color). If there is no significant growth inhibition or genetic instability even at 1000 $\mu M$ IPTG, the *lacI⁻* host strain WM1 can be used for subsequent experiments, making the addition of IPTG unnecessary for RNA-BP expression.

4. The degree of translational repression can be quantitated by performing $\beta$-galactosidase assays on extracts of cells containing the RNA-BP expression plasmid versus cells containing the control plasmid.

*$\beta$-Galactosidase Assays.* For each transformed strain, $\beta$-galactosidase assays should be performed in triplicate.

1. From plates containing IPTG at the optimal concentration, disperse three transformed colonies into separate culture tubes containing

1 ml of LB medium supplemented with ampicillin, chloramphenicol, and IPTG at the same concentration as in the plates. Grow overnight at 37°.

2. On the next day, inoculate tubes that each contain 1.5 ml of LB + antibiotics + IPTG with 5 $\mu$l of the overnight culture. Grow at 37° until the cells reach midlog phase ($OD_{600}$ 0.3–0.5), which should take ~3 hr.

3. Place the tubes on ice for 20 min.

4. Pour the cultures into polystyrene cuvettes. After wiping away any condensation on the outside of the cuvettes, measure $OD_{600}$ using a cuvette containing the growth medium as a blank.

5. Prepare several glass culture tubes, each containing 0.5 ml of *lacZ* assay buffer (60 m$M$ $Na_2HPO_4$, 40 m$M$ $NaH_2PO_4$, 10 m$M$ KCl, 1 m$M$ $MgSO_4$, pH 7.0) supplemented with 0.01% sodium dodecyl sulfate and 50 m$M$ 2-mercaptoethanol. Dispense two drops of chloroform into each tube. Add 0.5 ml of cell culture to each tube. Also prepare a blank tube to which 0.5 ml LB medium is added. Vortex each tube vigorously for 15 sec to permeabilize the cells and then incubate for 5 min in a 28° water bath.

6. Add 200 $\mu$l of *o*-nitrophenylgalactoside (ONPG) solution (4 mg/ml) to the tubes, noting the time at which the solution was added to each tube. Return the tubes to the water bath. Allow the ONPG hydrolysis reaction to proceed until the color in the tube becomes yellow. To stop the reaction, add 0.5 ml of 1 $M$ sodium carbonate solution to the tubes and vortex briefly, noting the time at which each reaction was stopped. If no significant color develops within 3–5 hr, stop the reaction in the same manner. Transfer the solutions to microfuge tubes, spin in a microfuge for 5 min at 14,000 rpm, and pour the supernatants into polystyrene cuvettes, taking care to avoid the transfer of chloroform droplets.

7. Measure $OD_{420}$, and then subtract the reading obtained for the LB blank reaction.

8. The $\beta$-galactosidase activity of each culture (in Miller units) = $(2000 \times OD_{420})/(\Delta T \times OD_{600})$, where $\Delta T$ is the time interval of the ONPG hydrolysis reaction in minutes.[14]

9. Calculate the repression ratio. $R$ is the average $\beta$-galactosidase activity of cells containing pACYC184 or a similar control plasmid divided by the average $\beta$-galactosidase activity of cells that contain the RNA-BP plasmid.

[14] J. H. Miller, *in* "Experiments in Molecular Genetics." Cold Spring Harbor Laboratory Press, Cold Spring Harbor, NY, 1972.

(Note: The Rev-expressing plasmid pREV1 and the reporter plasmid pLACZ-IIB can be used as controls to test for translational repression in WM1 cells. The repression ratio for this pair of plasmids should be about 30–40.[3])

If $R < 5$ and if the host bacteria can tolerate more of the RNA-BP without exhibiting signs of growth inhibition or genetic instability (see "Visualizing Translational Repression"), try repeating the assays using cells grown in the presence of a somewhat higher IPTG concentration to increase RNA-BP expression and translational repression.

Once repression by the RNA-BP has been achieved, it is important to confirm that the reduction in $\beta$-galactosidase synthesis is due specifically to binding of the RNA-BP to its target in the reporter mRNA and not to an indirect effect. This is demonstrated most readily by showing that a mutation in the RNA target that should disrupt protein binding also results in a loss of translational repression.

A potential complication of using WM1/F′ cells as the host strain for controlled expression of a toxic RNA-BP is that the measured repression ratio may vary somewhat from experiment to experiment due to the sensitivity of RNA-BP production to the precise concentration of IPTG in the culture medium. In such cases, it may be advantageous to dispense with WM1/F′ cells and IPTG once they have been used to show that an intermediate cellular concentration of the RNA-BP can repress translation without impairing cell growth. One strategy that can be used to achieve this objective is to randomize the 5′-untranslated region of the RNA-BP gene in the vicinity of the Shine–Dalgarno element or to randomize certain promoter base pairs and then to transform the resulting combinatorial plasmid library into WM1 cells (*lacI⁻*) that contain a reporter plasmid. By screening for transformants that generate large white colonies on X-Gal plates in the absence of *lac* repressor, plasmid variants can be identified that produce the RNA-BP at a fixed intermediate level that is nontoxic yet adequate to inhibit translation of the reporter mRNA.

## Strategies and Applications

Once the new reporter and RNA-BP plasmids have been constructed and optimal conditions for translational repression by the RNA-BP have been determined (repression ratio $\geq 5$), a variety of genetic analyses can be pursued, as follows.

### Defining Features of RNA Target Important for RNA-BP Binding

If the RNA target has not been characterized in detail, the genetic method outlined here should be useful for identifying the features of the target that are important for RNA-BP binding. A simple way to achieve this is by randomly mutagenizing the region of the reporter plasmid that encodes the RNA target. Small regions of the reporter plasmid ($\leq$50 bp) are most simply mutagenized by oligonucleotide-directed mutagenesis or PCR mutagenesis (see the methods described earlier for construction of the reporter plasmid) using a degenerate oligonucleotide that is "doped" with a small percentage of the non-wild-type nucleotides. To minimize the incorporation of multiple mutations, the mutagenic frequency should be kept at a low level ($\leq$1 mutation per RNA target).

The mutagenized plasmid mixture is transformed into competent cells containing the wild-type RNA-BP plasmid and plated on LB medium containing chloramphenicol, ampicillin, X-Gal, and, if appropriate, IPTG. Transformants with a defect in translational repression are identified by their enhanced blue color, indicative of increased $\beta$-galactosidase synthesis. Plasmid DNA is extracted from candidate colonies and retransformed into WM1, with selection only for ampicillin resistance to isolate the reporter plasmid. The observed defects in translational repression must be confirmed by cotransforming an appropriate strain with each mutant reporter plasmid and either pACYC184 or the RNA-BP plasmid and then measuring $\beta$-galactosidase activity and repression ratios. This precaution is important to weed out mutations that increase *lacZ* expression without affecting translational repression. Mutations that impair repression will indicate which features of the RNA target are important for RNA-BP binding.

Some calculations are necessary to estimate the number of transformants that must be screened to achieve adequate coverage. In general, if $N$ positions are mutagenized to the three non-wild-type nucleotides with equal probability, the number of possible single mutants is $3N$. If the fraction of transformants containing single mutations is $f$, it can be shown that in a pool of $3N/f$ transformants, about 63% of the possible mutations will be represented. If coverage of >90% of the possible mutants is desired, it is recommended that at least $10N/f$ colonies be screened. In practice, even the best methods of random mutagenesis do not give equal representation of all mutations. Therefore, if certain mutations are expected to affect binding but are not isolated by screening, these mutations should be introduced separately into the reporter plasmid and screened for their phenotype.

One important practical aspect of screening is to plate an optimal number of transformants on each indicator plate. If too few colonies are present on each plate, a large number of plates will be required, whereas if too many colonies grow on each plate, they may not be well separated and colony color phenotypes may not be clearly distinguishable. In general, for plates that are 8.5 cm in diameter, 500–1000 colonies per plate is a reasonable density. Poorly separated colonies with interesting phenotypes can be isolated by restreaking. In order to achieve the optimal colony density, we usually plate various volumes of the transformation culture (0.1, 1, and 10 $\mu$l) and incubate the plates overnight at 37°. Meanwhile, the rest of the transformation culture is stored at $-80°$ in 15–20% glycerol. The colonies are counted on the next day to obtain an estimate of the volume of the transformation culture required to give 500–1000 colonies, although it may be necessary to plate somewhat more than the calculated volume to compensate for the loss of cell viability that may result from freezing and thawing the cultures.

## Defining Features of RNA-BP Important for Binding to RNA Target

*Defective RNA-BP Variants.* A similar strategy can be used to define elements in the RNA-BP that are important for target recognition. To do so, the RNA-BP gene is mutagenized and transformed into a strain containing the reporter plasmid, and the transformants are screened for changes in *lacZ* repression. When little information is available about the RNA-binding domain of the RNA-BP, it is usually preferable to mutagenize the whole protein. However, if the role of only certain amino acid residues is being investigated, specific codons may be mutagenized.

Mutation of small regions of the RNA-BP can be accomplished by oligonucleotide-directed mutagenesis using a doped mutagenic oligonucleotide and single-stranded plasmid DNA, which is obtained by taking advantage of the presence of an f1 replication origin in pREV1.[10,11] Mutagenesis of the entire RNA-BP or large regions thereof is best achieved by error-prone PCR.[15] In either case, the mutagenic frequency should be kept at a low level ($\sim$1 mutation per RNA-BP gene). Higher mutagenic frequencies require extra steps to distinguish which of the mutations are responsible for the observed phenotype. The error-prone PCR protocol of Cadwell and Joyce[15] is reported to introduce approximately one mutation per 300 bp of amplified DNA, but our experience is that the mutation rate can be higher. Because this mutagenesis frequency is likely to be proportional to the number of rounds of DNA replication, which in turn should be inversely related to the initial concentration of the DNA template, the easiest way to increase or decrease the mutagenic frequency may be to vary the amount of template DNA added to the PCR reaction over a few orders of magnitude. The optimal mutagenic frequency is the lowest that yields an adequate number of mutants defective in translational repression.

In general, because PCR mutagenesis overwhelmingly results in single base-pair substitutions at any codon position, only a subset (typically 4–7)

---

[15] R. C. Cadwell and G. F. Joyce, *PCR Methods Appl.* **2**, 28 (1992).

of the 19 possible amino acid substitutions are accessible. Therefore, the complexity of such libraries is approximately six times the number of codons, or about $2N$, where $N$ is the number of nucleotides in the mutagenized region. Thus, it is desirable to screen at least $6N-10N$ colonies, assuming that a large fraction of the PCR-mutagenized library contains one mutation.

From a practical standpoint, it is best to construct a library whose complexity significantly exceeds the number of colonies that will be screened. Because this number might be in the thousands, it may sometimes be necessary to generate a bacterial library of high complexity by transforming the mutagenesis mix into high-efficiency electrocompetent cells (e.g., DH12S cells obtained from BRL, Gaithersburg, MD) rather than CCMB-competent WM1/F' cells. In either case, small aliquots of the transformation culture should be plated on chloramphenicol-containing plates (to obtain an estimate of the total number of transformants), whereas the bulk of the culture is plated on a separate chloramphenicol plate to generate a library. If, after overnight incubation at 37°, the calculated number of transformants on the library plate is deemed to be sufficient, that plate is flooded with a few milliliters of sterile 0.85% (w/v) NaCl solution, the colonies are resuspended by gentle scraping, and the cell suspension is recovered. A portion of this suspension is stored as a bacterial library by freezing the cells at $-80°$ after adding glycerol to 15–20% (v/v), and the remainder is used to prepare a plasmid DNA library by plasmid extraction.

To screen for mutants defective in RNA binding, competent cells containing the reporter plasmid are transformed with a small portion of the mutagenized RNA-BP plasmid library. After measuring the number of transformants per microliter in the resulting mixture of transformed cells (see "Testing for Translational Repression"), aliquots of the transformation mix are plated on LB medium containing chloramphenicol, ampicillin, X-Gal, and IPTG (if needed) to produce 500–1000 colonies per plate (see earlier discussion). Several plates may be required. The colonies are screened for elevated *lacZ* expression (increased blue color), and the $\beta$-galactosidase activity of selected candidates is quantified further by spectrophotometric assays of cell extracts, which are performed in parallel with assays of cells containing the wild-type RNA-BP expression plasmid or pACYC184. Mutant RNA-BP plasmids with an interesting phenotype are purified by extracting plasmid DNA, retransforming it into WM1/F', selecting transformants that are chloramphenicol resistant and ampicillin sensitive, and preparing plasmid DNA. These plasmids are then transformed into cells containing the reporter plasmid to reconfirm that the mutant RNA-BPs are defective for translational repression.

In addition to protein mutations that weaken binding, a second class of RNA-BP mutations expected to impair *lacZ* repression comprises those that destabilize the protein in *E. coli*. Prior to embarking on a detailed analysis of potentially interesting RNA-BP variants, it is important to discriminate between mutants that bind less tightly and those that accumulate at lower levels in *E. coli*. This is accomplished most simply by quantitative Western blot analysis of cell extracts from strains that express the wild-type RNA-BP or each candidate variant, using antibodies to the RNA-BP or to an epitope tag fused to it. Western blot analysis should also reveal any truncated RNA-BP variants resulting from nonsense mutations.

Once nondestabilizing RNA-BP mutations that cause translational repression defects have been identified, it may be useful to confirm their RNA-binding phenotype *in vitro*.[3,16] To do so, genes encoding the wild-type RNA-BP and interesting mutants are subcloned into an expression vector that allows their overexpression and rapid purification. We have had considerable success using *E. coli* plasmid vectors that contain a T7 RNA polymerase promoter to overproduce RNA-BPs as hexahistidine-tagged proteins.[3] The RNA required for the *in vitro*-binding assays can be synthesized by *in vitro* transcription,[17] taking advantage of the T7 RNA polymerase promoter in the reporter plasmid.

*Identification of RNA-BP Suppressor Variants.* As an alternative to screening for RNA-BP variants impaired for binding, a similar genetic strategy can be used to screen for RNA-BP variants that bind with enhanced affinity to defective RNA targets. RNA mutations that impair translational repression must first be identified (see earlier discussion). A set of roughly 5–10 defective RNA targets, each containing a point mutation or compensatory mutations in a base pair, should be an adequate starting point for screening RNA-BP suppressor mutants.

The strategy for identifying suppressor mutants is similar to the one used for identifying RNA-BP variants that are defective in binding. First, the mutant reporter plasmids are individually transformed into the appropriate host strain. Each of the resulting strains is then transformed with a plasmid library encoding mutant RNA-BPs. Approximately $6N$–$10N$ colonies (see earlier discussion) of each transformed cell culture are screened on X-Gal indicator plates. This time, however, colonies are screened for *diminished β-galactosidase activity* (a lighter blue color) relative to colonies expressing the wild-type RNA-BP in the same strain with the same mutant reporter gene. The suppressor phenotype of interesting candidates is confirmed by quantitative β-galactosidase assays. Western blot analysis can be used to ensure that suppression is not due to an increase in the cellular concentration of the RNA-BP.

Suppressor RNA-BPs can be characterized further by determining their allele specificity. To do so, purified plasmid DNAs encoding the mutant

---

[16] K. B. Hall and W. T. Stump, *Nucleic Acids Res.* **20,** 4283 (1992).
[17] J. F. Milligan and O. C. Uhlenbeck, *Methods Enzymol.* **180,** 51 (1989).

RNA-BP, the wild-type RNA-BP, or no RNA-BP are transformed into strains containing either the wild-type reporter plasmid or any of several different mutant reporter plasmids. $\beta$-Galactosidase assays are then performed for each reporter mutant to determine how well each RNA-BP variant represses *lacZ* translation as compared to the wild-type RNA-BP. It may be useful to confirm the specificity of important mutants by *in vitro* $K_d$ measurements.

Two classes of RNA-BP variants are likely to be identified based on their allele-specific pattern of suppression. One class are those that partially suppress the defects of most, if not all, mutant RNA targets. This phenotype may be due to a nonspecific increase in RNA-binding affinity or, less interestingly, to an increase in the cellular concentration of the RNA-BP. Mutants in the former category may help identify surfaces of the RNA-BP that contact the target in a manner that is not RNA sequence specific, whereas mutants in the latter category can be identified by Western blot analysis and discarded.

The other class of mutants are altered specificity mutants that specifically suppress the defects of only one or a few RNA targets. Such allele-specific suppressor mutants should be particularly instructive, as they are expected to provide information about amino acid residues in the RNA-BP that lie in close proximity to specific RNA nucleotides. A reasonable interpretation of such a mutant is that the specificity of interaction is a consequence of a direct and favorable interaction of the mutated amino acid with the mutated nucleotide(s) of the RNA target whose defect it suppresses. The identification of altered specificity variants of Rev provided us with sufficient spatial constraints to allow a structural model for target recognition by Rev to be constructed.[3] A similar approach allowed the modeling of a nucleolin–RNA complex.[4]

In general, altered specificity suppressor variants are rare. Our experience with Rev suppressors suggests that binding defects caused by purine transition mutations (A-to-G or G-to-A) may have the greatest chance to be suppressed by amino acid substitutions in the RNA-BP, perhaps because these two bases are structurally similar to one another and present a greater number of functional groups for possible interaction with the RNA-BP than uracil and cytosine. Once an amino acid substitution with an altered specificity phenotype has been identified in an RNA-BP library mutagenized by error-prone PCR, our experience also suggests that additional altered specificity mutants can often be isolated by completely randomizing the corresponding codon and then screening genetically for new suppressor mutants in a set of several isogenic *E. coli* strains that each contain a different reporter gene variant.

*Estimating Dissociation Constants*

When a substantial number of interesting RNA or RNA-BP variants have been identified, purifying each mutant and measuring its affinity ($K_d$)

*in vitro* for several different targets can be laborious. In such a case, it may be desirable to estimate these dissociation constants by using the mathematical model for translational repression described earlier in Eqs. (1) and (2). This model predicts that the degree of translational repression $(R - 1)$ should be inversely related to the dissociation constant $(K_d)$. We have tested this relationship for a number of mutant forms of Rev, nucleolin, and their RNA targets, and we find a good correlation between these two parameters (see later). The possibility of estimating relative $K_d$ values by measuring repression ratios can greatly expedite the detailed analysis of both RNA and protein mutants.

A good procedure to use when employing this strategy is to first purify the wild-type protein, the wild-type RNA target, and about six protein and/ or RNA mutants that repress *lacZ* to varying extents that span the range observed for the entire set of mutants. Dissociation constants are then measured *in vitro* for this set of purified proteins and RNAs, and resulting data are plotted as a graph of $\log(K_d)$ versus $\log(R - 1)$. A best-fit line is calculated by least-squares analysis and is used to estimate $K_d$ values for the other mutants on the basis of their repression ratios.

Troubleshooting

The genetic method described here is potentially applicable to any RNA-BP. This method is simple to use, and the time required to make the necessary plasmid constructs and test for translational repression should be less than 2 weeks in the hands of an experienced investigator. However, there may be some instances in which translational repression is not achieved. Some likely explanations for such a failure to achieve translational repression and possible solutions are as follows.

1. The RNA-BP cannot be expressed in *E. coli* at adequate levels. For many RNA-BPs, a cellular concentration typical of the average *E. coli* protein should be sufficient to achieve significant translational repression of a *lacZ* reporter containing a high-affinity RNA target. Our experience with RNA-BPs from diverse organisms is that most can be expressed in *E. coli* at levels sufficient to cause translational repression. However, it is possible that an adequate level of expression may not be achieved for certain RNA-BPs, even in the *lacI⁻* host strain WM1. Some sense of the cellular abundance of an RNA-BP can be obtained by Western blot analysis of bacterial extracts. Because the signals in pREV1 for RNA-BP expression are already quite strong, the most practical way to increase RNA-BP production may be to express the protein from a plasmid with a higher copy number (>50 per cell). However, to maintain compatibility with pLACZ-IIB, any such plasmid must not contain a ColE1-type replication origin. Equation 1 suggests that an alternative way to increase the repression ratio may

be to reduce the intrinsic affinity of the reporter mRNA for ribosomes, which compete with the RNA-BP for binding. Such a reduction could be achieved by mutating the rudimentary Shine-Dalgarno element of the reporter (5'-AG-3') so as to further diminish its already meager complementarity to the 3'-terminal segment of 16S rRNA (5'-CCUC-CUUA-3'), taking care not to abolish translation altogether. This strategy is likely to be most effective when there is already a modicum of translational repression that merely needs to be boosted in magnitude to facilitate genetic screening.

2. Failure of an RNA structure to form. Many RNA-BPs recognize features of primary, secondary, and tertiary structure in their RNA targets. Failure to achieve translational repression may be due to the failure of the required structure to form or to the formation of a competing structure. In many instances, this problem can be overcome by proper design of the target RNA so as to increase the stability of the required structure or minimize the potential for competing structures. For example, in our studies of Rev we added two G-C base pairs to the bottom of RRE stem–loop IIB to favor the formation of this stem–loop structure (Fig. 3).

3. The RNA target may not be defined adequately. If the RNA target inserted in the reporter transcript is too large, RNA-BP binding may not sterically hinder ribosome binding and translation initiation. The preferred, although somewhat laborious, solution is to first conduct studies to define a minimal high-affinity RNA target. A second possible solution is to make a nested series of unidirectional deletions upstream of the S/D element in the *lacZ* reporter plasmid, such that sequences at the 3' end of the RNA target are removed incrementally. Some of these deletion variants may retain the minimal RNA-BP-binding site while bringing it sufficiently close to the translation initiation region to allow translational repression.

## Application of Genetic Methods to Rev and Nucleolin

In cells infected by HIV-1, the Rev protein facilitates the nuclear export of unspliced or partially spliced messages containing its RNA target (the RRE). Although Rev has been considered a paradigm for the family of arginine-rich RNA-BPs, little was known about the basis for its RNA-binding specificity until recently.[3,18,19]

We have applied the genetic methods described here to identify and

[18] J. L. Battiste, H. Mao, N. S. Rao, R. Tan, D. R. Muhandiram, L. E. Kay, A. D. Frankel, and J. R. Williamson, *Science* **273**, 1547 (1996).

[19] X. Ye, A. Gorin, A. E. Ellington, and D. J. Patel, *Nature Struct. Biol.* **3**, 1026 (1996).

characterize novel variants of Rev that can bind with enhanced affinity to mutant targets. Two classes of Rev variants were identified: enhanced affinity variants that partially suppress defects in most mutant RRE targets tested and altered specificity variants that suppress defects in only one or a very few mutant elements. The suppressor phenotypes of these Rev variants are summarized in Fig. 4A. These Rev variants provided genetic evidence for direct contacts between specific Rev amino acids and nucleotides in its high-affinity RNA target, RRE stem–loop IIB, and this information was used to construct a structural model for how Rev docks with this stem–loop.[3] This genetically based model proved to be congruent with structures determined contemporaneously by nuclear magnetic resonance.[18–20]

A similar strategy has been employed to examine RNA recognition by nucleolin, an RNA-BP that participates in ribosomal RNA maturation.[4] Nucleolin, which is a member of the RRM family of RNA-BPs, uses two RRM domains (also referred to as RBD domains) to bind with high affinity to an RNA stem–loop called the NRE. Genetic screening identified nucleolin variants with enhanced affinity or altered specificity mutations that improve binding to one or more members of a set of mutant NREs (Fig. 4B). As in the case of Rev, these data allowed the three-dimensional structure of the nucleolin–NRE complex to be modeled.

Data obtained for mutant forms of Rev, nucleolin, and their RNA targets have also been used to test the quantitative relationship between the repression ratio $(R)$ and $K_d$ (Fig. 5). For 22 combinations of Rev and stem–loop IIB mutants and for seven combinations of nucleolin and NRE mutants, there is a linear relationship between $\log(K_d)$ and $\log(R - 1)$, as predicted by Eqs. (1) and (2). This relationship holds for $K_d$ values spanning more than four orders of magnitude. For 90% of the mutants, the $K_d$ value can be estimated to within a factor of three of its actual value from the observed repression ratio and the best-fit straight line calculated by least-squares analysis. These findings suggest the general utility of repression ratio measurements in *E. coli* as a rapid means for estimating relative $K_d$ values for mutant forms of RNA-BPs and their targets.

## Conclusions

To facilitate studies of RNA-BPs, we have devised a genetic strategy of broad applicability. This strategy should be useful for (i) identifying mutations in the RNA or the RNA-BP that affect binding affinity and/or specificity, (ii) obtaining quantitative estimates of relative binding affinities from repression ratios measured easily in *E. coli,* and (iii) developing models

[20] A. D. Ellington, F. Leclerc, and R. Cedergren, *Nature Struct. Biol.* **3**, 981 (1996).

A

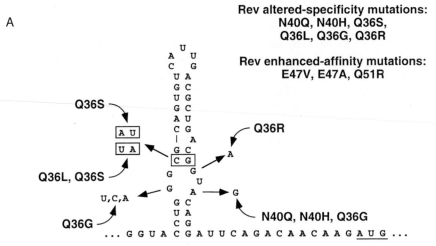

B

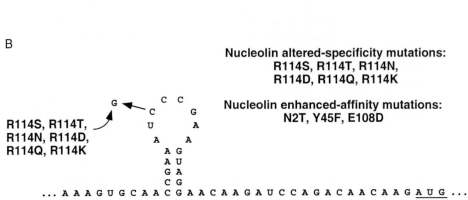

FIG. 4. Summary of Rev and nucleolin suppressor variants and their phenotypes. (A) Rev and stem–loop IIB. (B) Nucleolin and the NRE hairpin. Protein variants that suppress defects in their respective RNA targets are identified. Amino acid substitutions in the suppressor mutants are indicated, using a code that specifies the identity and position of the wild-type and mutant amino acid. The suppressor variants are classified into two groups based on their allele specificity. Enhanced affinity RNA-BP mutants improve repression of a variety of mutant RNA targets, whereas altered specificity protein variants are allele specific in their effect. Each RNA target is shown in the context of the translation initiation region of the reporter transcript, whose initiation codon is underlined. The NRE hairpin is drawn with two noncanonical base pairs in its stem, as proposed by Ghisolfi-Nieto et al.[27] Point mutations and compensatory base pair substitutions that can be suppressed by altered specificity protein variants are shown (straight arrows), as are the suppressing amino acid mutations (curved arrows). Other mutations in stem–loop IIB and NRE failed to yield allele-specific Rev or nucleolin suppressors.

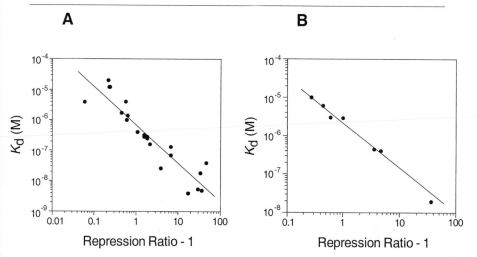

Fig. 5. Correlation of $K_d$ values with repression ratios. (A) A set of 22 Rev variants and/ or stem–loop IIB mutants were tested for binding *in vitro* ($K_d$) and translational repression in *E. coli*,[3] and data were plotted on a logarithmic scale. (B) Binding and repression data for variants of nucleolin and NRE were plotted in a similar fashion.[4] Note that 1 has been subtracted from the repression ratio, as a repression ratio of 1.0 indicates a lack of detectable binding. Also shown in each case is the best-fit straight line calculated by linear regression, which gives a close fit to data. The slope of this line is −1.25 for Rev and −1.1 for nucleolin, values close to the slope of −1.0 predicted by Eq. (1).

for RNA–protein interaction on the basis of the altered specificity or enhanced affinity of RNA-BP suppressor mutants thereby identified and characterized. Using this approach, translational repression of modified *lacZ* transcripts has been achieved not only with Rev and nucleolin, but also with other heterologous RNA-BPs, including U1A,[3] ribosomal protein S15 (C. Jain and J. Belasco, unpublished data), and the *Drosophila* Pumilio protein (S. Karudapuram and J. Belasco, unpublished data). In principle, it should be possible to adapt this method for cloning cDNAs encoding RNA-BPs that bind to specific RNA sequences. Other genetic approaches for studying RNA-BPs have also been reported.[21-26] We believe that to-

[21] M. P. MacWilliams, D. W. Celander, and J. F. Gardner, *Nucleic Acids Res.* **21**, 5754 (1993).

[22] I. A. Laird-Offringa and J. G. Belasco, *Proc. Natl. Acad. Sci. U.S.A.* **92**, 11859 (1995).

[23] D. J. SenGupta, B. Zhang, B. Kraemer, P. Pochart, S. Fields, and M. Wickens, *Proc. Natl. Acad. Sci. U.S.A.* **93**, 8496 (1996).

[24] U. Putz, P. Skehel, and D. Kuhl, *Nucleic Acids Res.* **24**, 4838 (1996).

gether these genetic methods will greatly expedite the analysis of RNA-BPs and allow important questions regarding this class of proteins to be answered in a relatively short period of time.

## Acknowledgment

The support of a research grant from the National Institutes of Health (GM55624) is gratefully acknowledged.

[25] K. Harada, S. S. Martin, and A. D. Frankel, *Nature* **380,** 175 (1996).
[26] E. Paraskeva, A. Atzberger, and M. W. Hentze, *Proc. Natl. Acad. Sci. U.S.A.* **95,** 951 (1998).
[27] L. Ghisolfi-Nieto, G. Joseph, F. Puvion-Dutilleul, F. Amalric, and P. Bouvet, *J. Mol. Biol.* **260,** 34 (1996).

# [22] RNA Challenge Phages as Genetic Tools for Study of RNA–Ligand Interactions

*By* Daniel W. Celander, Kristine A. Bennett, Derrick E. Fouts, Erica A. Seitz, and Heather L. True

## Introduction

Significant effort has been devoted to devising genetic approaches for the detection and characterization of RNA–protein interactions.[1-4] The RNA challenge phage system represents one method that enables direct genetic selection for a specific RNA–protein interaction in bacteria.[2] This system is based on the developmental program of the temperate bacteriophage P22 following infection of its natural host *Salmonella typhimurium.* Two sets of genes determine the developmental fate of bacteriophage P22 in an infected *S. typhimurium* host cell (Fig. 1a). These genes reside in two genetically distinct loci within the bacteriophage genome termed the immunity regions, *immC* and *immI.* The expression of the *c2* gene product, which resides in *immC,* causes the bacteriophage to establish and maintain a dormant, lysogenic state in the infected host cell. The c2 protein represses

[1] D. Peabody, *J. Biol. Chem.* **265,** 5684 (1990).
[2] M. P. MacWilliams, D. W. Celander, and J. F. Gardner, *Nucleic Acids Res.* **21,** 5754 (1993).
[3] K. Harada, S. S. Martin, and A. D. Frankel, *Nature* **380,** 175 (1996).
[4] D. J. SenGupta, B. Zhang, B. Kraemer, P. Pochart, S. Fields, and M. Wickens, *Proc. Natl. Acad. Sci. U.S.A.* **93,** 8496 (1996).

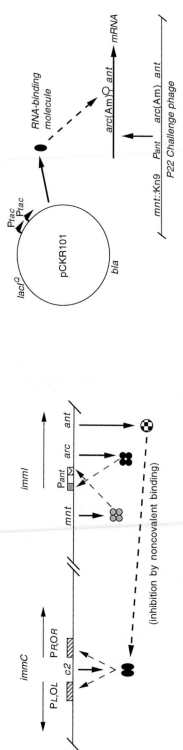

FIG. 1. (a) Life cycle of bacteriophage P22. Solid lines refer to expression of regulatory gene products. Dashed lines refer to negative regulatory pathways. For a more detailed discussion, see the description in the text. (b) RNA challenge phage development. The modified P22 phage contains an amber mutation in *arc* [*arc*(Am)] and a kanamycin resistance cassette inserted into the *mnt* gene (*mnt*::Kn9). The region encoding the ribosome-binding site of the *ant* gene is substituted by a sequence encoding an RNA sequence (shown by the lollipop) recognized by an RNA-binding molecule. The RNA-binding molecule is produced from an expression vector (e.g., pCKR101). The relative amounts of c2 and Ant proteins synthesized in an infected cell will dictate whether the phage becomes a lysogen or undergoes lytic growth. If a sufficient amount of the RNA-binding molecule is present in the cell, it will saturate its specific binding site on the newly synthesized *ant* mRNA transcripts following infection and inhibit Ant translation. Repression of Ant protein synthesis allows c2 protein-mediated establishment of lysogeny, which results in the formation of a kanamycin-resistant lysogen. (Adapted from M. P. MacWilliams, D. W. Celander, and J. F. Gardner, *Nucleic Acids Res.* **21**, 5754 (1993).)

lytic gene functions by binding to two DNA operator regions.[5] The *immI* region contains two additional repressor proteins termed Mnt and Arc and an antirepressor protein called Ant. Expression of these genes occurs from two divergently transcribed operons. One transcription unit consists of $P_{mnt}$ and the *mnt* gene. The second unit consists of $P_{ant}$ and the *arc* and *ant* genes.[5] The Mnt and Arc proteins selectively repress *ant* gene expression by binding to two different DNA operator sequences overlapping $P_{ant}$.[5–7] During the early phase of infection, $P_{ant}$ is utilized and a burst of Arc and Ant synthesis results. Arc represses transcription from $P_{ant}$ by binding to an overlapping operator site, $O_{arc}$. Although Arc also represses its own expression through binding $O_{arc}$, transcription of *mnt* from $P_{mnt}$ is stimulated, leading to the synthesis of Mnt protein. Mnt binds to $O_{mnt}$ to prevent further expression from $P_{ant}$, and a stable lysogen can form. In addition, Mnt binding to $O_{mnt}$ activates transcription from $P_{mnt}$, leading to continued expression of Mnt and maintenance of the lysogenic state.[8]

Repression of *ant* gene expression by Mnt is critical for maintaining the lysogenic state because Ant is able to inactivate c2 repressor function by binding noncovalently to the protein.[9] Without the functional c2 repressor, lytic gene functions would be fully expressed and the lytic growth phase would ensue. Although *ant* gene expression does not influence the developmental fate in the wild-type P22 context during normal infection,[5] phages that contain an *arc*(Am) mutation dramatically overproduce Ant after infection,[10] which can lead to lytic gene functions being expressed. The *arc*(Am)-bacteriophage can form lysogens if the transcription of *ant* is prevented.[11] The relative amounts of c2 and Ant proteins synthesized in an infected cell will dictate whether an *arc*(Am)-bacteriophage becomes lysogenic or lytic.

RNA challenge phages are modified versions of P22 in which posttranscriptional regulatory events controlling the expression of *ant* determine the developmental fate of the phage. The modified phage encodes a disruption of *mnt* with a kanamycin resistance gene cassette and an *amber* muta-

[5] M. Susskind and P. Youderian, *in* "Lambda II" (R. W. Hendrix, J. W. Roberts, F. W. Stahl, and R. A. Weisberg, eds.), p. 347. Cold Spring Harbor University Press, Cold Spring Harbor, NY, 1983.

[6] R. T. Sauer, W. Krovatin, J. DeAnda, P. Youderian, and M. M. Susskind, *J. Mol. Biol.* **168,** 699 (1983).

[7] A. K. Vershon, P. Youderian, M. M. Susskind, and R. T. Sauer, *J. Biol. Chem.* **260,** 12124 (1985).

[8] A. K. Vershon, S. M. Liao, W. R. McClure, and R. T. Sauer, *J. Mol. Biol.* **195,** 311 (1987).

[9] M. M. Susskind and D. Botstein, *J. Mol. Biol.* **98,** 413 (1975).

[10] M. M. Susskind, *J. Mol. Biol.* **138,** 685 (1980).

[11] N. Benson, P. Sugiono, S. Bass, L. V. Mendelman, and P. Youderian, *Genetics* **114,** 1 (1986).

tion within *arc* (Fig. 1b). Sequence-specific RNA-binding activities can be detected using derivatives of P22 that have RNA target sequences substituted for the *ant* 5'-untranslated leader sequence region. The bacteriophage P22$_{R17}$ is a derivative of P22 in which the chosen developmental pathway is regulated by the R17/MS2 coat protein interacting with its RNA target site located in the *ant* mRNA.[2] Lysogenic development of the phage relies on R17/MS2 coat protein expression in the susceptible host cell and the availability of a suitable coat protein-binding site encoded by the phage genome. The system was used successfully to identify novel RNA ligands that display reduced affinity for the R17/MS2 coat protein,[2] to select for suppressor coat proteins that recognize mutant RNA ligands,[12] and to characterize translationally repressive RNA structures.[13]

Some natural RNA-binding proteins bind to RNA structures that can be modeled as artificial translational operator sites in bacteria. These proteins recognize unique structural features of their RNA ligands with little concern for primary sequence content. Members of this class of RNA-binding proteins include the Tat and Rev proteins of human immunodeficiency virus type 1 (HIV-1)[14] and the R17/MS2 coat protein.[15] We have demonstrated that artificial RNA operator sites modeled on the consensus sequence of the R17/MS2 coat protein-binding site are very effective regulatory sequences in the RNA challenge phage system.[16] Thus, it is quite likely that RNA challenge phages will be useful for characterizing RNA–ligand interactions for other systems.

This article describes molecular genetic methods for the handling and propagation of bacteriophage P22 reagents and detailed protocols for the construction and evaluation of RNA challenge phages that encode specific RNA-binding sites. In addition, we describe procedures whereby one can use mutant phages to rapidly identify features of sequence and structure within the target RNA sites of P22 phage derivatives that contribute to efficient RNA–ligand interaction. Finally, we outline the use of randomized gene libraries to select for mutant proteins that display novel RNA–ligand-binding properties. Most of these techniques were adapted from those originally published by our group and have undergone modification to enhance their application to problems under investigation in the laboratory.

[12] S. Wang, H. L. True, E. M. Seitz, K. A. Bennett, D. E. Fouts, J. F. Gardner, and D. W. Celander, *Nucleic Acids Res.* **25,** 1649 (1997).
[13] D. E. Fouts and D. W. Celander, *Gene* **210,** 135 (1998).
[14] M. J. Gait and J. Karn, *Trends Biochem. Sci.* **18,** 255 (1993).
[15] G. W. Witherell, J. M. Gott, and O. C. Uhlenbeck, *Prog. Nucleic Acid Res. Mol. Biol.* **40,** 185 (1991).
[16] D. E. Fouts, H. L. True, and D. W. Celander, *Nucleic Acids Res.* **25,** 4464 (1997).

Materials

All reagents used in this work should be of the highest quality obtainable. All biochemical solutions and biological media should be prepared using sterile water; water is initially deionized using a Millipore (Bedford, MA) Milli-Q Plus water purification system. Table I outlines the bacterial cell strains, bacteriophages, and plasmids, as well as their intended application, for procedures described in this article.

*Biochemical Media and Reagents*

TE buffer: 0.010 $M$ Tris–HCl (pH 8.0), 0.001 $M$ $Na_2EDTA$

10× TBE buffer: 1 $M$ Tris base, 1 $M$ boric acid, 0.020 $M$ $Na_2EDTA$

10× thermal POL buffer: 0.20 $M$ Tris–HCl (pH 8.8 at 25°), 0.10 $M$ KCl, 0.10 $M$ $(NH_4)_2SO_4$, 0.02 $M$ $MgSO_4$, 1.0% (v/v) Triton X-100, and 1.0 m$M$ dNTPs

5× PNK buffer: 0.25 $M$ Tris–HCl (pH 7.6), 0.05 $M$ $MgCl_2$, 0.025 $M$ dithiothreitol (DTT), 0.5 m$M$ spermidine hydrochloride, and 0.5 m$M$ $Na_2EDTA$

2× stop buffer: 8 $M$ urea, 0.010 $M$ $Na_2EDTA$, 0.04% (w/v) bromphenol blue, 0.04% (w/v) xylene cyanole FF, 0.5× TBE

10× *Taq*-EP buffer: 0.10 $M$ Tris–HCl (pH 8.3), 0.50 $M$ KCl, 0.070 $M$ $MgCl_2$, 0.005 $M$ $MnCl_2$, 0.1% (w/v) gelatin, 2.0 m$M$ dGTP, 2.0 m$M$ dATP, 10.0 m$M$ dCTP, and 10.0 m$M$ dTTP

TEN buffer: 0.010 $M$ Tris–HCl (pH 8.0), 0.001 $M$ $Na_2EDTA$, and 0.15 $M$ NaCl

*Biological Media and Reagents*

LB media: 10 g Bacto-tryptone (Difco, Sparks, MD), 5 g yeast extract (Difco), and 10 g NaCl per liter

T-top agar: 10 g Bacto-tryptone, 10 g NaCl, and 7 g Bacto-agar (Difco) per liter

This agar is dissolved into solution with heating; aliquots (100 ml) are dispensed into 250-ml bottles prior to autoclave-sterilization. The T-top agar will solidify on cooling. We generally melt the T-top agar by heating the T-top agar for 1 min in a microwave oven. The liquefied T-top agar solution is then maintained at 55° with the use of a water bath.

LB agar plates: 10 g Bacto-tryptone, 5 g yeast extract, 10 g NaCl, and 15 g Bacto-agar per liter

For the phage work described in this article it is critically important that dehumidified LB-agar plates be used in conjunction with T-top agar

TABLE I
BIOLOGICAL REAGENTS USED

| Biological reagent | Relevant characteristics | Specific application | Source or reference |
|---|---|---|---|
| **Bacterial strain** | | | |
| *E. coli* | | | |
| CJ236 | F' [pCJ105 (Cm$^R$)] *dut-1 ung-1 thi-1 relA1* | Generation of phagemid templates | 18 |
| DH5α | F$^-$, φ80d*lacZ*ΔM15, *recA1, endA1, gyrA96, thi*-1, *hsdR17* (r$_k^-$, m$_k^+$), *supE44, relA1, deoR, Δ(lacZYA-argF)*U169 | Propagation of recombinant plasmids | D. Hanahan |
| *S. typhimurium* | | | |
| MS1883 | LT2 *leuA414* Fels$^-$ *hsdSB* (r$^-$m$^+$) *supE40* | Propagation and purification of bacteriophage P22 and challenge phage derivatives; titering phage | M. Susskind |
| MS1582 | LT2 *leuA414*(Am) Fels$^-$ *supE40 ataP*::[P22 *sieA44* 16(Am)*H1455* Tpfr49] | Counterselection of P22 (O$_{mnt}^+$) derivatives so that only P22 (O$_{mnt}^-$) phage progeny form plaques | M. Susskind |
| MS1868 | LT2 *leuA414*(Am) Fels$^-$ *hsdSB* (r$^-$m$^+$) *endE40* | Recipient used for RNA challenge phage assays | M. Susskind |
| **Bacteriophage** | | | |
| P22 *mnt*::Kn9 | *arc*$^+$ O$_{mnt}^+$ (Kan$^R$) | Bacteriophage P22 derivative used for performing RNA challenge phage constructions | 19 |
| P22$_{R17}$ | P22 *mnt*::Kn9 O$_{mnt}$::*Sma*I-*Eco*RI *arc*(Am) O$^+$$_{R17}$ replicase::*ant* (Kan$^R$) | Control RNA challenge phage; this phage only forms lysogens in MS1868[pR17coat] recipients | 2 |
| **Plasmid** | | | |
| pCKR101 | pBR322 *lacI*$^Q$, P$_{tac}$, multiple cloning site (Amp$^R$) | Plasmid vector used for IPTG-regulated expression of heterologous RNA-binding activity | B. Swalla |
| pφGen1 | pPY190 *mnt*::Kn9 *arc*(Am) O$_{mnt}^-$ *ant'* (KanR) | Recombination plasmid used to shuttle DNA-encoded RNA target site onto P22 *mnt*::Kn9 | 19 |
| pR17coat(+) | pCKR101, R17 coat gene (sense orientation) ~465 bp of 5'-untranslated leader sequence | Control plasmid; this plasmid will only direct lysogen formation for P22$_{R17}$ challenge phage | 2 |

overlays. Dehumidified plates can be prepared by incubating plates inverted and partially open in a 42° incubator for ~45 min. A plate is considered to be sufficiently dehumidified if the LB-agar surface possesses a slightly wrinkled line appearance without any visible moisture or condensation evident.

λ-Ca(II) buffer: 0.010 $M$ Tris–HCl (pH 8.0), 0.010 $M$ MgCl$_2$, and 0.005 $M$ CaCl$_2$

## Methods

### Plasmid Requirements for RNA Challenge Phage Experiments

Two different types of plasmids are required for RNA challenge phage work. The first plasmid encodes the RNA-binding activity (e.g., protein). We have used a derivative of pCKR101 for our work (Fig. 1b).[2] This plasmid contains a multiple cloning site downstream from a P$_{tac}$ fusion promoter to permit expression of the inserted gene sequence and a copy of the $lacI^Q$ gene to allow for IPTG-inducible control of the desired gene. The plasmid also encodes a copy of the $bla$ gene to enable antibiotic selection of the plasmid in $S.$ $typhimurium$ hosts. Other expression vectors may also be used for this purpose. Because RNA challenge phage assays rely on selection of kanamycin-resistant lysogens, it is important to avoid using plasmid expression vectors that encode genes for kanamycin resistance. If the RNA-binding protein is derived from a eukaryotic source, a conventional bacterial ribosome-binding site element should be installed four to seven nucleotides upstream of the open reading frame. This modest modification usually can be done using standard DNA amplification techniques prior to cloning the desired gene into the expression vector.

The second type of plasmid serves as the recombination substrate that is needed for constructing the particular RNA challenge phage encoding the desired RNA target site (Fig. 2a). The DNA encoding the desired RNA sequence may be incorporated into the recombination plasmid substrate using either standard recombinant DNA procedures[17] or oligonucleotide-directed mutagenesis procedures.[18] We have used the latter technique to construct most of our recombination templates. Our plasmid of choice for this purpose is pφGen1 (Fig. 2a).[19] This plasmid encodes a modified portion

[17] F. M. Ausubel, R. Brent, R. E. Kingston, D. D. Moore, J. G. Seidman, J. A. Smith, and K. Struhl (eds.), $in$ "Short Protocols in Molecular Biology," 2nd Ed. Wiley, New York, 1992.
[18] T. A. Kunkel, K. Bebenek, and J. McClary, $Methods$ $Enzymol.$ **204**, 125 (1991).
[19] D. E. Fouts and D. W. Celander, $Nucleic$ $Acids$ $Res.$ **24**, 1582 (1996).

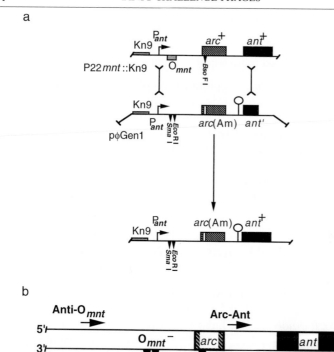

Fig. 2. (a) RNA challenge phage construction pathway. The RNA challenge phage is obtained following homologous recombination between the pφGen1 derivative and P22 *mnt*::Kn9 in *S. typhimurium*. P22 *mnt*::Kn9 *arc*(Am) progeny phages that contain the RNA-binding site are verified by PCR RFLP and sequence analyses. For simplicity, only progeny phages acquiring the RNA target site are shown in the recombination reactions. (b) Oligonucle-otide primer binding sites within the P22 *immI* region. The relative locations of primer annealing are illustrated for Anti-O*mnt*, Arc-Ant, and Intra-Ant. Arrows associated with each primer indicate the direction of DNA synthesis following annealing. The *Bso*FI (*Fnu*4HI) restriction site polymorphism and the corresponding nucleotide sequence information in the vicinity of the *arc*(Am) can be determined with the use of the primers Anti-O*mnt* and Intra-Ant. The nucleotide sequence encoding the *ant* mRNA leader sequence region, which would contain the acquired RNA target sequence, can be analyzed with the use of the primers Arc-Ant and Intra-Ant.

of the P22 *immI* region and an M13 *ori* sequence to allow for single-stranded phagemid production.

*Oligonucleotide Design and P22-Specific Primers.* The ideal primer for introducing target sequences into the P22 bacteriophage must bind specifically to one location on the phagemid template. An idealized oligonucleotide primer is shown:

5'-**GGCTTCGGTTGTCAG**-(oligo sequence of interest)-**AACCAACATGAATAG**-3'

Boldface nucleotides indicate regions of complementarity with the p$\phi$Gen1 phagemid template. In most of our work with RNA challenge phages, we have positioned the desired RNA sequence target immediately 5' of the *ant* ribosome-binding site so that the RNA–ligand interaction results in occlusion of ribosome engagement on the *ant* mRNA. A unique restriction site may be included very close to or within the target sequence to facilitate the identification of plasmids containing the oligonucleotide sequence of interest. Other primers that are useful for PCR and cycle sequencing include:

Anti-O$_{mnt}$    (5'-GATCATCTCTAGCCATGC-3')
Arc-Ant     (5'-CCAACTGCGGTAACAGTCAG-3')
Intra-Ant    (5'-GCGGTAAGAACATGCTGTC-3')

The relative locations where these primers bind to the bacteriophage P22 *immI* region are illustrated in Fig. 2b.

*PCR/RFLP Analysis.* We have found it extremely valuable to verify the status of our plasmid recombination substrates and challenge phage isolates using PCR/RFLP analysis. We recommend the inclusion of a unique restriction site in the mutagenesis oligonucleotide. The ability to screen for the presence of an acquired restriction site using this strategy will greatly reduce the time and effort required to identify candidate p$\phi$Gen1 derivatives that encode the desired RNA sequence. We generally use Arc-Ant and Intra-Ant as primers in PCR assays with plasmid DNA to verify oligonucleotide insertions in p$\phi$Gen1. The length of the PCR product generated for the p$\phi$Gen1 lacking oligonucleotide modification is 132 bp. More importantly, PCR/RFLP studies with candidate P22 challenge phage plaques permit rapid confirmation of the desired *arc*(Am) allele within the *immI* region as a *Bso*FI (*Fnu*4HI) restriction site is lost at the location of the amber mutation with the *arc* gene (Fig. 2b). We use Anti-O$_{mnt}$ and Intra-Ant as primers in PCR assays with phage to verify the presence of the *arc*(Am) allele in the desired recombinant P22 challenge phage. Lengths of the PCR product generated from both P22*arc$^+$* and P22*arc*(Am) phage may be identical (410 bp); however, only the P22*arc$^+$* phage would be sensitive to cleavage by *Bso*FI to yield two cleavage products (127 and 283 bp).

PCR assays are done with either phage suspension [$\sim$10$^8$ plaque-forming units (pfu)] or plasmid DNA ($\sim$10$^8$ molecules) in 100-$\mu$l reaction volumes containing 1$\times$ thermal POL buffer, 0.4 $\mu M$ each of the two relevant primers, and 0.5 unit Vent DNA polymerase (New England Biolabs, Beverly, MA). The PCR products (10 $\mu$l of the PCR reaction) are subsequently digested with restriction endonucleases (0.25–1 U/$\mu$l) for 1 hr according to the manufacturer's instructions and analyzed by agarose gel electrophoresis.

*5'-$^{32}$P-Labeling of Oligonucleotides.* Oligonucleotides are routinely 5'-$^{32}$P-labeled in a 10-$\mu$l reaction containing 1× PNK buffer, 2.5 $\mu M$ oligonucleotide (Operon, Inc., Alameda, CA), 2.5 $\mu M$ [$\gamma$-$^{32}$P]ATP (6000 Ci/mmol; NEN-DuPont, Boston, MA), and 10 U T4 polynucleotide kinase (New England Biolabs). The reactions are incubated at 37° for 30 min followed by incubation at 65° for 20 min to inactivate the kinase. The final oligonucleotide concentration is adjusted to 1 $\mu M$ using water.

*Dideoxynucleotide Cycle Sequencing.* The buffer components are separated from the amplified DNA products using diafiltration in a 30,000 NMWL Ultra Free-MC ultrafiltration unit (Millipore, Inc.) according to the manufacturer's instructions. The retentate containing the DNA is exchanged into TE buffer and adjusted to a final volume of 20 $\mu$l. Chain termination sequencing reactions are performed with the PCR products using 5'-$^{32}$P-labeled Intra-Ant or Arc-Ant primers and the asymmetric cycle sequencing procedure as originally described by Sears *et al.*[20]

## Construction of RNA Challenge Phage

An RNA target site is introduced into the challenge phage through homologous recombination in *S. typhimurium.* Alleles of the *arc* gene are used to identify P22 derivatives that contain the desired RNA-binding site. We have described an efficient means whereby RNA challenge phages are constructed using a single homologous recombination between a donor plasmid and a recipient P22 phage in the bacterial host cell (Fig. 2a).[19] Inserts ranging from 15 to 600 bp have been successfully introduced into challenge phage backgrounds. The construction of the RNA challenge phage involves the following steps, all of which are routinely accomplished in about 5 days.

*Transformation of Salmonella Recipient Strains.* Electrocompetent *S. typhimurium* strains are prepared using procedures suitable for generating competent *Escherichia coli* strains.[17] We usually transform competent MS1883 recipient cells with the desired p$\phi$Gen1 derivative using a Bio-Rad (Hercules, CA) Gene Pulser electroporator set at the following parameters: 1.8 kV, 400 $\Omega$, and 25 $\mu$F. Following recovery of the cells in 1 ml of LB media for 1 hr at 37°, cells are then spread or streaked onto LB-agar

---

[20] L. E. Sears, L. S. Moran, C. Kissinger, T. Creasey, H. Perry-O'Keefe, M. Roskey, E. Sutherland, and B. E. Slatko, *Biotechniques* **13**, 626 (1992).

plates supplemented with 100 $\mu$g/ml ampicillin and incubated at 37° until colonies are obtained.

*Infection of MS1883 [pφGen1] with Modified Bacteriophage P22.* To obtain recombinant P22 that encodes the desired RNA sequence, *in vivo* bacteriophage × plasmid homologous recombination reactions must be performed. A single MS1883[pφGen1] transformant obtained from the transformation plate is grown to saturation in 2 ml LB media containing ampicillin with aeration at 37° overnight. The cells are diluted 1:100 into fresh LB media containing ampicillin and grown at 37° with aeration until a cell density of $2 \times 10^8$ cells/ml is achieved ($OD_{600} = \sim0.3$–0.5). An aliquot of the culture (0.10 ml) is transferred to a sterile 13 × 100-mm glass culture tube, and $\sim$0.02 ml of diluted bacteriophage P22 *mnt*::Kn9 stock ($\sim1 \times 10^{10}$ pfu/ml; $2 \times 10^8$ phage) is added to yield a multiplicity of infection of 10 phage to each cell. The inoculated culture is incubated at room temperature for 30 min without agitation to allow for phage adsorption. The culture is combined with 2 ml of fresh LB media and incubated with aeration at 37° for 6 hr.

The inoculated cell culture usually contains stringy fragments of lysed cell debris following lytic growth of the phage. About 300 $\mu$l of chloroform is added to the cell suspension, and the mixture is vortexed briefly to lyse any viable cells that remain following infection. An aliquot of the mixture (1.5 ml) is added to microfuge tubes and clarified by centrifugation at 16,000g at room temperature for 2 min. The supernatant is transferred to a sterile microfuge tube and $\sim$50 $\mu$l of chloroform is added to prevent spoilage of the recombination lysate by contaminating bacterial growth. The recombination lysate can be stored safely at 4° for a period of 1–3 months.

*Identification of Progeny Phage That Encodes $O_{mnt}^-$ Allele.* The purpose of the following procedure is to isolate single phage plaques that encode the $O_{mnt}^-$ allele. The recombination lysate prepared in the preceding section will consist of a mixture of phage–parental bacteriophage (P22 *mnt*::Kn9) and recombinant progeny phage that contain various regions of homology with the recombination plasmid substrate, pφGen1. The growth of the parental phage and any progeny phage that carry the $O_{mnt}^+$ allele can be inhibited using as a passaging recipient a strain that constitutively expresses the *mnt* gene product. We routinely use the MS1582 cell strain as this strain harbors a defective P22 prophage that effectively prevents lytic infection for P22 derivatives that encode the $O_{mnt}^+$ allele. Only phage carrying the $O_{mnt}^-$ allele, such as those acquired from recombination with pφGen1, will grow lytically in MS1582 because the resident Mnt protein cannot bind to this mutant operator site to repress *ant* transcription.

This isolation procedure is accomplished using the following phage passaging experiment. An MS1582 culture is grown to saturation in 2 ml

LB media with aeration at 37° overnight. The cells are diluted 1:100 into fresh LB media and grown at 37° with aeration until a cell density of $2 \times 10^8$ cells/ml is achieved ($OD_{600} = \sim 0.3-0.5$). Aliquots of the culture (0.10 ml) are transferred to sterile $13 \times 100$-mm glass culture tubes, and aliquots (10 $\mu$l) of a serially diluted recombination lysate are used to inoculate the cultures. We usually prepare a dilution series of the recombination lysate in $\lambda$-Ca(II) buffer and inoculate four individual MS1582 cultures with recombination lysate(s) either undiluted or diluted 10-, 100-, or 1000-fold in $\lambda$-Ca(II) buffer. We have found that this range of lysate dilution will yield at least one plate that contains independent plaques. The phage–MS1582 cultures are incubated without agitation at room temperature for 30 min. Following this adsorption period, 2.5 ml of liquefied T-top agar is added to each phage–MS1582 culture, and the mixture is vortexed quickly and thoroughly and then poured rapidly onto a dehumidified LB plate. Care is taken to distribute the T-top agar evenly across the plate, and we have found that this is best done simply by tilting the plate at various angles in one's hand. The T-top agar will cool rapidly on the LB-agar surface, which may lead to an uneven, often ugly, T-top agar overlay if this procedure is not done quickly; therefore, it may take some practice to initially develop a proficient level of dexterity with the technique. The T-top agar overlay will usually solidify at room temperature after 20 min. The plates are then incubated at 37° for 5–12 hr to allow for plaque development. The candidate plaques will become evident as the lawn of MS1582 cells grows over this period. For plaque isolation, we usually select plates that contain numerous, well-separated plaques.

*Purification of arc(Am) Bacteriophages.* As far as repressor activity of *ant* transcription is concerned, the Arc protein is not as efficient as the Mnt protein; consequently, an $O_{mnt^-}$ phage that encodes both a functional *arc* gene and an $O_{arc^+}$ allele will form plaques on a MS1582 lawn. Phages that encode a functional *arc* allele will form turbid plaques as the Arc protein only partially represses Ant synthesis. Phages that encode an *arc*(Am) allele will produce abundant levels of Ant, resulting in the formation of clear plaques. Turbid plaques and clear plaques display subtle differences in their morphological attributes that tend to confuse students who are inexperienced in phage genetics. Examples of these different plaque morphologies are illustrated in Fig. 3. The different plaque morphologies are clearly revealed if one prepares a control LB-agar plate that contains a P22 $O_{mnt^-}$ *arc*(Am) phage and a P22 $O_{mnt^-}$ *arc*$^+$ phage on the same *Salmonella* strain lawn. Because we intend to use the *ant* gene as a biological reporter gene assay for monitoring RNA–ligand interactions in bacteria and because the *arc*$^+$ allele would reduce the sensitivity of this assay greatly, we select the phage encoding the *arc*(Am) allele.

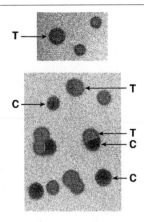

F$_{\mathrm{IG}}$. 3. Morphology of bacteriophage P22 plaques. Plaques displaying either a turbid morphology (T) or a clear morphology (C) are illustrated. The P22 *mnt*::Kn9 O$_{mnt^-}$ *arc*$^+$ phage displays a turbid plaque morphology that is characterized by a cloudy interior surrounded by a dark outer ring. The very center of the turbid plaques is customarily dark, giving the plaque a "bull's-eye" appearance when the plate is held in front of a light source, although this feature is not evident here. The P22 *mnt*::Kn9 O$_{mnt^-}$ *arc*(Am) phage displays a clear plaque morphology that lacks both a cloudy interior and a bull's-eye appearance. In this photograph, clear plaques display dark regions in which no bacterial growth is evident; however, clear plaques will appear more translucent when the plate is held in front of a light source. The light gray background surrounding the plaques depicts the background lawn of bacterial cells.

Clear plaques are individually picked from the developed MS1582 lawn using wooden toothpicks and streaked sequentially onto two lawns of MS1582 cells and then onto two lawns of MS1883 cells to ensure phage purity. These lawns are prepared by mixing 100 $\mu$l of 2 × 10$^8$ cells/ml culture with 2.5 ml of liquefied T-top agar, followed by rapidly distributing the mixture evenly onto dehumidified LB plates. The plates are allowed to solidify at room temperature for 20 min. We routinely mark each T-top agar overlay plate into quadrants and carefully apply a single phage-containing streak from a toothpick-selected plaque onto each quadrant for a total of four independent plaques per plate. Individual plaques are obtained within each quadrant by applying light pressure to the T-top agar surface to spread the original streak line within a given quadrant using two to three fresh sterile round toothpicks (or rounded-end wooden dowels) successively. The T-top agar surface can be torn easily if too much pressure is applied to the surface during the streaking procedure, which can interfere with subsequent plaque purification efforts from the affected quadrants. We generally purify 12 clear plaques for subsequent analysis, which means we will use 12 individual plates for preparing the bacterial lawns in a typical purification effort. The plaques usually develop within 6 hr, thereby making it possible to streak purify any given phage plaque on three separate bacterial lawns in 1 day.

A purified bacteriophage plaque is harvested for subsequent analysis by coring a plug through both the T-top agar overlay and the underlying LB-agar layer in the region surrounding the isolated plaque with a sterile Pasteur pipette. The phage from the agar plug is liberated by vortexing the agar plug in $H_2O$ (50 $\mu$l) for 15 sec, followed by incubation at room temperature for 30 min. These phage suspensions are subjected to PCR RFLP analyses to characterize any restriction site polymorphism within the regions encoding the arc(Am) allele and the acquired RNA target site. The DNA region encoding the RNA target site of promising arc(Am) phage isolates is then sequenced using the primers Intra-Ant and Arc-Ant. The isolated phage plugs or their aqueous suspensions can be stored at 4°; however, we recommend amplifying the phage soon after individual plaques have been isolated and characterized in this fashion.

*Large-Scale Production of Phage Lysate.* The production of a high-titer lysate is important for use in challenge phage assays. MS1883 cells are grown in 10 ml of LB with aeration at 37° until saturation. An aliquot (200 $\mu$l) of cell culture is added to a microfuge tube (1.5 ml) along with an isolated phage contained within an agar plug and incubated at room temperature for 30 min. During this incubation period, the remainder of the 10-ml saturated culture is combined with 200 ml LB media in a 1-liter flask and incubated with aeration at 37° for 30 min. The inoculated cell culture is then added to the 200-ml culture and incubated with aeration at 37° for 6 hr. Chloroform (10 ml) is added to the culture and the phases are mixed well. The supernatant is transferred to a sterile 250-ml polypropylene centrifuge tube using a serological pipette. Cellular debris is removed by centrifugation at 10,400$g$ at 4° for 15 min. The supernatant is transferred to four Oakridge style tubes (38 ml capacity) and centrifuged in a fixed angle rotor at 38,700$g$ at 4° for 1 hr. The resulting pellets are resuspended in $\lambda$-Ca(II) buffer (10 ml) and placed into sterile glass screw-cap tubes (16 × 125 mm). Chloroform (1 ml) is added to the suspension to prevent bacterial contamination, and the phage suspension is stored at 4° until use. This procedure will reproducibly yield high titers of phage suspensions ($\sim 10^{11}$ to $10^{12}$ pfu/ml). Although the phage titer will remain stable for several months, we recommend checking the titer periodically for phage stocks kept in storage for longer than 1 year. We would advise against using tubes composed of polypropylene for the long-term storage of phage suspensions as the phage will absorb to the surfaces of such tubes, resulting in significant variations in phage titers over time.

*Determination of P22 Bacteriophage Titer.* A phage titer must be done to accurately determine the concentration of bacteriophage in the large-scale lysate. This important parameter will be used to determine the appropriate amount of phage required for the challenge phage assays. The phage lysate is diluted serially 10-fold using $\lambda$-Ca(II) buffer. Twenty microliters of $10^{-7}$, $10^{-8}$, $10^{-9}$, and $10^{-10}$ dilutions is spotted onto a freshly prepared lawn of MS1883. One counts the appropriate dilution that yields well-resolved and abundant plaques following growth of the bacterial lawn. The phage titer is calculated with the following equation:

$$\text{pfu/ml} = [\text{average number of plaques}/0.02 \text{ ml}] \times \text{dilution factor}$$

A reliable titer value is determined from obtaining plaque counts from replicate lawn spottings and from using at least two independently prepared serial dilutions of phage stocks.

*RNA Challenge Phage Assays*

Bacteriophage P22 derivatives are used for infecting MS1868 strains that encode either a wild-type or a mutated version of the protein of interest. Electrocompetent MS1868 cells are transformed with the plasmid encoding the RNA-binding protein. Transformants of the recipient strain are grown until the culture density is $\sim 5 \times 10^8$ cells/ml. The cells are diluted $1:100$ into fresh media and grown with aeration at $37°$ for 2 hr. After this incubation period, the cells are diluted $1:4$ into fresh media containing a range of [IPTG] (typically 0–1 m$M$) to allow for the induced expression of the RNA-binding protein. The culture growth is continued at $37°$ until the cell density is about $5 \times 10^8$ cells/ml. A 100-$\mu$l aliquot of these cultures (i.e., $5 \times 10^7$ cells) is inoculated with the RNA challenge phages at a multiplicity of infection of $\sim 20$–25 (i.e., $\sim 10^9$ phage). High multiplicities of infection favor lysogen formation due to the increased copy number of the P22 *c2* gene; however, concomitant expression of *ant* will negate c2 protein function, leading to lytic growth. Following phage adsorption at room temperature for 20 min, the infected cells are plated onto LB-agar plates containing 100 $\mu$g/ml ampicillin and 40 $\mu$g/ml kanamycin and the appropriate concentration of IPTG. Uninfected cells are also plated onto LB-agar plates containing ampicillin (and IPTG, when relevant) to determine the viable cell number.

Because lysogenization frequencies can be monitored over a range spanning 7 orders of magnitude, we have found it necessary to prepare dilutions of the infected cell culture prior to plating the cells onto selective LB-agar plates. We generally apply 20 $\mu$l of a suitable dilution of the culture onto the selective LB-agar plate. A typical challenge phage selection plate will contain as many as 18–24 individually spotted samples, which reflects a collection of six 10-fold serial dilutions (e.g., $10^{-2}$, $10^{-3}$, $10^{-4}$, $10^{-5}$, $10^{-6}$, and $10^{-7}$) for three to four different infected cultures. These dilution series are usually performed with LB selective media using conventional polypropylene microtubes. For those who intend to perform a lot of these experiments, we recommend the use of 96-well microtiter plates and multichannel pipette dispensers for preparing the dilution series and for applying the diluted cultures onto the selective LB-agar plates. The applied spottings are allowed to dry completely before the plates are incubated at 37° for 8–12 hr.

The frequency of lysogenization (expressed as percentage lysogeny at a given IPTG concentration) is calculated as the number of colonies per milliliter obtained on LB-agar plates containing ampicillin and kanamycin divided by the number of viable colonies per milliliter obtained on LB-agar plates containing ampicillin, multiplied by 100%. Standard errors that are typically associated with these assays relate to variations in micropipetting and to modest fluctuations observed in spotting culture dilutions onto the selective plates. The assays are usually performed in triplicate to permit a statistical treatment of data. Data are plotted as the log(% lysogeny) versus log([IPTG]), as illustrated in Fig. 4.

We usually include two types of control infection experiments in parallel with the described RNA challenge phage assays. One positive control exper-

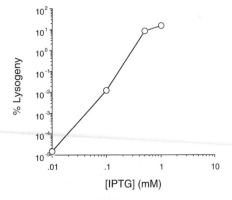

FIG. 4. Example of RNA challenge phage data plot. MS1886[pR17coat(+)] recipients were treated with increasing concentrations of isopropylthiogalactoside (IPTG) to induce the expression of the R17/MS2 coat protein. The recipients were infected with P22$_{R17}$ at a multiplicity of infection of 20–25 and then later plated onto LB-agar plates containing kanamycin. The coat protein directs repression of Ant biosynthesis, thereby enabling the challenge phage to establish a lysogenic state. Because the P22$_{17}$ phage encodes a kanamycin resistance cassette, colony growth results on selective LB-agar media. The frequency of lysogenization for a recipient cell culture following infection by a particular phage can be determined by simply counting the number of colonies obtained under selective conditions and by applying the formula described in the text. One representation of these data is illustrated.

iment is to monitor lysogen formation of MS1868[pR17 coat(+)] recipients following infection with P22$_{R17}$ phage (Table I); normally, the R17/MS2 coat protein will direct high frequencies of lysogenization ($\sim$20%) for this phage because it encodes the RNA-binding site to allow for coat protein-mediated repression of Ant biosynthesis (Fig. 4).[2] A series of negative control experiments would monitor lysogen formation of MS1868 recipient cells following infection with heterologous challenge phages. For example, the P22$_{R17}$ should undergo lytic development in any recipient cell that fails to express the R17/MS2 coat protein.[2]

## Isolation of Phages with Mutations in RNA Target Site

The procedure for mutagenizing phage has been published previously[2]; we present a modification of the procedure in this article. RNA challenge phages with mutations in the RNA target site are identified as clear plaques when plated on a lawn of JG1201 expressing the cognate RNA-binding protein. JG1201 has two important features that make it useful for identifying phage mutants.[2] First, it contains a plasmid (pMUC12) that expresses the *mucAB* gene products whose functions are to increase the frequency of error-prone repair of DNA that has been damaged by ultraviolet (UV) light. Phage can also form a clear plaque morphology if the genome encodes a defective *c2* gene, as the c2 protein serves as the primary repressor required for lysogeny. These phage mutants are discounted as viable candidates by virtue of a second attribute of JG1201: a resident, replication-defective copy of a P22 prophage that expresses a functional c2 repressor protein. Any phage that carries a defective *c2* gene and a nonmutated version of the encoded RNA target site will form turbid plaques on this strain background.

Briefly, 100 $\mu$l of 1 $\times$ 10$^{10}$ pfu/ml of phage containing the RNA target site is spotted on the bottom of a plastic petri dish. The sample is exposed to UV light in a UV cross-linker at an energy of 12,000 $\mu$J/cm$^2$. After exposure, the phage suspension is diluted serially with $\lambda$-Ca(II) buffer. One hundred microliters of an IPTG-induced JG1201 culture (2 $\times$ 10$^8$ cells/ml) is infected with 10 $\mu$l of 10$^0$–10$^{-3}$ dilutions of mutagenized phage at room temperature without agitation for 20 min. Approximately 2.5 ml of liquefied T-top agar is mixed rapidly with the cell–phage suspension, and the mixture is distributed evenly onto the surface of dehumidified LB-agar plates containing IPTG (300 $\mu$g/ml), ampicillin (100 $\mu$g/ml), and spectinomycin (100 $\mu$g/ml). After the T-top agar overlays have solidified, the plates are incubated at 37° for 4–8 hr. Clear plaques are streaked onto two lawns of JG1201 expressing the RNA-binding protein and onto two lawns of MS1883. Individual phage clones are subjected to PCR and cycle sequencing procedures to determine the identity of the mutation(s). Challenge phage assays are usually conducted with the isolated mutant phage to evaluate the manner in which the RNA mutation(s) disturbs protein–RNA interactions, as revealed by protein-mediated repression of Ant biosynthesis.

## Isolation of Suppressor Proteins

Ultraviolet mutagenesis and error-prone PCR are two common methods used to create random mutations within a protein of interest. Mutagenesis of protein open reading frames by ultraviolet light is done in the same fashion as described for the generation of mutant phages. For mutagenesis of protein open reading frames by error-prone PCR,[21] the entire gene or region of the gene to be mutated is amplified by PCR in a modified reaction cocktail that promotes base misincorporation during chain elongation by *Taq* DNA polymerase. Briefly, 200 p$M$ DNA template, 300 n$M$ (each) primers, and 5 U *Taq* DNA polymerase are combined in a 100-$\mu$l reaction volume containing 1× *Taq*-EP buffer; the error-prone DNA amplification procedures are carried out using a thermal cycling program similar to conventional PCR procedures used for the primer pair–DNA template of interest. The frequency of error incorporation in amplified templates can be evaluated with the use of restriction endonucleases. The PCR products are extracted with chloroform/isoamyl alcohol, ethanol precipitated, and inserted into a plasmid for expression. The ligated plasmids may be transformed directly into electrocompetent *S. typhimurium* MS1868 recipients. Suppressors are identified from this library as lysogens of MS1868 following infection with an RNA challenge phage containing a mutant RNA target site. The lysogens are identified on LB-agar plates containing kanamycin. The plasmid DNA is usually recovered from these lysogens following their expansion in liquid LB cultures containing ampicillin and kanamycin.

## Preparation of Salmonella Total Soluble Protein Lysates

Low-level expression of the desired RNA-binding protein in MS1868 recipients is often responsible for the low frequencies of lysogenization observed in challenge phage assays. If the intracellular expression level of the protein of interest is questionable, Western blots should be conducted to determine the level of expression. Of course, the utility of this method would rely on the availability of specific antisera to the protein of interest. We describe our procedure for preparing soluble protein lysates from *S. typhimurium* cells.[12] MS1868 transformants are grown in 2 ml LB media containing ampicillin with aeration at 37° to saturation. The cells are diluted

[21] R. C. Cadwell and G. F. Joyce, *PCR Methods Appl.* **2,** 28 (1992).

1:100 into fresh LB media containing ampicillin and grown with aeration at 37° for 2 hr. After this incubation period, the cells are diluted 1:4 into fresh LB media containing ampicillin supplemented with IPTG at the appropriate concentration and grown at 37° for 1 hr to allow induction of the recombinant protein. Cell cultures (3 ml) are harvested into 1.5-ml microcentrifuge tubes by centrifugation at 16,000g at room temperature for 30 sec. The pellets are resuspended into TEN buffer (200 μl) and frozen at −20°. The cells are thawed at 37° and phenylmethylsulfonyl fluoride is added to a final concentration of 0.5 m$M$. The cells are lysed on ice by sonication using a microtip ultrasonic cell disruptor at 50% maximum power (1-min burst–30-sec rest; the process is repeated a total of three times). Cellular debris is pelleted by centrifugation at 16,000g at 4° for 15 min. The supernatant is transferred to a fresh microcentrifuge tube, and the total protein concentration is determined in duplicate using a Bio-Rad DC protein assay with bovine serum albumin as a standard.

## Acknowledgments

We thank E. Brusca for comments on this article and B. Swalla for providing an updated map of pCKR101. This work was supported from grants provided by the American Foundation for AIDS Research and the National Institutes of Health.

# [23] Screening RNA-Binding Libraries Using Tat-Fusion System in Mammalian Cells

By Stephen G. Landt, Ruoying Tan, and Alan D. Frankel

## Introduction

The development of genetic assays to monitor protein–protein, DNA–protein, and RNA–protein interactions has greatly facilitated structure/function studies and cloning of interacting partners. For RNA–protein interactions, several assays have been described that utilize in vivo reporters to monitor effects on translation or transcription[1–7] (see articles 20–22, 24, 25,

---

[1] C. Jain and J. G. Belasco, Cell **87**, 115 (1996).
[2] K. Harada, S. S. Martin, and A. D. Frankel, Nature **380**, 175 (1996).
[3] D. J. SenGupta, B. Zhang, B. Kraemer, P. Pochart, S. Fields, and M. Wickens, Proc. Natl. Acad. Sci. U.S.A. **93**, 8496 (1996).
[4] H. Kollmus, M. W. Hentze, and H. Hauser, RNA **2**, 316 (1996).

27 in this volume). Two advantages of *in vivo* reporter systems are that they (1) can be used for moderate to large scale screening and (2) can monitor interactions in biologically meaningful environments. This article describes a screening method for studying RNA–protein interactions in mammalian cells that appears to be relatively adaptable and may be particularly suitable for studying mammalian complexes that may, for example, require posttranslation modification or multiple cellular components for binding.

## Overview of the Method

The screening method, called the Tat-fusion system, is based on the transcriptional activation properties of human immunodeficiency virus type 1 (HIV-1) Tat.[7] Tat is an unusual transcription factor that operates by binding to an RNA hairpin, known as TAR, located at the 5′ end of the HIV-1 transcripts and enhances the efficiency of transcriptional elongation from the HIV-1 LTR promoter. A short, arginine-rich RNA-binding domain in Tat recognizes a bulge region in TAR, while a cellular protein, cyclin T1, binds to TAR as part of a high-affinity ternary complex with Tat and recognizes the adjacent TAR loop.[8] Tat also enhances transcription when delivered to the RNA via a heterologous RNA–protein interaction in place of the Tat–cyclin T1–TAR interaction.[7,9–14] In the Tat-fusion system, a library is fused to the activation domain of Tat and RNA-binding domains are identified by their ability to activate an HIV-1 LTR reporter in which TAR is replaced by an RNA site of interest. The system appears to accommodate a wide variety of interactions, operates in different cell types, and, in many cases, shows high levels of activation.

The Tat-fusion system uses two plasmids (Fig. 1): a Tat-fusion expressor plasmid and a reporter plasmid expressing green fluorescent protein (GFP)[15] from a modified HIV-1 LTR. Reporter cell lines (typically HeLa)

[5] S. Wang, H. L. True, E. M. Seitz, K. A. Bennett, D. E. Fouts, J. F. Gardner, and D. W. Celander, *Nucleic Acids Res.* **25,** 1649 (1997).

[6] E. Paraskeva, A. Atzberger, and M. W. Hentze, *Proc. Natl. Acad. Sci. U.S.A.* **95,** 951 (1998).

[7] R. Tan and A. D. Frankel, *Proc. Natl. Acad. Sci. U.S.A.* **95,** 4247 (1998).

[8] P. Wei, M. E. Garber, S. M. Fang, W. H. Fischer, and K. A. Jones, *Cell* **92,** 451 (1998).

[9] C. Southgate, M. L. Zapp, and M. R. Green, *Nature* **345,** 640 (1990).

[10] M. J. Selby and B. M. Peterlin, *Cell* **62,** 769 (1990).

[11] S. J. Madore and B. R. Cullen, *J. Virol.* **67,** 3703 (1993).

[12] R. Tan, L. Chen, J. A. Buettner, D. Hudson, and A. D. Frankel, *Cell* **73,** 1031 (1993).

[13] L. Chen and A. D. Frankel, *Biochemistry* **33,** 2708 (1994).

[14] W. S. Blair, T. B. Parsley, H. P. Bogerd, J. S. Towner, B. L. Semler, and B. R. Cullen, *RNA* **4,** 215 (1998).

[15] A. B. Cubitt, R. Heim, S. R. Adams, A. E. Boyd, L. A. Gross, and R. Y. Tsien, *Trends Biochem. Sci.* **20,** 448 (1995).

A

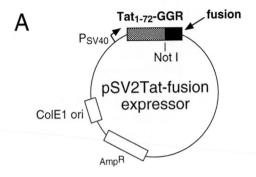

```
M   E   P   V   D   P   R   L   E   P   W   K   H   P   G
ATG GAA CCG GTG GAC CCG CGG CTG GAG CCT TGG AAG CAC CCC GGG

S   Q   P   K   T   A   C   T   N   C   Y   C   K   K   C
TCC CAG CCA AAG ACC GCG TGC ACC AAT TGC TAC TGT AAG AAG TGC

C   F   H   C   Q   V   C   F   I   T   K   A   L   G   I
TGT TTC CAC TGT CAG GTG TGC TTC ATC ACC AAG GCC CTA GGA ATC

S   Y   G   R   K   K   R   R   Q   R   R   R   P   P   Q
TCA TAT GGC CGC AAA AAA CGA CGT CAG AGG CGT CGA CCA CCT CAG

G   S   Q   T   H   Q   V   S   L   S   K   Q   G   G   R
GGA TCC CAG ACC CAC CAG GTC TCT CTG AGC AAG CAG GGC GGC CGC
                                                  Not I

X
TGA GTGAGTGA CTCGAG ACTAGT
              Xho I  Spe I
```

B

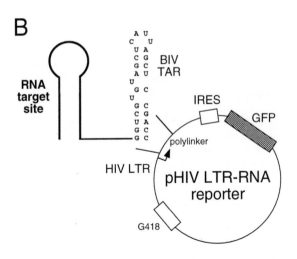

```
HIV LTR    Afl II  Spe I      Nhe I      Hind III   Eco RI
ACTGGGTCTCTCTGGCTTAAGGACTAGTCTAGCTAGCTAGAAGCTTCCCGGAATTC
    +1
```

are generated containing the stably integrated reporter, and the Tat-fusion library is introduced into cells by bacterial protoplast fusion, which delivers library members to cells in a nearly clonal manner.[7] Cells expressing high levels of GFP (as a result of RNA binding and Tat activation) are isolated by fluorescence-activated cell sorting (FACS) and plasmid DNA is recovered. Plasmids are reintroduced into bacteria, protoplasts are generated, and fusion and sorting are repeated until the population is highly enriched for functional activators. Individual positive clones are then tested for the desired RNA-binding specificity by measuring the activation of several mutant or unrelated RNA reporters. Clones that show the proper specificity are sequenced and characterized further. The overall screening protocol is outlined in Fig. 2.

## Considerations for Tat Fusions

RNA-binding domains are fused either to the C terminus of the Tat activation domain (following amino acid 48) or to the complete first exon (following amino acid 72), which includes the nuclear localization sequence and TAR-binding domain (amino acids 49–57; Fig. 1A). Fusing to Tat1–72 will usually ensure proper localization and allows the activity of the Tat activation domain to be assessed, as fusions to Tat1–72 are expected to be active on the wild-type HIV-1 LTR unless they are poorly expressed or disrupt the Tat–TAR interaction. It is possible that retaining the TAR-binding domain will cause nonspecific binding to some RNA reporters, but this has not yet been observed. RNA-binding domains may also be appended to the N terminus of Tat, although at least one fusion in this configuration shows reduced (~3- to 5-fold) activity.

---

FIG. 1. Diagrams of Tat-fusion expressor and RNA reporter plasmids. (A) The pSV2Tat-fusion expressor plasmid uses an SV40 early promoter to drive expression of Tat1–72[22] containing a C-terminal Gly-Gly-Arg extension. The plasmid contains a NotI cloning site at the end of the Tat gene to generate fusions and a ColE1 origin for chloramphenicol amplification. The sequence of the Tat gene is shown, with the overlapping nuclear localization sequence and RNA-binding domain (amino acids 49–57) underlined, and cloning sites are indicated. (B) The pHIV LTR-RNA reporter plasmid uses the HIV-1 LTR to drive expression of the GFP gene, with translation initiated internally from an encephalomyocarditis virus (EMC) IRES element.[23] The GFP used is an S65T variant that also contains codons optimized for expression in human cells.[19] The vector (derived from pcDNA3; Invitrogen) contains a neomycin resistance gene for selecting stable cell lines with G418 and a polylinker (sequence shown below) for cloning an RNA target site. In general, a BIV TAR element is also placed downstream of the RNA target site for monitoring activation, as described in the text.

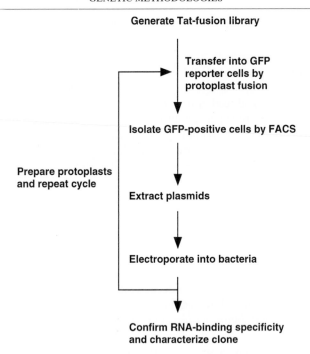

FIG. 2. Overall strategy for screening libraries with the Tat-fusion system.

Results from several groups suggest that functional Tat fusions can be generated with a variety of RNA-binding domains, including the bacteriophage MS2 coat protein, HIV-1 Rev protein and related arginine-rich RNA-binding peptides, BIV Tat RNA-binding domain, U1A RNP domain, and KH domains[7,9-13] (H. Peled-Zehavi and A. D. Frankel, unpublished results). In our experience, all fusions to HIV-1 Tat have activated well through their cognate RNA reporters, with the exception of a fusion to the iron response element (IRE)-binding protein, IRP-1.[16] It is unclear why this particular fusion was inactive. Additional new data suggest that Tat can function when fused to different components of a multiprotein complex (H. Peled-Zehari and A. D. Frankel, unpublished results). A functional fusion has also been generated between the poliovirus 3C protein and the equine infectious anemia virus (EIAV) Tat protein[14] and this RNA–protein complex also may require cooperative binding of an additional cellular protein.

To date, the RNA-binding domains fused to Tat have had affinities from subnanomolar (U1A) to >10 n$M$ (MS2 and several RNA-binding peptides). The upper and lower limits to affinity are not yet known, and

[16] T. A. Rouault, C. K. Tang, S. Kaptain, W. H. Burgess, D. J. Haile, F. Samaniego, O. W. McBride, J. B. Harford, and R. D. Klausner, *Proc. Natl. Acad. Sci. U.S.A.* **87,** 7958 (1990).

as with other *in vivo* systems, a strict correlation may not exist between affinity and activation levels when comparing different RNA reporters or fusions.[3] Any of a number of variables may affect the readout *in vivo,* including protein expression levels, protein stability, mislocalization, or structural incompatibilities between fused partners. However, for any given reporter we have observed an excellent correlation between *in vitro* RNA-binding affinities and levels of activation (for examples, see Refs. 12 and 13).

At least two other difficulties may be envisioned for the Tat-fusion system. First, large domains fused to Tat might sterically hinder the formation of transcription elongation complexes. Preliminary experiments with cDNA libraries, however, indicate that a substantial fraction of large fusion proteins still activate through HIV-1 TAR (S. G. Landt and A. D. Frankel, unpublished results). Second, it is possible that endogenous nuclear proteins or RNAs might compete for RNA binding and thereby squelch activation. This has not been a problem for the U1A–U1 interaction (the protein and RNA are highly abundant nuclear components) and, in fact, it generally has been difficult to use untethered RNA-binding domains as dominant negative inhibitors of Tat fusions. This likely reflects a requirement for multiple interaction surfaces to achieve Tat activation.

The vector shown in Fig. 1A is used for cloning fusions to Tat1–72. Tat is expressed from a simian virus 40 (SV40) early promoter and contains a C-terminal Gly-Gly-Arg linker and codons optimized for mammalian expression. *Not*I and *Xho*I or *Spe*I sites are used for cloning, and termination codons have been placed in all three reading frames immediately downstream of Tat for insertion of libraries.

## Considerations for RNA Reporters

The reporter plasmid (Fig. 1B) contains the HIV-1 LTR followed by a polylinker for cloning RNA target sites (in place of TAR) followed by an internal ribosome entry site (IRES) and the GFP gene. An IRES was included because structured RNA elements, or the binding of proteins to these elements, can block cap-mediated translation when located near the 5' end of a transcript.[17] By initiating translation internally, it also is possible to clone RNAs containing AUG codons upstream of GFP without competing for initiation. The vector also contains the neomycin resistance gene for selecting stable reporter-containing cell lines using G418.

RNA sites are best cloned as close as possible to the 5' end of the transcript, using the Afl II site in the polylinker (Fig. 1B). We typically include the bovine immunodeficiency virus (BIV) TAR hairpin (the binding site for

[17] R. Stripecke, C. C. Oliveira, J. E. McCarthy, and M. W. Hentze, *Mol. Cell. Biol.* **14,** 5898 (1994).

BIV Tat; see Ref. 17a) immediately downstream of the RNA target site for three reasons. First, because Tat activates transcription through an elongation mechanism, it is possible that RNAs beyond a certain length are not transcribed before transcription terminates and thus no RNA target is synthesized to recruit Tat.[18] By placing BIV TAR downstream of the target site, transcription past the site can be monitored by testing activation through BIV TAR using the HIV-BIV Tat65–81 fusion protein to deliver the HIV-1 Tat activation domain.[13] We have observed levels of GFP activation that are readily detectable by FACS when BIV TAR is located more than 400 nucleotides from the 5' end. Thus, it appears that RNAs at least 300–400 nucleotides in length can be accommodated by the Tat-fusion system. Second, BIV TAR is useful for identifying appropriate reporter cell lines for screening libraries. Individual clones are transfected with the HIV-BIV Tat65-81 protein to test whether the reporter can be activated, and clones showing the highest levels of activation are chosen for screening. Third, activation through BIV TAR provides a positive control to estimate protoplast fusion efficiency and to provide a standard for setting FACS sorting windows.

GFP and Reporter Cell Lines

GFP was chosen as the reporter to allow library screening by FACS. We have examined several GFP variants that display different fluorescence intensities in order to optimize the signal from Tat activation. Previous vectors, which did not contain an IRES, used a S65A GFP variant[15] that produced the best signal to noise ratio on Tat activation.[7] An S65T variant that also contains codons optimized for expression in human cells (EGFP[19]) produced a high background in the absence of Tat and therefore was unsuitable for the system. In the context of the IRES-containing vector, however, the EGFP variant produces a low background and is highly activatable by Tat. Future improvements may use other GFP variants with different fluorescence properties.

To date, all library experiments have been performed using stable HeLa cell lines that contain the appropriate reporter. An example of activation of a U1 RNA reporter cell line by a Tat–U1A fusion is shown in Fig. 3. Preliminary experiments suggest that the reporter plasmid may instead be introduced transiently along with the Tat-fusion plasmid by protoplast fusion (A. J. Lynn and A. D. Frankel, unpublished results). In principle, this would eliminate the need to select stable cell lines, but further experiments are needed to define a working protocol. In generating stable reporters, we have observed large differences in the levels of Tat activation between

[17a] C. A. Smith, L. Chen, and A. D. Frankel, *Methods Enzymol.* **318,** [28], (2000) (this volume).
[18] M. J. Selby, E. S. Bain, P. A. Luciw, and B. M. Peterlin, *Genes Dev.* **3,** 547 (1989).
[19] J. Haas, E.-C. Park, and B. Seed, *Curr. Biol.* **6,** 315 (1996).

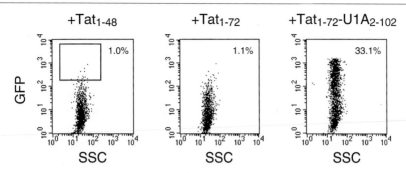

FIG. 3. Activation of a U1 RNA reporter by a Tat-U1A fusion protein. Protoplasts were prepared with the Tat expressors shown and were fused to HeLa cells containing a stably integrated HIV-1 LTR-U1 RNA reporter containing hairpin II of U1 snRNA. Tat1–48 contains the activation domain alone whereas Tat1–72 includes the nuclear localization sequence and HIV-1 TAR RNA-binding domain. Cells were sorted by FACS 2 days after protoplast fusion, and plots show GFP intensity on the y axis and side scatter (a measure of cell granularity) on the x axis. Only live cells that exclude propidium iodide are plotted, and numbers indicate the percentage of sorted cells found in the inset shown on the left-hand side.

clones, as monitored through the BIV Tat–TAR interaction. We therefore advise testing at least 10 clones for activation and choosing those that show high activities, low fluorescence in the absence of Tat, and good growth rates. We have used the Tat-fusion system in other cell types, including mouse NIH 3T3 and Chinese hamster ovary (CHO) cells, and in principle, screens may be performed using any cell type that supports Tat activation through a heterologous RNA–protein interaction. This may be advantageous if, for example, endogenous cell-type specific proteins are recruited to an RNA target site to assemble a high-affinity complex. Protoplast fusion efficiencies and plasmid recovery differ substantially among cell types, and each protocol must be optimized before initiating a large-scale screen.

## Procedure: Generating Reporter Cell Lines

1. Grow HeLa cells in six-well plates to ~50% confluence in Dulbecco's modified Eagles medium (DMEM) containing 10% fetal bovine serum (FBS). Transfect cells with 0.5–2 μg reporter plasmid per well (typically using lipofectin according to the manufacturer's instructions; GIBCO/BRL, Gaithersburg, MD). In one well, cotransfect the reporter along with 100 ng pSV2HIV-BIV Tat65–81 (which expresses the BIV Tat RNA-binding domain fused to the HIV-1 Tat activation domain). After 2 days, examine GFP expression of the cotransfected control by fluorescence microscopy or FACS.

2. Trypsinize and dilute transfected reporter cells to 40,000 cells/ml in DMEM containing 10% FBS and an appropriate concentration of

G418. Add 100 $\mu$l cells to 100 $\mu$l medium in one column of a 96-well plate and serially dilute (by twofold) into subsequent columns. Prepare four plates to ensure growth of a sufficient number of clones. To determine the appropriate G418 concentration for selection, treat untransfected HeLa cells with G418 and use the lowest concentration required to kill the entire population (typically 0.8–1 mg/ml).

3. Grow cells for 1–3 weeks, identify single colonies that appear normal and grow well, and trypsinize and expand cell lines. Test each for background GFP expression and Tat activation (by transfecting with 100 ng pSV2HIV-BIV Tat65–81) by fluorescence microscopy or FACS. Prepare frozen stocks of appropriate lines and maintain growth in the presence of G418.

### Delivering the Library into Reporter Cells

We have observed that Tat-fusion libraries can be delivered into HeLa cells in a nearly clonal manner by bacterial protoplast fusion.[7] Delivering many copies of a single library member into a HeLa cell may allow detection of weak signals that otherwise would be diluted in a population of cotransfected plasmids and substantially reduces the background from cotransfected library members. Plasmids are amplified in *Escherichia coli* with chloramphenicol (to ~2000–3000 copies/cell using the ColE1 origin), protoplasts are generated by treating with lysozyme and EDTA, and cell fusion is mediated by polyethylene glycol (PEG).[20,21] Under the conditions described, 10–20% of HeLa cells typically become fused. It is not known how many plasmids enter a target cell nucleus, but plasmids can be recovered from sorted cells at sufficient levels for electroporation, even 2 days after fusion (see later). Several steps in the procedure appear to be particularly critical: treating bacteria with chloramphenicol to amplify the plasmid, optimizing the time of lysozyme treatment to ensure efficient protoplasting, handling bacteria gently to avoid lysis, and minimizing the time of PEG treatment to maintain HeLa cell viability.

### Procedure: Protoplast Preparation

1. Transform Tat-expressor plasmids into *E. coli* DH-5$\alpha$ or DH-10B cells, grow cultures overnight in LB/ampicillin (100 $\mu$g/ml), and inoc-

[20] R. M. Sandri-Goldin, A. L. Goldin, M. Levine, and J. C. Glorioso, *Mol. Cell. Biol.* **1**, 743 (1981).

[21] M. Rassoulzadegan, B. Binetruy, and F. Cuzin, *Nature* **295**, 257 (1982).

[22] A. D. Frankel and C. O. Pabo, *Cell* **55**, 1189 (1988).

[23] R. P. Jackson and A. Kaminski, *RNA* **1**, 985 (1995).

ulate 50 ml LB/amp with 0.5 ml culture. For libraries, transformed colonies should be scraped from plates in LB medium (5 ml per 100-mm plate) and stored as glycerol stocks at $2 \times 10^9$ cells/ml [in 50% (v/v) LB, 32.5% (v/v) glycerol, 50 m$M$ MgSO$_4$, 12.5 m$M$ Tris–HCl, pH 8.0.] Use 1-ml aliquots to inoculate 50 ml LB/amp cultures.

2. Grow cells by shaking at 37° to OD$_{600}$ = 0.7–0.8 (typically 3–5 hr), add 12.5 mg chloramphenicol, and grow an additional 12–16 hr.

3. Harvest cells in 50-ml tubes by centrifuging at 4° for 10 min at 2000$g$. Resuspend in 10 ml 50 m$M$ Tris–HCl, pH 8.0, centrifuge, and resuspend in 2.5 mL chilled 20% (w/v) sucrose, 50 m$M$ Tris–HCl, pH 8.0.

4. To prepare protoplasts, add 500 $\mu$l 5 mg/ml lysozyme (Boehringer-Mannheim) prepared freshly in 250 m$M$ Tris–HCl, pH 8.0, and incubate on ice for 5 min. Add 1 ml 250 m$M$ EDTA, pH 8.0, and incubate an additional 5 min on ice. Dilute cells slowly with 1 ml 50 m$M$ Tris–HCl, pH 8.0, and incubate at 37°. Monitor cells by phase-contrast microscopy to ensure complete protoplasting while minimizing cell lysis. Bacteria are converted from highly motile rod shapes to stationary circular protoplasts. Typically >90% conversion occurs within 15–20 min.

5. Dilute slowly with 20 ml prewarmed serum-free DMEM containing 10 m$M$ MgCl$_2$ and 10% sucrose to terminate protoplasting. Medium must be added slowly to prevent cell lysis. Viscous remnants of the cell wall and membrane may be seen. Mix suspension gently but thoroughly by inverting two to three times and incubate at room temperature for 15 min to allow bacteria to recover. Protoplasts should be used within 30 min, before the cell wall begins to regenerate.

Procedure: Protoplast Fusion

1. Plate HeLa reporter cells in six-well plates at ~30–40% confluence in DMEM lacking G418 and grow overnight to ~80–90% confluence. This density is optimal for protoplast fusion and does not result in substantial cell–cell fusion.

2. Wash each well with 2 ml serum-free DMEM, add 2 ml protoplast suspension, and centrifuge at room temperature for 8–10 min at 1650$g$. Handle plates very carefully to avoid dislodging the bacteria. Aspirate the solution and add 2 ml 45–50% polyethylene glycol (PEG; either PEG1000 or PEG1500) diluted in serum-free DMEM. Store PEG solutions at 4° in the dark to prevent photooxidation.

3. Aspirate PEG solution after 2 min and wash three times with 4 ml serum-free DMEM prewarmed to 37°. It is important to treat with

PEG for ≤2 min to minimize cell toxicity. Wash cells carefully but relatively vigorously to remove the PEG completely (which is viscous and difficult to aspirate) and residual protoplasts. Add 2 ml DMEM containing 10% FBS and supplemented with 30 μg/ml kanamycin to prevent growth of any remaining bacteria. Following fusion, HeLa cells often exhibit membranous protrusions and other morphological changes and require about 1 day to return to normal growth.

Cell Sorting and Plasmid Recovery

The Tat-fusion system utilizes transient transfection (via protoplast fusion) and recovery of plasmid DNA from sorted recipient cells. Other methods have been considered in which positive members of a library are recovered by selecting for cell lines containing replicating vectors or integrated plasmids. However, transient methods offer the advantages that they can be performed relatively rapidly, may be used with a variety of cell types, and may introduce fewer selection biases.

In order to recover sufficient plasmid with the transient procedure, it is important to minimize the time between protoplast fusion and plasmid isolation, before plasmids are completely degraded. Generally, we incubate cells for 2 days after fusion to accumulate enough GFP for detection, although plasmid yields may be two to three times higher after just 1 day. Thus, it is desirable to sort cells as early as possible, depending on the level of activity expected from the library.

The most important factor in FACS sorting is to set an appropriate limit for GFP expression. Because the level of activity expected for any particular RNA–protein interaction is unknown, the cutoff generally should be set conservatively to ensure that all positive members of a library are retained. The level of activation through BIV TAR may be used as an approximate guide for any given reporter, setting the window below this value. It is worth sacrificing enrichment for yield, especially during the early rounds of a screen when library complexity is high. It may be desirable to sort cells at several different stringencies during later rounds, when the distribution of positives becomes more obvious.

It has been difficult to obtain high plasmid yields from sorted cells, and the success of the method relies on using alkaline cell lysis and preparing highly electrocompetent cells. We have found that the commonly used Hirt method for preparing cell extracts gives relatively low plasmid yields, whereas the alkaline procedure yields about one plasmid (one transformed colony) per sorted cell. Although not ideal, this yield is sufficient for library screening. We also have used polymerase chain reaction to recover library inserts from sorted cells but have observed an unacceptable bias against large fragments.

Procedure: FACS Sorting

1. Two days after protoplast fusion, wash cells once with 2 ml PBS per well, add 1 ml EDTA dissociation buffer (GIBCO/BRL), and incubate at 37° for 5–10 min. Transfer to a 6-ml conical tube, centrifuge at 4° for 5 min at 1650$g$, remove supernatant, and wash cells twice with 2 ml phosphate-buffered saline (PBS). Resuspend in PBS containing 5% cell dissociation buffer, 0.3% FBS, and 1 $\mu$g/ml propidium iodide to a final concentration of $10^6$ cells/ml (~1 ml/well).

2. Determine the GFP sorting window (at 510 nm) by first scanning 10,000 reporter cells fused to protoplasts containing an inactive plasmid to define the negative population and, if possible, to protoplasts containing an active plasmid that approximates the expected activity. For reporters with a downstream BIV TAR element, activation by pSV2HIV-BIV Tat65–81 provides a reasonable standard. Exclude dead cells from the sorted population by the uptake of propidium iodide (monitored at 620 nm) and exclude cell and bacterial debris by forward scatter, which provides a measure of cell size. Sort positive cells into a tube containing 500 $\mu$l PBS containing 5% cell dissociation buffer and 0.3% FBS. Using a FACStar$^{Plus}$ cell sorter (Becton-Dickinson, San Jose, CA), ~2500 cells may be sorted per second.

Procedure: Plasmid Recovery

1. Mix sorted cells with 20,000 untransfected HeLa cells and microfuge for 5 min at 4°.

2. Resuspend pellet in 10 $\mu$l TE (10 m$M$ Tris–HCl, pH 8.0, 1 m$M$ EDTA) containing 0.2 mg/ml tRNA. Lyse cells by adding 20 $\mu$l 1% SDS, 0.2 $N$ NaOH and incubate on ice for 5 min. Add 15 $\mu$l 3 $M$ sodium acetate (pH 4.8), incubate on ice for an additional 10 min, and microfuge for 5 min at 4°. Transfer supernatant to a fresh tube and extract with an equal volume of phenol/chloroform (1 : 1).

3. Add 1 $\mu$l 20 mg/ml glycogen (Sigma, St. Louis, MO) and precipitate DNA with 0.1 volume 3 $M$ sodium acetate (pH 4.8) and 3 volumes ethanol at −70° for 1 hr. Microfuge for 30 min at 4°, wash pellet once with 70% ethanol, air dry, and resuspend in 1–5 $\mu$l distilled water.

Procedure: Preparation of Electrocompetent Cells
and Electroporation

A relatively standard procedure is used to prepare electrocompetent cells, with the addition of an extra water wash step to generate high competency and resuspension of cells at high concentration to enhance recovery of small amounts of plasmid.

1. Inoculate 500 ml LB with an overnight *E. coli* DH-5α culture. Grow at 37° with shaking to $OD_{600} \approx 0.5$ and chill in ice water for 15 min. To achieve high electrocompetence, it is critical that all subsequent steps be performed at 0° and that all solutions, containers, and rotors be thoroughly prechilled to 0°.

2. Centrifuge cells for 20 min at 4000$g$, resuspend in 500 ml 1 m$M$ HEPES, pH 7.0, and recentrifuge for 20 min. Repeat HEPES wash, and then wash twice with 250 ml deionized water. It is critical to perform four wash steps to achieve sufficient competency. Cell pellets may be rather loose after the water washes, and decanting or aspirating should be performed carefully to minimize cell loss.

3. Resuspend cells in 100 ml 10% (v/v) glycerol and centrifuge for 10 min at 4000$g$. Centrifuging in conical tubes helps minimize loss of the cell pellet. Estimate the pellet volume and resuspend in 0.5 volume 10% (v/v) glycerol (typically ~500 μl). Cell concentrations should be ~3 × $10^{11}$/ml. Fresh electrocompetent cells give the highest transformation efficiencies. Aliquots (100 μl) may be frozen on dry ice, typically with a 2- to 3-fold decrease in competency. When tested with small amounts of supercoiled pUC19 DNA (0.1–10 pg), ~3–10 × $10^{10}$ transformants typically are obtained with frozen cells.

4. Electroporation conditions will vary and should be optimized for highest efficiency. For a typical procedure, add 1 μl recovered DNA to 50 μl competent cells in a chilled 1-mm electroporation cuvette (Bio-Rad, Hercules, CA). Electroporate at 1.8 kV, 25 μF, 200 Ω (using a Bio-Rad Gene Pulser). Immediately following electroporation, add 1 ml room temperature Superbroth (32 g Bacto-tryptone, 20 g yeast extract, 5 g NaCl, 500 μL 10 $N$ NaOH, in 1 liter), transfer to a culture tube, and incubate at 37° for 1 hr with shaking. Plate cells on LB/amp and grow at 37°.

Closing Points

The Tat-fusion system provides a useful tool for structure/function studies of RNA–protein complexes and screening libraries in mammalian cells. Several applications of the method currently are being investigated, including screening cDNA libraries and examining the assembly of multiprotein complexes. Future technical improvements may include (1) delivering the reporter and Tat-fusion plasmids simultaneously, eliminating the need to select stable reporter cell lines, (2) performing screens simultaneously with a nonspecific RNA reporter to sort out the population of nonspecific activators, and (3) using other types of reporters, such as proteins that can be identified using cell-permeant fluorescent substrates or probes, cell-surface proteins that allow positive cells to be separated using antibody-coated beads, or other GFP variants with different emission properties.

### Acknowledgments

This work was supported by a predoctoral fellowship from the Howard Hughes Medical Institute to S.G.L. and by grants from the National Institutes of Health.

## [24] Frameshifting Assay to Characterize RNA–Protein Interactions in Eukaryotic Cells

By Heike Kollmus and Hansjörg Hauser

### Introduction

Most assay systems currently used for the demonstration of RNA–protein interaction are carried out *in vitro*. Typical applications are outlined elsewhere in this volume. However, it is not clear to what extent the results reflect the *in vivo* situation, as neither the structure of RNA nor the folding of the binding protein can be simulated reliably *in vitro*. In addition, the natural surroundings of a living cell add additional variations and factors that influence the interactions. We have quantified ribosomal frameshifting efficiency in mammalian cell lines, a translational process that requires a secondary structure at the frameshift site of the mRNA. We could show that the protein binding to such secondary structures enhances frameshifting efficiency in proportion to the amount of bound protein. In this way, an assay for RNA–protein interaction, which plays a central role in gene regulation,[1] has been created. We have devised an assay to quantitatively determine such interactions *in vivo* with high reliability as outlined here.

Ribosomal frameshifting is an unusual translational recoding mechanism in which ribosomes shift in the reading frame in $-1$ or $+1$ direction during the elongation of translation. Frameshifting has been described in bacteria, yeast, and higher organisms and is extremely low in the absence of special frameshifting sequences that enhance efficiency (programmed

[1] J. E. G. McCarthy and H. Kollmus, *Trends Biochem. Sci.* **20,** 191 (1995).

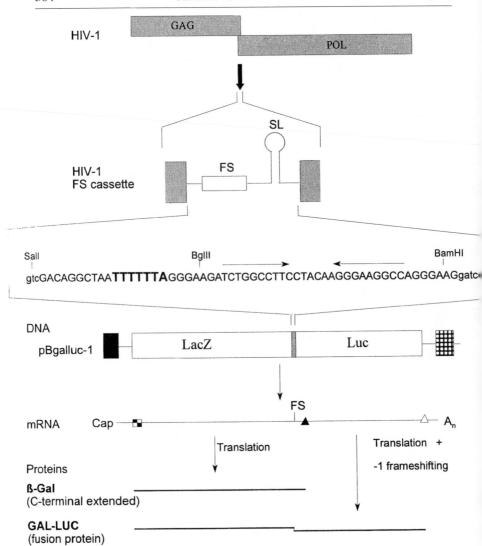

Fig. 1. GAL–LUC expression system for the quantification of ribosomal frameshifting in permanent cell lines exemplified by the HIV-1 *gag-pol* sequence. To quantify frameshifting efficiency in mammalian cells, an expression system was established in which the retroviral genes *gag* and *pol* are replaced by the genes encoding the β-galactosidase (lacZ) and firefly luciferase (luc), respectively. The intergenic region between the two reporter enzymes comprises the HIV-1 frameshifting cassette with the HIV-1 slippery sequence (TTTTTTA) and the 3′ adjacent stem–loop structure (indicated by inverted arrows) flanked by the restriction sites *Bgl*II and *Bam*HI. Transcription of the GAL–LUC fusion protein expression vector is driven by an SV40 promotor (black box) and an SV40 polyadenylation signal (hatched box).

frameshifting).[2,3] Programmed $-1$ frameshifting is an essential step in the replication cycle of retroviruses. It determines the amount of catalytic Pol proteins, which are needed in lower quantities than structural Gag proteins.[4,5] Two RNA sequence elements are involved in this process: A heptanucleotide slippery sequence, which induces a low level of frameshifting, and an RNA structure, a stem–loop, or a pseudoknot, which acts as an enhancer of the slippery sequence located upstream.[4,5] Because the established methods only allowed the determination of frameshifting efficiency *in vitro*, we established a sensitive *in vivo* frameshifting assay. Two reporter enzymes—the bacterial $\beta$-galactosidase (lacZ) and the firefly luciferase (luc)[6]—are fused either in frame or by a frameshift $-1$ (Fig. 1). Sequences to be tested for their ability to mediate frameshifting are inserted between the reading frames of the two enzymes. Translation of the RNA derived from this construct yields mainly a C-terminally extented $\beta$-galactosidase protein. The $\beta$-galactosidase–luciferase fusion protein (GAL–LUC) is expressed via frameshifting. Both $\beta$-galactosidase proteins show enzymatic activity, which is indistinguishable from the wild-type protein. The same is true for luciferase activity in the GAL–LUC fusion protein. Thus, frameshifting activity can be measured by the determination of luciferase activity relative to the control reporter construct where the luciferase is fused in frame to the reading frame of the $\beta$-galactosidase. Because our assay system is based on a bifunctional fusion protein, the activity of the upstream encoded enzymatic activity serves as an internal standard. This permits a highly accurate determination of frameshifting as compared to earlier assays that required a cotransfection of another reporter construct.[7,8] The $\beta$-galactosidase activity not only reflects transfection efficiency, but also

[2] R. F. Gesteland and J. F. Atkins, *Annu. Rev. Biochem.* **65,** 741 (1996).
[3] P. J. Farabaugh, *Microbiol. Rev.* **60,** 103 (1996).
[4] D. L. Hatfield, J. G. Levin, A. Rein, and S. Oroszlan, *Adv. Virus Res.* **41,** 193 (1992).
[5] T. Jacks, *Curr. Top. Microbiol. Immunol.* **157,** 93 (1990).
[6] H. Reil, H. Kollmus, U. H. Weidle, and H. Hauser, *J. Virol.* **67,** 5579 (1993).
[7] M. Cassan, V. Berteaux, P. O. Angrand, and J. P. Rousset, *Res. Virol.* **141,** 597 (1990).
[8] H. Reil and H. Hauser, *Biochim. Biophys. Acta* **1050,** 288 (1990).

The resulting mRNA harbors an in-frame stop codon relative to the start codon of the $\beta$-galactosidase (filled triangle). Translational termination leads to a C-terminally extended $\beta$-galactosidase. Translation with a ribosomal frameshift in $-1$ direction allows synthesis of the fusion protein GAL–LUC since the coding region of the luciferase is fused in $-1$ direction in relation to the start codon. The efficiency of ribosomal frameshifting is quantified by the measurement of luciferase expression in relation to luciferase expression of an expression plasmid in which the luciferase is fused in frame. The expression of this control plasmid is designated 100%.

provides an internal control for transcriptional and translational rates and for vector stability. This test system is sensitive enough to determine frame-shifting efficiency that is above the background of 0.05%.

The assay system was used to investigate the influence of the primary and secondary RNA structure on *in vivo* −1 frameshifting efficiency of the *gag-pol* HIV-1 and the *gag-pro* HTLV-2 regions.[6,9] It was modified to investigate frameshifting in +1 direction by changing the reading frame of luciferase in relation to that of β-galactosidase. It was employed to characterize a natural mutant of the thymidine kinase (TK) gene of the herpes simplex virus (HSV).[10] The assay system is very versatile and can be used *in vitro* with the bacteriophage T7 promoter or T3 promoter.[6,10] Alternatively, a composite promoter leading to identical transcripts in mammalian cells and *in vitro* can be used to confirm *in vitro* results by *in vivo* data.[11] In addition, the influence of human immunodeficiency virus type 1 (HIV-1) infection on frameshifting efficiency[12] can be determined to evaluate the role of transacting factors such as viral proteins or modified tRNAs.

Regulation of the *in vivo* frameshifting efficiency by a transacting factor, an RNA-binding protein, was demonstrated.[13] The stem–loop structure from HIV-1 (Fig. 1) was replaced by the iron-responsive element (IRE) from the ferritin mRNA (Fig. 2). This stem–loop structure binds iron regulatory proteins (IRPs), depending on the iron supply of the cell.[14–16] We demonstrated that frameshifting can be regulated reversibly. Under conditions where IRPs bind to the stem–loop structure, the frameshifting efficiency was enhanced severalfold (Fig. 3A). The influence of the RNA–protein interaction on frameshifting efficiency is thought to promote a ribosomal pause, which enhances the probability of the ribosome to shift in −1 direction[13] (Fig. 4). Alternatively, the protein might stabilize the RNA structure and thereby increase the frameshifting rate. Supporting evidence comes from data showing that stable RNA structures such as a pseudoknot can cause a translational pause.[17,18] In this way, the frameshift-ing assay offers an experimental tool to test the stability of secondary

---

[9] H. Kollmus, A. Honigman, A. Panet, and H. Hauser, *J. Virol.* **68**, 6087 (1994).

[10] B. C. Horsburgh, H. Kollmus, H. Hauser, and D. M. Coen, *Cell* **86**, 949 (1996).

[11] W. Dirks, F. Schaper, and H. Hauser, *Gene* **149**, 389 (1994).

[12] H. Reil, M. Hoexter, D. Moosmayer, G. Pauli, and H. Hauser, *Virology* **205**, 371 (1994).

[13] H. Kollmus, M. Hentze, and H. Hauser, *RNA* **2**, 316 (1996).

[14] R. D. Klausner, T. A. Rouault, and J. B. Harford, *Cell* **72**, 19 (1993).

[15] Ö. Melefors and M. W. Hentze, *Blood Rev.* **7**, 251 (1993).

[16] K. Pantopoulos, N. K. Gray, and M. W. Hentze, *RNA* **1**, 155 (1995).

[17] C. Tu, T.-H. Tzeng, and J. A. Bruenn, *Proc. Natl. Acad. Sci. U.S.A.* **89**, 8636 (1992).

[18] P. Somogyi, A. J. Jenner, I. Brierley, and S. C. Inglis, *Mol. Cell. Biol.* **13**, 6931 (1993).

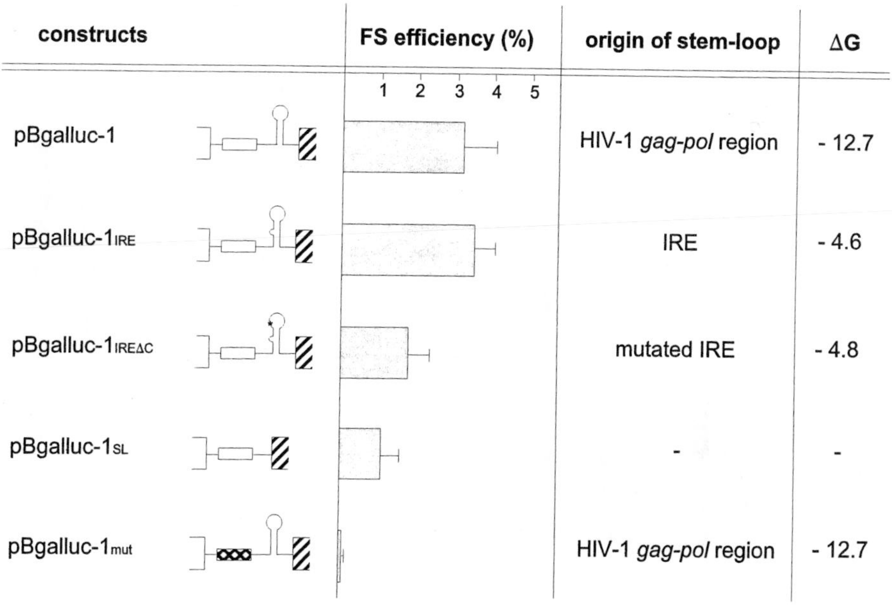

| constructs | FS efficiency (%) | | | | | origin of stem-loop | ΔG |
|---|---|---|---|---|---|---|---|
| | 1 | 2 | 3 | 4 | 5 | | |
| pBgalluc-1 | | | | | | HIV-1 *gag-pol* region | - 12.7 |
| pBgalluc-1IRE | | | | | | IRE | - 4.6 |
| pBgalluc-1IREΔC | | | | | | mutated IRE | - 4.8 |
| pBgalluc-1SL | | | | | | - | - |
| pBgalluc-1mut | | | | | | HIV-1 *gag-pol* region | - 12.7 |

FIG. 2. Secondary structures determine the efficiency of frameshifting. *In vivo* frameshifting efficiency was determined in transfectants with the indicated plasmids. Values are obtained from six independent transient transfection experiments in BHK-21 cells. Standard deviations are indicated by error bars. The relevant parts of the plasmids with the frameshift cassettes are symbolized on the left. The β-galactosidase open reading frame is depicted as an open box upstream, a hatched box downstream shows the luciferase open reading frame, an unfilled box shows the HIV-1 heptanucleotide, and a cross-hatched box indicates the mutated heptanucleotide. The free energies of the secondary structures (ΔG) were calculated with the computer program of M. Zuker and P. Stiegler, *Nucleic Acids Res.* **9**, 133 (1981). The extent of frameshifting was determined as described in the text.

structures as well as to investigate RNA–protein interactions under *in vivo* conditions.

The structure–function relationship of secondary structures mediates stop codon suppression of the selenoprotein biosynthesis.[19,20] These seleno-cysteine insertion sequences (SECIS[21]) of eukaryotes are located in the 3′ UTR and function via interactions with an RNA-binding protein(s) that has not been identified so far.[22,23]

[19] H. Kollmus, L. Flohé, and J. E. G. McCarthy, *Nucleic Acids Res.* **24**, 1195 (1996).
[20] H. Kollmus, J. E. G. McCarthy, and L. Flohé, *Z. Ernährungswiss.* **37**, 114 (1998).
[21] M. J. Berry, L. Banu, J. W. Harney, and P. R. Larsen, *EMBO J.* **12**, 3315 (1993).
[22] S. C. Low and M. Berry, *Trends Biochem. Sci.* **21**, 203 (1996).
[23] A. Böck, K. Forchhammer, J. Heider, and C. Baron, *Trends Biochem. Sci.* **16**, 463 (1991).

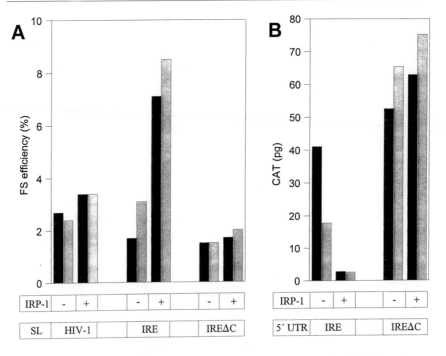

FIG. 3. Frameshifting efficiency is enhanced by RNA–protein interaction. (A) Dependence of the frameshifting efficiency by the RNA-binding protein IRP-1. BHK-21 cells were transfected transiently with 1.5 $\mu$g of the reporter constructs pBgalluc-1, pBgalluc-1$_{IRE}$, or pBgalluc-1$_{IRE\Delta C}$ and with or without the same amount of the IRP-1 expression plasmid pSG5-IRF. The medium was replaced 24 hr before harvesting the cells and supplemented with 100 $\mu M$ hemin (+ Fe, ■) or 100 $\mu M$ desferrioxamine (− Fe, ▨). (B) Control of translation initiation by iron ions in transiently transfected cells in dependence of IRP-1. BHK-21 cells were transfected with the CAT plasmids IRE.CAT and IRE.$\Delta$CAT. The cells were treated as described in (A).

This article describes a procedure to analyze frameshifting efficiency *in vivo* and gives a detailed description of our protocol that depends on an RNA–protein interaction (IRE/IRPs).

## Procedures

### Principle of Assay System GALLUC

The principle of the *in vivo* frameshifting assay is outlined in Fig. 1. It shows the reporter construct pBgalluc-1, which was used to investigate the

**High iron**                                    **Low iron**

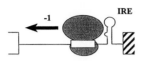
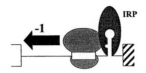

FIG. 4. Model for the enhancement of ribosomal frameshifting by protein binding to a downstream secondary structure. In iron-deficient cells, IRPs are converted into their apoprotein forms. This enables IRPs to bind to the IRE stem–loop, which is located 3′ adjacent to the slippery sequence of the frameshift cassette. Because ribosomal progression is impeded by the RNA/protein complex, they stall at the slippery sequence. The probability of ribosomes to change the reading frame is enhanced (right site). Consequently, frameshifting efficiency is higher when compared to the situation where protein binding to the IRE is abolished by iron loading of the cells (left site). The symbols correspond to those of Fig. 2.

HIV-1 frameshift cassette.[6] It encodes the bacterial $\beta$-galactosidase to which the firefly luciferase is fused in $-1$ frame. Translation of the mRNA derived from the expression construct yields mainly $\beta$-galactosidase protein. The fusion protein GAL–LUC with both enzymatic activities is expressed by frameshifting. To calculate frameshifting efficiency, a control plasmid is needed in which the luciferase is fused in frame in relation to the coding region of the $\beta$-galactosidase.

The luciferase expression of this plasmid is set to 100%, whereby the corrected luciferase expression of the frameshifting constructs gives the mediated frameshifting efficiency. The synthesized fusion protein can be detected by Western blotting to confirm measured data[6,9] (see later for details). To investigate the influence of an RNA–protein interaction on frameshifting, the HIV-1 stem–loop structure is replaced by the RNA sequence to be analyzed. Depending on the rate and strength of protein binding to the inserted element, an enhancement of the frameshifting rate is expected.

By comparing the frameshift efficiency of the heptanucleotide alone[6] (pBgalluc-1$_{SL}$) with that of the heptanucleotide in combination with the secondary structure, the protein binding efficiency can be determined. In this way, the ability of the heterologous stem–loop structure to enhance frameshifting is controlled (Fig. 2). Not every secondary structure can substitute the respective wild-type structure as shown for the HTLV-2 heptanucleotide.[9,13] Another useful tool is a known mutant of the stem–loop structure to be tested to measure frameshift efficiency without protein binding. In our experiments we used the IRE from the 5′ UTR of human ferritin H-chain mRNA,[24] which forms a moderately stable stem–loop structure with a bulge in the stem. A single nucleotide deletion in the loop of the IRE (IRE$\Delta$C) serves as a negative control. It displays similar thermal stability, but IRPs cannot bind.[25–27] A construct with a mutated slippery sequence is used to determine the background expression level of the system (pBgalluc-1$_{mut}$, Fig. 2).

*Plasmids*

The luciferase gene is fused in $-1$ direction in relation to the AUG of the $\beta$-galactosidase. Appropriate control plasmids carry luciferase gene in frame. The plasmid pBgalluc-1[6] or its derivative pBGLB is the starting point for cloning the test sequences.[19] In the latter a *Bam*HI site in the N-terminal region of the $\beta$-galactosidase is mutated to replace the intergenic region between the two reporter genes. Frameshift cassettes tested for their ability to mediate ribosomal frameshifting comprise a shifty heptanucleotide and an adjacent stem–loop structure. The cassettes are inserted between the restriction sites *Sal*I and *Bam*HI[6] (Fig. 1). The replacement of the HIV-derived sequence by any other stem–loop structure through the *Bgl*II and the *Bam*HI site is possible (Fig. 1). Plasmids containing the cloned DNA sequences are grown and purified by common techniques.[28] We recommend using double cesium chloride gradients or DNA purification columns (Qiagen, Hilden, Germany), as the purity of the DNA is of great importance for transfection efficiency.

*Cell Culture and Gene Transfer*

A difference in efficiency from the same frameshift cassette could not be detected in cell lines from different species and tissues.[6] Thus, it is convenient to use cell lines that efficiently express foreign genes, such as baby hamster kidney (BHK)-21 cells (ATCC, Rockville, MD, CCL10) or 293, adenovirus type 5-transformed human embryonal kidney cells (ATCC, CRL 1573) as recipients. This ensures a level of luciferase expression significantly above background levels, even in the case of low frameshifting

---

[24] M. W. Hentze, S. W. Caughman, T. A. Rouault, J. G. Barriocanal, A. Dancis, J. B. Harford, and R. D. Klausner, *Science* **238**, 1570 (1987).

[25] E. A. Leibold and H. N. Munro, *Proc. Natl. Acad. Sci. U.S.A.* **85**, 2171 (1988).

[26] T. A. Rouault, M. W. Hentze, S. W. Caughman, J. B. Harford, and R. D. Klausner, *Science* **241**, 1207 (1988).

[27] B. Goossen, S. W. Caughman, J. B. Harford, R. D. Klausner, and M. W. Hentze, *EMBO J.* **9**, 4127 (1990).

[28] J. Sambrook, E. F. Frisch, and T. Maniatis, "Molecular Cloning: A Laboratory Manual." Cold Spring Harbor Laboratory, Cold Spring Harbor, NY, 1989.

efficiencies.[6,9] The cell lines are cultivated in Dulbecco's modified Eagle's medium (DMEM) supplemented with 10% fetal calf serum. Transfections are carried out using calcium phosphate techniques[29] or other DNA transfer methods (e.g., transfection kit Superfect, Fa. Qiagen, Hilden, Germany). A transient gene transfer on single wells of a six-well plate is sufficient for performing the enzymatic frameshifting assay. For Western blotting analysis of the reporter proteins, pools of stable transfectants have to be used in order to detect the underexpressed GAL–LUC fusion protein.

*Transient Transfections.* Seed $1 \times 10^5$ cells per well in duplicate the day before transfection. For calcium phosphate precipitation, 1.5 $\mu$g reporter plasmid and 1.0 $\mu$g carrier DNA (preferably isolated from the recipient cell line) are cotransfected using standard protocols.[29] For other than the previous mentioned cell lines or other gene transfer methods, the optimal DNA amount must be determined prior to the assay. Transient expression is measured in cell extracts prepared from cells harvested 48 hr after transfection. A relevant RNA-binding protein can be cotransferred using the same amount of DNA as the reporter plasmid. For determination of IRP-induced frameshifting, we contransfected the expression plasmid pSG5hIRF, which encodes IRP-1.[13]

*Stable Transfections.* Stable transfectants are obtained by seeding 3–$4 \times 10^5$ cells per culture flask (25 cm$^2$) and transferring 5 $\mu$g expression plasmid, 0.5 $\mu$g puromycin resistance gene (e.g., pSpac$\Delta$p$^6$), or 0.5 $\mu$g G418 resistance plasmid (e.g., pAG60$^{30}$) and 5 $\mu$g of carrier DNA. The selection of stable transfectants is initiated 48 hr after transfection by medium, which is supplemented with 5 $\mu$g/ml of puromycin or 700 $\mu$g/ml of G418 for 293 cells and 5 $\mu$g/ml of puromycin or 1000 $\mu$g/ml of G418 for BHK-21 cells. After 1–2 weeks of selection, the clones are pooled and assayed for luciferase and $\beta$-galactosidase activity or protein. Cells without antibiotic resistance plasmids serve as a negative control.

*Conditions for Determination of Frameshifting Efficiency Depending on Iron Amount in Cell*

The two cytoplasmic proteins IRP-1 and IRP-2 bind to IRE in iron-deficient cells, whereas IRE binding of both IRPs is switched off in iron-repleted cells.[15,16] To manipulate the iron content in the culture medium, hemin chloride (Sigma) is added as an iron source and desferrioxamine mesylate (Sigma) as an iron chelator, both in concentrations of 100 $\mu M$.

[29] M. Wigler, S. Silverstein, L.-S. Lee, A. Pellicer, Y.-C. Cheng, and R. Axel, *Cell* **11**, 223 (1977).

[30] F. Colbère-Garapin, F. Horodniceanu, P. Khourilsky, and A. C. Garapin, *J. Biol. Chem.* **150**, 1 (1981).

In transient assays these additions are given 24 hr after transfection. The cells are incubated another 24 hr.[13] To monitor the effect of these compounds to the cells, control CAT expression plasmids (IRE.CAT and IREΔ.CAT,[13,24]) are transfected in parallel. A CAT mRNA harboring the IRE or its mutant structure (IREΔC) in the 5' UTR is encoded by these plasmids. IRP binding under iron depletion leads to strong inhibition of the CAT translation from the mRNA with the wild-type IRE. In contrast, the mutant IRE containing mRNA is not affected significantly (Fig. 3B[24,31]).

*Preparation of Cell Extracts*

The following method for the preparation of cell extracts not only measures the reporter enzymes β-galactosidase, luciferase, and chloramphenicol acetyltransferase (CAT) in the same buffer, but also uses these samples for protein measurements and Western blotting. The cells are detached from the culture plates by a short incubation with TEN (40 m$M$ Tris–HCl, pH 7.5; 1 m$M$ EDTA, 150 m$M$ NaCl) at 37° after rinsing the cells twice with phosphate-buffered saline (PBS: 137 m$M$ NaCl, 2.7 m$M$ KCl, 8 m$M$ $Na_2HPO_4$, 1.47 m$M$ $KH_2PO_4$; pH 7.0). The cells are pelleted at 1000 rpm in an Eppendorf centrifuge, and the supernatants are removed. At this stage, the cell pellets can be frozen if they are not used immediately. For the preparation of extracts, the cell pellets are resuspended in 250 μl 250 m$M$ Tris–HCl (pH 7.5), subjected to three cycles of freeze-thawing, and centrifuged for 10 min at 4° at 14,000 rpm in a bench-top centrifuge. The supernatants are transferred into reaction tubes.

*Measuring Reporter Genes*

*Luciferase.* For the measurement of luciferase activity, 20–60 μl of the cell extract is mixed with 350 μl of reaction buffer (25 m$M$ glycylglycerine, pH 7.8; 5 m$M$ ATP, 15 m$M$ $MgSO_4$). The light emission (light units/10 sec) of the reaction is measured after injection of 100 μl luciferin (0.2 m$M$; Sigma) in a Berthold Biolumat (Berthold Biolumat, Bad Wildbad, Germany).

*β-Galactosidase.* To assay β-galactosidase activity, 20–60 μl of cell extract is incubated with 1 μl of reaction buffer (60 m$M$ $Na_2PO_4$, 40 m$M$ $NaH_2PO_4$, 10 m$M$ KCl, 1 m$M$ $MgCl_2$, 50 m$M$ 2-mercaptoethanol) and 200 μl of substrate (2 mg/ml *o*-nitrophenyl-β-D-galactopyranoside in 60 m$M$ $Na_2PO_4$, 40 m$M$ $NaH_2PO_4$) at 37° until a visible yellow color change is observed. The reaction is stopped by the addition of 500 μl of 1 $M$ $NaCO_3$. The colorimetric change is measured in a spectrophotometer at 405 nm.

---

[31] N. K. Gray and M. W. Hentze, *EMBO J.* **13**, 3882 (1994).

*CAT.* The protein amount of CAT enzyme in the cell extracts is determined by the ELISA kit from Boehringer/Mannheim.
Every enzyme assay should be performed in duplicate.

## Western Blotting

An analysis of the cellular products from transfected cells by Western blotting can be performed, although we demonstrated earlier that the activity of the β-galactosidase enzyme is insensitive to fusions of proteins to its C terminus.[6,9] After determination of the supernatant protein concentration by the Bradford method,[32] aliquots equivalent to 150 μg of total protein are separated by tricine–sodium dodecyl sulfate (SDS) gel electrophoresis.[33] The gel has to be equilibrated in a transfer buffer[34] (25 mM Tris, 192 mM glycine, 15% methanol) before the proteins are transferred onto filters in a wet blotting chamber (Immobilon, Millipore, Bedford, MA). For the immunological detection of the fusion protein GAL–LUC and the C-terminally extended β-galactosidase, the membranes are incubated with a luciferase antiserum (Promega, Madison, WI) or monoclonal antibodies against β-galactosidase (Promega), respectively, according to standard protocols.[35] The immobilized antibodies are detected by a second alkaline phosphatase-conjugated antibody (Dianova, Hamburg, Germany), which is identified by an alkaline phosphatase assay system (Bio-Rad, München, Germany).

## Calculation of Frameshifting Efficiency

The luciferase activity of cells transfected with the different reporter constructs has to be normalized to its respective β-galactosidase activity. This is to compensate for the variable transfection efficiency which, for instance, is caused by different DNA quality or by the extent of expression in stable transfectants. The luciferase activity of cells transfected with the construct pBgalluc0 containing the luciferase in frame with respect to the β-galactosidase reading frame serves as a reference value (100%). Frameshifting efficiency is calculated by relating the corrected luciferase activity to that of the in-frame fusion, as the β-galactosidase activity from different fusion protein constructs is indistinguishable. We recommend using results from at least six independent transfection experiments for calculating the mean value with standard deviations.

[32] M. M. Bradford, *Anal. Biochem.* **72,** 248 (1976).
[33] H. Schägger and G. von Jagow, *Anal. Biochem.* **166,** 368 (1987).
[34] H. Towbin, T. Staehelin, and J. Gordon, *Proc. Natl. Acad. Sci. U.S.A.* **76,** 4350 (1979).
[35] M. S. Blake, *Anal. Biochem.* **136,** 175 (1979).

Conclusion

The test system provides the prerequisite not only for measuring the effect of overexpressed RNA-binding proteins on frameshifting efficiency, but also for screening endogenous RNA-binding proteins. Furthermore, mutants of secondary structures can be designed and tested for their ability to bind proteins *in vivo* and *in vitro*. The principle of this *in vivo* frameshifting assay is also applicable to other fusion proteins, e.g., a fusion of *Renilla* and firefly luciferase as published previously.[36]

Acknowledgment

The method presented here is based on earlier work to which H. Reil, A. Honigman, and M. Hentze have contributed.

[36] G. Grentzmann, J. A. Ingram, P. J. Kelly, R. F. Gesteland, and J. A. Atkins, *RNA* **4,** 479 (1998).

# [25] Translational Repression Assay Procedure: A Method to Study RNA–Protein Interactions in Yeast

By Efrosyni Paraskeva and Matthias W. Hentze

Introduction

The biochemical analysis of RNA–protein interactions can, in some cases, be difficult or even impossible due to the low abundance of a given cell type or tissue of interest. To circumvent such limitations, alternative strategies for the identification, cloning, and study of RNA–protein interactions have been devised based on prokaryotic and yeast genetics (see also other articles, this volume).

In eukaryotes, proteins that bind to specific sites near the cap structure of an mRNA inhibit the stable association of the small ribosomal subunit and repress its translation. This kind of translational repression mechanism was first described and studied in detail for the ferritin mRNA (reviewed in Hentze and Kühn[1]), which contains an iron responsive element (IRE) close to the cap structure of the mRNA. The iron regulatory protein-1 (IRP-1) binds to the IRE and inhibits the ferritin mRNA translation. Interestingly, translational repression is also achieved by RNA-binding proteins

[1] M. W. Hentze and L. C. Kühn, *Proc. Natl. Acad. Sci. U.S.A.* **93,** 8175 (1996).

with functions unrelated to eukaryotic translation when their cognate-binding site is introduced into a cap proximal position of a reporter mRNA. This kind of translational inhibition was shown to operate *in vitro* as well as *in vivo*, both in mammalian cells and in yeast.[2]

Based on this principle, we have developed a translational repression assay procedure (TRAP) as a method to study RNA–protein interactions in the cytoplasm of the yeast *Saccharomyces cerevisiae*.[3,4] TRAP offers a new strategy to clone RNA-binding proteins for which little else than the binding region is known. It can also be used to delineate the RNA sequence or protein domains required for binding and to study effectors, including pharmacological agents, of RNA–protein interactions *in vivo*.

## Principle and Components of System

TRAP is based on the diminished translation of a reporter mRNA into an indicator protein. This reduction occurs when a protein binds to its cognate binding sequence that has been introduced into the 5′ UTR of the reporter mRNA (see Fig. 1).

Specifically, yeast cells are transformed with two plasmids: (a) a plasmid for the expression of the RNA-binding protein (or a cDNA expression library) from a galactose inducible promoter and (b) a plasmid for the expression of the reporter protein from a constitutively active promoter.

The binding site for the RNA-binding protein of interest is cloned into the 5′ UTR of the mRNA encoding the reporter protein. Expression of the reporter in glucose-containing medium (i.e., when the promoter driving the expression of the RNA-binding protein is turned off) confers an easily detectable phenotype to the cells. After induction of the expression of the RNA-binding protein (following replacement of glucose by galactose in the growth medium), the translation of the reporter mRNA is repressed (due to formation of the RNA–protein complex at the 5′ UTR) and the cells lose the reporter phenotype.

TRAP uses the green fluorescent protein (GFP), particularly the S65T mutant of GFP, as a reporter.[5,6] The levels of GFP expression in the living

[2] R. Stripecke, C. C. Oliveira, J. E. G. McCarthy, and M. W. Hentze, *Mol. Cell. Biol.* **14,** 5898 (1994).

[3] E. Paraskeva, A. Atzberger, and M. W. Hentze, *Proc. Natl. Acad. Sci. U.S.A.* **95,** 951 (1998).

[4] M. W. Hentze and E. Paraskeva, United Kingdom Patent Application No. 9801631.4, "Method of Isolating RNA-Binding Compounds" (1998).

[5] M. Chalfie, Y. Tu, G. Euskirchen, W. W. Ward, and D. C. Prasher, *Science* **263,** 802 (1994).

[6] R. Heim, A. B. Cubitt, and R. Y. Tsien, *Nature* **373,** 663 (1995).

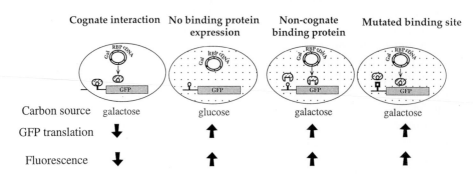

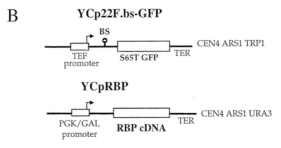

FIG. 1. Schematic representation of TRAP. (A) The principle of TRAP. Yeast cells are transformed with the GFP indicator plasmid and the RNA-binding protein (RBP) expression plasmid. TRAP: In galactose medium, expression of the RBP is induced. The RNA–protein interaction at the 5′ end of the GFP mRNA represses its translation and the fluorescence of the cell is reduced (cognate interaction). Controls: GFP mRNA translation yields high fluorescence levels when expression of the RBP is repressed in glucose medium (no RNA-binding protein expression) or when the expressed protein cannot interact with the GFP mRNA (noncognate-binding protein or mutated binding site). (B) Description of the plasmids used. The GFP reporter plasmid YCp22F.bs-GFP contains the GFP S65T open reading frame and a TRP1 selection marker. The binding sites (BS) are cloned into the *Afl*II site. Plasmid YCpRBP is used for the galactose-inducible expression of the RNA-binding protein and contains a URA3 selection marker. From E. Paraskeva, A. Atzberger, and M. W. Hentze, *Proc. Natl. Acad. Sci. U.S.A.* **95,** 951 (1998).

yeast cells can be followed by flow cytometry. This allows the detection of cells in which the translational repression of GFP, caused by the expression of the specific RNA-binding protein, has led to decreased levels of fluorescence. In addition, these cells can be isolated by fluorescence-activated cell sorting (FACS). A schematic representation of TRAP is shown in Fig. 1A. We have utilized the haploid *S. cerevisiae* strain RS453 (*ade 2-1, trp 1-1, leu 2-3, his 3-11, ura 3-52, lys, can 1-100*). The two plasmids used are centromeric and are stably maintained at one to two copies per cell. Plasmid YCp22F.bs-GFP contains the GFP S65T mutant open reading frame and harbors the TRP1 selection marker.[3] GFP expression is driven by the translation elongation factor-1 (TEF-1) promoter.[7] Plasmids for the expression of the RNA-binding proteins (YCpRBP) harbor the strong, galactose-inducible PGK/GAL fusion promoter and the URA3 selection marker.[8]

Using TRAP to Study RNA–Protein Interactions

TRAP is suitable for the study of RNA–protein interactions with affinities spanning a useful physiological range. It has been used successfully when the binding affinities were in the micromolar to the nanomolar range, namely for the binding of iron regulatory protein (IRP)-1 to IREs (reported affinity 1.0–10 $nM^{9–11}$), the binding of the spliceosomal protein U1A to loop 2 of U1 snRNA (reported affinity 0.02–80 $nM^{12–14}$), and the binding of the bacteriophage MS2 coat protein to the MS2 replicase mRNA (with reported affinities 0.02–0.1 $nM$ and 0.1–1 $\mu M$, respectively, for the strongest and weakest binding sites we tested[15,16]).

The following steps are used to study an RNA–protein interaction pair by TRAP.

1. The RNA sequence that binds the protein of interest is cloned into the reporter gene plasmid YCp22F.bs-GFP. A convenient site is *Afl*II, located 9 nucleotides downstream from the transcription start site and 32 nucleotides upstream from the GFP translation initiation codon. A similar plasmid that harbors the TRP selection marker but does not express GFP is used as control for the background levels

[7] C. Guthrie and G. R. Fink, "Guide to Yeast Genetics and Molecular Biology" (1991).
[8] C. C. Oliveira, J. J. van den Heuvel, and J. E. G. McCarthy, *Mol. Microbiol.* **9,** 521 (1993).
[9] E. A. Leibold, A. Laudano, and Y. Yu, *Nucleic Acids Res.* **18,** 1819 (1990).
[10] A. J. E. Bettany, R. S. Eisenstein, and H. N. Munro, *J. Biol. Chem.* **267,** 16531 (1992).
[11] B. R. Henderson, E. Menotti, C. Bonnard, and L. C. Kühn, *J. Biol. Chem.* **269,** 17481 (1994).
[12] C. Lutz-Freyermuth, C. C. Query, and J. D. Keene, *Proc. Natl. Acad. Sci. U.S.A.* **87,** 6393 (1990).
[13] T. H. Jessen, C. Oubridge, C. Hiang Teo, C. Pritchard, and K. Nagai, *EMBO J.* **10,** 3447 (1991).
[14] W. T. Stump and B. K. Hall, *RNA* **1,** 55 (1995).
[15] P. T. Lowary and O. C. Uhlenbeck, *Nucleic Acids Res.* **15,** 10483 (1987).
[16] G. W. Witherell, J. M. Gott, and O. C. Uhlenbeck, *Prog. Nucleic Acids Res. Mol. Biol.* **40,** 185 (1991).

of cell fluorescence. The cDNA encoding the RNA-binding protein (RBP) is subcloned into the galactose-inducible YCpRBP expression plasmid. A schematic representation of plasmids YCp22F.bs-GFP and YCpRBP is shown in Fig 1B.

Concerning the suitability of a particular RNA-binding sequence for TRAP, the following considerations apply: (a) The GFP mRNA is transcribed by the TEF promoter. An advantage of using an RNA polymerase II promoter is that essentially any RNA sequence for the binding site can be used. This is important because a number of examples of RNA-binding proteins, especially proteins playing a role in development or cell differentiation, bind to sequences that contain long runs of uridines. Such sequence motifs can lead to premature transcription termination by an RNA polymerase III promoter. (b) The specific principle of TRAP requires that the introduction of the RNA binding region into the 5' UTR does not interfere with the GFP expression per se. Therefore, binding regions that harbor inhibitory open reading frames may not represent suitable target sequences. (c) However, the presence of an AUG within the binding sequence does not necessarily represent a problem. We have cloned such a sequence into the 5' UTR of the GFP indicator mRNA in a way that the AUG was introduced in frame with the GFP open reading frame. Translation initiated at the AUG of the binding sequence led to the expression of a fully functional, N-terminally extended GFP (unpublished observations). (d) Highly structured binding sites might interfere with the translation of the GFP mRNA, even in the absence of a specific binding protein. To avoid such problems, the minimal binding region should ideally be defined to reduce the probability of introducing inhibitory features into the 5' UTR.

2. Yeast cells are cotransformed[17] with the YCpRBP and the YCp22F.bs-GFP or the control plasmid, and double transformants are selected on plates lacking tryptophan and uracil at 30°.

3. Single colonies are expanded in selective liquid medium (2-ml cultures) containing either 2% glucose or 2% galactose as the carbon source. The cultures are grown at 30° for 14 hr to a density of $OD_{600}$ 0.5–1 and the GFP fluorescence levels of the cells are then assessed by flow cytometry.

We have used a Facscan (Becton-Dickinson) flow cytometer, which uses an argon ion laser fixed at 488 nm for excitation and a 530-nm bandpass filter for collection of the emitted fluorescence. Debris is excluded from

---

[17] H. Ito, Y. Fukuda, K. Murata, and A. Kimura, *J. Bacteriol.* **153**, 163 (1983).

the analysis by gating on the forward scatter versus side scatter parameters. The relative fluorescence intensity of single cells can be monitored this way.

Cells that have been grown in glucose and express GFP from the YCp22F.bs-GFP plasmid display high fluorescence, whereas cells that carry the control plasmid display low fluorescence. Induction of the cognate RBP by galactose decreases the mean fluorescence measurably to an intermediate value. This reduction of the mean fluorescence value is manifested by a leftward shift of the fluorescence distribution curve toward lower values.

Reduction of fluorescence levels requires the presence of the wild-type binding site and the cognate-binding protein. It is not seen with a mutated binding site or a noncognate-binding protein. Figure 2 shows results obtained when the the IRE-IRP-1 interaction was studied by TRAP.

For a given RNA-binding protein the degree of repression (monitored as a drop in fluorescence intensity) correlates with its binding affinity.[3] For example, mutations of the MS2 coat protein RNA-binding site that were shown previously to decrease the affinity of the RNA–protein interaction resulted in a smaller reduction of the mean fluorescence value. However, different RNA–protein interactions cannot be compared quantitatively because other aspects, such as the stability of the protein or its mRNA, also affect the result (see later).

## Cloning by TRAP

The initial steps for cloning an RBP are similar to those performed for the study of a known RNA–protein interaction. Instead of the cloned cDNA for the RNA-binding protein, a cDNA library from RNA isolated from a cell type or tissue in which the RNA-binding protein is expressed is subcloned to plasmid YCpRBP to generate an expression library. Then the procedure is as follows:

1. RS453 cells are transformed with plasmid YCp22F.bs-GFP, which contains the RNA-binding site, and transformants are selected on plates lacking tryptophan. Positive colonies are expanded in liquid medium cultures. The fluorescence of cells carrying the YCp22F.bs-GFP plasmid is checked by FACS. This is necessary to ensure that (a) the binding site does not interfere with GFP translation and (b) no yeast protein is interacting nonspecifically with the binding site.

2. These cells are transformed with the cDNA library. Double transformants are selected on plates lacking tryptophan and uracil and pooled in liquid selective medium containing glucose. In our hands, approximately 60 $\mu$g plasmid DNA yields 3–7 × 10^6 transformants.

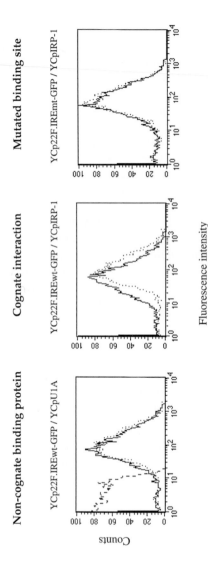

Fig. 2. Analysis of fluorescence of the IRE/IRP-1 interaction with TRAP. Fluorescence of GFP-expressing cells is specifically reduced by cognate RNA–protein interactions. Cells cotransformed with a GFP indicator plasmid carrying an IRE and either plasmid YCpIRP-1 or YCpU1A were grown in glucose- or galactose-containing medium. The fluorescence of cells grown in glucose (dotted lines) or galactose (solid lines) media was analyzed by flow cytometry. Control cells not carrying a GFP plasmid are represented by the hatched line. Fluorescence drops when expression of the cognate-binding protein is induced by the addition of galactose. Expression of a noncognate-binding protein does not affect fluorescence levels. No change of fluorescence is observed with a mutated binding site present within the 5'UTR of GFP mRNA. Fluorescence intensity is plotted against counts (number of cells analyzed per channel of fluorescence). From E. Paraskeva, A. Atzberger, and M. W. Hentze, *Proc. Natl. Acad. Sci. U.S.A.* **95**, 951 (1998).

3. The cultures are expanded in selective medium containing galactose and grown for 16–20 hr at 30° to induce expression of the binding proteins. Cells are sorted by FACS according to their GFP fluorescence levels. Populations of cells that display low fluorescence levels are recovered in glucose-containing medium, expanded in galactose medium, and sorted for a second time.

We have analyzed and isolated the cells on a Facs Vantage (Becton-Dickinson) cell sorter. Excitation was at 488 nm and the emitted fluorescence was collected with a 530-nm bandpass filter. Sorting gates were set using forward scatter width versus side scatter height signals, to exclude debris and clumps, and forward scatter height versus fluorescence height signal, to sort cells that fell within a certain fluorescence intensity.

4. The enrichment for cells expressing the cognate RNA-binding protein is usually tested after two rounds of sorting, but the actual number of rounds needed may well vary with different interaction pairs. One parameter that can affect the efficiency of sorting is the abundance of the RNA-binding protein cDNA in the expression library. Enrichment in RNA-binding activity in extracts prepared from the starting cultures and the cultures of the sorted cells can be tested, for example, by gel retardation or UV cross-linking assays. If sorting is successful, extracts from galactose-induced cultures of cells selected for low fluorescence levels will show specific RNA-binding activity compared to the those from the initial cultures (see also Fig. 3 and the comments section).

5. Cultures enriched in cells expressing the cognate RNA-binding protein are subsequently regrown in galactose medium and sorted further into regions of cells with different fluorescence. The sorted cells are recovered, and enrichment in cells expressing the protein of interest is tested again, as described earlier. We have consistently found that cells expressing the cognate RNA-binding protein are found among cells with low but not minimal fluorescence (see Fig. 3).

6. When satisfactory enrichment is achieved, cells are plated in -ura, -trp selective plates, and single colonies are picked and grown in liquid culture. Analysis of extracts from single cell cultures in galactose-containing medium (e.g., by bandshift or UV cross-linking) leads to the identification of single cell clones that express a repressory RNA-binding protein.

7. The RNA-binding protein expressing plasmid YCpRBP is recovered from positive clones and further specificity controls are performed. These include checking the interaction of the protein with the cognate, mutated, reverse, or unrelated RNA-binding sequences. A se-

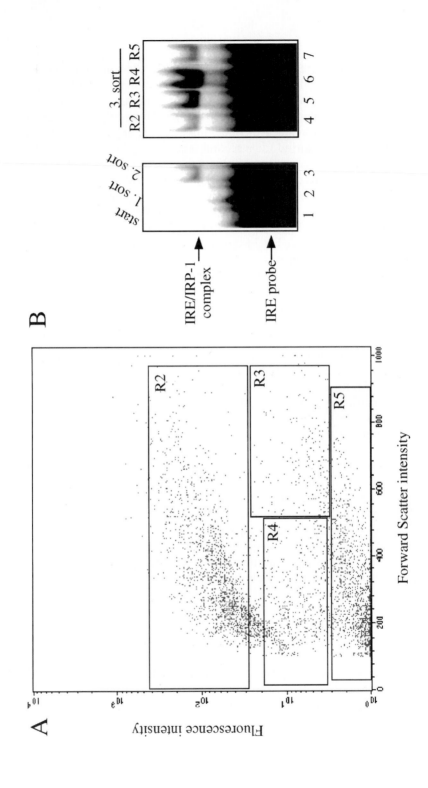

quence-specific RNA-binding protein should only interact with the cognate-binding site. Comparison with functionally characterized mutant binding sites is highly recommended at this point.

TRAP has been used successfully for recloning IRP-1 cDNA. Instead of an expression library, we mixed plasmids YCpIRP-1 and YCpU1A that express proteins IRP-1 and U1A, respectively, in a 1:500,000 ratio and used this DNA to transform RS453 cells carrying YCp22F.IREwt-GFP. Cells that displayed low fluorescence levels were sorted three times in total. Figure 3 shows the dot–plot profile and sorting of cells that have been grown in galactose medium according to their fluorescence levels. The greatest enrichment in IRE-binding activity was achieved in cells in region R4. When single colonies from the sorted cells were analyzed, 1 out of 20 expressed IRP-1. Thus, TRAP allowed successful recloning of the cDNA encoding a cognate RNA-binding protein from a 500,000-fold background.[3]

Comments

The specific features of TRAP make it useful for studying RNA-binding proteins. In particular, it combines the use of a eukaryotic host cell with the advantages implicit in assaying cytoplasmic (rather than nuclear) expression and the possibilities afforded by successive rounds of enrichment. TRAP allows the study of RNA-binding proteins in their native form without having to fuse additional protein domains that might interfere with their RNA-binding affinity or specificity. Although TRAP detects RNA-binding proteins in the cytoplasm, it also allows the study of nuclear proteins. Overexpression allows the accumulation of sufficient amounts of newly synthesized protein in the cytosol, which can affect the translation of the GFP indicator mRNA. TRAP only requires two components: the RNA-binding protein expression plasmid (or a suitable cDNA expression library) and the indicator mRNA. This simplicity minimizes the potential for inad-

FIG. 3. Recovery of IRP-1 expressing cells from a transformation with mixed plasmid DNA. RS453 cells carrying plasmid YCpIREwt-GFP were transformed with a DNA mixture containing plasmids YCpIRP-1 and YCpU1A in a ratio of 1:500,000. Transformants were pooled and grown in galactose medium. Cells with low fluorescence were sorted three times sequentially. (A) The third sorting of cells into populations R2–R5. (B) Total extracts were analyzed in a gel retardation assay with 1 ng IRE probe (lanes 1–7). The extract of cells sorted twice shows IRE-binding activity (lane 3). IRE-binding activity is below the detection limit in the extract from pooled transformants (lane 1) or cells sorted once (lane 2). The highest IRE-binding activity of extracts from cultures R2–R5 is present in extract R4 (lane 6). From E. Paraskeva, A. Atzberger, and M. W. Hentze, *Proc. Natl. Acad. Sci. U.S.A.* **95,** 951 (1998).

vertent nonspecific effects on the assay, a consideration that may be particularly relevant when studying pharmacological effectors of RNA–protein interactions.

Furthermore, the utilization of FACS permits the rapid processing of large numbers of independent clones and the recovery of living cells after sorting. This way, multiple rounds of sorting can be performed while monitoring the enrichment of the population in cells expressing the protein of interest. The highest level of specific enrichment was found, in all cases tested so far, in fractions of cells displaying reduced but not minimal fluorescence. The minimally fluorescent population is possibly contaminated by cells that may harbor mutations in the GFP cDNA or may have lost the GFP-expressing plasmid (even during culture in selective medium). However, the design of TRAP allows counterselection against nonspecific, constitutive loss of GFP fluorescence because the cells that display reduced fluorescence due to the RNA-binding protein interaction shift to and can be collected from the high fluorescent pool following incubation in glucose-containing media.

The minimal affinity required for an RNA–protein interaction to be studied by TRAP is currently unknown, but will certainly depend on case-specific parameters such as the stability of the corresponding RNA-binding protein and the maximal level of its overexpression in yeast. In addition, because all of the proteins tested so far bind to hairpin structures, we cannot predict how proteins binding to unstructured RNA motifs will perform. Finally, like other strategies for the detection of RNA–protein interactions in heterologous hosts, TRAP requires that the protein under investigation can bind to RNA as a monomer or homopolymer.

We have so far only tried using TRAP in yeast, but the same principle can be applied to other eukaryotes, including mammalian cells. This way the utility of TRAP can be expanded for the study of RNA-binding proteins that cannot be produced in a functional form in yeast, i.e., in cases where the interaction is influenced by posttranslational modifications, such as regulated phosphorylation.

# [26] Analysis of Low-Abundance Ribonucleoprotein Particles from Yeast by Affinity Chromatography and Mass Spectrometry Microsequencing

*By* Scott W. Stevens

## Introduction

In eukaryotic cells, numerous ribonucleoprotein particles (RNPs) carry out a wide variety of functions, including mRNA splicing,[1] protein synthesis,[2] protein secretion,[3] and tRNA,[4] rRNA,[5] and snRNA[6] processing. These particles are composed of at least one molecule of RNA and a variable number of proteins. Their cellular location varies from the cytoplasm as in the case of the ribosome, the nucleus as in the case of mRNA splicing machinery, or the nucleolus as in the case of tRNA and rRNA processing. There is a large (>1000-fold) difference in the copy number between yeast and mammalian mRNA splicing RNPs,[7] making the biochemical preparation of large amounts of these RNPs from yeast technically challenging. As a result, most of the work done in preparative purification of splicing RNPs has been performed using mammalian extracts.

The yeast *Saccharomyces cerevisiae* has been widely used for the genetic analysis of RNPs. There is a high degree of compositional as well as mechanistic conservation observed between RNPs in yeast and mammals. The determination of proteins associated with yeast RNPs provides a comparison with their mammalian counterparts and a tool for the dissection of their structure and biochemical functions in a system amenable to genetic manipulation. A rapid, highly efficient yet gentle means of preparatively purifying these large complexes provides for the recovery of sufficient amounts of protein to analyze by mass spectrometry microsequencing.[8] Completion of the yeast genome sequencing project and the recent explosion of genomic sequencing of other organisms will provide biologists with

---

[1] J. P. Staley and C. Guthrie, *Cell* **92,** 315 (1998).
[2] H. F. Noller, *Annu. Rev. Biochem.* **60,** 191 (1991).
[3] H. Lutcke, *Eur. J. Biochem.* **228,** 531 (1995).
[4] T. H. Reilly and M. E. Schmitt, *Mol. Biol. Rep.* **22,** 87 (1996).
[5] E. S. Maxwell and M. J. Fournier, *Annu. Rev. Biochem.* **64,** 897 (1995).
[6] K. T. Tycowski, Z. H. You, P. J. Graham, and J. A. Steitz, *Mol. Cell* **2,** 629 (1998).
[7] J. A. Wise, D. Tollervey, D. Maloney, H. Swerdlow, E. J. Dunn, and C. Guthrie, *Cell* **35,** 743 (1983).
[8] M. T. Davis and T. D. Lee, *J. Am. Soc. Mass Spec.* **8,** 1059 (1997).

the means for the rapid identification of the open reading frames coding for the proteins identified by these methods.

Several methods have been used previously to identify RNP components in yeast. Methods such as classical genetic screens, synthetic lethality analysis, identification of second-site suppressors, and two-hybrid analysis have given important if not incomplete insight into the composition of many RNPs. Most recently, the direct biochemical purification and identification of proteins associated with the yeast U4/U6 · U5 small nuclear RNP (snRNP),[9] the U1 snRNP,[10] and the snR30[11] and snR42[12] small nucleolar (sno)RNPs have been achieved through the use of affinity chromatography.

The U4/U6 · U5 snRNP from yeast is used due to its large size and potential for new proteins. This snRNP is involved in the pre-mRNA splicing pathway and contains factors thought to promote the catalysis of the splicing reaction. In previous RNP purifications from yeast, Lührmann and colleagues took advantage of the presence of a trimethylguanosine (TMG) moiety present in a number of yeast snRNAs to purify the entire set of TMG-containing snRNAs and snRNPs.[13] They then used an epitope tag inserted into an integral protein of the U1 snRNP to isolate it from the bulk TMG containing RNPs.[10] This method was also employed to "background subtract" the U1 snRNP for the isolation of snR30 and snR42 snRNPs by classical biochemical methods.

In this work, two protein affinity tags are inserted into the coding sequence of the core spliceosomal snRNP protein D3, which allows serial purification on two affinity matrices. Preparative glycerol gradient purification is then used to effect the separation of the splicing snRNPs, which exist as six discrete entities: U1, U2, U5, U6, U4/U6, and U4/U6 · U5. This method was sufficient to enrich for the U4/U6 · U5 snRNP for analysis of the associated polypeptides by electrospray tandem mass spectrometry microsequencing and should be generally applicable to other low-abundance RNPs from any genetically manipulable organism. A series of vectors that aid in the use of these affinity chromatography tags in *S. cerevisiae* are also presented.

[9] S. W. Stevens and J. Abelson, *Proc. Natl. Acad. Sci. U.S.A.* **96,** 7226 (1999).

[10] G. Neubauer, A. Gottschalk, P. Fabrizio, B. Seraphin, R. Lührmann, and M. Mann, *Proc. Natl. Acad. Sci. U.S.A.* **94,** 385 (1997).

[11] B. Lubben, P. Fabrizio, B. Kastner, and R. Lührmann, *J. Biol. Chem.* **270,** 11549 (1995).

[12] N. J. Watkins, A. Gottschalk, G. Neubauer, B. Kastner, P. Fabrizio, M. Mann, and R. Lührmann, *RNA* **4,** 1549 (1998).

[13] P. Fabrizio, S. Esser, B. Kastner, and R. Lührmann, *Science* **264,** 261 (1994).

## Selection of Proper Tagged Protein and Affinity Matrices

The best protein to epitope tag is one that is known to associate tightly with the complex of interest and has been shown to tolerate modification or insertion at either the amino terminus or the carboxy terminus. The selection of the protein to epitope tag is crucial as disruption of its structure may affect its function or its interaction with other proteins.

mRNA splicing snRNPs in all eukaryotes contain a set of canonical proteins termed the Sm core. These polypeptides, termed B (and B' in metazoans), D1, D2, D3, E, F, and G, assemble at a consensus RNA sequence located in the U1, U2, U4, and U5 snRNAs (U6 snRNA does not contain this sequence but contains its own set of core proteins[14,15]). Because purifying spliceosomal snRNPs away from nonspliceosomal TMG-containing snRNPs would provide a purification enrichment, the SmD3 protein was chosen due to its ability to quantitatively precipitate pre-mRNA splicing snRNPs from a yeast extract in small-scale preparations.[16] The next consideration was to determine the best affinity tags to use. Ideally, affinity tags allow purification in a way that is economical, quantitative, and preserve the native structure and composition of the complex.

The use of metal affinity chromatography has been widely used for the affinity purification of countless proteins and complexes from a wide variety of organisms.[17] The small size of the polyhistidine epitope, the relatively inexpensive Ni-NTA resin (Qiagen, Valencia, CA) resin, and gentle elution protocol make it ideal for large-scale use as an affinity tag. Purification of polyhistidine-containing complexes from whole cell yeast extracts using up to 16 histidine residues in the SmD3 polypeptide has proven to be very poor as there are numerous proteins from yeast that contain polyhistidine sequences and outcompete the much larger and likely sterically challenged RNPs (unpublished observations, 1998). A second affinity tag that allows purification directly from a yeast-splicing extract with a low background of protein binding, which is of reasonable cost and allows for the gentle elution of the RNPs, would make an ideal first-step purification complemented by Ni-NTA chromatography as a second step. A polyoma antigen epitope[18] was chosen (termed Py) for which the monoclonal antibody-producing hybridoma line is available. This antibody has been shown previously to work well in the purification of proteins directly from yeast extract with low background.[19]

[14] M. Cooper, L. H. Johnston, and J. D. Beggs, *EMBO J.* **14,** 2066 (1995).
[15] A. E. Mayes, L. Verdone, P. Legrain, and J. D. Beggs, *EMBO J.* **18,** 4321 (1999).
[16] J. Roy, B. H. Zheng, B. C. Rymond, and J. L. Woolford, *Mol. Cell. Biol.* **15,** 445 (1995).
[17] S. A. Lopatin and V. P. Varlamov, *Appl. Biochem. Micro.* **31,** 221 (1995).
[18] K. R. Schneider, R. L. Smith, and E. K. O'Shea, *Science* **266,** 122 (1994).
[19] P. L. Raghunathan and C. Guthrie, *Science* **279,** 857 (1998).

## Construction of Epitope-Tagged Gene

To generate an SmD3 gene that contains the polyhistidine tag followed by two tandem repeats of the Py epitope, the polymerase chain reaction (PCR) is used to generate two fragments, as shown in Fig. 1. PCR fragment SmD3-A contains 300 bp of sequence upstream of the SmD3-coding sequence with an *Xho*I site engineered before the stop codon of the SmD3 gene. PCR fragment SmD3-B contains an *Xho*I site and the polyhistidine and Py epitopes, which are incorporated into the sequence of the D3-B oligonucleotide (Fig. 1) as well as 200 bp of DNA downstream of the SmD3-coding sequence. The two PCR fragments are digested with *Xho*I, ligated together, and reisolated. The resulting ligation product is then digested with *Bam*HI and *Sal*I restriction endonucleases, which cut at sequences incorporated into the D3-5 and D3-3 oligonucleotides to create convenient

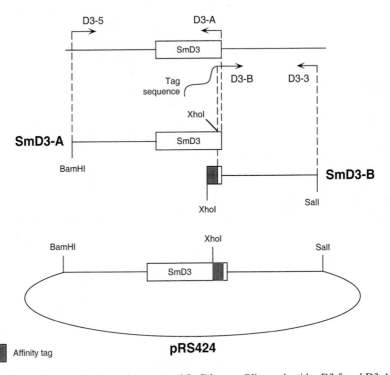

FIG. 1. Construction of the epitope-tagged SmD3 gene. Oligonucleotides D3-5 and D3-A are used to amplify the open reading frame and promoter region of the gene; D3-B and D3-3 are used to incorporate the affinity tag sequence and amplify the stop codon and the terminator of the gene. These fragments are digested with *Xho*I, ligated, reisolated, and digested with *Bam*HI and *Sal*I for insertion into the *Bam*HI and *Sal*I sites in the yeast vector pRS424.

sites for insertion into the pRS series of yeast vectors.[20] The epitope-tagged SmD3 fragment is ligated into pRS424, a high copy TRP1 marked vector, and the proper SmD3-coding sequence and in-frame epitope are confirmed by DNA sequencing.

## Complementation of SmD3 Knockout Strain

The carboxyl-terminal epitope-tagged SmD3 gene (SmD3-28) is introduced into yeast strain JWY2445[16] [*MAT* **a** *his1 leu2-1 trp1-Δ101 ura3-52 smd3Δ1∷LEU2* + YCp50-SMD3 ] by the lithium acetate transformation procedure.[21] Counterselection of the wild-type SmD3-URA3 plasmid is performed on 5-fluoroorotic acid as described previously.[22] The epitope tag has no effect on the growth rate of cells; however, a doubly tagged (N and C terminus) SmD3 gene almost triples the wild-type growth rate (data not shown). Extracts are prepared from both C-terminally tagged and N- and C-terminally tagged strains and show identical immunoprecipitation of the spliceosomal RNPs as determined by silver staining of RNAs (data not shown). Because of the dramatically shorter doubling time and identical results, the SmD3-28 strain is used for further experiments.

## Yeast Growth and Extract Preparation

A large-scale (200 liter) fermentation is performed in YPD to OD of 3.5 (absorbance measured at 600 nm). This growth routinely yields 1.2 kg yeast cells. The cells are processed immediately by resuspension in 1.2 liters buffer A (10 m$M$ HEPES, pH 7.9, 10 m$M$ KCl) and frozen in liquid nitrogen by extruding slowly through a syringe. Extracts are prepared in a Waring blendor from 150 g frozen cell suspension with frequent addition of liquid nitrogen. Cells are blended at maximum speed for 15 min per 150 g cell suspension. Frozen cell powder is stored overnight at −20° and thawed in the morning at room temperature. The addition of KCl to 200 m$M$ with stirring at 4° is followed by the addition of 2-mercaptoethanol to 10 m$M$, phenylmethylsulfonyl fluoride (PMSF) to 0.4 m$M$, leupeptin to 1 μg/ml, and pepstatin A to 1 μg/ml. The extract is stirred gently at 4° for 30 min, then spun at 17,000 rpm in an SS34 rotor for 40 min. The supernatant is then spun at 37,000 rpm in a Ti60 ultracentrifuge rotor for 1 hr. The splicing

[20] T. W. Christianson, R. S. Sikorski, M. Dante, J. H. Shero, and P. Hieter, *Gene* **110,** 119 (1992).
[21] R. D. Geitz and R. H. Scheistel, *Methods Mol. Cell. Biol.* **5,** 255 (1995).
[22] J. D. Boeke, J. Trueheart, G. Natsoulis, and G. R. Fink, *Methods Enzymol.* **154,** 164 (1988).

extract is pooled and dialyzed against 3 × 2 liter volumes buffer D [20 m$M$ HEPES, pH 7.9, 50 m$M$ KCl, 8% (v/v) glycerol, 10 m$M$ 2-mercaptoethanol, 0.4 m$M$ PMSF] for 1.5 hr each.

## Affinity Chromatography of Spliceosomal snRNPs

The $\alpha$-Py antibody is grown from cell line AK6967 in GIBCO (Grand Island, NY) serum-free hybridoma medium II with no antibiotics or supplements 3 days past confluence. Hybridoma cell supernatants are spun to remove cell debris, filtered, and stored at $-20°$ with 0.02% sodium azide until use. Antibody is coupled to protein G-Sepharose (Pharmacia, Piscataway, NJ.) (5 ml resin per 200 ml antibody supernatant). After incubating the resin with antibody for 2 hr at room temperature, the resin is washed at room temperature once in 8 resin volumes of phosphate-buffered saline (PBS), pH 7.4, and again in 0.2 $M$ sodium borate, pH 9.0. The beads are resuspended in 8 resin volumes of borate buffer, and solid dimethyl pimelimidate is added to 20 m$M$ to cross-link the antibody to the protein G resin. The resin is gently rocked for 30 min at room temperature. The resin is pelleted and resuspended in 40 ml 0.2 $M$ ethanolamine, pH 8.0, and allowed to gently rock at room temperature for 2 hr to quench the cross-linking reaction. The resin is then washed three times in 8 resin volumes of PBS and stored in PBS with 0.02% (w/v) sodium azide at 4°.

Whole cell yeast splicing extract (150 ml) is passed over a 1-ml bed volume of $\alpha$-Py antibody resin at 5 ml per hour at 4°. The extract is then washed with 15 column volumes of buffer D50 or D250 (buffer D + 50 m$M$ or 250 m$M$ KCl). Elution of the bound material is affected by very slow (1 ml per hour) incubation with buffer D50 or D250 containing 100 $\mu$g/ml competitor peptide (peptide sequence EYMPME). The yield from this column is generally about 300 $\mu$g protein as determined by the Bradford protein assay (Bio-Rad, Hercules, CA) and eluted reproducibly in the second and third column volumes of elution buffer. Figure 2A shows the protein eluted from this column. Sm proteins migrate near the bottom of the gel, and the epitope-tagged SmD3 protein is seen as an intensely stained band of 18kDa. A protein of about 85 kDa is seen in Fig. 2, which appears in the flow-through on chromatography on Ni-NTA agarose (see later). The contaminant is also seen when the KCl concentration is adjusted to 300 m$M$, indicating a relatively strong interaction with the antibody. Although the yeast genome does not appear to have any natural epitopes of exact sequence for this antibody (unpublished observations, 1998), this and other minor contaminants indicate the utility of a second affinity purification step to eliminate nonspecifically bound proteins. The identity of the ~85 kDa protein is determined by mass spectrometry to be the yeast HIS4 gene

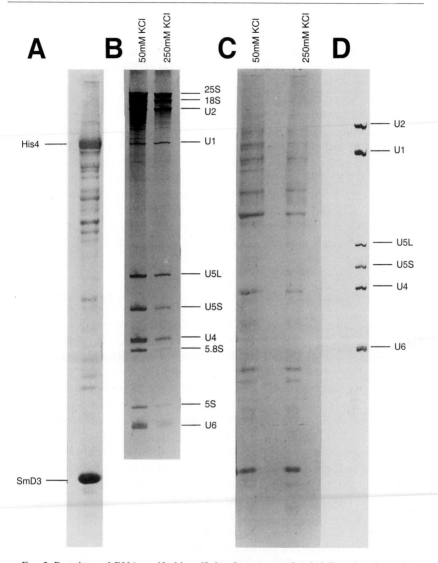

F IG. 2. Proteins and RNA purified by affinity chromatography. (A) Proteins eluted from the α-Py affinity matrix. Locations of the SmD3 band and the contaminating His4 band are shown. (B) RNA extracted from the α-Py affinity matrix eluate. U1, U2, U4, U5L, U5S, and U6 snRNAs, as well as contaminating 25S, 18S, 5.8S, and 5S rRNAs, are noted. (C) Proteins eluted from the Ni-NTA affinity matrix at 50 and 250 mM KCl. Note the absence of the His4 contaminant. (D) RNA extracted from the 50 mM KCl Ni-NTA affinity matrix eluate. Locations of U1, U2, U4, U5L, U5S, and U6 snRNAs are shown. Note the absence of contaminating rRNAs.

product, an 88-kDa polypeptide that contains no obvious epitope for this antibody by sequence analysis. RNA extracted from α-Py antibody affinity chromatography eluted in D50 and D250 is shown in Fig. 2B. 25S, 18S, 5.8S, and 5S rRNAs are significant contaminants at this stage of the purification, but are eliminated in the Ni-NTA chromatography step.

For the second affinity purification step, the α-Py antibody eluate is adjusted to 10 m$M$ imidazole, pH 7.0, and passed over a 0.5-ml Ni-NTA agarose column. The column is washed with 20 column volumes of buffer D50 or D250 containing 10 m$M$ imidazole and eluted with buffer D50 or D250 + 100 m$M$ imidazole. The yield of protein from the second column is generally 125–175 µg. Representative proteins from purification performed at either 50 or 250 m$M$ KCl seem to be essentially identical at this stage (Fig. 2C) and the loss of the HIS4 protein and a few other minor bands can be seen. RNA extracted from the D50 Ni-NTA elution is shown in Fig. 2D. The splicing snRNAs are essentially free from rRNA after this chromatography step.

## Glycerol Gradient Separation of Spliceosomal snRNPs

Because spliceosomal snRNPs exist as discrete entities U1, U2, U5, U6, U4/U6, and U4/U6 · U5, it is therefore necessary at this point to utilize other biochemical methods to isolate them individually. One method that has worked for others for the purification of mammalian snRNPs is Mono Q or Mono S chromatography.[11,23,24] Because the elution of many of these yeast snRNPs is effected only at high salt concentrations (unpublished observations, 1998), it is possible that some weakly associated snRNP proteins may be lost. A more gentle method for separating these complexes is glycerol gradient centrifugation. Figure 3 shows a profile of snRNAs extracted from fractions of a 10–30% glycerol gradient enriched for the U4/U6 · U5 snRNP. The 25S U4/U6 · U5 snRNP reproducibly sedimented primarily in fractions 22–25 with small amounts of contaminating U1 and U2 snRNPs. By analyzing serially diluted glycerol gradient fractions and performing a densitometric analysis of snRNAs in U4/U6 · U5 fractions, U4/U6 · U5 snRNAs are shown to be present at a 50- to 100-fold molar excess over U1 or U2 snRNAs. This large excess should allow the distinction of U4/U6 · U5 proteins in these fractions from U1 or U2 snRNP proteins.

The U4/U6 · U5 snRNP exists in equilibrium with the U5, U6, and U4/U6 snRNPs, which makes its purification more difficult than the relatively

[23] P. Bringmann and R. Lührmann, *EMBO J.* **5,** 3509 (1986).
[24] M. Bach, P. Bringmann, and R. Lührmann, *Methods Enzymol.* **181,** 232 (1990).

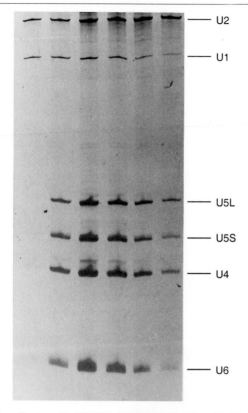

— U2

— U1

— U5L

— U5S

— U4

— U6

FIG. 3. Glycerol gradient-purified U4/U6 · U5 snRNP. Fractions 20–25 of a 10–30% glycerol gradient are shown. U4, U5L, U5S, and U6 snRNAs are noted as are contaminating U1 and U2 snRNAs.

more abundant U1 snRNP. Due to the low abundance of the U4/U6 · U5 snRNP in these extracts, 16 10–30% linear glycerol gradients are run, each containing 50–75 μg protein eluted from the Ni-NTA column. Fractions 22–25 from each gradient are pooled, extracted with phenol/chloroform/ isoamyl alcohol (25:24:1), and precipitated with 5 volumes of acetone from the organic phase as described previously.[3] Precipitated proteins are washed twice with 80% ethanol and redissolved in 5× SDS–PAGE gel-loading buffer and electrophoresed on a 12% high-TEMED SDS–poly-acrylamide gel. The presence of high amounts of TEMED allows for better separation of the smaller proteins in a gel of this construction. Figure 4 shows proteins isolated from these fractions.

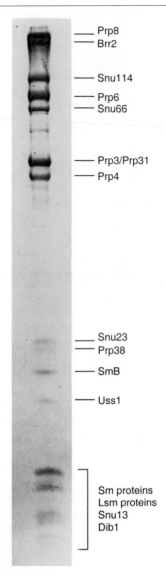

FIG. 4. U4/U6·U5 snRNP proteins. Fractions 20–25 from 16 glycerol gradients were extracted as described in the text and electrophoresed on a 12% high-TEMED SDS–PAGE gel. Identities of the proteins were determined by mass spectrometry microsequencing.

Mass Spectrometry Microsequencing Analysis of U4/U6 · U5 Proteins

The identification of small quantities of protein has been achieved by the use of electrospray tandem mass spectrometry (ES/MS/MS).[8,25] This very sensitive method uses a liquid chromatography system attached to a triple quadrupole tandem mass spectrometer to separate polypeptides derived from a trypsin digest of the protein. Identified peptides of interest are then collided with high energy gas to produce both N-terminally and C-terminally truncated fragments whose masses are determined in the second mass spectrometer. For a more detailed description of these procedures, refer to Volume 193 of this series.[25a]

Each band from the gel is excised carefully with a fresh scalpel and stored at $-20°$. The gel slice is then destained, reduced, and alkylated in gel and digested overnight with trypsin. Peptides are eluted from the matrix, desalted, and analyzed by ES/MS/MS as described previously.[8] These data are compared against a nonredundant database of computer-generated trypsin digest profiles of all proteins in a nonredundant database to assign a best fit. All peptides that produce good fragmentation during high-energy gas collision are analyzed in this way, and a consensus is drawn as to the best fit for these data. Because a peptide mass and MS/MS profile is rarely unique in the database, several peptide spectra are usually required to confirm the identity of the protein. For very small proteins, there are often not very many peptides generated during digestion with trypsin, and other methods such as $^{18}O$ labeling during digestion can aid in the correct assignment of a peptide to the parent protein. In the case of the Prp3/Prp31 overlapping band, each protein had a number of peptides whose sequences allowed their correct identification, which is evidence of the robustness of the procedure. The identity of the proteins from this study are shown in Fig. 4. A more detailed description of the U4/U6 · U5 proteins is presented elsewhere. Proteins termed LSM are U6-specific Sm proteins that were not even visible by Coomassie blue staining, yet were identified by mass spectrometry.

Vectors for Epitope Tagging Amino or Carboxyl
    Terminus of Proteins in Yeast

To make this procedure generally applicable in yeast, a series of vectors allow any gene to be epitope tagged on either the amino terminus (N-terminal) or the caboxyl terminus (C-terminal). These vectors are based

[25] D. C. Stahl, K. M. Swiderek, M. T. Davis, and T. D. Lee, *J. Am. Soc. Mass Spec.* **7,** 532 (1996).
[25a] J. A. McCloskey (ed.), *Methods Enzymol.* **193** (1990).

## pNHPxxx series

Ndel   Ncol   Sall
EcoRI  Nhel  AccI  XhoI  ApaI

```
 *  H  H  H  H  H  H  H  M  E  Y  M  P  M  E  M  E  Y  M  P  M  E
ATGCATCATCATCATCATCACCACCACATGGAATATATGCCAATGGAAATGGAATATATGCCAATGGAAcatatggaattcccatggctagcgtcgacctcgaggggggcccggtacc
TACGTAGTAGTAGTAGTAGTGGTGGTGGTACCTTATATACGGTTACCTTTACCTTTATATACGGTTACCTTgtatacctaagggtaccgatcgcagctggagctcccccgggccatgg
```

## pCHPxxx series

SacI  SacII  NotI  XbaI  SpeI
          Eagl      BamHI

```
                                        H  H  H  H  H  H  H  H  M  E  Y  M  P  M  E  M  E  Y  M  P  M  E  *
gagctccaccggtggcggccgctctagaactagtggatccAAGAGAAGGctcgagCATCATCATCATCATCACCACCACATGGAATATATGCCAATGGAAATGGAATATATGCCAATGGAAtga
ctcgaggtggccaccgccggcgagatcttgatcacctaggTTCTCTTCCgagctcGTAGTAGTAGTAGTAGTGGTGGTGTACCTTATATACGGTTACCTTTACCTTTATATACGGTTACCTTact
```

Fig. 5. Relevant sequences of the pNHPxxx and pCHPxxx series of epitope-tagging vectors. The sequence of epitope tag regions and polylinker regions of vectors is shown with available restriction endonuclease cleavage sites. Note that not all restriction endonuclease sites are available in every vector due to their presence in the various yeast auxotrophic marker genes (refer to Table I). The start and stop codons are represented above the sequences by an asterisk.

on the pRS series of vectors[20] and allow their selection using auxotrophic markers. Additionally, versions of the N-terminal and C-terminal constructions were placed into a pUC-based vector to facilitate cloning and subcloning when sites of interest are not available in the yeast vectors. Figure 5 shows the relevant sequences of these vectors, and Table I shows the features of each of the vectors, including the sites available in their polylinkers. Because epitope tagging of proteins can lead to deleterious effects at one or both of the termini, it is wise to construct both N- and C-terminally tagged versions of a given gene and select the one that generates the least interference with respect to growth rate and function.

Vectors created for N-terminal epitope tagging include the SmD3 promoter and start codon, the polyhistidine sequence followed by two tandem repeats of the Py epitope, and a polylinker. The gene of interest can be created by PCR to include an appropriate restriction site and must also include the terminator of the gene of interest, which is generally contained in the 200–300 bp downstream of the stop codon. Vectors created for C-terminal epitope tagging include a polylinker followed by the polyhistidine sequence and two tandem repeats of the Py epitope with the SmD3 terminator following the stop codon. Again, generation of the epitope-tagged gene may be accomplished using PCR by amplifying the open reading frame and the promoter region of the gene of interest. In general, 200–300 bp upstream of the start codon has worked well in these studies.

TABLE I
FEATURES OF NHP AND CHP SERIES OF VECTORS[a]

| Vector | Auxotrophic marker | Copy number | Location of affinity tag | Cloning sites available |
|---|---|---|---|---|
| pNHP316 | URA | Low | N-terminal | *Eco*RI, *Nhe*I, *Sal*I, *Xho*I, *Apa*I |
| pNHP413 | HIS | Low | N-terminal | *Eco*RI, *Nco*I, *Sal*I, *Xho*I, *Apa*I |
| pNHP414 | TRP | Low | N-terminal | *Nde*I, *Eco*RI, *Nco*I, *Nhe*I, *Sal*I, *Xho*I, *Apa*I |
| pNHP424 | TRP | High | N-terminal | *Nde*I, *Eco*RI, *Nco*I, *Nhe*I, *Sal*I, *Xho*I, *Apa*I |
| pNHP425 | LEU | High | N-terminal | *Nde*I, *Nco*I, *Nhe*I, *Sal*I, *Xho*I, *Apa*I |
| pCHP316 | URA | Low | C-terminal | *Sac*I, *Not*I, *Xba*I, *Spe*I, *Bam*HI |
| pCHP413 | HIS | Low | C-terminal | *Sac*I, *Not*I, *Xba*I, *Spe*I, *Bam*HI |
| pCHP414 | TRP | Low | C-terminal | *Sac*I, *Not*I, *Spe*I, *Bam*HI |
| pCHP424 | TRP | High | C-terminal | *Sac*I, *Not*I, *Spe*I, *Bam*HI |
| pCHP425 | LEU | High | C-terminal | *Sac*I, *Not*I, *Spe*I, *Bam*HI |

[a] N-terminal (NHP) and C-terminal (CHP) HIS-Py-tagged vectors are based on the pRS series of vectors. Auxotrophic markers, copy number of the plasmid, and cloning sites available in the vectors are shown. Refer to Fig. 5 for the sequences of the polylinkers.

Polylinker sites were chosen to allow the same oligonucleotides for the construction of bacterial expression in the pET series of vectors.[26]

## Conclusion

This article has shown that the preparative purification of very low-abundance RNPs from yeast in quantities necessary for identification of the associated proteins is possible using affinity chromatography and mass spectrometry. The general applicability of these methods to any genetically manipulable organism makes the purification of RNPs and other protein complexes possible where conventional methods have failed in the past. The relatively low cost of the resins and general applicability to almost any protein known to be a component of an RNP or other large multiprotein complex make this method an attractive alternative to more conventional biochemical purifications. The gentle elution protocol helps ensure that the integrity and composition of the complexes will remain as close to native as possible. With the rapid advances in genomics, the ability to quickly identify the open reading frames of the purified proteins will help accelerate studies on their structure and function.

## Acknowledgments

I especially thank John Abelson—in whose laboratory this work was performed—for support and guidance, John Woolford for providing the SmD3 knockout strain, Amy Kistler and Christine Guthrie for providing the hybridoma cell line, and the members of John Abelson's and Christine Guthrie's laboratories for helpful advice. I am also very grateful to Tony Moreno, Mary Young, and Terry Lee at the City of Hope for mass spectrometry expertise. This work was supported by NIH Grant GM32627 to John Abelson, CA33572 to the City of Hope Beckman Research Institute Mass Spectrometry Facility, and an American Cancer Society postdoctoral fellowship (PF4447) to S.S.

---

[26] F. W. Studier, A. H. Rosenberg, J. J. Dunn, and J. W. Dubendorff, *Methods Enzymol.* **185,** 60 (1990).

# [27] Yeast Three-Hybrid System to Detect and Analyze RNA–Protein Interactions

*By* Beilin Zhang, Brian Kraemer, Dhruba SenGupta, Stanley Fields, and Marvin Wickens

## Introduction

The interactions of RNAs with proteins are critical for a wide variety of cellular processes, ranging from translation to the proliferation of certain viruses. For this reason, several methods have been developed to facilitate the analysis of RNA–protein interactions by genetic means.[1–8]

This article focuses on one such genetic method, the yeast three-hybrid system.[9] In this method, an RNA–protein interaction in yeast results in transcription of a reporter gene. The interaction can be monitored by cell growth, colony color, or the levels of a specific enzyme. Because the RNA–protein interaction of interest is analyzed independent of its normal biological function, a wide variety of interactions are accessible. Among the applications of the method are the discovery of proteins that bind to and regulate specific, previously known RNA sequences. In addition, it may also be possible to identify the previously unknown RNA partner of an RNA-binding protein.[10] These applications complement an array of biochemical strategies, including RNA affinity chromatography. The three-hybrid system, like other genetic strategies, has the attractive feature that a clone encoding the protein of interest is obtained directly in the screen.

This article emphasizes the use of the three-hybrid system to find the partner in an RNA–protein interaction when only one of the components

[1] M. P. MacWilliams, D. W. Celander, and J. F. Gardner, *Nucleic Acids Res.* **21,** 5754 (1993).

[2] R. Stripecke, C. C. Oliveira, J. E. McCarthy, and M. W. Hentze, *Mol. Cell. Biol.* **14,** 5898 (1994).

[3] I. A. Laird-Offringa and J. G. Belasco, *Proc. Natl. Acad. Sci. U.S.A.* **92,** 11859 (1995).

[4] C. Jain and J. G. Belasco, *Cell* **87,** 115 (1996).

[5] K. Harada, S. S. Martin, and A. D. Frankel, *Nature* **380,** 175 (1996).

[6] E. Paraskeva, A. Atzberger, and M. W. Hentze, *Proc. Natl. Acad. Sci. U.S.A.* **95,** 951 (1996).

[7] R. Tan and A. D. Frankel, *Proc. Natl. Acad. Sci. U.S.A.* **95,** 4247 (1998).

[8] U. Putz, P. Skehel, and D. Kuhl, *Nucleic Acids Res.* **24,** 4838 (1996).

[9] D. J. SenGupta, B. Zhang, B. Kraemer, P. Pochart, S. Fields, and M. Wickens, *Proc. Natl. Acad. Sci. U.S.A.* **93,** 8496 (1996).

[10] D. SenGupta, M. Wickens, and S. Fields, *RNA,* in press.

is known. We describe how to perform such screens and discuss some of
the strengths and limitations of the method. The system is still at a relatively
early stage; this article is provided for those who wish to use it as is or
might wish to improve it.

## Principles of Method

The general strategy of the three-hybrid system is diagrammed in Fig.
1. DNA-binding sites are placed upstream of a reporter gene in the yeast
chromosome. A first hybrid protein consists of a DNA-binding domain
linked to an RNA-binding domain. The RNA-binding domain interacts
with its RNA-binding site in a bifunctional ("hybrid") RNA molecule. The
other part of the RNA molecule interacts with a second hybrid protein
that consists of another RNA-binding domain linked to a transcription
activation domain. When this tripartite complex forms at a promoter, even
transiently, the reporter gene is turned on. This expression can be detected
by phenotype or simple biochemical assays.

The specific molecules used most commonly for three-hybrid screens
at the time of writing are depicted in Fig. 2. The DNA-binding site consists
of a 17 nucleotide recognition site for the *Escherichia coli* LexA protein
and is present in multiple copies upstream of both *HIS3* and *LacZ* genes.
Hybrid protein 1 consists of LexA fused to bacteriophage MS2 coat protein,
a small polypeptide that binds as a dimer to a short stem–loop sequence
in its RNA genome. The hybrid RNA (depicted in more detail in Fig. 5)
consists of two MS2 coat protein-binding sites linked to the RNA sequence

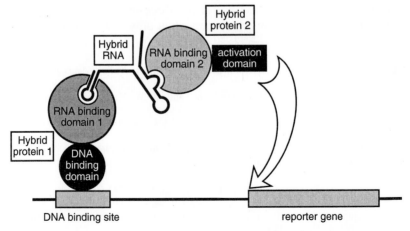

Fig. 1. General scheme of the three-hybrid system. Adapted from Harada *et al.*[5]

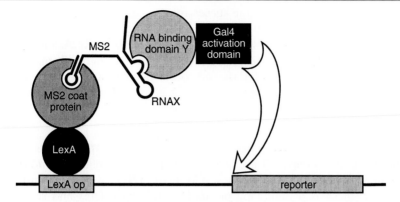

FIG. 2. One array of specific components used commonly in the three-hybrid system. Both *lacZ* and *HIS3* are present in strain L40coat under the control of lexA operators and so can be used as reporters. For simplicity, the following features are not indicated. Multiple LexA operators are present (four in the *HIS3* promoters and eight in the *lacZ* promoter), and the LexA protein binds as a dimer. The hybrid RNA contains two MS2-binding sites, and the MS2 coat protein binds as a dimer to a single site.

of interest, X. Hybrid protein 2 consists of the transcription activation domain of the yeast Gal4 transcription factor linked to an RNA-binding protein, Y.

Although the components depicted in Fig. 2 are used most commonly, other RNAs and proteins can be substituted. For example, the histone mRNA 3′ stem loop and the protein to which it binds (SLBP) can replace the MS2 components (BZ and MW, unpublished).

The three-hybrid approach has many of the same strengths and limitations of the two-hybrid system to detect protein–protein interactions. By introducing libraries of RNA or protein, cognate partners can be identified. As in two-hybrid screens, the challenge then becomes to identify those molecules whose interaction is biologically relevant. For this purpose, mutations in the known RNA or protein component are very useful. The system also makes it possible to identify those regions of an RNA or protein that are required for a known interaction and to test combinations of RNA and protein to determine whether they interact *in vivo*.

Perspective

The three-hybrid system is still in an early stage of development. Many known combinations of RNA and protein partners have been tested, with a very high frequency of success. However, the number of library screens reported to date is small: three successful screens of cDNA libraries have been reported,[11–13] and we have completed a single RNA library screen.

The main thrust of this review is the screening of RNA and protein libraries. However, the system has a number of other potential applications (see "Prospects"). We offer the protocols here as a starting point for investigators wanting to use the system as is or to develop it further.

Plasmid Vectors

Several plasmids have been constructed to express hybrid RNA sequences and are depicted in Fig. 3. An activation domain plasmid that is useful in screening RNA libraries, pACTII/CAN, is also shown in Fig. 4. Each of these is a multicopy plasmid containing origins and selectable markers for propagation in yeast and bacteria. They are described next.

pIII/MS2-1 and pIII/MS2-2 (Fig. 3A) are yeast shuttle vectors derived from pIIIEx426RPR, which was developed by Good and Engelke.[14] Sequences to be analyzed are inserted at the *Sma*I/*Xma*I site. pIII/MS2-1 and pIII/MS2-2 differ only in the relative position of the *Sma*I/*Xma*I site and the MS2-binding sites. Both plasmids carry a *URA3* marker and produce hybrid RNAs from the yeast RNase P RNA (*RPR1*) promoter, an RNA polymerase III promoter. The *RPR1* promoter was chosen for two reasons. First, it is efficient, directing the synthesis of up to several thousand molecules per cell. Second, transcripts produced from this promoter presumably do not enter the pre-mRNA processing pathway and may not leave the nucleus. An alternative method using an RNA polymerase II promoter to generate the hybrid RNA has also been described.[8]

pIIIA/MS2-1 and pIIIA/MS2-2 (Fig. 3B) are similar to pIII/MS2-1 and pIII/MS2-2, but carry the yeast *ADE2* gene (in addition to *URA3*). The *ADE2* gene is exploited in screening for RNA-binding proteins (see later). The two plasmids differ from one another only in the relative position of the *Sma*I/*Xma*I site and the MS2-binding sites.

The vector pACTII-CAN (Fig. 4) has been generated using the pACTII backbone[15] by inserting the yeast *CAN1* gene at the unique *Sal*I site of

[11] F. Martin, A. Schaller, S. Eglite, D. Schumperli, and B. Muller, *EMBO J.* **16,** 769 (1997).
[12] Z. F. Wang, M. L. Whitfield, T. C. I. Ingledue, Z. Dominski, and W. F. Marzluff, *Genes Dev.* **10,** 3028 (1996).
[13] B. Zhang, M. Gallegos, A. Puoti, E. Durkin, S. Fields, J. Kimble, and M. P. Wickens, *Nature* **390,** 477 (1997).
[14] P. D. Good and D. R. Engelke, *Gene* **151,** 209 (1994).
[15] C. Bai and S. J. Elledge, *in* "The Yeast Two-Hybrid System" (P. L. Bartel and S. Fields, eds.), p. 11. Oxford Univ. Press, London, 1997.

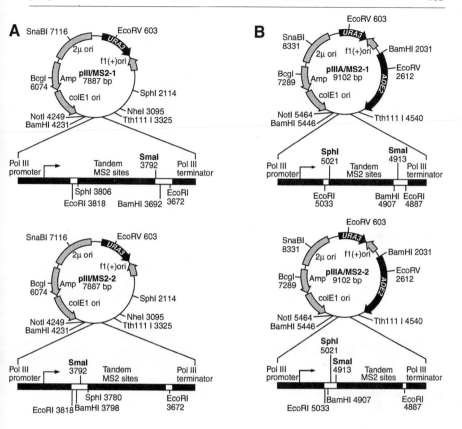

Fig. 3. Plasmid vectors used to express hybrid RNAs in the three-hybrid system. Restriction sites indicated either are unique or can be used to verify the presence of various elements on the plasmids. Sites suitable for insertion of sequences of interest are in bold.

pACTII. The protein sequence to be tested is inserted at the *Sfi*I, *Nco*I, or *Sma*I/*Xma*I sites. The presence on the plasmid of the *CAN1* gene, which encodes an arginine permease, causes cells to die in media containing the arginine analog canavanine. Thus, canavanine selection leads to the loss of this plasmid in a yeast strain that carries a chromosomal *can1* allele, which confers resistance. In this fashion, canavanine can be used to select for loss of the activation domain plasmid, whereas 5-FOA can be used to select for loss of the hybrid RNA plasmid.

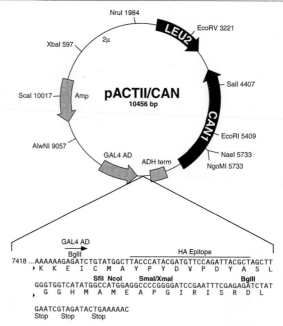

Sfil, Ncol, Smal/Xmal sites are unique and can be used to insert fragment. Bglll also can be used, but will result in loss of the HA epitope.

Fig. 4. Vector used to express the protein "bait" in an RNA library screen. Restriction sites either are unique or can be used to verify the presence of various elements on the plasmid. Sites suitable for insertion of the "bait" protein are in bold.

## Hybrid RNAs

### General Considerations

The hybrid RNA and the tripartite nature of the system present unique considerations for three-hybrid vs two-hybrid screens. Among these are issues concerning the length of the RNA that can be analyzed, the ability of cDNA/activation domain proteins to activate transcription independent of binding to the RNA, and of RNAs to activate transcription independent of the cDNA/activation domain protein. These issues are discussed in the relevant sections.

RNA sequences to be tested are inserted into the unique SmaI/XmaI site of any of the four RNA vectors depicted in Fig. 3. The hybrid RNA molecule that is transcribed *in vivo* from one of these plasmids consists of the sequence of interest, X, linked to two MS2 coat protein-binding sites,

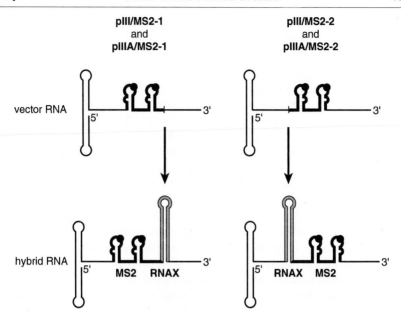

FIG. 5. Schematic diagram of the secondary structure of hybrid RNA molecules. Thin lines, RNase P RNA 5′ leader and 3′ trailer sequences; bold lines, tandem MS2 recognition sites, including a point mutation that increases affinity (black dot); gray lines, inserted RNA sequence, depicted as a stem loop.

and RNase P RNA 5′ leader and 3′ trailer sequences (Fig. 5). The MS2-binding sites are present in two copies because binding to coat protein is cooperative. Each site contains a single nucleotide change that enhances binding.[16]

In our experience, most RNA inserts of suitable size produce RNAs of comparable and high abundance. RNA abundance may be important in determining the level of signal produced from the reporter gene: transferring the hybrid RNA gene from normal high-copy vectors to low-copy vectors reduces levels of *LacZ* expression substantially.

The relative order of the RNA sequence of interest and the MS2 sites can affect signal strength. In the few cases that have been tested systematically, both orientations yield activation and are specific. However, in the IRE/IRP interaction, placing the IRE upstream of the MS2 sites results in two- to threefold more transcription than the alternative arrangement. Although RNA-folding programs can be used to determine whether one arrangement is more likely to succeed, the accuracy of their predictions *in*

[16] P. T. Lowary and O. C. Uhlenbeck, *Nucleic Acids Res.* **15,** 10483 (1987).

*vivo* is problematic. In most cases tested that we have performed, we have placed the RNA sequence of interest upstream of the MS2 sites.

## Limitations

RNA sequences to be analyzed are restricted in two respects at present. First, runs of four or more U's in succession can terminate transcription by RNA polymerase III. Second, typically, RNA inserts of lengths less than 150 nucleotides yield higher signals: longer inserts commonly reduce the level of activation of the reporter (see "General Considerations"). In principle, both of these limitations may be overcome by using a different polymerase, such as that of a bacteriophage polymerase.

*Runs of U's.* Four or more U's in succession can function as RNA polymerase III terminators and so can prevent production of the desired hybrid RNA. The efficiency of termination at oligouridine tracts is context dependent, however, so that it may be worth testing whether a suspect RNA sequence will function in the system. Northern blotting should always be performed to make certain that the hybrid RNAs are expressed at high levels. For long runs of U's, it may be necessary to eliminate the terminator by mutagenesis.

*Size of the RNA Element.* The size of the RNA insert appears to be an important determinant of three-hybrid activity. In reconstruction experiments using known RNA–protein partners, RNA sequences that are 30–100 nucleotides in length (e.g., TAR, IRE) typically yield substantial and specific reporter activation. We have investigated the effects of additional RNA sequences flanking a known protein-binding site. The addition of heterologous sequences to the IRE caused a reduction in the IRE-IRP1 three-hybrid signal. The effect was proportional to the size of the insertion, with the addition of 150–200 nucleotides commonly leading to almost complete abolition of the IRE-IRP1 three-hybrid signal. Similarly, we have tested the effect of an additional natural RNA sequence on the ability to detect the interaction of the yeast Snp1 protein, a homolog of the mammalian U1-70K protein, with the loop I of yeast U1 RNA. Insertion of even the RNA sequences that normally flank loop I, but are not part of the Snp1 binding site, decreased the three-hybrid signal due to Snp1–U1 interaction. Taken together, these experiments suggest that a minimal binding site is preferable and that the optimal length of the RNA insert is less than 150 nucleotides.

## Yeast Strains

The yeast reporter strain L40coat is derived from L40-ura3 (a gift of T. Triolo and R. Sternglanz, Stony Brook, NY). The genotype of the strain is *MATa, ura3-52, leu2-3,112, his3Δ200, trp1Δ1, ade2, LYS2::(lexA op)-HIS3, ura3::(lexA-op)-lacZ, LexA-MS2 coat (TRP1)*. The strain is auxotrophic for uracil, leucine, adenine, and histidine. Each of these markers is exploited in the three-hybrid system. Both the *HIS3* and the *lacZ* genes have been placed under the control of *lexA* operators, and hence are reporters in the three-hybrid system. A gene encoding the LexA-MS2 coat protein fusion has been integrated into the chromosome.

Strain R40coat is identical to L40coat, but of the opposite mating type.

A canavanine-resistant derivative of L40coat, L40coat-can, carries a *can1* allele. This strain can be rendered canavanine sensitive by transformation with a plasmid such as pACTII/CAN, and it becomes resistant again on loss of the plasmid when plated on media containing canavanine.

Testing Known or Suspected Interactors

The protocol to test a known or suspected RNA–protein interaction is straightforward. The RNA sequence of interest and the gene encoding the RNA-binding protein are cloned into an RNA plasmid and an activation domain vector, respectively. These are then introduced into yeast by transformation, and the level of expression of a reporter gene is determined. Figure 6 depicts such an analysis of the IRE–IRP1 interaction[9,17] using selection for *HIS3* to monitor the interaction. The results shown demonstrate that each segment of the system is required for activation of the reporter, which is monitored by growth on selective media.

This approach can be useful in testing candidate interactors obtained by other means. For example, an interaction between mouse telomerase RNA and a newly cloned telomerase protein was confirmed in this fashion (TP1).[18] Similarly, binding of yeast ribosomal protein S9 with *CRY2* pre-mRNA, deduced by genetic experiments, has been demonstrated using the system (S. Fewell and J. Woolford, personal communication).

In principle, the protocol also provides a facile way to delineate the portions of the RNA or protein that are required for an interaction. Regions required for RNA binding of FBF1 to its RNA target were inferred through such an analysis.[13] In such experiments, it is important to confirm that protein or RNA is present in cases in which the interaction no longer appears to function. This confirmation can be accomplished by either Northern or Western blotting. In principle, it should be possible to identify single amino acids and nucleotides that affect the interaction.

[17] B. Zhang, B. Kraemer, D. SenGupta, S. Fields, and M. Wickens, *in* "The Yeast Two-Hybrid System" (P. L. Bartel and S. Fields, eds.), p. 298. Oxford Univ. Press, 1997.
[18] L. Harrington, T. McPhail, V. Mar, W. Zhou, R. Oulton, M. B. Bass, I. Arruda, and M. O. Robinson, *Science* **275,** 973 (1997).

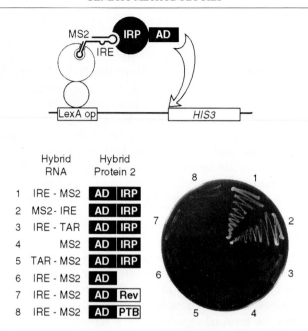

FIG. 6. The IRE/IRP1 interaction monitored by activation of the *HIS3* gene. Plasmids encoding the indicated hybrid RNAs and activation domain fusions were transformed into strain L40coat. After selecting transformants for the presence of the plasmids, colonies were restreaked onto media selecting for expression of *HIS3*. Only the two combinations that should lead to IRE/IRP1 interactions (lines 1 and 2) yield growth.

## Finding Protein Partner for Known RNA Sequence

The three-hybrid system can be used to identify a protein partner of a known RNA sequence. A library containing many cDNAs is introduced into a yeast strain that carries a plasmid that encodes the RNA sequence of interest as a hybrid RNA. From such screens emerge cDNAs that are capable of activating the reporter, some of which require the hybrid RNA to do so. As in other such methods, secondary and tertiary screens to eliminate several categories of false positives are crucial.

If the RNA–protein interaction is particularly strong, as in the case of SLBP and its stem–loop target in histone mRNA, then the initial selection can demand high levels of expression of the reporter gene.[9,11,12] A stringent selection eliminates weak activators. However, a less stringent selection is often preferable because the "strength" of the interaction being sought is not known. In turn, this reduced stringency leads to a higher background and an increased need for subsequent screens.

The following protocol has been used to clone novel RNA-binding proteins.

## Step 1. Introduce the RNA Plasmid and cDNA Library

The host strain, L40coat, is normally transformed with the RNA plasmid first. Cells containing the RNA plasmid are then transformed with a cDNA library fused with a transcription activation domain. (Hybrid RNAs are rarely toxic to the host cell, unlike some hybrid protein baits used in two-hybrid screens; as a result, there is no need to cotransform both plasmids.) The transformation mix is plated out on media lacking both leucine (selecting for the cDNA plasmid) and histidine (selecting for *HIS3* gene expression). Maintenance of the RNA plasmid (i.e., selection for *ADE2* or *URA3*) is not demanded. This permits cells that can activate *HIS3* without the RNA to lose the RNA plasmid, permitting the colony color screen (see "Step 2").

*Protein Libraries.* Activation domain libraries prepared in *LEU2* vectors can be used with the hybrid RNA plasmids and yeast strains described earlier. Many such activation domain libraries have been prepared for use with the two-hybrid system. These may be obtained from individual laboratories and several commercial sources.

*3-Aminotriazole in Initial Selection.* We typically add 3-aminotriazole (3AT), a competitive inhibitor of the *HIS3* gene product, to select for stronger interactions. Some RNAs activate reporter genes weakly on their own, and many proteins appear to activate the reporter slightly, independent of the hybrid RNA; both situations yield "false positives." To eliminate weak activation by the RNA "bait," 3AT should be titrated using a strain carrying only the RNA plasmid prior to undertaking the initial transformation. To diminish the number of false-protein positives ("RNA-independent" positives; see later), concentrations of 3AT in the range of 2 to 5 m*M* are a good starting point; they offer a reasonable balance between suppressing the background and permitting "real" positives to grow. *In vitro* data on the affinity of the RNA–protein interaction may be valuable in determining the concentration of the 3AT that should be used. In one of the screens that yielded SLBP, whose interaction with its RNA target is particularly robust *in vitro,* 25 m*M* 3AT was included, reducing the background greatly.[11]

## Step 2. Eliminate RNA-Independent False Positives by Colony Color

Two classes of positives are obtained from the initial transformation. One class of transformants requires the hybrid RNA to activate *HIS3*; these are termed "RNA dependent." A second class of positives activates

*HIS3* with or without the hybrid RNA; these are termed "RNA independent." RNA-independent positives can carry proteins that may bind to the
promoter regions of the reporter genes, or proteins that interact directly
with the LexA–MS2 coat protein fusion. The RNA-independent class of
transformants can be very abundant and can account for more than 95%
of the total number of colonies.

To facilitate eliminating RNA-independent false positives, we exploit
the *ADE2* gene on the RNA plasmid. The host strain is an *ade2* mutant.
When the level of adenine in the medium becomes low, cells attempt to
synthesize adenine *de novo* and accumulate a red purine metabolite due
to lack of the *ADE2*-encoded enzyme. This accumulation renders the cell
pink or red in color. In contrast, cells carrying the wild-type *ADE2* gene
are white.

In the initial transformation, selection is imposed only for activation of
*HIS3*, not for maintenance of the RNA plasmid. For RNA-dependent
positives, selection for *HIS3* selects indirectly for the RNA plasmid, which
carries the *ADE2* gene; thus these transformants are white and must remain
so if they continue to grow in the absence of exogenous histidine. However,
RNA-independent positives do not require the RNA plasmid to activate
the *HIS3* gene and so can lose that plasmid, which they do with a frequency
of a few percent per generation. These false positives therefore yield pink
colonies or white colonies with pink sectors.

The initial transformation plates are usually incubated at 30° for a week.
This duration allows positives to accumulate *HIS3* gene product and grow
and also provides enough time for the color to develop. If the pink color
is not strong after a week, incubation at 4° overnight sometimes helps. Pink
or pink-sectored colonies are discarded, and the uniformly white colonies
are picked for further analysis. We usually pick all the white colonies
(typically a few large, and many small, colonies) and patch onto media
selecting again for both the cDNA plasmid and the RNA plasmid. Most
of the small white colonies turn out to be RNA independent and fail to
grow. White colonies that are able to grow on the selective media are
subject to further analysis.

The identification of RNA-dependent positives by colony color is not
perfect; many, but not all, RNA-independent positives can be identified
and discarded. A majority of the white colonies may still be RNA independent. It is important to rigorously eliminate the remaining RNA-independent activators from among the white colonies (Steps 4 and 5) before
recovering plasmids in *E. coli*.

*Step 3. Assay β-Galactosidase Activity*

To corroborate that the white colonies contain cDNAs that activate
through the three-hybrid system, the level of expression of the *lacZ* gene

is monitored. In strain L40coat, the *lacZ* gene is integrated into the chromosome and placed under the control of LexA-binding sites.

β-Galactosidase can be assayed by measuring the conversion of a lactose analog to a chromogenic or luminescent product. This assay can be performed using colonies permeabilized on either a filter or a cell lysate. The filter assay yields qualitative results, whereas the liquid assay is more quantitative.

### Qualitative (Filter) Assays[19]

1. Restreak colonies from Step 2 onto the appropriate selective media (SD-Leu-Ura) and grow overnight.
2. Replica colonies onto plate-sized nitrocellulose filters or filter papers (Whatman, Clifton, NJ, 3MM).
3. Immerse filter in liquid nitrogen for 20 sec.
4. Allow filter to thaw on bench top (approximately 2 min).
5. Prepare petri dish-sized circles of 3MM Whatman paper, place in petri dish, and saturate with Z buffer (60 m$M$ $Na_2HPO_4$, 40 m$M$ $NaH_2PO_4$, 10 m$M$ KCl, 1 m$M$ $MgSO_4$, 50 m$M$ 2-mercaptoethanol, pH 7.0), supplemented with 300 $\mu$g/ml X-Gal. The X-Gal should be added fresh. Remove excess buffer.
6. Overlay filter onto Whatman paper and seal dish with parafilm.
7. Incubate 30 min to overnight at 30°. Examine the filters regularly.

A strong interaction (such as that between IRE and IRP) should turn blue within 30 min. With protracted incubation, weak interactions eventually yield a blue color. For this reason, it is important to examine the filters periodically to determine how long it takes for the color to develop.

*Quantitative (Liquid) Assays.* The specific activity of β-galactosidase can be determined in yeast cell lysates using any of a variety of substrates. Colorimetric assays using ONPG[20] or CPRG[21] are common; CPRG is more sensitive, but also more expensive. Alternatively, luminescent substrates provide high sensitivity yet are relatively inexpensive. The following protocol uses a luminescent substrate, Galacton-Plus (Tropix, Bedford, MA). The assay requires an instrument to detect luminescence. The following protocol was designed for a Monolight 2010 luminometer (Analytic Luminescent Laboratories, San Diego, CA). Certain details of the assay, such as sample volumes, will vary with the instrument used.

1. Inoculate 5-ml cultures of selective media in triplicate for each interaction to be tested. Grow overnight.

---

[19] L. Breeden and K. Nasmyth, *Cold Spring Harb. Symp. Quant. Biol.* **50,** 643 (1985).
[20] J. H. Miller, "Experiments in Molecular Genetics." Cold Spring Harbor Laboratory, Cold Spring Harbor, NY, 1972.
[21] K. Iwabuchi, B. Li, P. Bartel, and S. Fields, *Oncogene* **8,** 1693 (1993).

2. Inoculate fresh selective media to an OD of 0.1.

3. Grow to midlog phase (OD ~0.8).

4. Pellet ~1.0 OD units worth of cells for each culture.

5. Resuspend pellet in 100 $\mu$l of lysis buffer (100 m$M$ potassium phosphate, pH 7.8, 0.2% Triton X-100).

6. Lyse by freeze-thaw. This lysis requires three sequential cycles of freezing in liquid nitrogen for 10 sec followed by incubation at 37° for 90 sec.

7. Vortex each tube briefly.

8. Pellet in microfuge at 12,000g.

9. Collect supernatant for luminometer assays. If necessary, samples may be frozen at −70°.

10. Add 10 $\mu$l of lysate to luminometer tube. For strong interactions, it may be necessary to dilute the lysate to keep the assay in the linear range.

11. Dilute Galacton reagent 1 : 100 in reaction buffer (100 m$M$ sodium phosphate, 1 m$M$ magnesium chloride, pH 8.0). Add 100 $\mu$l to each luminometer tube.

12. Incubate at 25° for 60 min.

13. Measure luminescence as directed by luminometer manufacturer. Use light emission accelerator II from Tropix.

14. Measure protein concentration in lysates by Bradford or equivalent assay.

15. Normalize light emission by protein concentration, yielding a specific activity.

### Step 4. Cure RNA Plasmid and Test Again

Most but not all of the RNA-independent false positives are eliminated by the colony color screen in Step 2. To ensure that positives are genuinely RNA dependent, the RNA plasmid is removed by counterselection against *URA3*. Expression of the reporter gene is then monitored. Candidates that fail to activate the reporter genes are analyzed further.

*URA3 Counterselection.* To select for cells that have lost the RNA plasmid, cells are plated on media containing 5-fluoroorotic acid (5-FOA). 5-FOA is converted by the *URA3* gene product to 5-fluorouracil, which is toxic. Cells lacking the *URA3* gene product can grow in the presence of 5-FOA if uracil is provided, whereas cells containing the *URA3* gene product cannot.

1. Replica plate the positives from Step 2 to SD-Leu plates. Let the cells grow for a day, allowing cells to lose the plasmid.

2. Replica plate onto SD-Leu + 0.1% 5-FOA plates. Incubate at 30° for a few days.

3. Cells that grow can be streaked on a SD-Ura plate to confirm the loss of the RNA plasmid. A single pass through 5-FOA counterselection is usually sufficient.

4. Assay $\beta$-galactosidase activity.

The *ADE2* marker on the RNA plasmid can be useful in monitoring the loss of the RNA plasmid. Cells that lose the plasmid will turn pink in a few days. If the number of the positives is small, then the use of 5-FOA, which is quite expensive, can be avoided. Cells can simply be grown in rich media overnight and then spread onto SD-Leu plates. After a few days, some of the colonies become pink or show pink sectors. Uniformly pink colonies, which have lost the RNA plasmid, can be reassayed for $\beta$-galactosidase activity.

## Step 5. Determine Binding Specificity Using Mutant and Control RNAs

To test RNA-binding specificity, reintroduce plasmids encoding various hybrid RNAs into the strains that have been cured of their original RNA plasmid. If the number of positives is small, the various RNA plasmids can be introduced by transformation. Otherwise, mating is used to introduce the plasmids. Strain R40coat can be used for this purpose. A sample protocol for the mating assay follows.

1. Grow lawns of separate R40coat transformants carrying a specific hybrid RNA plasmid (e.g., mutant vs wild type) on SD-Ura plates.

2. Replica plate the grid of Ura colonies from the 5-FOA plate to a YPD plate.

3. Replica plate the lawn from each R40coat strain to the same YPD plate.

4. Incubate the plates overnight at 30° to allow mating.

5. Replica plate to SD-Leu-Ura plate to select for diploids.

6. Assay $\beta$-galactosidase activity.

*RNAs to be Used in Specificity Tests. Ab initio,* positives that have survived Step 4 carry proteins that bind preferentially to the hybrid RNA relative to cellular RNAs. Although these positives recognize some features on the hybrid RNA, these need not be the ones that are recognized by the biologically relevant factor(s). Therefore, the ideal controls are subtle (e.g., point) mutations that affect the biological functions or interaction in a sequence-specific manner. In some cases, an antisense bait can be used to control for secondary structure or base composition.

*Fraction of All RNA-Dependent Positives that Are "Correct."* The fraction of positives that are physiologically relevant is unpredictable at the outset, as it is a function of the abundance of the protein (in a random library screen), strength of the interaction, and level of 3AT used in the initial transformation, among other parameters. The following provides guidelines based on a published screen.

We screened a *Caenorhabditis elegans* cDNA activation domain library using a portion of the *fem-3* 3'UTR as bait, leading to the identification of a regulatory protein, FBF.[13] In this screen, 5 m$M$ 3AT was used in the initial transformation, selecting for reasonably strong interactions. In total, 5,000,000 transformants were analyzed, yielding approximately 5000 His⁺ Leu⁺ colonies. Of these, 100 were white. Sixty of these activated the *lacZ* gene. Twelve of the 60 proved to be genuinely RNA dependent, and 3 of the 12 displayed the appropriate binding specificity. Each of these was FBF, the genuine regulator. Thus, RNA-dependent positives were 0.2% of the total number of colonies obtained in the initial transformation. Importantly, 25% of all RNA-dependent positives were FBF. In screens that identified the protein (SLBP) that binds to the 3' end of histone mRNAs,[11,12] "correct" positives represented 70 to 100% of the total of the total His⁺ LacZ⁺ transformants.

*What if No Subtle RNA Mutations Are Available?* Additional screens must be devised to identify those positives that are correct. Clearly, functional tests are ideal, but in many organisms and systems these are time-consuming and labor-intensive. The sequence of the cDNAs may be directly informative by comparison to known RNA-binding proteins or by comparison to the predicted molecular weight of the expected protein (based on, for example, UV cross-linking). Because each case is idiosyncratic, no general discussion will be offered here. However, we caution that such secondary screens are critical.

### Step 6. Identify Positive cDNAs

cDNA plasmids that display the predicted RNA-binding specificity are recovered from the yeast cells and introduced into *E. coli* by transformation. The yeast cells can contain multiple cDNA plasmids, only one of which encodes the protein that binds to the RNA. Thus, plasmids should be isolated from multiple *E. coli* transformants and reintroduced into yeast to ensure that the correct plasmid has been obtained. A sample protocol follows. All steps are performed at room temperature.

1. Use a toothpick to transfer a yeast colony to 50 $\mu$l lysis solution (2% Triton X-100, 1% sodium dodecyl sulfate, 100 m$M$ NaCl, 10 m$M$ Tris–Cl, pH 8.0, 1 m$M$ EDTA).
2. Add 50 $\mu$l phenol/chloroform and ca. 0.1 g acid-washed glass beads.
3. Vortex at high speed for 2 min.

4. Spin at high speed in a microcentrifuge for 5 min.
5. Transfer the supernatant to a clean tube and ethanol precipitate the DNA. Wash with 70% (v/v) ethanol and dry the pellet.
6. Resuspend the pellet in 10 $\mu$l $H_2O$. Use 1 $\mu$l to transform electrocompetent *E. coli.*

For convenience, to determine how many plasmids are present in each yeast transformant, we often perform polymerase chain reactions (PCR) using lysed yeast colonies. To do so, we use primers that flank the inserts. The following is a protocol for yeast colony PCR.[22]

1. Touch the yeast colony with a sterile disposable pipette tip.
2. Rinse the tip in 10 $\mu$l incubation buffer [1.2 $M$ sorbitol, 100 m$M$ sodium phosphate, pH 7.4, and 2.5 mg/ml zymolyase (ICN Biomedicals, Inc., Costa Mesa, CA)] by pipetting up and down several times. Incubate at 37° for 5 min.
3. Remove 1 $\mu$l for a 20-$\mu$l PCR reaction. If desired, the PCR product can be purified by chromatography or purification from a gel and sequenced directly.

## Step 7. Functional Tests or Additional Screens

Almost invariably, additional steps will be needed to identify those positives that are biologically meaningful. As stated earlier (see Step 5), those screens are idiosyncratic, depending on the interaction and organisms studied. In a screen with an RNA "bait," it is not surprising that one might identify irrelevant RNA-binding proteins, as well as the legitimate interactor.

## Finding RNA Partner for Known RNA-Binding Protein

The three-hybrid system can be used to identify a natural RNA ligand for a known RNA-binding protein by screening an RNA library with a protein/activation domain fusion as "bait." Although only one such screen has been performed to date,[10] the objective is sufficiently general that we include discussion of our experience here.

A strain that carries a plasmid expressing an activation domain fusion with a known RNA-binding protein is transformed with a library of plasmids expressing hybrid RNAs, each composed of the MS2 coat protein-binding sites fused to an RNA element. In our initial experiments, the library

[22] M. Ling, F. Merante, and B. H. Robinson, *Nucleic Acids Res.* **23,** 4924 (1995).

consisted of fragments of yeast genomic DNA that are transcribed along with the coat protein-binding sites. We demonstrated that we could identify a fragment of the U1 RNA that binds to the yeast Snp1 protein in this type of search.[10] This hybrid RNA library should be useful in assigning RNA ligands to RNA-binding proteins of *Saccharomyces cerevisiae,* particularly those that have been classified as RNA binding solely on the basis of primary sequence data.

A three-hybrid search to identify RNAs follows much the same logic as that described earlier to identify proteins. One particular class of false positives that must be eliminated consists of RNAs that are able to activate expression of the reporter genes on their own without interacting with the protein fused to the activation domain (i.e., "protein-independent" positives).[10] A step must be included to classify RNAs as protein dependent, and thus worthy of additional analysis, or protein independent, and thus of no further interest in this context. Additionally, the protein-dependent class of RNAs must be tested with the activation domain vector and other RNA-binding protein fusions to identify those RNAs that are specific to the protein of interest.

### Step 1. Introduce Activation Domain Plasmid and Hybrid RNA Library

1. Transform L40coat with the plasmid expressing the RNA-binding protein as an activation domain hybrid, based on either the pACTII or the pACTII-CAN vector.
2. Transform cells from a single colony carrying this plasmid with the RNA library, selecting on media lacking tryptophan, leucine, uracil, and histidine and containing 0.5 m$M$ 3AT. Without any 3AT, most of the transformants will grow on a plate lacking histidine.

*Construction of RNA Library.* The yeast RNA library used in our experiments was constructed as follows.

Chromosomal DNA from *S. cerevisiae* is partially digested with the following four enzymes, listed with their recognition sequences: *Mse*I (TTAA), *Tsp*509I (AATT), *Alu*I (AGCT), and *Rsa*I (GTAC). The digests are pooled, and fragments in the size range of 50 to 150 nucleotides are purified from a preparative agarose gel. The ends of the digested DNA are filled in with the Klenow fragment of DNA polymerase I where required. Plasmid pIII/MS2-2 is digested with *Sma*I, treated with calf intestine phosphatase, and ligated to the blunt-ended genomic DNA fragments. The ligations are used to transform electrocompetent HB101 *E. coli.* DNA fragments cloned at the *Sma*I site of the pIII/MS2-2 are expressed such that the RNA sequence corresponding to the yeast genomic fragment is positioned 5' to the MS2 coat protein-binding sites within the hybrid RNA. More than 1.5 million *E. coli* transformants are obtained, pooled, and used to prepare plasmid DNA for the RNA library.

### Step 2. Screen for Activation of Second Reporter Gene

To ensure that activation of *HIS3* is not spurious, the level of expression of the *LacZ* gene is monitored.

1. Patch individual colonies that grew in the library transformation onto plates lacking leucine, tryptophan, and uracil.
2. Carry out filter β-galactosidase assays as described earlier.

### Step 3. Eliminate Protein-Independent False Positives

The identification of protein-independent false positives requires two successive steps. First, the activation domain plasmid must be removed from the strain. Then the level of LacZ expression must again be determined. The colonies that are LacZ$^+$ are cured of the activation domain plasmid by one of the following methods. If the plasmid is derived from pACTII, the plasmid is cured by growing a transformant overnight in YPD media and then plating for single colonies on SD-ura to select for the RNA plasmid. These colonies are then replica plated onto an SD-leu plate to determine which of the colonies lack the *LEU2* marker on the activation domain plasmid. If the activation domain plasmid is derived from pACTII/CAN, the plasmid can be cured by patching colonies onto a canavanine plate. For complete curing of the plasmid, a second patching onto a canavanine plate is generally required. Canavanine is prepared as a stock solution of 20 mg/ml, which is filter sterilized. The selective plates are SD-arg with 60 μg/ml canavanine.

Assay cells cured of the activation domain plasmid for β-galactosidase activity by filter assays, using the protocols described earlier. Colonies that have lost activity on loss of the activation domain plasmid contain possible RNA ligands of interest.

*How Common Are Protein-Independent Activators?* Protein-independent activators include Hybrid RNAs that activate transcription when bound to a promoter. In our experiments to date, the frequency of such "activating RNAs" in a genomic library is rather high. For example, 92% of all His$^+$ LacZ$^+$ positives obtained after selection in 0.5 m$M$ 3AT were protein independent.

*Step 4. Determine Binding Specificity*

As in cDNA library screening, specificity tests are necessary in secondary screens to narrow down the number of candidates. Of course, the more subtle the mutation used, the better.

Transform cells from Step 3 that contain an RNA plasmid and appear to be protein dependent for three-hybrid activity with control activation domain plasmids such as pACTII and an IRP1-activation domain fusion in pACTII. An RNA ligand that is specific for the RNA-binding protein used in the library screen should not produce a three-hybrid signal with these control plasmids.

*Step 5. Sequence RNAs of Interest*

Determine the sequence of the RNA ligand and test the binding by alternative methods such as *in vitro* binding.

*Fraction of Protein-Dependent Activators That Are "Correct".* In our experience with yeast Snp1 protein as "bait," we screened $2.5 \times 10^6$ transformants and obtained 13 that were protein dependent. Of these, the strongest by far was the appropriate segment of U1snRNA. Some of the other positives have weak sequence similarity to the relevant region of U1 and are now being analyzed further.

Prospects

The main focus of this review has been the use of the three-hybrid system to identify partners in an RNA–protein interaction. In principle, with minimal modifications, the system may be modified to detect factors that enhance or prevent an RNA–protein interaction, to identify RNA ligands that enhance or prevent a protein–protein interaction, and to detect an RNA–RNA interaction. Such applications have not yet been reported. As stated earlier, we have written this article in part to facilitate the development of the system by others, as well as to facilitate its use as is. In that spirit, we have included screening of both protein and RNA libraries, although relatively few such experiments have yet been published.

One asset of the three-hybrid method, shared with other genetic approaches to finding protein partners, is that a DNA clone of the interacting protein is obtained. With the proliferation of sequence databases and genomic information, a small bit of sequence information may be enough to help determine whether that protein is a legitimate partner or to shed light on its function. In some organisms, obtaining the clone immediately enables direct tests of its function *in vivo*. In others, alternative screening steps will be needed to identify meaningful mates from among the assembled suitors.

## Acknowledgments

We are grateful to the media laboratory of the biochemistry department of the University of Wisconsin for help with figures. We also appreciate the helpful comments and suggestions of members of both Wickens' and Fields' laboratories and communications of results from several laboratories prior to publication. Work in our laboratories is supported by grants from the NIH and NSF. S.F. is an investigator of the Howard Hughes Medical Institute.

## Section V

## Protein Engineering Methodologies Useful for RNA–Protein Interaction Studies

# [28] Using Peptides as Models of RNA–Protein Interactions

*By* COLIN A. SMITH, LILY CHEN, and ALAN D. FRANKEL

## Introduction

Some RNA–protein interactions can be modeled using relatively short peptides that bind RNAs specifically and mimic many characteristics of larger complexes. For our purposes, we arbitrarily distinguish RNA-binding peptides from other types of RNA-binding domains by considering only peptides <30 amino acids in length that do not fold into stable tertiary structures, although some peptides can form weak secondary structures on their own. The small size of these peptides allows one to dissect RNA–protein contacts relatively systematically and to engineer novel interactions with a desired RNA site. Most methods described in this article are adaptations of more general methods used to study RNA–protein complexes, and we attempt to highlight those aspects that pertain particularly to peptides.

## Overview

This article discusses general considerations for identifying and designing an RNA-binding peptide, methods for characterizing synthetic peptides, use of gel-shift assays for measuring RNA–peptide-binding constants, use of RNase footprinting, peptide affinity columns, and chemical modification interference to define the binding site and potential RNA–peptide contacts, and use of circular dichroism and absorption spectroscopy to help characterize peptide and RNA structure and binding-induced conformational changes. We use two well-characterized RNA–peptide complexes as examples: the bovine immundeficiency virus (BIV) TAR–Tat complex and the human immundeficiency virus (HIV-1) RRE–Rev complex (Fig. 1). In both cases, arginine-rich peptides recognize short hairpins containing asymmetric bulges with subnanomolar dissociation constants ($K_d$ values).

The BIV Tat protein binds to its RNA target, the transactivation response element (TAR), located at the 5' end of the viral transcripts and activates transcription. A 14 amino acid arginine-rich peptide corresponding to the RNA-binding domain of BIV Tat binds specifically to BIV TAR, and a combination of spectroscopic, biochemical, and transcriptional activation experiments have shown that the peptide makes a set of specific amino

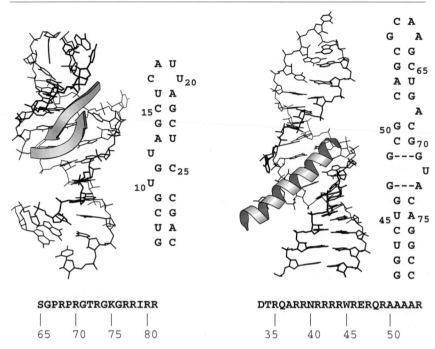

```
            A  U                          C  A
         C     U20                     G     A
            U  A                          C  G
         15 C  G                          G  C65
            G  C                          A  U
            A  U                          C  G
            U                          50 G  C  A
               G  C25                       C  G70
         10 U                             G---G
            G  C                                   U
            C  G                          G---A
            U  A                          G  C
            G  C                       45 U  A75
                                          C  G
                                          U  G
                                          G  C
                                          G  C
```

SGPRPRGTRGKGRRIRR                        DTRQARRNRRRRWRERQRAAAAR
 |    |    |    |                          |    |    |    |
 65   70   75   80                         35   40   45   50

Fig. 1. BIV TAR–Tat (left) and HIV-1 RRE–Rev (right) RNA–peptide complexes as determined by NMR.[3,4,7,8] The BIV Tat peptide adopts a $\beta$-hairpin conformation on specific RNA binding whereas the Rev peptide must be in an $\alpha$-helical conformation prior to binding. Also shown are the RNA hairpin secondary structures and peptide sequences, with previously reported numbering.[3,7]

acid–nucleotide contacts and adopts a $\beta$-turn conformation on RNA binding.[1–4] TAR forms a pseudo-continuous A-form helix with two unstacked bulged nucleotides. The HIV-1 Rev protein binds to its RNA target, the Rev response element (RRE), located in the viral *env* gene and facilitates transport of unspliced and partially spliced mRNAs from the nucleus to the cytoplasm. A 17 amino acid arginine-rich peptide from Rev binds specifically to a high-affinity binding site in the RRE (stem–loop IIB), and spectroscopic, biochemical, and transcriptional activation experiments have identified a set of specific amino acid–nucleotide contacts and have shown that the peptide

[1] L. Chen and A. D. Frankel, *Biochemistry* **33,** 2708 (1994).
[2] L. Chen and A. D. Frankel, *Proc. Natl. Acad. Sci. U.S.A.* **92,** 5077 (1995).
[3] J. D. Puglisi, L. Chen, S. Blanchard, and A. D. Frankel, *Science* **270,** 1200 (1995).
[4] X. Ye, R. A. Kumar, and D. J. Patel, *Chem. Biol.* **2,** 827 (1995).

binds the RRE in an $\alpha$-helical conformation.[5–8] The RRE contains two purine–purine base pairs (G : G and G : A) that help create a widened major groove that accommodates the $\alpha$ helix. In the RRE–Rev complex, both the peptide and RNA structures are stabilized on binding, and the entire complex may be viewed as a single folding unit.[9]

## Peptide Design and Characterization

To date, peptides that bind specifically to RNAs have been derived from proteins containing an arginine-rich RNA-binding motif or having highly localized clusters of basic amino acids, such as in viral coat proteins.[10] These peptides show a wide diversity of structures, including $\alpha$ helices, bent helices, and $\beta$ turns.[3,4,7,8,11–14] It is expected that other types of RNA-binding peptides (perhaps rich in aromatic residues, for example) will be identified in other proteins through genetic and structure–function analyses. Specific RNA-binding peptides have also been found by screening arginine-rich combinatorial libraries[15–17] and, in principle, it may be possible to identify peptides targeted toward many types of RNA-binding sites.

Once a specific RNA-binding peptide sequence has been identified, it is important to consider the surrounding context in which the peptide is placed in order to provide a proper structural framework. For example, the Rev peptide must be at least partially $\alpha$ helical in order to dock correctly against the RRE.[6] The helical conformation of a peptide may be stabilized by including favorable charge interactions at the two ends of the helix dipole (the alignment of peptide bonds in a helix generates a partial positive charge at the N terminus and a partial negative charge at the C terminus).

[5] J. Kjems, A. D. Frankel, and P. A. Sharp, *Cell* **67,** 169 (1991).

[6] R. Tan, L. Chen, J. A. Buettner, D. Hudson, and A. D. Frankel, *Cell* **73,** 1031 (1993).

[7] J. L. Battiste, H. Mao, N. S. Rao, R. Tan, D. R. Muhandiram, L. E. Kay, A. D. Frankel, and J. R. Williamson. *Science* **273,** 1547 (1996).

[8] X. Ye, A. Gorin, A. D. Ellington, and D. J. Patel, *Nature Struct. Biol.* **3,** 1026 (1996).

[9] R. Tan and A. D. Frankel, *Biochemistry* **33,** 14579 (1994).

[10] P. Ansel-McKinney, S. W. Scott, M. Swanson, X. Ge, and L. Gehrke, *EMBO J.* **15,** 5077 (1996).

[11] R. Tan and A. D. Frankel, *Proc. Natl. Acad. Sci. U.S.A.* **92,** 5282 (1995).

[12] Z. Cai, A. Gorin, R. Frederick, X. Ye, W. Hu, A. Majumdar, A. Kettani, and D. J. Patel, *Nature Struct. Biol.* **5,** 203 (1998).

[13] P. Legault, J. Li, J. Mogridge, L. E. Kay, and J. Greenblatt, *Cell* **93,** 289 (1998).

[14] M. A. Weiss, *Nature Struct. Biol.* **5,** 329 (1998).

[15] K. Harada, S. S. Martin, and A. D. Frankel, *Nature* **380,** 175 (1996).

[16] K. Harada, S. S. Martin, R. Tan, and A. D. Frankel, *Proc. Natl. Acad. Sci. U.S.A.* **94,** 11887 (1997).

[17] R. Tan and A. D. Frankel, *Proc. Natl. Acad. Sci. U.S.A.* **95,** 4247 (1998).

Thus, helical peptides should be synthesized with a neutral (acetyl) or negatively charged (succinyl) blocking group at the N terminus and a neutral amide group at the C terminus.[18] Positively or negatively charged amino acid side chains may also be positioned near the appropriate ends of the peptide sequence to help stabilize the helix dipole. A helical conformation may be stabilized further by adding several helix-promoting alanine residues at either or both peptide termini (typically four alanines are used to generate one helical turn).[18] Extending the peptide helix may be undesirable in cases where the extension generates steric clashes with the RNA. In general, if a peptide sequence is derived from the middle of a protein, it is reasonable to include an acetylated N terminus and amidated C terminus to mimic the normally adjacent peptide bond. It can also be useful to include an aromatic residue (tryptophan, tyrosine, or phenylalanine), if accommodated by the sequence, to allow peptide quantitation by absorption spectroscopy. A cysteine may be added to allow coupling to a solid support for the preparation of affinity columns. Nonnatural amino acids may be incorporated for structure–function analyses as desired.

Synthetic peptides should be purified to homogeneity by high-performance liquid chromatography (HPLC). Typically, peptides are eluted from a $C_4$ reversed-phase column using an acetonitrile gradient (0.2%/min) in 0.1% trifluoroacetic acid. Basic peptides <20 amino acids in length usually elute below 20% acetonitrile. Following lyophilization, peptides should be stored in water at $-80°$ and diluted as needed just before use. Peptide identity should be confirmed by mass spectrometry (electrospray is most often used and typically is performed by a local protein analysis facility). An accurate determination of peptide concentration is essential for most experiments, and the best method is quantitative amino acid analysis, also typically performed by a local facility. For peptides containing aromatic residues, absorption spectra may be used to estimate concentrations using the following extinction coefficients: tryptophan, 5700 $M^{-1}$ cm$^{-1}$ at 280 nm; tyrosine, 1300 $M^{-1}$ cm$^{-1}$ at 280 nm; and phenylalanine, 200 $M^{-1}$ cm$^{-1}$ at 250 nm.[19] For peptides without aromatic residues, absorbance at 230 nm (not a distinct peak) may be used, assuming a molar extinction coefficient per residue of 200 $M^{-1}$ cm$^{-1}$. It is also possible to estimate peptide concentrations and assess purity by native polyacrylamide gel electrophoresis. Peptides are stained with Coomassie blue and compared to a set of known standards, preferably using peptides of similar composition.

[18] S. Marqusee, V. H. Robbins, and R. L. Baldwin, *Proc. Natl. Acad. Sci. U.S.A.* **86,** 5286 (1989).
[19] C. R. Cantor and P. R. Schimmel, "Biophysical Chemistry." Freeman, New York, 1980.

Procedure: Peptide Quantitation by Native Gel Electrophoresis

1. Prerun a 15% polyacrylamide gel (0.8 mm thickness, 37.5 : 1 mono-acrylamide : bisacrylamide, 30 m$M$ sodium acetate, pH 4.5) for 45 min at room temperature in 30 m$M$ sodium acetate, pH 4.5. At this pH, most peptides will be positively charged and electrodes are reversed from typical gels, running from anode to cathode.
2. Load a series of peptide dilutions, estimated to be 0.5–5 $\mu$g, in running buffer with 10% glycerol, along with a standard peptide series, and electrophorese for 2–3 hr at 150 V.
3. Stain with Coomassie blue [0.006% (w/v) Coomassie brilliant blue in 10% (v/v) acetic acid, 5% (v/v) methanol] for 1 hr and destain (10% acetic acid, 5% methanol).
4. Visually estimate peptide concentration by comparing band intensities to the standards.

Measurement of Binding Affinities

The most commonly used method to measure binding constants of nucleic acid–protein complexes is the gel-shift assay (see Refs. 20 and 21 for more details). In general, the method works well for RNA–peptide complexes with $K_d$ values in the nanomolar range, provided that the binding reaction is at equilibrium and there is minimal dissociation during electrophoresis. For short peptides, it is important that the gel be run at low temperature to minimize dissociation. For weaker binding peptides, dissociation rates can be significant, although relative binding constants still can be derived.[22,23] Binding specificity typically is measured in competition experiments with unlabeled specific and nonspecific RNAs or by measuring affinities directly with radiolabeled wild-type and mutant RNAs. Because RNA structures are diverse and can provide many types of binding surfaces, it generally is best to compare mutants of similar structure to assess specificity. Adding competitor RNAs (such as tRNA) often helps minimize other nonspecific effects (binding to tubes, etc.), but may also decrease apparent binding constants, and it generally is easiest to measure relative, rather than absolute, binding constants. Filter-binding assays with short peptides are not advised because binding to the filter often competes with binding to the RNA.

[20] J. Carey, *Methods Enzymol.* **208,** 103 (1991).
[21] D. L. Black, R. Chan, H. Min, J. Wang, and L. Bell, *in* "RNA : Protein Interactions: A Practical Approach" (C. W. J. Smith, ed.), p. 109. Oxford Univ. Press, Oxford, 1998.
[22] K. M. Weeks and D. M. Crothers, *Biochemistry* **31,** 10281 (1992).
[23] K. S. Long and D. M. Crothers, *Biochemistry* **34,** 8885 (1995).

Binding conditions should be optimized for each RNA–peptide complex, in particular varying the type of buffer and pH, and the concentrations of monovalent salt, divalent metal ion (RNA folding is often dependent on $Mg^{2+}$), and RNA competitor. For the BIV TAR–Tat peptide complex, optimal conditions include Tris buffer, NaCl, and no divalent cation,[1] whereas conditions for the RRE–Rev peptide complex include HEPES buffer, KCl, and $MgCl_2$.[6] To measure apparent dissociation constants, the peptide concentration is varied and $K_d$ values are determined by quantitating free and bound RNAs and fitting to standard binding equations or estimated from the peptide concentration required to shift 50% of the RNA into the complex. RNA concentrations should be significantly lower than the $K_d$ (at least 10-fold), often requiring that RNAs are radiolabeled to high specific activity; if the RNA concentration approaches the $K_d$, binding will largely reflect the RNA:peptide stoichiometry. Representative gel-shift assays are shown with BIV TAR RNA and Tat peptides (Fig. 2). The

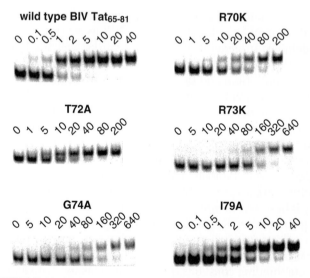

Fig. 2. BIV TAR binding by BIV Tat peptide mutants. Gel-shift assays were performed with a BIV Tat peptide corresponding to residues 65–81. Peptide and RNA were incubated together for 30 min on ice in 10-$\mu$l reactions containing 10 m$M$ Tris–HCl (pH 7.5), 70 m$M$ NaCl, 0.2 m$M$ EDTA, 5% glycerol, and 25 $\mu$g/ml yeast tRNA. To determine relative binding affinities, 0.1–0.25 n$M$ radiolabeled BIV TAR RNA was titrated with the indicated concentrations (n$M$) of peptides. RNA–peptide complexes were resolved on 10% polyacrylamide/0.5× TBE gels run at 200 V for 2.5 hr at 4°. In this experiment, apparent $K_d$ values (0.75–160 n$M$) were estimated visually as the peptide concentrations required to shift 50% of the unbound RNA into the complex.

wild-type peptide binds with an apparent $K_d \approx 0.75$ n$M$, whereas mutation of any of several key amino acids decreases affinity 3- to 200-fold.

Procedure: Gel-Shift Assay

Internally labeled RNAs are transcribed *in vitro*, purified, and annealed. The *in vitro* transcription conditions described have worked well for a variety of different RNA hairpins ~30–50 nucleotides in length but can be optimized as needed. Labeled RNAs are incubated with peptides at different concentrations, and bound and unbound RNAs are resolved on native polyacrylamide gels.

Materials

10× T7 buffer: 800 m$M$ HEPES–KOH, pH 8.1; 50 m$M$ dithiothreitol (DTT); 10 m$M$ spermidine; 0.01% Triton X-100

RNA elution buffer: 600 m$M$ sodium acetate, pH 6.0; 1 m$M$ EDTA; 0.01% sodium dodecyl sulfate (SDS).

10× renaturation buffer: 200 m$M$ Tris–HCl, pH 7.5; 1 $M$ NaCl

4× binding buffer (BIV complex): 40 m$M$ Tris–HCl, pH 7.5; 280 m$M$ NaCl; 0.8 m$M$ EDTA; 100 $\mu$g/ml yeast tRNA; 20% glycerol

Radioactive CTP: [$\alpha$-$^{32}$P]CTP (NEN, 3000 Ci/mmol, 10 mCi/ml, 3.3 $\mu M$). A higher specific activity label may be required for subnanomolar affinity interactions.

10× TBE: 900 m$M$ Tris, 900 m$M$ boric acid, 10 m$M$ EDTA

Transcription and RNA Purification

1. Anneal single-stranded DNA template (5 $\mu M$) with equimolar T7 promoter primer in 10 m$M$ Tris, pH 7.5, 100 m$M$ NaCl by heating to 65° for 5 min and slow cooling to room temperature.

2. Prepare transcription reaction (typically 40 $\mu$l) containing 1× T7 buffer, 80 mg/ml PEG-8000, 8 m$M$ GTP, 8 m$M$ UTP, 4 m$M$ ATP, 100 $\mu M$ unlabeled CTP, 0.6 $\mu M$ radiolabeled CTP, 40 m$M$ MgCl$_2$, 900 unit/ml human placental RNase inhibitor, 100 $\mu$g/ml T7 RNA polymerase, and 500 n$M$ annealed template. Incubate at 37° for 2 hr. Precipitated magnesium pyrophosphate may be visible.

3. Add an equal volume of deionized formamide to the 40-$\mu$l transcription reaction and load in one or two lanes of 15% acrylamide/8 $M$ urea gel (1.5 mm thickness, 19:1 monoacrylamide:bisacrylamide, 1× TBE). Unlabeled RNAs (~5 $\mu$g) of similar length can be run in adjacent lanes as standards.

4. Visualize the RNA by UV shadowing (at 254 nm) against a fluorescent TLC plate or by overlaying a brief autoradiogram. Excise the desired RNA band, crush gel, add 600 $\mu$l RNA elution buffer, and shake for 2 hr. Microfuge debris and transfer 400 $\mu$l supernatant to a fresh tube.

5. Precipitate the RNA with 2.5 volumes of ethanol, incubate on ice for 10 min, microfuge for 10 min at 4°, and wash pellet with 200 $\mu$l 75% ethanol. Dry pellet and dissolve in 100 $\mu$l water.

6. Count 1 $\mu$l to determine RNA concentration (see equation). For a typical 30 nucleotide hairpin, a 40-$\mu$l transcription reaction yields 3–30 pmol of RNA.

7. Renature the RNA (5–20 n$M$) in renaturation buffer by heating to $\geq$65° and slow cooling to room temperature. Store overnight at $-80°$. Overnight incubation often enhances gel shift quality.

### Quantitation of Radiolabeled RNA

Assuming a specific activity for labeled CTP of 3000 Ci/mmol and a concentration of 3.3 $\mu M$, and diluting 1 : 170 with unlabeled CTP, the following equation can be used to estimate the RNA concentration:

$$\frac{\text{measured dpm}}{\mu\text{l sample}} \times \frac{1\,\mu\text{Ci}}{10^6\,\text{dpm}} \times \frac{1\,\text{pmol}}{3\,\mu\text{Ci}} \times \frac{170}{\text{No. of C's in transcript}} = \mu M\,\text{RNA}$$

### Gel Shift

1. Prepare a 10% polyacrylamide gel (0.8 mm thickness, 37.5 : 1 monoacrylamide : bisacrylamide, 0.5$\times$ TBE) and chill to 4° overnight. Prerun for 1 hr in 0.5$\times$ TBE running buffer before loading samples.

2. Dilute renatured RNA in 2$\times$ binding buffer to ~2 n$M$. This concentration may vary depending on the binding affinity and competitor concentration, but a 1 n$M$ final RNA concentration is typical.

3. Serially dilute peptide in water and mix 5 $\mu$l RNA with 5 $\mu$l peptide. Incubate a minimum of 10 min on ice. Initially, incubation times should be varied to ensure that equilibrium has been reached.

4. Load binding reactions onto gel at 4°. Load bromphenol blue and xylene cyanole in a free well to monitor gel migration.

5. Electrophorese at 200 V for 3 hr (bromphenol blue migrates ~10 cm), dry gel under vaccuum for 30 min at 80°, and expose to film or phosphorimaging plate.

6. Determine apparent $K_d$ by quantitating the fraction of RNA free and bound and curve fitting or estimate from the peptide concentration that shifts 50% of the free RNA into the complex.

Footprinting

RNase and chemical footprinting are good methods to help localize sites of peptide binding. Nucleotides generally are protected from chemical or enzymatic attack in the presence of peptide, although sometimes enhanced accessibility is observed, typically due to conformational changes in the RNA. Several sequence-specific or structure-specific RNases are available: RNase A cleaves 3' to pyrimidines, RNase CL3 cleaves 3' to cytidines, RNase T1 cleaves 3' to guanines, RNase PhyM cleaves 3' to adenines and uridines, RNase T2 cleaves relatively nonspecifically at single-stranded regions, and RNase CV cleaves specifically at double-stranded regions. Several chemical reagents can be used to modify RNAs for footprinting: dimethyl sulfate (DMS) preferentially methylates N-7's of guanines, diethyl pyrocarbonate (DEPC) carbethoxylates N-7's of adenines, and hydroxyl radicals generated by reaction with Fe(II)–EDTA attack riboses. For most chemical reactions, sites of modification are identified using a mild base, typically aniline, to initiate strand cleavage or by primer extension using reverse transcriptase. For short RNAs that do not include primer-binding sites, end-labeled RNAs are used and cleavage sites are identified on sequencing gels, with alkaline hydrolysis ladders and RNase sequencing reactions as markers. It is often helpful to perform footprinting experiments at several peptide : RNA stoichiometries to confirm the patterns of protection or enhancement. In general, RNase footprinting is substantially simpler than chemical modification and is a reasonable choice to define the boundaries of peptide binding. Footprinting with peptides is similar to footprinting with proteins, and detailed descriptions of chemical and RNase methods can be found elsewhere.[24,25,25a,25b] Chen and Frankel[1] provide an example of RNase mapping of the BIV TAR–Tat peptide complex that defines the boundaries of a peptide-binding site.

Chemical Modification Interference

Chemical modification interference is an important method to help pinpoint specific nucleotides and chemical groups involved in peptide binding. RNAs are modified prior to binding, free and bound RNAs are sepa-

[24] C. Ehresmann, F. Baudin, M. Mougel, P. Romby, J. P. Ebel, and B. Ehresmann, *Nucleic Acids Res.* **15,** 9109 (1987).
[25] C. Merryman and H. F. Noller, *in* "RNA : Protein Interactions: A Practical Approach" (C. W. J. Smith, ed.), p. 237. Oxford Univ. Press, Oxford, 1998.
[25a] H. Moine, B. Ehresmann, C. Ehresmann, and P. Romby, in "RNA Structure and Function" (R. W. Simons, and M. Grunberg-Manago, eds.), p. 77. Cold Spring Harbor Lab. Press, NY, 1998.
[25b] C. Brunel and P. Romby, *Methods Enzymol.* **318,** [1], (2000) (this volume).

rated, and sites whose modification prevents binding are identified as missing bands on sequencing gels. Most sites of interference represent direct RNA–peptide contacts, although some modifications may inhibit binding indirectly by introducing a neighboring bulky group or altering the RNA structure. Several methods are available for separating the free and bound RNAs, and gel shift is the method used most commonly. We have found that peptide affinity columns are particularly useful for studying RNA–peptide complexes. Using a salt gradient to elute RNAs with increasing affinities, it is possible to examine the relative importance of each modified position by analyzing different fractions.[26] Following separation, modified sites are detected as blocks to primer extension by reverse transcriptase or by aniline-promoted cleavage. Details of the chemical modification, cleavage, and primer extension reactions have been described.[24,25] In one example with the BIV TAR–Tat interaction, we have used ethylnitrosourea (ENU) to ethylate phosphates, dimethyl pyrocarbonate (DMPC) to carbmethoxylate N-7's of adenines, hydrazine (HZ) to cleave pyrimidine rings, and DMS to methylate N-7's of guanines and have used a peptide affinity column to assess sites of interference (Fig. 3). Modification of G9, G11, A13, G14, C15, U16, G22, U24, and C25, and alkylation of phosphates 3′ to G9 and G22, interfered strongly with peptide binding, correlating closely to contacts observed in nuclear magnetic resonance (NMR) models of the BIV TAR–Tat complex.[3,4]

## Procedure: Coupling Peptide to Solid Support

A peptide containing a reactive cysteine (best located at the N or C terminus) may be coupled to $\omega$-aminohexyl agarose activated with the bifunctional cross-linker sulfo-SMCC.[1] Solutions used in this procedure should be deoxygenated thoroughly (by purging with argon or helium) to increase coupling efficiency, and metal ions should be kept to a minimum, using a chelating resin if highly deionized solutions are not available.

## Materials

Sulfo-SMCC: 4-(N-maleimidomethyl)cyclohexane-1-carboxylic acid 3-sulfo-N-hydroxysuccinimide ester (Pierce Chemical Co., Rockford, IL)
$\omega$-aminohexyl agarose (Sigma Chemical Corp., St. Louis, MO)
50 m$M$ sodium phosphate buffer (pH 7.4)

---

[26] J. Tao and A. D. Frankel, *Biochemistry* **35**, 2229 (1996).

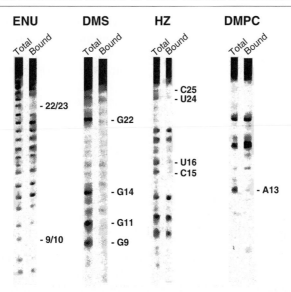

FIG. 3. Chemical modification interference of BIV Tat peptide binding to BIV TAR. Approximately 250–500 ng ($2 \times 10^6$ cpm) 5'-end-labeled BIV TAR RNA was mixed with 10 $\mu$g unlabeled RNA and modified at less than one site per molecule with ethylnitrosourea (ENU), dimethyl pyrocarbonate (DMPC), hydrazine (HZ), or dimethyl sulfate (DMS). Modification reactions were carried out under denaturing conditions, and RNA was ethanol precipitated twice and renatured. A mixture of 5 $\mu$g modified RNA and 5 $\mu$g nonspecific RNA (in this case a mutant of the related HIV-1 TAR site having a deletion of a 3-nucleotide bulge) was applied to a BIV Tat peptide affinity column, and RNAs were eluted at 4° with a salt gradient. Nonspecific RNA and RNAs modified at positions that interfere with binding eluted at 430 n$M$ NaCl, whereas specifically bound RNAs eluted at 900 n$M$ NaCl. Eluted RNAs were ethanol precipitated, cleaved at the modified positions by aniline, and analyzed on 17–20% polyacrylamide/8 $M$ urea sequencing gels. The cleavage patterns of specifically bound fractions were compared to total modified RNAs not loaded on the column, and positions whose modification interferes with specific binding are indicated. Numbering refers to the nucleotide positions shown in Fig. 1, and for ENU interference, the two nucleotides surrounding the phosphates are indicated.

1. Wash 1 ml $\omega$-aminohexyl agarose with 30 ml phosphate buffer and centrifuge for 3 min. Remove supernatant and repeat wash three times.
2. Add 5 ml 2.5 m$M$ sulfo-SMCC in phosphate buffer and stir for 30 min at room temperature. Sulfo-SMCC is unstable in water and should be prepared just prior to use.
3. Wash agarose with 30 ml phosphate buffer and centrifuge for 3 min. Remove supernatant and repeat wash three times.
4. Add ~1 mg reduced peptide in 2 ml phosphate buffer, transfer to a small glass vial, and stir for 2 hr at room temperature. Note that the concentration of coupled peptide will determine the elution proper-

ties of the affinity column and may be varied depending on the RNA-binding affinity. The concentrations given have been used for a 17 amino acid BIV Tat peptide.

5. Add 20 $\mu$l 500 m$M$ DTT to quench the reaction and stir for an additional 30 min.
6. Wash the resin twice with phosphate buffer as in step 1.

### Procedure: Fractionation of Modified RNAs Using Peptide Affinity Column

Labeled modified RNAs are loaded onto a peptide affinity column along with unlabeled nonspecific RNA that serves as a marker. RNAs are eluted in a salt gradient, cleaved with aniline, and visualized by denaturing gel electrophoresis and autoradiography. All column steps are performed at 4°.

1. Wash the peptide affinity column (0.5 ml) with 5 column volumes starting buffer (typically in 100 m$M$ NaCl).
2. Apply a mixture of unlabeled nonspecific RNA ($\sim$5 $\mu$g) and modified labeled specific RNA to the column. The best nonspecific RNA is a mutant RNA related in structure to the wild-type site.
3. Elute RNAs using a 100-ml salt gradient (typically 100 m$M$ to 1 $M$ NaCl) at 5 ml/min and collect 1-ml fractions. The gradient may be altered depending on the binding affinity, concentration of coupled peptide, and desired separation.
4. Monitor elution of the nonspecific RNA by measuring UV absorbance at 260 nm, and monitor elution of the labeled RNA using a Geiger counter. Because the RNA has been modified at a frequency of <1 site per molecule, the bulk of the labeled RNA will bind specifically and elute as a single peak late in the gradient.
5. Precipitate fractions with ethanol, cleave under conditions appropriate for each modification reaction, and analyze on a polyacrylamide sequencing gel. In the example shown in Fig. 3, the specifically bound fraction is compared to the total modified RNA prior to fractionation. Fractions eluting between nonspecific and specific positions of the gradient may be analyzed to assess the importance of particular positions in intermediate affinity complexes (see Ref. 26).

### Circular Dichroism

Circular dichroism (CD) spectroscopy, which monitors the absorption of left and right circularly polarized light, can provide preliminary information about peptide structure and conformational changes in peptide–RNA complexes. Base stacking interactions in nucleic acid helices generally give rise to CD ellipticity near 260 nm, whereas structured peptides (especially $\alpha$ helices) often have characteristic CD signals in the far-UV region.[27] Although peptides rarely form stable structures, the propensity of a peptide to adopt a particular conformation may be assessed by recording CD spectra in the absence of RNA. The arginine-rich, RNA-binding domain of HIV-1 Rev forms a marginally stable $\alpha$ helix, with characteristic CD minima at 208 and 222 nm (Fig. 4A). The fraction of $\alpha$ helix may be estimated from the intensity of the signal at 222 nm, and the fraction of $\beta$ sheet in a peptide may also be estimated, although its spectrum is less well defined.[28] In the example shown, the Rev helix was partially stabilized (to ~50%) by adding a succinyl group to the N terminus and four alanines and an amide group to the C terminus.[6] The helical conformation is further stabilized on RRE binding, as seen by a decrease in ellipticity at 222 nm when RNA is added.[9] It is possible to estimate the peptide $\alpha$-helix content even in the presence of RNA because the RRE contributes little CD signal at 222 nm. Additional information about the RNA may be obtained at higher wavelengths (260–280 nm) where peptides generally contribute little to the CD signal. In the RRE–Rev complex, the RNA undergoes a conformational change on binding, as seen by a change in the CD signal near 280 nm. The CD difference spectrum (in which the spectrum of the free RNA has been subtracted from the spectrum of the complex) is shown in Fig. 4B. It is interesting that, in this case, both the peptide and RNA structures are stabilized on binding, indicating a coupled folding event. The spectra shown were recorded on an Aviv 62DS spectropolarimeter in a 1-cm path length cuvette; details of the measurement will vary according to the instrument used. In general, CD-transparent reagents such as potassium phosphate buffer and fluoride salts should be used, with typical peptide concentrations of 10–20 $\mu M$ and RNA concentrations of 1–5 $\mu M$. Spectra of peptides and RNA–peptide complexes should be recorded at low temperature (4°) as these short structures usually are unstable. CD melting curves can provide further proof of peptide structure, as indicated by a cooperative transition.

## Absorbance Melting Curves

Peptide binding may stabilize or reorganize an RNA structure, and in such cases UV absorption spectroscopy can be a useful method to monitor the interaction. Changes in base pairing or stacking may increase or de-

[27] G. D. Fasman, "Circular Dichroism and the Conformational Analysis of Biomolecules." Plenum Press, New York, 1996.
[28] Y. H. Chen, J. T. Yang, and K. H. Chau, *Biochemistry* **13**, 3350 (1974).

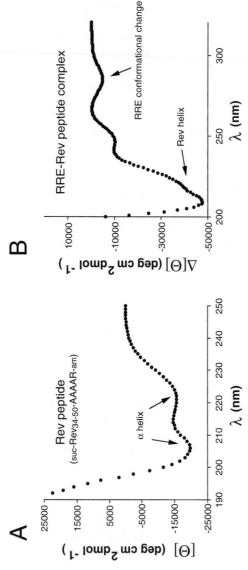

FIG. 4. CD spectrum of an α-helical Rev peptide (A) and difference CD spectrum of an RRE–Rev peptide complex (B). The Rev peptide shown in (A) contains a succinyl group at its N terminus and alanine residues and an amide group at its C terminus to help stabilize the helical conformation.[6] The peptide displays a characteristic helical spectrum with double minima at 208 and 222 nm and contains ~50% helix at 4° as estimated from the mean molar ellipticity at 222 nm.[28] In (B), spectra of the unbound RRE and RRE–Rev peptide complex (1:1 stoichiometry) were recorded, and the free RNA spectrum was subtracted from the spectrum of the complex. The signal near 222 nm largely reflects the helical peptide conformation (the RNA contributes little ellipticity at the wavelength), whereas the signal near 280 nm in the difference spectrum largely reflects changes in RNA base stacking on peptide binding.

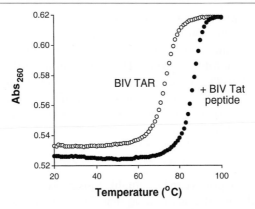

FIG. 5. UV absorbance melting curves of BIV TAR and the RNA–peptide complex. In this experiment, melting curves were recorded with an Aviv 14DS spectrophotometer equipped with a thermoelectric cell holder. RNA samples (2 $\mu M$) were renatured prior to melting experiments, and absorbance was monitored at 260 nm using a heating rate of 1°/min. The RNA–peptide complex was at a 1:1 stoichiometry.

crease absorbance at 260 nm, and in some complexes, increased RNA thermostability may be observed. At least in part, effects on thermostability probably reflect favorable electrostatic interactions with the peptide, analogous to ion shielding, and may result from specific or nonspecific contacts. Highly basic peptides are the most likely to cause increased thermostability, which will be observed only if the RNA unfolds at a lower temperature than the peptide. This is the case in the BIV TAR–Tat interaction, in which the peptide undergoes a disorder-to-order transition on RNA binding and requires the folded RNA scaffold to maintain its $\beta$-hairpin structure.[2,3] A set of thermodynamic values ($\Delta G$, $\Delta H$, and $\Delta S$) can be calculated from absorbance melting curves[29,30] and contributions of peptide interactions to the stability may be estimated.

In the example shown in Fig. 5, BIV TAR shows a single cooperative UV-melting transition with $T_m$ of 74°, $\Delta H$ of $-73$ kcal mol$^{-1}$, and $\Delta S$ of 210 cal K$^{-1}$ mol$^{-1}$.[2] In the presence of the BIV Tat peptide, the stability of the RNA increases, with $T_m$ of 88°, $\Delta H$ of $-93$ kcal mol$^{-1}$, and $\Delta S$ of 260 cal K$^{-1}$ mol$^{-1}$. Some of the stabilization may be attributed to an ionic effect, as the addition of 1 m$M$ Mg$^{2+}$ alone increases the $T_m$ by 5–6°.[2] Melting curves typically are monitored at 260 nm, with heating and equilibration times ~1 min/°. $T_m$ values may be determined by calculating the first derivative of the melting curve, and for unimolecular RNAs, $\Delta G$, $\Delta H$,

[29] J. D. Puglisi and I. Tinoco, Jr., *Methods Enzymol.* **180,** 304 (1989).
[30] M. T. Record, Jr., J. H. Ha, and M. A. Fisher, *Methods Enzymol.* **208,** 291 (1991).

and $\Delta S$ values may be estimated by calculating the concentrations of folded and unfolded molecules at each temperature, assuming a two-state transition and using baseline absorbance values (baselines must be well defined) for the folded and unfolded states.[29] The equilibrium constant can be calculated, $K = (f)/(1 - f)$, where $f$ is the fraction of bases paired, and values for $\Delta H$ and $\Delta S$ can be derived from van't Hoff plots ($\ln K$ versus $1/T$) using the standard equations $d \ln K/d(1/T) = -\Delta H/R$, $\Delta G = -RT \ln K$, and $\Delta S = (\Delta H - \Delta G)/T$.[29]

## Closing Points

RNA-binding peptides can be valuable tools in examining RNA–protein interactions and have the advantages that they are easy to synthesize, can be mutated systematically, and generally are amenable to NMR and other spectroscopic techniques. We have attempted to emphasize aspects of commonly used techniques that apply especially to RNA–peptide complexes, and we refer the reader to other articles in this volume that describe genetic, NMR, and other biophysical methods that also may be useful.

## Acknowledgments

This work was supported by a postdoctoral fellowship from the American Cancer Society to C.A.S. and by grants from the National Institutes of Health.

# [29] Analysis of RNA–Protein Cross-Link Sites by Matrix-Assisted Laser Desorption/Ionization Mass Spectrometry and N-Terminal Microsequencing

*By* Bernd Thiede and Brigitte Wittmann-Liebold

N-terminal sequencing and matrix-assisted laser desorption/ionization mass spectrometry (MALDI-MS) are well-established techniques for the analysis of peptides and proteins. An automatic N-terminal microsequence analysis is used routinely to determine the sequence of peptides and proteins by which the common 20 amino acids can be detected after Edman degradation by on-line reversed-phase high-performance liquid chromatography (RP-HPLC). Modified amino acids often can be observed by an unusual retention time or due to a gap within the sequence, but the nature of the modification cannot be determined by Edman chemistry. Mass spectrometry, however, is a modern tool for the determination of modified amino acids. MALDI-MS is not restricted to the analysis of peptides and proteins and has also been applied successfully to oligosaccharides, DNA, and RNA.[1] Furthermore, MALDI-MS can be used to obtain the sequence of such biomolecules by fragmentation with postsource decay (PSD) or in combination with chemical or enzymatic degradation.[2] An approach was developed using the features of these two methods to localize RNA–protein cross-link sites to single amino acids and nucleotides.

The complex ribonucleoprotein particles of the 30S[3] and 50S[4] ribosomal subunits of *Escherichia coli* are used as starting material. The general approach is presented in Fig. 1 with the 30S subunit as an example. The subunits were cross-linked by UV treatment alone or in combination with 2-iminothiolane. Free rRNA and rRNA cross-linked with ribosomal proteins are separated from the non-cross-linked proteins after precipitation and size-exclusion chromatography. Endoprotease digestion, followed again by precipitation and size-exclusion chromatography, produces rRNA cross-linked with ribosomal peptides and removes non-cross-linked peptides. Subsequently, oligoribonucleotide–peptide complexes are obtained after ribonuclease T1 digestion. An additional second treatment with the same endoprotease is performed to ensure complete digestion of the proteins, which may be protected by the large rRNA at the first digestion. The generated complexes are purified and separated by reversed-phase HPLC, and the presence of cross-links is identified by their absorbance at 220 and 260 nm.

The oligoribonucleotide–peptide complexes are analyzed by the combination of N-terminal sequencing and MALDI-MS. The complexes show a gap within the amino acid sequence at the cross-linked amino acid. By comparison with the Swiss-Protein database the sequence of the peptide is assigned to the parent ribosomal protein. Three MALDI-MS spectra are recorded to obtain the sequence of the oligoribonucleotide moiety and to determine the cross-linked nucleotide. Mass analyses are performed of the total complex, the complex after partial hydrolysis from the 5′ and 3′ end by treatment with aqueous ammonium hydroxide, and the complex after digestion with 5′ → 3′-phosphodiesterase to determine the 5′ end.

The precise localization of the RNA–protein contact sites permits arranging the known protein structures of ribosomal proteins into three-

[1] R. Kaufmann, *J. Biotechnol.* **41,** 155 (1995).

[2] M. Mann and G. Talbo, *Curr. Opin. Biotechnol.* **7,** 11 (1996).

[3] H. Urlaub, B. Thiede, E.-C. Müller, R. Brimacombe, and B. Wittmann-Liebold, *J. Biol. Chem.* **272,** 14547 (1997).

[4] B. Thiede, H. Urlaub, H. Neubauer, G. Grelle, and B. Wittmann-Liebold, *Biochem. J.* **334,** 39 (1998).

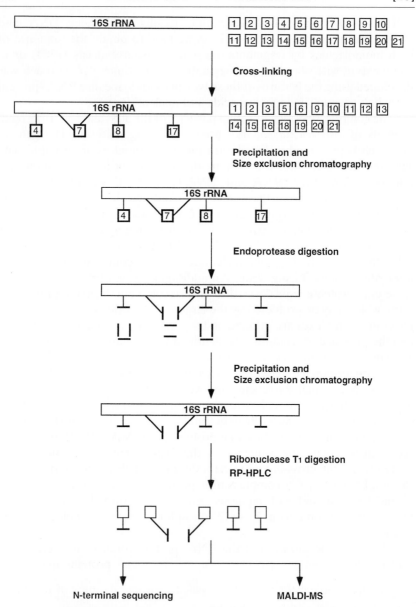

FIG. 1. Schematic presentation of the approach for the analysis of RNA–protein contact sites derived after cross-linking illustrated with the 30S ribosomal subunit of *E. coli*.[3] The ribosomal proteins are labeled 1 to 21 for S1 to S21. Empty boxes denote fragments of the rRNA cross-linked to short peptides.

dimensional molecular models of the rRNA by these fixing points.[5,6] Furthermore, an RNA–protein recognition pattern can be identified using the described approach. Cross-linking data of the ribosome reveal mainly hairpin loops and internal loops as secondary structure motifs of the rRNAs for protein recognition, and predominately loop structures of the ribosomal proteins are identified as RNA-binding sites. Furthermore, RNA-binding motifs are found within or close to the identified peptide stretches of the proteins, such as the K homology (KH) motif and the zinc-finger domain.[7]

This article describes and discusses the application of RP-HPLC, MALDI-MS, and N-terminal sequencing for the analysis of oligoribonucleotide–peptide complexes.

## RP-HPLC

### Procedure

One hundred fifty $A_{260}$ units (starting material for cross-linking) dissolved in 1 ml 25 m$M$ Tris–HCl, pH 7.8, and 2 m$M$ EDTA are injected into the HPLC system equipped with a $C_{18}$ column (250 $\times$ 4 mm; 300 Å). An isocratic elution with 90% buffer A (0.1% trifluoroacetic acid in water) and 10% buffer B (0.085% trifluoroacetic acid in acetonitrile) at a flow rate of 0.5 ml/min is performed until the absorbance returns to baseline, indicating removal of the free oligoribonucleotides. Subsequently, a linear gradient within 240 min from 10% buffer B to 45% buffer B at a flow rate of 0.5 ml/min is applied. Fractions containing an absorbance at 220 and 260 nm are evaporated using a Speed-Vac concentrator Savant, Life Sciences International GmbH, Frankfurt, Germany. The fractions are dissolved in 0.1% trifluoroacetic acid/49.9% water/50% acetonitrile for analyses by N-terminal sequencing and MALDI-MS.

### RP-HPLC Analysis

The oligoribonucleotide–peptide complexes are clearly identified due to the same intensity of the peaks at 220 and 260 nm. Peptide peaks are observed as well, but with no or a significantly lower intensity at 260 nm compared to the absorbance at 220 nm. The purity of the cross-links can be a severe problem for further analysis. Rechromatography of the cross-

[5] F. Mueller and R. Brimacombe, *J. Mol. Biol.* **271,** 545 (1997).
[6] I. Tanaka, A. Nakagawa, H. Hosaka, S. Wakatsuki, F. Mueller, and R. Brimacombe, *RNA* **4,** 542 (1998).
[7] H. Urlaub, V. Kruft, O. Bischof, E.-C. Müller, and B. Wittmann-Liebold, *EMBO J.* **14,** 4578 (1995).

links with the linear gradient as described earlier after precipitation with 0.1 volume of 3 $M$ sodium acetate, pH 5.5, and 2 volumes of ethanol at $-20°$ with 25 $\mu$g/500 ml glycogen as carrier[7] is performed if a cross-link cannot be identified unequivocally by N-terminal sequence analysis and MALDI-MS. Nevertheless, the loss of material is about 50% compared to the first HPLC run.

### MALDI-Mass Spectrometry

*Procedure*

Mass spectra are recorded with a MALDI time-of-flight (TOF) mass spectrometer (TofSpec, Fisons Instruments, Manchester, UK) equipped with a 337-nm nitrogen laser at 22 kV acceleration voltage. The dried-droplet sample preparation method is used: mix 0.7 $\mu$l matrix (saturated solution of $\alpha$-cyano-4-hydroxycinnamic acid in 0.1% trifluoroacetic acid/49.9% water/50% acetonitrile, v/v) with 0.5 $\mu$l sample solution on the sample holder. Spectra with the same quality can be obtained as well using the thin-layer preparation technique: saturated solution of $\alpha$-cyano-4-hydroxy-cinnamic acid in acetone/nitrocellulose (10 mg/ml acetone) (4:1), 2,5-dihydroxybenzoic acid (10 mg/ml in 70% aqueous 0.1 trifluoroacetic acid and 30% acetonitrile), and 2,4,6-trihydroxyacetophenone [20 mg/ml in methanol/100 m$M$ ammonium citrate (2:1)].[8] Spectra can be recorded in the positive as well as in the negative ion mode.

Alkaline hydrolysis is performed by adding 20 $\mu$l of aqueous ammonium hydroxide, pH 10.5, to the vacuum-dried sample and incubating for 15 min at 95°.

Hydrolysis with 5' → 3'-phosphodiesterase (calf spleen) is performed by adding 2 mU enzyme and 20 $\mu$l 1 m$M$ Tris–HCl, pH 7.8, to the vacuum-dried sample and incubating for 1 hr at 37°.

Samples are dried in the vacuum after hydrolysis and dissolved in 0.1% trifluoroacetic acid/49.9% water/50% acetonitrile, v/v, for mass analyses.

*Mass Analysis of Total Complex*

Mass spectra of cross-links with 2-iminothiolane show masses of the total complex and the corresponding peptide plus 2-iminothiolane spacer (101 Da). Therefore, the oligoribonucleotide composition can be calculated directly from the mass spectrum of 2-iminothiolane cross-links. In contrast,

[8] O. N. Jensen, S. Kulkarni, J. V. Aldrich, and D. F. Barofsky, *Nucleic Acids Res.* **24,** 3866 (1996).

only the mass of the total complex is observed by UV cross-links. Nevertheless, the oligoribonucleotide composition of UV cross-links can be calculated as well due to the simultaneously performed N-terminal peptide sequence analysis whereby the mass of the peptide can be determined. The accuracy of the composition analysis is dependent on the resolution of the mass spectrometer used. A mass accuracy better than 1 Da is usually obtained even with a MALDI-TOF instrument that employs continuous extraction and low resolution due to the fact that mass spectra can be calibrated by the mass of the corresponding peptide or by an internal standard. In the majority of mass spectra the cross-linked complex shows additional mass peaks of 22 Da (sodium ion), 26 Da (magnesium ion), or 24 Da (a mixture of sodium and magnesium ions). These adduct ions clearly help differentiate between peaks of cross-links and non-cross-linked peaks of peptides. In principle, the occurrence of other peptides within the same HPLC fraction was not a problem in mass analysis, but sometimes these peptides may have suppression effects on the complex and hence the sample must be purified further.

Mass analysis of the total complex enables determination of the oligoribonucleotide composition of different oligoribonucleotide–peptide cross-links in the ribosome, but this information is not sufficient to determine the cross-linking sites within large rRNAs. For this reason the oligoribonucleotides are partially hydrolyzed.

### Mass Analysis after Partial Hydrolysis with Ammonium Hydroxide

Mass ladders from the 5′ and 3′ end of the oligoribonucleotide moiety of the oligoribonucleotide–peptide complexes can be generated by partial alkaline hydrolysis. Aqueous ammonium hydroxide at pH 10.5 is selected as reagent due to its volatility. Most of the mass spectra show all possible fragments (Fig. 2). Therefore, it is essential to assign all peaks differing from the peaks of the mass spectrum of the total complex. Some characteristic properties help in analyzing the mass spectra. Higher peak intensities are obtained for products that are cleaved from the 3′ end in comparison to cleavage products from the 5′ end. The loss of G (363.2 Da) from the 3′ end due to the specificity of ribonuclease T1 is observed in any case. No further G could be detected within the sequences and the other nucleotides had lower mass differences (A, 329.2 Da; C, 305.2 Da; U, 306.2 Da). Therefore, products due to hydrolysis from the 5′ end have higher masses than those from the 3′ end if they have the same number of nucleotides. Analysis of mass spectra can be complicated due to the adduct ions as mentioned earlier and the additional mass of water (+18 Da) in some peaks. We are not able to distinguish between C and U using a mass spectrometer with a

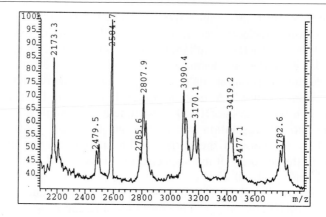

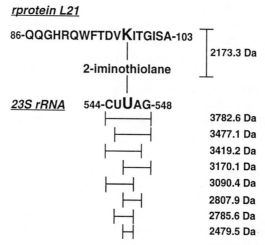

FIG. 2. MALDI mass spectrum of an oligoribonucleotide–peptide complex of the 50S ribosomal subunit after alkaline hydrolysis. The deduced oligoribonucleotide and amino acid sequence is presented.

low resolution (about 200) and continuous extraction, but with new MALDI instruments with delayed extraction,[9] this should not be a problem.

*Mass Analysis after Partial Hydrolysis with 5′ → 3′-Phosphodiesterase*

The phosphodiesterase is an exonuclease that attacks the 5′-terminal OH group and releases 3′ mononucleotides. The enzyme must be applied

[9] R. S. Brown and J. J. Lennon, *Anal. Chem.* **67,** 1998 (1995).

if the nucleotide sequence cannot be determined unequivocally by the alkaline hydrolysis. In our examples the phosphodiesterase was able to partially hydrolyze nucleotides from the 5' end, but not at the cross-linking site and the neighboring nucleotide. Therefore, sequence information is only obtained if the cross-linking position is not at the 5' end or next to it. However, in combination with alkaline hydrolysis, the whole oligoribonucleotide sequence and the single cross-linking position can be determined. The longer the sequence, the easier is the assignment to the rRNA sequence.

## N-Terminal Sequencing

### Procedure

The dissolved sample is applied onto a BioBrene precoated filter and introduced in an automatic N-terminal sequencer (Procise, Applied Biosystems, Foster City, CA). The N-terminal sequence analysis is performed as described by the manufacturer in the pulsed-liquid mode.

### N-Terminal Sequence Analysis

The N-terminal sequence analysis of oligoribonucleotide–peptide complexes is used to determine the peptide as well as the cross-linking position. A gap within the sequence defines the cross-linked amino acid. Lysine is found to be the cross-linked amino acid in 2-iminothiolane cross-links, whereas tyrosine or methionine is observed in UV cross-links. Analysis of the amino acid sequence and the cross-linking position is relatively simple if only one peptide is detected. With a mixture of two peptides, often one sequence dominates. Therefore, the analysis is simple due to the fact that only ribosomal proteins from one subunit of a defined organism must be considered. Nevertheless, N-terminal sequences of three or more peptides within one sample are difficult to interpret and precipitation and rechromatography as described earlier are recommended.

## Conclusion and Outlook

This article describes the techniques in use in our laboratory for the analysis of RNA–protein cross-links in the ribosome after reduction in size to oligoribonucleotide–peptide complexes. These techniques enabled the precise determination of RNA–protein contact sites within the ribosome for single molecules at the RNA and protein level with high sensitivity. This approach promises to be applicable for other RNA–protein complexes. Other techniques, such as nanoelectrospray ionization tandem mass spectrometry (nanoESI-MS/MS),[10] may replace using two different techniques. In fact, preliminary results demonstrate that the oligoribonucleotide moiety of the complexes can be sequenced using nanoESI-MS/MS, whereas the peptide fragments showed low intensities.

[10] M. Wilm and M. Mann, *Anal. Chem.* **68,** 1 (1996).

## [30] *In Vitro* Reconstitution of 30S Ribosomal Subunits Using Complete Set of Recombinant Proteins

*By* Gloria M. Culver and Harry F. Noller

### Introduction

The ability to reconstitute functional 30S ribosomal subunits from individually purified components has greatly aided the study of the structure, function, and assembly of this complex ribonucleoprotein particle.[1,2] However, obtaining large amounts of highly pure components, especially ribosomal proteins, can be costly, laborious, and fraught with difficulties. To this end, we have developed methods to overexpress and purify large quantities of each of the *Escherichia coli* small-subunit ribosomal proteins from plasmid-borne copies of the ribosomal protein genes. Because standard reconstitution procedures resulted in very inefficient reconstitution using these recombinant proteins, we have developed an ordered assembly protocol that results in efficient *in vitro* reconstitution of functional 30S subunits using natural 16S rRNA and recombinant small-subunit ribosomal proteins. The ordered assembly procedure is based on the addition of recombinant proteins following the pathway of *in vitro* assembly implied by the assembly map for the 30S subunit.[1,2] Therefore, large amounts of pure, individual small-subunit ribosomal proteins can be isolated for a variety of uses, and the potential exists for preparing large amounts of 30S subunits containing mutant or individually modified components, such as Fe(II)-derivatized proteins.[3]

[1] S. Mizushima and M. Nomura, *Nature* **226,** 1214 (1970).
[2] W. A. Held, S. Mizushima, and M. Nomura, *J. Biol. Chem.* **218,** 5720 (1973).
[3] G. M. Culver and H. F. Noller, *Methods Enzymol.* **318** [31] (2000) (this volume).

Reagents

All glassware, plasticware, reagents, and buffers should be handled with care to avoid nuclease contamination.[3]

Preparation of 16S rRNA follows a standard procedure that has been described previously.[4]

*Escherichia coli* strain BL21 (DE3) and the vector pET24b are from Novagen. All chemicals are purchased from Sigma Chemical Company (St. Louis, MO) except where noted.

Urea, ultrapure (ICN, Costa Mesa, CA).

Methanol, glacial acetic acid, and formamide (Fisher Scientific, Santa Clara, CA); formamide is super pure grade.

Nikkol (octaethylene glycol mono-*n*-dodecyl ether; $C_{12}E_8$) (Calbiochem, La Jolla, CA).

Isopropyl-$\beta$-D-1-thiogalactopyranoside (IPTG) (Quantum Biochemicals, Montreal, Canada).

Bradford reagent (Bio-Rad, Hercules, CA): Microassay procedure is followed as outlined in the company's literature.

Dialysis tubing [3500 molecular weight cutoff; Fisher Scientific Spectra/Por]: soak in water and then in appropriate dialysis buffer prior to use.

Centricon 3 and Centricon 100 ultraconcentrators and Microcon 100 microconcentrators (Amicon, Danvers, MA): Pretreated with 5% Tween 20 (Fisher Scientific; 200 or 50 $\mu$l, respectively), washed extensively with water, and finally washed in appropriate buffer prior to use; once wetted, the filters should never be allowed to dry.

Acrylamide : Bisacrylamide (29 : 1) (Amresco): Ultrapure grade; referred to later as polyacrylamide (29 : 1). Store at $4°$.

Buffers

Preparation of all the buffer components follows standard procedures as outlined in detail in Sambrook *et al.*[5]

Buffer A: 80 m$M$ $K^+$–HEPES (pH 7.6), 20 m$M$ $MgCl_2$, 330 m$M$ KCl, and 0.01% Nikkol (v/v). Store at $4°$.

Buffer B: 20 m$M$ Tris–HCl (pH 7.0), 20 m$M$ KCl, 6 $M$ urea, and 6 m$M$ 2-mercaptoethanol (BME). Store at $4°$.

Buffer B HIGH: 20 m$M$ Tris–HCl (pH 7.0), 1 $M$ KCl, 6 $M$ urea, and 6 m$M$ BME. Store at $4°$.

[4] D. Moazed, S. Stern, and H. F. Noller, *J. Mol. Biol.* **187,** 399 (1986).
[5] J. Sambrook, E. F. Fritsch, and T. Maniatis, *in* "Molecular Cloning" (N. Ford, ed.), p. B.1. Cold Spring Harbor Press, Cold Spring Harbor, NY, 1989.

Buffer C: 20 m$M$ sodium acetate (pH 5.6), 20 m$M$ KCl, 6 $M$ urea, and 6 m$M$ BME. Store at 4°.

Buffer C HIGH: 20 m$M$ sodium acetate (pH 5.6), 1 $M$ KCl, 6 $M$ urea, and 6 m$M$ BME. Store at 4°.

Buffer D: 80 m$M$ K$^+$–HEPES (pH 7.6), 20 m$M$ MgCl$_2$, 1 $M$ KCl, and 6 m$M$ BME. Store at 4°.

Buffer E: 20 m$M$ K$^+$–HEPES (7.6), 20 m$M$ KCl, and 6 m$M$ BME. Store at 4°.

Buffer F: 20 m$M$ K$^+$–HEPES (pH 7.6) and 20 m$M$ MgCl$_2$. Store at 4°.

Buffer G: 50 m$M$ K$^+$–HEPES (pH 7.6), 100 m$M$ KCl, and 10 m$M$ MgCl$_2$. Store at 4°.

Buffer H: 80 m$M$ K$^+$–HEPES (pH 7.6), 20 m$M$ MgCl$_2$, and 0.01% Nikkol. Store at 4°.

SDS gel sample buffer: 0.12 $M$ Tris–HCl (pH 6.8), 4% sodium dodecyl sulfate (SDS), 20% glycerol, 8 $M$ urea, 750 m$M$ BME, and 0.05% bromphenol blue. Store at −20°.

SDS gel running buffer: 25 m$M$ Tris–base, 192 m$M$ glycine, and 0.01% SDS. Store at room temperature.

4% stacking gel: 4% polyacrylamide (29:1), 125 m$M$ Tris–HCl (pH 6.8), and 0.1% SDS. Store at 4°.

12% resolving gel: 12% polyacrylamide (29:1), 375 m$M$ Tris–HCl (pH 8.8), and 0.1% SDS. Store at 4°.

Methylene blue stain: 0.5 $M$ sodium acetate (pH 5.2) and 0.04% (w/v) methylene blue. Store at room temperature.

Formamide loading dye: 95% formamide, 20 m$M$ EDTA (pH 8.0), 0.05% xylene cyanol, and bromphenol blue. Store at −20°.

Coomassie blue stain: 40% (v/v) methanol, 10% (v/v) acetic acid, and 0.1% (w/v) Coomassie blue R-250. Store at room temperature.

Coomassie destain: 40% (v/v) methanol and 10% (v/v) acetic acid. Store at room temperature.

1× TBE: 0.09 $M$ Tris–borate and 0.002 $M$ EDTA (pH 8.0). Store at room temperature.

3× sucrose gradient buffer: 60 m$M$ K$^+$–HEPES (pH 7.6), 60 m$M$ MgCl$_2$, and 1 $M$ KCl. Store at 4°.

Luria-Bertani medium (LB): 10% Bacto-tryptone, 10% Bacto-yeast extract, and 5% NaCl. Autoclave for 25 min at 15 psi on liquid cycle. Store at room temperature.

LB agar plates: 10% Bacto-tryptone, 10% Bacto-yeast extract, 5% NaCl, and 15% Difco agar. Autoclave for 25 min at 15 psi on liquid cycle, cool to 55°, add kanamycin to a final concentration of 50 $\mu$g/ml, and pour into presterilized polystyrene petri plates (100 × 15 mm). Store plates at 4°.

Kanamycin stock: 50 mg kanamycin in 1 ml water. Store at −20°.
In all cases, the BME is added just prior to use.

## Equipment and Supplies

Pharmacia (Piscataway, NJ) LKB FPLC automated system with Controller LCC-501 Plus, Pumps P-500, UV-M II UV monitor, Frac-100 fraction collector, optical unit with Hg lamp, 5 m$M$ flow cell and a 280-nm filter, FPLC director and FPLC assistant
FPLC (fast protein liquid chromatography) columns and accessories: Pharmacia Resource S, 6-ml column, Resource Q, 6-ml column, and 10-ml superloop
Branson Sonifier 450 with microtip attachment
Bausch and Lomb refractometer, ABBE-3L
Isco density gradient fractionator model 183 with accompanying UV monitor ($A_{254}$) and chart recorder
Hoefer Gradient Former
Beckman induction drive centrifuge
Beckman JA-10 and JA-20 centrifuge rotors
Corning (Corning, NY) Corex Rubber sleeve adapter for JA-20
Beckman Ultraclear 14 × 89-mm centrifuge tubes
Beckman swinging bucket ultracentrifuge rotor SW41
Beckman L7-65 ultracentrifuge
Bio-Rad (Hercules, CA) Mini-PROTEAN II electrophoresis cell
Becton-Dickinson 50-ml Falcon tubes
Fisher brand sterile polystyrene petri plates

## Overexpression of Small Subunit Ribosomal Proteins

Wild-type clones of ribosomal proteins S2-S21 in the vector pET24b are carried in *E. coli* strain BL21(DE3), where the proteins can be overexpressed from an inducible T7 promoter on the plasmid.[6] For large-scale, individual overexpression of each of the small subunit ribosomal proteins, a 10-ml culture of LB containing 50 $\mu$g/ml kanamycin in water (10 $\mu$l of 50 mg/ml kanamycin stock in water) is inoculated by picking a single colony from a streak of the BL21(DE3) strain harboring the pET24b/ribosomal protein gene plasmid. (The original streak is prepared from a glycerol stock of the same strain on an LB agar plate containing 50 $\mu$g/ml kanamycin.) These cultures (10 ml) are grown overnight in a 37° room in a roller drum.

---

[6] F. W. Studier, A. H. Rosenberg, J. J. Dunn, and J. W. Dubendorff, *Methods Enzymol.* **185,** 60 (1990).

A 500-ml LB culture containing 50 $\mu g/ml$ kanamycin is inoculated with the 10-ml saturated overnight culture and is grown in a 37° room on a rotary shaker for approximately 3 hr until the $OD_{600}$ is between 0.4 and 1.0. To induce protein production, 5 ml of 100 m$M$ IPTG is added to the 500-ml culture (final IPTG concentration of 1 m$M$) and growth is continued at 37° on a rotary shaker for 4 hr. Cultures are split into two chilled 250-ml bottles and cells are pelleted by centrifugation in a JA-10 rotor at 5000 rpm for 20 min at 4°. The supernatant is discarded, the cell pellet is washed with 50 ml ice-cold buffer E, and cells are harvested by centrifugation at 5000 rpm for 20 min at 4°. Washed pellets are stored at −20°. Ribosomal proteins are overexpressed at different levels and have different solubilities; both of these properties are reflected in the final amount and concentration of isolated protein that is obtained (see Table II).

To analyze the level of overexpression of the individual proteins, 50 $\mu l$ of induced cell culture (prior to harvest and washing) is mixed with 50 $\mu l$ SDS gel sample buffer, and 30 $\mu l$ of this mixture is analyzed by SDS gel electrophoresis (4% stacking gel : 12% resolving gel, both containing 6 $M$ urea).[7] Gels (Mini-PROTEAN II, 1 mm thick) are run at 180 V for 1.5 hr in SDS gel running buffer at room temperature. The gels are stained with 250 ml Coomassie stain by shaking for 1 hr at room temperature and are destained by shaking with multiple washes with Coomassie destain (Fig. 1A). As a control, the BL21(DE3) strain harboring only the pET24b vector (no ribosomal protein gene insert) should be grown and induced identically to strains overexpressing the ribosomal proteins (Fig. 1A, lane 1).

## Disruption of Cells Containing Overexpressed Proteins

For disruption of cells and subsequent protein purification, it is convenient to work with as many as three to five protein samples at a time. Cell pellets from 500 ml of culture, containing the overexpressed proteins, are thawed at 4° in 15 ml of buffer E. Sonication of the mixture is done at 4° in 50-ml Falcon tubes using a microtip attachment at constant duty cycle with pulses of 30 sec on, 90 sec off for six cycles. Cell lysates containing overexpressed insoluble proteins do not appear to change significantly during sonication, whereas cell lysates containing overexpressed soluble proteins change appearance from cloudy to almost clear during the course of sonication. Postsonication centrifugation at 10,000 rpm in a JA20 rotor for 10 min at 4° either clears the protein-containing lysate (soluble) or pellets the protein-containing inclusion bodies (insoluble). Inclusion-body pellets can be distinguished from unlysed cells by their very distinct appear-

[7] U. K. Laemmli, *Nature* **227,** 680 (1970).

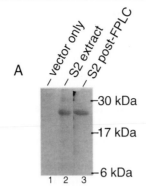

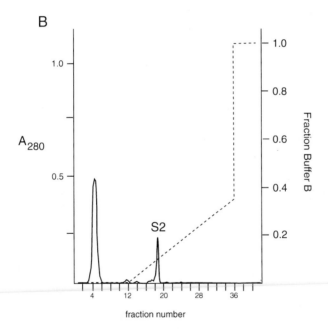

Fig. 1. Overexpression and purification of recombinant ribosomal protein S2. (A) Overexpressed and purified ribosomal protein S2: vector only, cells harboring pET24b; S2 extract, cells harboring ribosomal protein gene encoding S2 cloned into pET24b; S2 post-FPLC, FPLC-purified recombinant ribosomal protein S2. (B) Chromatogram of FPLC purification of ribosomal protein S2. Sample was loaded in buffer A (0% buffer B), and a linear 125-ml gradient of buffer B was introduced starting at fraction 12. The peak corresponding to S2 is indicated.

ance, which is very white and somewhat flaky, whereas cell pellets tend to have a yellowish hue and appear more uniform and smooth. Some of the ribosomal proteins are distributed between the soluble fraction and inclusion bodies, whereas others are found almost exclusively in the soluble fraction or in inclusion bodies as a result of overexpression (see Table I); the amount of protein found in the inclusion body pellet is generally a function of induction and growth time. Because subsequent purification is performed under denaturing conditions, a substantial portion of protein can generally be recovered from these pellets (see later).

## Purification of Recombinant Ribosomal Proteins

For soluble proteins (S4, S7, S10, S12–S16, S19, and S20; Table I), the cleared cell lysate (see earlier discussion) is placed in dialysis tubing (molecular weight cutoff 3500) and dialyzed overnight at 4° against these 1-liter changes of buffer B (S4, S7, S12–S16, S19, and S20) or buffer C (S10).

TABLE I

PROGRAMMING AND RUNNING FPLC FOR PURIFICATION OF
OVEREXPRESSED SMALL SUBUNIT RIBOSOMAL PROTEINS

| Time (min) | Function |
|---|---|
| 0.0 | Injection valve to inject position |
| | S2 is injected onto column from superloop |
| 0.0 | Fraction collector turned on |
| | 5-ml fractions |
| 0.0 | Concentration C HIGH-0% |
| 0.0 | Flow rate |
| | 1 ml/min |
| 15.0 | Injection valve to load position |
| | Superloop is bypassed |
| 23.0 | Flow rate |
| | 5 ml/min |
| 27.0 | Begin gradient |
| | Concentration buffer C HIGH-0% |
| 51.0 | End gradient |
| | Concentration buffer C HIGH-35% |
| 51.2 | Start high salt wash |
| | Concentration buffer C HIGH-100% |
| 53.0 | Fraction collector turned off |
| 55.0 | End high salt wash |
| | Concentration buffer C HIGH-100% |
| 55.2 | Reequilibrate system in low salt buffer |
| | Concentration buffer C HIGH-0% |
| 60.4 | End |

Inclusion body pellets containing overexpressed protein are resuspended in 12 ml ice-cold buffer B (S3, S5, S8, S9, S11, S18, and S21) or in 12 ml ice-cold buffer C (S2, S6, and S17) by pipetting up and down and gentle flicking of the tubes; this process can take up to 30 min and care should be taken not to lodge particles in the pipette. Some of these pellets are particularly hard to resuspend (S9, S11, and S18), even in buffer containing 6 *M* urea, and all of the particulate material may not go into solution. Thus, the yields and resulting protein concentrations can be lower for later proteins than for others, such as S6 (see Table II). The resuspended mixtures are transferred to dialysis tubing (molecular weight cutoff 3500) and dialyzed over-

TABLE II

PROPERTIES OF *E. coli* RIBOSOMAL PROTEINS S2–S21 RELEVANT TO PURIFICATION OF RECOMBINANT PROTEINS

| Protein | Solubility[a] | Buffer[b] | Elution[c] (KCl, m*M*) | Molecular weight[d] | Concentration of purified protein (mg/ml)[e] | Yield[f] |
|---------|---------------|-----------|------------------------|---------------------|----------------------------------------------|----------|
| S2  | I | C | 140 | 26,613 | 1.5  | 70   |
| S3  | I | B | 210 | 25,852 | 0.3  | 6.0  |
| S4  | S | B | 200 | 23,137 | 3.0  | 47   |
| S5  | I | B | 160 | 17,515 | 1.5  | 50   |
| S6  | I | C | 140 | 15,704 | 20.0 | 430  |
| S7  | S | B | 200 | 19,732 | 0.7  | 50   |
| S8  | I | B | 110 | 13,996 | 2.0  | 70   |
| S9  | I | B | 240 | 14,725 | 0.2  | 5.0  |
| S10 | S | C | 170 | 11,736 | 0.3  | 10   |
| S11 | I | B | 230 | 13,728 | 0.5  | 2.5  |
| S12 | S | B | 230 | 13,606 | 3.0  | 40   |
| S13 | S | B | 230 | 12,968 | 2.0  | 50   |
| S14 | S | B | 270 | 11,191 | 3.0  | 60   |
| S15 | S | B | 210 | 10,137 | 2.0  | 75   |
| S16 | S | B | 190 | 9,191  | 7.5  | 65   |
| S17 | I | C | 190 | 9,573  | 0.5  | 25   |
| S18 | I | B | 250 | 8,896  | 2.0  | 30   |
| S19 | S | B | 260 | 10,229 | 2.0  | 90   |
| S20 | S | B | 230 | 9,553  | 2.0  | 40   |
| S21 | I | B | 230 | 8,369  | 2.0  | 40   |

[a] Solubility of the majority of overexpressed protein in extract; I, insoluble; S, soluble.

[b] Buffer used for dialysis and FPLC column chromatography: (B) 20 m*M* Tris–HCl (pH 7.0), 20 m*M* KCl, 6 *M* urea, and 6 m*M* BME and (C) 20 m*M* sodium acetate (pH 5.6), 20 m*M* KCl, 6 *M* urea, and 6 m*M* BME.

[c] Concentration of KCl where the peak of overexpressed protein elutes from the column.

[d] From Giri *et al.*[19]

[e] Concentration of purified protein as determined by Bradford assay, from a standard preparation starting with a 500-ml culture, typical recoveries post-FPLC and dialysis.

[f] Yield, typical total yield, in milligrams, of purified protein obtained from a 500-ml culture.

night at 4° against two 1-liter changes of the same buffer. Dialyzed protein samples are centrifuged at 10,000 rpm for 15 min at 4° in a JA20 rotor to remove any insoluble or precipitated material.

An automated FPLC system is used for protein purification. Proteins S2–S5 and S7–S21 are purified at 4° by FPLC cation-exchange chromatography using a Resource S column. S6 is purified at 4° by FPLC chromatography on a Resource Q anion-exchange column, due to its unusually (for ribosomal proteins) low pI.[8] For each protein, the appropriate buffer for chromatography and the concentration of KCl at which each protein elutes from the column are given in Table II, along with other relevant information. The purification of ribosomal protein S2 is presented as an example and is described in detail later. The method described in Table I is used to program the FPLC system and to purify all of the small subunit ribosomal proteins. The purification scheme used here is generalized for purifying a complete set of small subunit ribosomal proteins and can be optimized to increase the yield of individual proteins.

Example: Purification of S2

Prior to chromatography, the FPLC system is equilibrated with buffer C. The system (including the Resource S column) is washed with 50 ml buffer C, followed by a 50-ml wash with buffer C HIGH, and is reequilibrated by washing with 100 ml low salt buffer (buffer C). The dialyzed, cleared S2 lysate (approximately 10 ml; Fig. 1A, lane 2) is loaded manually into the 10-ml Superloop. At time zero, the fraction collector turns on and is set to collect 5-ml fractions. At a flow rate of 1 ml/min, the S2 lysate is injected automatically into the Resource S column, followed by a 5-ml wash of the loop with buffer C. The injection loop is then isolated (closed off) from the rest of the system. The column is washed with an additional 8 ml buffer C at a flow rate of 1 ml/min. The flow rate is adjusted to 5 ml/min, and the low salt wash is continued with 20 ml buffer C. A 125-ml linear gradient from 20 to 350 m$M$ KCl is introduced starting in buffer C and titrating in buffer C HIGH, with an increase of approximately 13.7 m$M$ KCl/min. The gradient is immediately followed by a 19-ml wash with 100% buffer C HIGH (1 $M$ KCl). The system in then reequilibrated to low salt conditions by a 26-ml wash with buffer C and is thereby prepared for purification of the next protein. A representative chromatogram of S2 purification using this method is shown in Fig. 1B. Protein-containing fractions are identified by SDS gel electrophoresis: 20 $\mu$l of each 5-ml fraction is combined with 20 $\mu$l SDS gel sample buffer, electrophoresed on an SDS

---

[8] E. Kaltschmidt, *Anal. Biochem.* **43**, 25 (1971).

gel, and stained with Coomassie blue (see earlier discussion; Fig. 1A). Appropriate fractions are pooled, transferred to dialysis tubing (3500 molecular weight cutoff) and dialyzed at 4° against four 1-liter changes of buffer D, except for S10, which is dialyzed against four 1-liter changes of buffer D containing 4 $M$ urea (Fig. 1A, lane 3). Protein concentrations are determined by Bradford assay. Protein concentrations typically range from as little as 0.2 mg/ml for S9 to over 20 mg/ml for S6; typical observed yield and concentrations for each protein are given in Table II. Aliquots of the proteins are quick-frozen in liquid nitrogen and stored at −80°.

Avoidance of Nuclease Contamination

Contaminating nucleases in cellular extracts appear to bind to the Resource Q anion-exchange column and elute early in the salt gradient (approximately 50 m$M$ KCl). Therefore, proteins that have p$I$ values close to neutral, such as S2 (p$I$ 6.7), are purified at lower pH (buffer C) on a cation-exchange column (Resource S) to avoid possible contamination with this nuclease.[8] The following assay illustrates how such nucleases can be detected.

Although the proteins appear to be quite pure by SDS gel electrophoresis (Fig. 1A, lane 3), individual protein-containing fractions should be assayed for contamination with nuclease prior to proceeding with experiments. Ribosomal protein S2 will again be used as the reference protein in this experimental outline. To achieve an appropriate concentration of S2 for this assay, the 1.5-mg/ml stock of S2 is thawed on ice and a 2.7-$\mu$l aliquot of 1.5 mg/ml S2 is diluted with 37.3 $\mu$l buffer D to a final concentration of 0.1 mg/ml S2. This diluted sample is stored on ice until needed and then discarded. In a standard reaction, 1.5 $\mu$l of 14 $\mu M$ 16S rRNA (in buffer F) is incubated in 5 $\mu$l of buffer H at 42° for 15 min. To this reaction, 2 $\mu$l 0.1 mg/ml S2 is added along with 41.5 $\mu$l buffer A. The sample is mixed and incubated at 42° for 60 min. As a control, a sample of 16S rRNA should be treated identically as just described, except for the addition of 2 $\mu$l of buffer D in place of the diluted sample of S2. The integrity of the input RNA is assessed by preparing a sample of 1.5 $\mu$l of 14 $\mu M$ 16S rRNA diluted directly with 50 $\mu$l formamide loading dye. The incubated samples are combined with 50 $\mu$l formamide loading dye, and all samples are heated at 90° for 2 min. Samples (30 $\mu$l) are loaded on a 4% acrylamide 6 $M$ urea and 1× TBE gel (16 × 20 × 0.25 cm), which is run for 4 hr at 70 V at room temperature with an aluminum heat-diffusing plate. To visualize the RNA, the gel is transferred to a glass baking dish and 500 ml methylene blue stain is added, agitated for 10 min, and then destained by multiple rinses with water.

## Preparation of Recombinant Protein Mixtures

Prior to reconstitution, equimolar mixtures of pure recombinant proteins are prepared (Table III). Group I contains S4, S7, S8, S15, S17, and S20; group II contains S5, S6, S9, S11, S12, S13, S16, S18, and S19; and group III contains S2, S3, S10, S14, and S21. To achieve optimal protein concentrations while maintaining the ionic conditions required for reconstitution, the protein mixtures are prepared and then concentrated using Centricon 3 ultraconcentrators. For consistency and convenience, large

TABLE III
PREPARATION OF MIXTURES OF RIBOSOMAL PROTEINS FOR RECONSTITUTION

| Group | Protein | Total amount of protein added to pool ($\mu$g)[a] | Volume ($\mu$l)[b] | Final concentration (mg/ml)[c] |
|---|---|---|---|---|
| I | | | | 1.25 |
| | S4 | 345 | 115 | |
| | S7 | 296 | 423 | |
| | S8 | 210 | 105 | |
| | S15 | 152 | 76 | |
| | S17 | 144 | 288 | |
| | S20 | 143 | 71.5 | |
| | | | 1078.5 | |
| II | | | | 1.7 |
| | S5 | 263 | 175 | |
| | S6 | 236 | 11.8 | |
| | S9 | 221 | 1105 | |
| | S11 | 206 | 412 | |
| | S12 | 204 | 68 | |
| | S13 | 195 | 97.5 | |
| | S16 | 138 | 18.4 | |
| | S18 | 133 | 66.5 | |
| | S19 | 153 | 76.5 | |
| | | | 2030.7 | |
| III | | | | 1.2 |
| | S2 | 399 | 266 | |
| | S3 | 389 | 1297 | |
| | S10 | 176 | 587 | |
| | S14 | 168 | 56 | |
| | S21 | 126 | 63 | |
| | | | 2269 | |

[a] The pooled amount corresponds to 15 nmol.
[b] Based on concentrations given in Table II.
[c] Resulting concentration of mixture following Centricon 3 concentration as determined by Bradford assay.

batches of mixtures are generally prepared (on the order of 15-nmol equivalents), quick-frozen in liquid nitrogen, and stored in small aliquots (1-nmol equivalent) at −80°. Although the mixture of group I proteins tends not to require much concentration, it is generally treated identically to the other mixtures. Individual protein stock solutions (Table II) are thawed on ice, and the groups are prepared using equimolar amounts of each protein. Mixtures of groups I, II, and III are prepared as described in Table II and are placed in pretreated Centricon 3 ultraconcentrators at 4°. For group II and group III mixtures, because the volume of the mixtures is greater than the capacity of the Centricon 3, the mixture is prepared and 1.7 ml is placed into the Centricon 3 concentrators; additional material is added to the Centricon 3 concentrators after approximately 1 hr of centrifugation. Protein mixtures are concentrated in the Centricon 3 ultraconcentrators by centrifugation at 5000 rpm for approximately 3 hr (time varies depending on the starting concentration of the individual proteins) at 4° in a JA-20 rotor (with adapters). Proteins are recovered from Centricons by placing the sample chamber upside down in a retentate vial and centrifuging at 5000 rpm for 2.5 min at 4° in a JA-20 rotor (with adapters), after which the retentate vial is put on ice. Generally, less than 1 ml is recovered and concentrations are determined by Bradford assay of 10 $\mu$l recovered mixture. Concentrations are adjusted with buffer D to 1.25 mg/ml group I, 1.7 mg/ml group II, and 1.2 mg/ml group III. Mixtures are aliquoted, quick-frozen in liquid nitrogen, and stored at −80°; aliquots should be appropriately sized such that the mixtures are not thawed and refrozen.

## Reconstitution of 30S Subunits

Reconstitution of 30S particles is optimal using a fourfold molar excess of each purified recombinant protein relative to 16S rRNA. This ratio was determined empirically by titration over a range of 1 to 8 molar equivalents protein to RNA monitoring the efficiency of reconstitution by sucrose gradient sedimentation.[9] In a standard reconstitution, 40 pmol of 16S rRNA is used; therefore, 160 pmol of each protein must be added corresponding to 13.8 $\mu$g for group I, 18.7 $\mu$g for group II, and 13.4 $\mu$g for group III (Table III). Because these protein mixtures are stored in 1 $M$ KCl (buffer D) and reconstitution proceeds at 330 m$M$ KCl (buffer A), the KCl concentration must be readjusted after the addition of each mixture during the reconstitution reaction. In a standard reconstitution, 2.9 $\mu$l of 14 $\mu M$ 16S rRNA (in buffer F) is mixed with 5 $\mu$l of buffer H by pipetting up and

[9] G. M. Culver and H. F. Noller, submitted for publication.

down, followed by brief centrifugation and incubation at 42° for 15 min. Samples can be kept briefly at room temperature during subsequent additions of components and are always mixed by pipetting up and down and centrifugation at top speed for 5 sec prior to returning to incubation at 42°. To this reaction, 11 $\mu$l of group I proteins (1.25 $\mu$g/$\mu$l) is added along with 14 $\mu$l buffer H, and the reaction is incubated at 42° for 20 min. Group II proteins, 11 $\mu$l (1.7 mg/$\mu$l) and 22 $\mu$l buffer H are added and then incubated at 42° for 20 min. Finally, 11 $\mu$l group III proteins (1.2 mg/$\mu$l), and 23 $\mu$l buffer H are added and the reaction is incubated at 42° for 20 min.

Purification of 30S Subunits

Reconstituted 30S subunits can be analyzed and purified by sucrose gradient sedimentation. A 60% (w/v) sucrose solution (concentration determined using a refractometer) is prepared in water. For every six (11 ml) sucrose gradients needed, 50 ml each of a 10 and a 40% solution in 1× sucrose gradient buffer should be made. To make the 10% sucrose solution in 1× sucrose gradient buffer, mix 8.3 ml 60% sucrose with 16.7 ml 3× sucrose gradient buffer and 25 ml water. The 40% sucrose solution in 1× sucrose gradient buffer is made by mixing 33.3 ml 60% sucrose with 16.7 ml 3× sucrose gradient buffer. An 11-ml 10–40% sucrose gradient in 1× sucrose gradient buffer is prepared using a gradient former, with 5.5 ml 10% sucrose solution and 5.5 ml 40% sucrose solution, and dripped into a 14 × 49-mm ultraclear centrifuge tube, as described in detail elsewhere.[10] Sucrose gradients are balanced using 10% sucrose solution, stored at 4° prior to use, and can be stored for up to 24 hr. Reconstituted 30S subunits (100 $\mu$l in buffer A) are layered on top of the gradients, which are centrifuged in a SW41 rotor (32,000 rpm) for 15.5 hr at 4°. The gradient is fractionated and analyzed using a density gradient fractionator. Fractions containing 30S subunits are collected. Natural 16S rRNA [2.9 $\mu$l of 14 $\mu$M 16S rRNA (in buffer F) in 97.1 $\mu$l buffer A] and 30S subunits [2.9 $\mu$l of 14 $\mu$M 30S subunits (in buffer G) in 97.1 $\mu$l buffer A] should be run on sucrose gradients as sedimentation markers. Prior to use either for directed hydroxyl radical probing or for functional analysis, the reconstituted 30S subunits must be concentrated and the sucrose removed. Collected reconstituted 30S subunits, approximately 1 ml, are placed in a pretreated Centricon 100 ultraconcentrator, 1 ml buffer A is added to all samples, which are then centrifuged at 2400 rpm for 45–60 min at 4° in a JA-20 rotor (with adapters). Samples are washed three times by the addition of 2 ml buffer A followed by centrifugation at 2400 rpm

---

[10] C. Merryman and H. F. Noller, in "RNA : Protein Interactions: A Practical Approach" (C. J. W. Smith, ed.), Oxford Univ. Press, London, 1998.

for 30–45 min at 4° in a JA-20 rotor (with adapters) for each wash. Samples are recovered by placing the sample chamber upside down in a retentate vial and centrifuging for 2.5 min at 2400 rpm at 4° in a JA-20 rotor (with adapters). Generally, 50–100 $\mu$l is recovered and the volume of each sample is adjusted to 100 $\mu$l with buffer A for further use.

Alternatively, any unincorporated proteins can be removed from reconstituted 30S subunits by centrifugation in Microcon 100 microconcentrators. Reconstituted 30S subunit samples are placed on a pretreated Microcon 100 microconcentrator and centrifuged at 2000 rpm for 10 min at 4°. Samples are washed three times by the addition of 100 $\mu$l buffer A followed by centrifugation at 2000 rpm for 5–10 min at 4° for each wash. The samples are recovered by placing the sample chamber upside down in a clean tube and centrifuging for 2.5 min at 2000 rpm at 4°. The 30S subunits are kept on ice until used. This alternative purification can be done if the reconstituted 30S subunits are used in functional assays prior to sucrose gradient sedimentation or if sucrose gradient sedimentation is to be avoided. One example where a presucrose-gradient purification step in necessary is for the association of reconstituted 30S subunits with natural 50S subunits to form 70S ribosomes. Because subunit association proceeds at a lower salt concentration (100 m$M$ KCl; buffer G) than reconstitution (330 m$M$ KCl; Buffer A), it is possible that additional unreconstituted ribosomal proteins could bind nonspecifically to reconstituted 30S, natural 50S subunits, or the resulting 70S ribosomes; therefore, removal of unreconstituted proteins prior to lowering the salt concentration is necessary.

The 30S subunits reconstituted in this way are active in *in vitro* assays. After sucrose gradient purification, purified 30S subunits bind tRNA in a poly(U)-dependent manner, as measured by a filter-binding assay (Table

TABLE IV

RECOVERY, tRNA BINDING, AND POLYPHENYLALANINE SYNTHESIS
OF RECONSTITUTED 30S SUBUNITS[a]

| | | tRNA[Phe] binding activity (%) | | |
|---|---|---|---|---|
| 30S subunits | Recovery (%) | poly(U) dependent | poly(U) independent | poly(Phe) synthesis, activity (%) |
| Natural | — | 100 ± 3 | 20 ± 4 | 100 ± 2 |
| Recombinant | 45 | 48 ± 4 | 12 ± 3 | 34 ± 2 |

[a] Recovery is based on comparison of amount of input 16S rRNA to the amount of 30S subunits isolated and purified from sucrose gradients. For tRNA binding, 100% binding corresponds to 0.7 pmol tRNA[Phe] bound/pmol of 30S subunits. The method for tRNA binding is as described previously.[11,12] For polyphenylalanine synthesis, 100% activity corresponds to incorporation of 9.0 pmol of phenylalanine/pmol 30S subunits. The assay for polyphenylalanine synthesis was as described previously.[13]

IV).[11,12] The same 30S subunits can direct polyphenylalanine synthesis in the presence of 50S subunits, elongation factors, and poly(U) (Table IV), indicating that they are functional in subunit association and that the observed tRNA binding is productive.[13] Association of recombinant protein-reconstituted 30S subunits with 50S subunits to form 70S ribosomes can be observed directly by sucrose gradient sedimentation.[14]

## Summary

This system allows convenient purification of large quantities of all of the small subunit ribosomal proteins by overexpression from cloned genes. This not only allows large-scale reconstitution of 30S subunits from individual proteins, but also facilitates protein purification greatly. These proteins can be reconstituted into functional 30S subunits using an ordered assembly protocol based on the *in vitro* 30S assembly map. Reconstitution of 30S subunits using this system enables mutant or modified proteins, such as Fe(II)-BABE-derivatized proteins, to be incorporated into subunits for studying ribosome structure and function.[3,15–18]

## Acknowledgment

The authors thank John Diener for critical reading of the manuscript.

[11] M. Nirenberg and P. Leder, *Science* **145,** 1399 (1964).
[12] D. Moazed and H. F. Noller, *Cell* **47,** 985 (1986).
[13] P. Traub, S. Mizushima, C. V. Lowry, and M. Nomura, *in* "RNA and Protein Synthesis" (K. Moldave, ed.). Academic Press, London, 1981.
[14] B. Hapke and H. Noll, *J. Mol. Biol.* **105,** 97 (1976).
[15] G. M. Heilek, R. Marusak, C. F. Meares, and H. F. Noller, *Proc. Natl. Acad. Sci. U.S.A.* **92,** 1113 (1995).
[16] G. M. Heilek and H. F. Noller, *Science* **272,** 1659 (1996).
[17] G. M. Heilek and H. F. Noller, *RNA* **2,** 597 (1996).
[18] G. M. Culver and H. F. Noller, *RNA* **4,** 1471 (1998).

# [31] Directed Hydroxyl Radical Probing of RNA from Iron(II) Tethered to Proteins in Ribonucleoprotein Complexes

*By* Gloria M. Culver *and* Harry F. Noller

## Introduction

Characterizing the RNA environment surrounding specific proteins and ligands is of great interest in the study of complex RNAs or ribonucleoprotein particles (RNPs). One such example is the ribosome, for which classical techniques, such as footprinting and cross-linking, have been used to map ribosomal RNA (rRNA) elements that interact with or are in the vicinity of specific ribosomal proteins.[1–3] More recently, directed hydroxyl radical probing has been used as a direct and comprehensive method of mapping the rRNA environment in the vicinity of specific locations on the surface of ribosomal proteins or ligands.[4–10] In this approach, iron(II) is tethered to a single position on the surface of a protein via the linker 1-($p$-bromo-acetamidobenzyl)-EDTA (BABE) devised by Rana and Meares.[11] Fenton chemistry is initiated at the Fe(II) site, locally generating hydroxyl radicals that cleave the backbone of nearby RNA.[12] Sites of RNA cleavage are identified by primer extension as specific additional bands that are absent or reduced in control reactions. Generally, such control reactions include a cysteine-less protein that is treated identically to the cysteine-containing proteins to identify sites of nonspecific RNA cleavage.

There are a number of special advantages to this method. First, like cross-linking, this method provides direct information about the RNA envi-

---

[1] S. Stern, T. Powers, L. M. Changchien, and H. F. Noller, *Science* **244,** 783 (1989).

[2] T. Powers and H. F. Noller, *RNA* **1,** 194 (1995).

[3] R. Brimacombe, *Biochimie* **73,** 927 (1991).

[4] G. M. Heilek, R. Marusak, C. F. Meares, and H. F. Noller, *Proc. Natl. Acad. Sci. U.S.A.* **92,** 1113 (1995).

[5] G. M. Heilek and H. F. Noller, *Science* **272,** 1659 (1996).

[6] G. M. Heilek and H. F. Noller, *RNA* **2,** 597 (1996).

[7] K. S. Wilson and H. F. Noller, *Cell* **92,** 131 (1998).

[8] G. M. Culver and H. F. Noller, *RNA* **4,** 1471 (1998).

[9] K. R. Lieberman and H. F. Noller, *J. Mol. Biol.* **284,** 1367 (1998).

[10] G. M. Culver, G. M. Heilek, and H. F. Noller, *J. Mol. Biol.* **286,** 355 (1999).

[11] T. M. Rana and C. F. Meares, *Proc. Natl. Acad. Sci. U.S.A.* **88,** 10578 (1991).

[12] W. J. Dixon, J. J. Hayes, J. R. Levin, M. F. Weidner, B. A. Dombroski, and T. D. Tullius, *Methods Enzymol.* **208,** 380 (1991).

ronment in the vicinity of the probe, unlike chemical or enzymatic protection, which can be caused by both direct and indirect effects.[13] Second, the Fe(II) probe can be tethered systematically to many different positions across the surface of a protein, comprehensively exploring the RNA environment of the protein. Third, many positions in the RNA backbone are typically cleaved from a single tethering position, providing relatively large numbers of data points on RNA–protein proximities for each experiment. Fourth, distances between probe positions and target sites can be estimated from the strength of the observed RNA backbone cleavage.[14] This article uses probing of 16S rRNA from ribosomal protein S20 as an example.

## General Practices

Due to the widespread presence of ribonucleases and the sensitivity of RNA to nuclease attack, precautions should be taken to avoid contaminating nucleases. Gloves should be worn for all manipulations, and glassware should be baked at 200°. Any suspect plasticware can be washed with 95% (v/v) ethanol and autoclaved. When water is called for in the text it is understood to be distilled, deionized, and filtered using a Barnstead Nanopure water purification system or the equivalent. Generally, reagents are kept on ice prior to use. Reactions for which the total volume is less than 1.5 ml are performed in Eppendorf tubes. Any centrifugations in Eppendorf tubes are assumed to be in an Eppendorf-style microfuge at 13,500 rpm unless stated otherwise.

## Reagents

Preparation of 16S rRNA follows a standard procedure that has been described previously.[15]

Synthesis of BABE is facilitated by starting with a commercially available precursor, aminobenzyl-EDTA (Dojindo Labs), and then following the improved synthesis as described by Meares and co-workers, who also detail methods for purifying and determining the concentration of purified BABE.[11,16]

All chemicals are purchased from Sigma Chemical Company (St. Louis, MO) except where noted:

[13] C. Merryman and H. F. Noller, *in* "RNA:Protein Interactions: A Practical Approach" (C. J. W. Smith, eds.). Oxford Univ. Press, London, 1998.

[14] S. Joseph, B. Weiser, and H. F. Noller, *Science* **278,** 1093 (1997).

[15] D. Moazed, S. Stern, and H. F. Noller, *J. Mol. Biol.* **187,** 399 (1986).

[16] J. K. Moran, D. P. Greiner, and C. F. Meares, *Bioconj. Chem.* **6,** 296 (1995).

Nikkol (octaethylene glycol mono-$n$-dodecyl ether; $C_{12}E_8$) (Calbiochem, La Jolla, CA).

Ammonium iron(II) sulfate (hexahydrate) (Aldrich, Milwaukee, WI) in the form that is 99.997% pure.

7-Diethylamino-3-([4'-(iodoacetyl)amino]phenyl)-4-methylcoumarin (DCIA) (Molecular Probes, Eugene, OR).

Kethoxal (ICN, Costa Mesa, CA).

dNTP and ddNTP stocks (100 m$M$) Pharmacia (Piscataway, NJ).

Centricon 100 ultraconcentrators and Microcon 3 microconcentrators (Amicon, Danvers, MA): Pretreat with 5% Tween 20 (200 or 50 ml, respectively) for 10 min at room temperature, wash extensively with water, and finally wash and store in appropriate buffer prior to use. Once wetted, filters should not be allowed to dry.

Bradford reagent (Bio-Rad, Hercules, CA): Microassay procedure is followed as outlined in the company's literature.

Urea, ultrapure (ICN)

Acrylamide: Bisacrylamide (29:1) (Amresco): Ultrapure grade solution, referred to later as polyacrylamide (29:1). Store at 4°.

Phenol (Amresco).


Buffers

Preparation of all buffer components follows standard procedures as outlined in detail elsewhere.[17]

Buffer A: 80 m$M$ K$^+$–HEPES (pH 7.6), 20 m$M$ MgCl$_2$, 330 m$M$ KCl, and 0.01% Nikkol. Store at 4°.

Buffer D: 80 m$M$ K$^+$–HEPES (pH 7.6), 20 m$M$ MgCl$_2$, 1 $M$ KCl, and 6 m$M$ 2-mercaptoethanol (BME). Store at 4°; add BME just prior to use.

Buffer F: 20 m$M$ K$^+$–HEPES (pH 7.6) and 20 m$M$ MgCl$_2$. Store at 4°.

BABE modification buffer: 1 $M$ KCl, 80 m$M$ K$^+$–HEPES (pH 7.6) and 0.01% Nikkol. Store at 4°.

RNA extraction buffer: 0.3 $M$ sodium acetate (pH 6.0), 0.5% sodium dodecyl sulfate (SDS), and 5 m$M$ EDTA (pH 8.0). Store at room temperature; if precipitate forms, discard.

SDS gel sample buffer: 0.12 $M$ Tris–HCl (pH 6.8), 4% SDS, 20% glycerol, 8 $M$ urea, 750 m$M$ BME, and 0.05% bromphenol blue. Store at −20°.

[17] J. Sambrook, E. F. Fritsch, and T. Maniatis, in "Molecular Cloning" (N. Ford, ed.), p. B.1. Cold Spring Harbor Press, Cold Spring Harbor, NY, 1989.

SDS gel running buffer: 25 m$M$ Tris–Base, 192 m$M$ glycine, and 0.01% SDS. Store at room temperature.

4% stacking SDS gel: 4% polyacrylamide (29:1), 125 m$M$ Tris–HCl (pH 6.8), and 0.1% SDS. Store at 4°.

12% resolving gel: 12% polyacrylamide (29:1), 375 m$M$ Tris–HCl (pH 8.8), and 0.1% SDS. Store at 4°.

Coomassie blue stain: 40% (v/v) methanol, 10% (v/v) acetic acid, and 0.1% (w/v) Coomassie blue R-250. Store at room temperature.

Coomassie destain: 40% (v/v) methanol and 10% (v/v) acetic acid. Store at room temperature.

1× TBE, 0.09 $M$ Tris–borate: 0.002 $M$ EDTA (pH 8.0). Store at room temperature.

Also see primer extension section for an extensive set of buffers that are specific to that method.

### Equipment and Supplies

Beckman JA-20 centrifuge rotor
Corning Corex Rubber Sleeve Adaptor for JA-20
Bio-Rad Mini-PROTEAN II electrophoresis cell
Fotodyne Foto/Prep I transilluminator
IEC Clinical tabletop centrifuge No. 221
Bio-Rad Bio-Spin disposable chromatography columns

### Constructing Cysteine-Containing Mutants

Fe(II) is tethered to different positions on the surface of a protein using the bifunctional linker, BABE, which can be attached to proteins via the sulfhydryl groups of cysteine. Thus, unique cysteine residues are introduced at different sites on the surface of the protein by directed mutagenesis to allow Fe(II) tethering to different positions. Also, a version of the protein that is devoid of cysteines serves as a control for unintended derivatization of any noncysteine side chains. A few approaches have proven useful in selecting residues for cysteine substitution. If the three-dimensional structure of the protein has been determined, it can be used to select exposed surface residues and to ensure even distribution of probing sites over the surface of the protein. Importantly, phylogenetic amino acid sequence alignments, if available, are used to identify nonconserved, polar (potentially exposed) residues for cysteine substitution (see later).

*Escherichia coli* ribosomal protein S20 contains no cysteine residues and can thus be used as a cysteine-less control in directed hydroxyl radical probing experiments. The lack of cysteine residues necessitates introduction of all tethering sites by directed mutagenesis. Because the three-dimensional structure of S20 has yet to be determined, an amino acid sequence alignment of S20 proteins from five different organisms was used to select six sites for cysteine substitution.[18] The selected sites target nonconserved, mainly hydrophilic residues likely to be found on the surface of S20. The wild-type copy of the S20 gene was cloned into the pET24b vector (Novagen), which allows both overexpression of the recombinant protein and production of single-stranded DNA for use in site-directed mutagenesis.[8,19,20] Oligonucleotides used for mutagenesis are generally designed as 27-mers with the site of mutation (cysteine-encoding codon) at the center. Preparation of single-stranded DNA and mutagenesis follow standard techniques that have been described in detail previously.[21] The desired number of tethering sites depends on the size and shape of the protein and on the scope of the experiment.

Endogenous cysteines can also be used as tethering sites by eliminating all but one of the wild-type cysteine residues in a given protein. An amino acid sequence alignment provides a good basis for selecting appropriate amino acids to substitute for endogenous cysteine residues based on amino acids that are known to be tolerated in other organisms at the same positions.[7] Unless there is a physiologically essential cysteine residue, it is usually the case that all endogenous cysteine residues can be substituted to generate a cysteine-less control protein.

Activity of mutant proteins relative to wild-type protein should be assessed by enzymatic and/or binding assays. For small-subunit ribosomal proteins, footprinting 16S rRNA has been used as a sensitive test for RNA–protein interaction and can be performed both before and after derivatization with Fe(II)–BABE (see later).[13]

## Assessing Efficiency of Derivatization of Cysteine Residues

The accessibility of each introduced cysteine residue for derivatization can be assessed conveniently by reaction with the fluorescent reagent, 7-diethylamino-3-([4'-(iodoacetyl)amino]phenyl)-4-methylcoumarin (DCIA; Molecular Probes). DCIA reacts with cysteine sulfhydryl groups by a nucleophilic reaction similar to that of Fe(II)–BABE and can be used to estimate the efficiency of Fe(II)–BABE derivatization. Separation of unre-

[18] I. S. Mian, B. Weiser, and H. F. Noller, unpublished (1995).
[19] F. W. Studier, A. H. Rosenberg, J. J. Dunn, and J. W. Dubendorff, *Methods Enzymol.* **185,** 60 (1990).
[20] G. M. Culver and H. F. Noller, *Methods Enzymol.* **318** [30] (2000) (this volume).
[21] T. A. Kunkel, K. Bebenek, and J. McClary, *Methods Enzymol.* **204,** 125 (1991).

acted DCIA from DCIA-conjugated protein is achieved by SDS gel electrophoresis, allowing the relative reactivity of each cysteine-containing protein to be compared. Wild-type S20 is assayed in parallel with cysteine-containing S20 mutant proteins to assess nonspecific derivatization at positions other than cysteine. In a standard reaction, 5 $\mu$l S20 (2.0 mg/ml) is mixed with 5 $\mu$l 1.5 mM DCIA [in dimethylformamide (DMF)] and 40 $\mu$l BABE modification buffer. Reactions are incubated at 30° for 15 min, quenched by the addition of 3 $\mu$l BME (14.4 $M$) followed by the addition of 50 $\mu$l SDS gel sample buffer, and placed on ice until needed. For gel analysis, 40 $\mu$l of the reaction is heated (90° for 2.5 min), loaded on an SDS gel electrophoresis (Mini-PROTEAN, 1 mm thick; 4% stacking gel: 12% resolving gel, both including 6 $M$ urea), and run at 180 V for 1.5 hr in SDS gel running buffer at room temperature.[22] Gels are examined by UV illumination using a Foto/Prep I transilluminator (Fotodyne) on the analytic setting to visualize free DCIA, which migrates at the front, and DCIA-derivatized protein. Wild-type (cysteine-less) S20 is not labeled with DCIA and therefore does not fluoresce, whereas mutant cysteine-containing S20 proteins are all labeled with DCIA and fluoresce when illuminated. Gels are photographed at this stage prior to loss of fluorescent signal. Gels are then stained with 250 ml Coomassie blue stain by shaking for 1 hr at room temperature and destained by shaking with multiple washes with Coomassie destain to confirm equal loading of proteins.

The DCIA reactivity assay should be performed again after Fe(II)–BABE derivatization of the proteins to confirm cysteine modification. Derivatization of the cysteine sulfhydryl residue with Fe(II)–BABE should effectively compete for reaction with DCIA. Thus, fluorescent labeling should be reduced or abolished when DCIA is reacted with Fe(II)–BABE-derivatized protein compared to the observed reactivity of the unmodified proteins.

Preparation of Fe(II)–BABE Complex

Typically, approximately 70% of BABE is loaded with Fe(II), and this complex is freshly prepared prior to each use. In a standard reaction, 8 $\mu$l freshly prepared 50 mM $(NH_4)_2Fe(SO_4)_2 \cdot 6H_2O$, 4 $\mu$l 1 $M$ sodium acetate (pH 6.0), and 20 $\mu$l 28 mM BABE are mixed by gentle flicking of the tube and brief pulse centrifugation at room temperature in a total volume of 40 $\mu$l. The sample is incubated at room temperature for 30 min. As a chase, 2 $\mu$l 50 mM EDTA is added, and incubation is continued at room

---

[22] U. K. Laemmli, *Nature* **227,** 680 (1970).

temperature for another 15 min. The Fe(II)–BABE complex is stored on ice until use.

## Fe(II) Derivatization of Protein

Conditions are adjusted so that wild-type (cysteine-less) S20 does not react with Fe(II)–BABE under conditions where it reacts with cysteine residues in S20 mutant proteins, as demonstrated by DCIA reactivity. In a standard reaction, 14.3 $\mu$l of S20 (2.0 mg/ml) and 7 $\mu$l Fe(II)–BABE complex (see earlier discussion) are mixed with 78.7 $\mu$l modification buffer by pipetting and brief pulse centrifugation at room temperature and incubated at 37° for 30 min. To remove unreacted Fe(II)–BABE from Fe(II)–BABE-derivatized S20 [Fe(II)-S20], reactions are loaded onto pretreated Microcon 3 microconcentrators on ice and are centrifuged at 6500 rpm for 30 min at 4°. Samples are washed by the addition of 400 $\mu$l modification buffer followed by centrifugation at 6500 rpm for 90–120 min at 4°. The wash is repeated three additional times. After the final wash, samples are recovered from Microcons by placing the sample chamber upside down in a clean tube and centrifuging for 2.5 min at 2000 rpm at 4°; typically, 50 $\mu$l is recovered. Fe(II)-derivatized proteins are kept on ice until needed. Protein recovery and concentration are determined by Bradford assay of 5 $\mu$l of sample. Fe(II)-derivatized protein can be quick-frozen (approximately 10-mg aliquots) in liquid nitrogen and stored at −80° for several weeks; however, in most cases the modified protein is used immediately.

## Complex Formation of S20 with 16S rRNA

Ribosomal protein S20 is a primary binding protein; i.e., it interacts directly with 16S rRNA independent of the presence of other ribosomal proteins. Thus, the ability of S20 to yield a characteristic footprint on 16S rRNA can be used as an assay to test whether the introduction of cysteine residues or their modification with Fe(II)–BABE interferes with binding. The initial step for these experiments is formation of a complex of S20 and 16S rRNA. The ratio of protein to RNA required for complex formation is determined by the titration of wild-type unmodified protein; for recombinant S20, this has been found to be a fourfold molar excess of protein to RNA, as used in the experiments described later.[8] Complex formation using wild-type, underivatized S20, and 16S rRNA will be described; this same procedure is used for mutant proteins or Fe(II)-derivatized proteins. For complex formation, 2 $\mu$l 2.0 mg/ml S20 is added to 8 $\mu$l buffer D for a final concentration of 0.4 mg/ml (41.9 $\mu M$) S20. In a standard reaction, 2.9 $\mu$l of 14 $\mu M$ 16S rRNA is incubated in 5 $\mu$l of buffer A minus KCl at 42° for

15 min. To this reaction, 3.8 $\mu$l S20 (0.4 mg/ml in buffer D) is added along with 91.2 $\mu$l buffer A; the reaction is incubated at 42° for 60 min and then on ice for 10 min. The S20/16S rRNA complex can then be used directly in footprinting experiments to assess binding or it can be purified from free protein for directed hydroxyl radical probing (see later).

### Footprinting S20 on 16S rRNA

Full details of footprinting using base-specific probes and solution-based hydroxyl radicals have been described.[13] Here, use of the base-specific probe kethoxal is provided as an example. Solution-based hydroxyl radical probing using Fe(II)–EDTA cannot be used to footprint Fe(II)-derivatized proteins, as hydroxyl radicals would be generated from the tethered site as well as in solution. If solution-based hydroxyl radical probing is necessary, the protein of interest can be derivatized with BABE that has not been complexed with Fe(II). The S20/16S rRNA complex need not be purified for footprinting analysis. Controls for footprinting should include 16S rRNA incubated without kethoxal as well as uncomplexed 16S rRNA treated with kethoxal. After the complex is formed (see earlier discussion), 4 $\mu$l kethoxal [37 mg/ml, made by diluting 10 $\mu$l of a 54% stock (in ethanol) into 140 $\mu$l water] is added to the 100 $\mu$l complex sample, mixed by pipetting and brief pulse centrifugation at 4° and incubated at 0° for 60 min. Samples are adjusted to 25 m$M$ potassium borate by the addition of 5.2 $\mu$l 500 m$M$ potassium borate (pH 7.0) prior to RNA extraction (see later) to stabilize the kethoxal–guanine adduct.[23] Isolation of 16S rRNA from the complex and subsequent primer extension analysis are detailed later.

### Purification of Fe(II)–S20–16S rRNA Complex

Prior to site-directed hydroxyl radical probing, 16S rRNA–Fe(II)-S20 complexes must be purified from free Fe(II)-S20. A 1-ml Sephacryl S-200 spin column is prepared in a Bio-Spin disposable column and equilibrated in buffer A by loading 100 $\mu$l buffer A onto the column and centrifuging at 4° in a clinical tabletop centrifuge at three-quarter maximum speed for 3.5 min until exactly 100 $\mu$l is recovered; this typically takes three to four equilibration spins. Once the column is equilibrated, the sample is loaded onto the column, which is subsequently centrifuged at 4° in a clinical table-top centrifuge at three-quarter maximum speed for 3.5 min. The 16S

[23] M. Staehlin, *Biochim. Biophys. Acta* **31,** 448 (1959).
[24] S. Mizushima and M. Nomura, *Nature* **226,** 1214 (1970).
[25] W. A. Held, S. Mizushima, and M. Nomura, *J. Biol. Chem.* **218,** 5720 (1973).

rRNA–Fe(II)-S20 complex appears in the first eluant, whereas free Fe(II)-S20 does not appear in the eluant until three or four additional spins are performed. The recovered 100-$\mu$l sample, containing the purified 16S rRNA–Fe(II)-S20 complex, is kept on ice until needed.

### Reconstitution of 30S Subunits Containing Fe(II)-Derivatized Protein

Fe(II)-derivatized S20 can be incorporated into complete 30S subunits for site-directed hydroxyl radical probing. The procedure for reconstitution and purification of 30S subunits using recombinant proteins is described in detail elsewhere and requires that the small subunit ribosomal proteins be divided into three groups (groups I–III) for serial addition. Because S20 belongs to group I along with S4, S7, S8, S15, and S17, an equimolar mixture of S4, S7, S8, S15, and S17 lacking S20 (group I-S20) is made following the same procedure outlined in the accompanying article and summarized in Table I.[20] To reconstitute 30S subunits containing Fe(II)-S20, the detailed procedure is initiated as described except in place of the standard addition of a fourfold molar excess of the group I mixture, a fourfold molar excess of both Fe(II)-S20 and group I-S20 mixtures is added. The remaining steps of the reconstitution are followed as described.

Competition experiments can be performed to control for specificity of observed 16S rRNA cleavages in 30S subunits. Wild-type S20 can be used

TABLE I

PREPARATION OF PROTEIN MIXTURES TO ALLOW INCLUSION OF Fe(II)-DERIVATIZED S20

| Group | Assembly map[a] | Amount to concentrate (mg/15,000 pmol) | mg/pmol of mixture |
|---|---|---|---|
| I | | | 0.086 |
| | S4 | 345 | |
| | S7 | 296 | |
| | S8 | 210 | |
| | S15 | 152 | |
| | S17 | 144 | |
| | S20 | 143 | |
| I-S20 | | | 0.0765 |
| | S4 | 345 | |
| | S7 | 296 | |
| | S8 | 210 | |
| | S15 | 152 | |
| | S17 | 144 | |
| Fe(II)-S20 | | | 0.0095 |

[a] From Refs. 24 and 25.

as a competitor of incorporation of Fe(II)-S20 into 30S subunits. During reconstitution with Fe(II)-S20, an additional fourfold molar excess of wild-type, underivatized S20 is included along with group I-S20 and Fe(II)-S20. If Fe(II)-S20 occupies its correct position in the 30S subunits, cleavage of 16S rRNA in the presence of excess competitor should be reduced accordingly.

The functional integrity of reconstituted 30S subunits containing Fe(II)-derivatized protein should be assayed prior to probing. Association of 30S subunits containing Fe(II)-derivatized protein with natural 50S subunits to form 70S ribosomes demonstrates that the derivatized 30S subunits are functional and, in addition, allows probing of 70S ribosomes.[8] Template-directed tRNA binding and poly(U)-directed polyphenylalanine synthesis can also be used to monitor the functional integrity of the Fe(II)-derivatized particles.

## Hydroxyl Radical Probing

Fe(II)-derivatized RNPs, 30S subunits, and 70S ribosomes must be purified prior to directed hydroxyl radical probing; the Fe(II)-S20–16S rRNA complex can be purified as described (see "Purification of Fe(II)-S20–16S rRNA Complex"), and 30S subunits and 70S ribosomes can be purified using sucrose gradient centrifugation as described in detail elsewhere.[13,20] Note that it is important to exclude sucrose before initiating hydroxyl radical formation due to quenching of the free radicals by sucrose. Once purified, Fe(II)-S20-containing RNPs, 30S subunits, and 70S ribosomes are all treated identically. Reagents used in the hydroxyl radical probing reactions are freshly prepared just prior to use. To each ice-cold 100-$\mu$l sample, 2 $\mu$l 250 m$M$ ascorbic acid and 2 $\mu$l 1.25% $H_2O_2$ are added sequentially to the side of the Eppendorf tube. The probing reaction is initiated by brief pulse centrifugation at 4° to mix the reactants. Reactions are returned to ice and incubated for 10 min. The reaction is quenched by the addition of 20 $\mu$l 20 m$M$ thiourea. RNA is isolated immediately following probing as described later.

## Isolation of RNA after Probing

For both footprinting and directed hydroxyl radical probing experiments, 16S rRNA is recovered by the addition of 250 $\mu$l ethanol and 10 $\mu$l 3 $M$ sodium acetate (pH 5.2). Samples are mixed by vortexing, quick-frozen in a dry ice/ethanol bath for 10 min, and centrifuged for 10 min at 4°. Ethanol is decanted carefully by slowly pouring it out of the tube, taking care to pour away from the pellet and not to disturb it. The pellet (slight residual ethanol can be ignored at this stage) is resuspended in 200 $\mu$l RNA extraction buffer. RNA extraction can be performed at room temperature until the final precipitation steps. Samples are shaken on an Eppendorf mixer for 5 min, followed by the addition of 200 $\mu$l water-saturated phenol (Tris-buffered, pH 7.5) and continued shaking for 5 min. Samples are centrifuged for 5 min, and the phenol phase is carefully removed from the bottom of the tube using a gel-loading tip on a Pipetman. Phenol extraction is repeated two additional times. After the last phenol phase has been removed, 200 $\mu$l chloroform is added, followed by vortexing for 30 sec and centrifugation for 2 min. Chloroform is removed from the bottom of the tube and the chloroform extraction is repeated. To precipitate extracted RNA, 600 $\mu$l ice-cold 95% (v/v) ethanol is added to samples, mixed by gentle vortexing, placed in a dry ice/ethanol bath for 10 min, and centrifuged for 10 min at 4°. Ethanol is carefully and slowly poured out of the tube. Pellets are washed by the addition of 500 $\mu$l ice-cold 70% (v/v) ethanol, which is mixed by gently flicking the tubes followed by centrifugation for 1 min at 4°. Ethanol is carefully and slowly poured from the tube, samples are pulse centrifuged, and any residual ethanol is removed using a fine (gel loading) pipette tip [e.g., Rainin #HR-250 (Emeryville, CA)], taking care to avoid disturbing the pellet. Pellets are dried under vacuum for 5 min and then are resuspended in 40 $\mu$l water. Assuming quantitative recovery, the resulting RNA samples should be approximately 1 $\mu M$.

## Analysis of Probing Data by Primer Extension

Primer extension is an extremely sensitive technique for detecting breaks in the RNA backbone. Synthetic DNA oligomers anneal to target RNA template sequences and prime DNA polymerization by reverse transcriptase. The primers are generally 17 nucleotides long and have homology to a single region of the target RNA. Each primer can be used to analyze approximately 200 nucleotides of target RNA. Primers bind to the target RNA with different affinities, necessitating that optimal primer concentrations be determined empirically; these typically fall within the range from 0.01 to 0.5 $\mu M$. Primer stocks are prepared in water and stored at $-20°$. Dideoxy sequencing reactions of template RNA should accompany primer extension analysis of modified samples.

## Reagents and Buffers

All reagents are kept on ice during use and stored at either $-20°$ or $-80°$. A large (60 cm $\times$ 20 cm $\times$ 0.25 mm) sequencing gel is used to resolve primer extension samples.[13]

4.5× hybridization buffer: 225 m$M$ K$^+$–HEPES (pH 7.0) and 450 m$M$ KCl.

ddNTP stocks: Both 1.5 and 67 $\mu M$ stocks each of all four individual nucleotides (pH 7.5).

10× extension buffer: 1.3 $M$ Tris–HCl (pH 8.5), 100 m$M$ MgCl$_2$, and 100 m$M$ dithiothreitol (DTT).

dNTP-T: 110 m$M$ dATP, 110 m$M$ dCTP, 110 m$M$ dGTP, 6 m$M$ dTTP, and [$\alpha$-$^{32}$P]dTTP (NEN; 10 mCi/ml).

Reverse transcriptase, avian myeloblastosis virus (AMV), 24640 units/ml (Seikagku American Inc.).

Chase: 1 m$M$ dATP, 1 m$M$ dCTP, 1 m$M$ dGTP, and 1 m$M$ dTTP.

Precipitation mix: 67% ethanol and 85 m$M$ sodium acetate (pH 6.5).

Primer extension loading buffer: 7 $M$ urea, 0.1× TBE, 0.03% xylene cyanol, and bromphenol blue.

RNA for primer extension is generally at 1 $\mu M$.

Primers are at a concentration range of 0.01–0.5 $\mu M$ (see earlier discussion).

Amersham Hyperfilm

Whatman 3CHR chromatography paper

This protocol is sufficient for 18 primer extension samples (including sequencing lanes). To allow for error, mixtures are prepared in excess of the required amount for 18 samples as noted.

Hybridization mixture is prepared (enough for 22 samples) by mixing 22 $\mu$l 4.5× hybridization buffer and 22 $\mu$l primer (see earlier discussion). To each 2.5-$\mu$l sample of 1 $\mu M$ RNA, 2 $\mu$l of hybridization mixture is added and mixed by brief vortexing and pulse centrifugation at room temperature. Samples are placed in an open metal rack and incubated in a water bath at 95° for 90 sec. As the end of the incubation approaches, hot water (95°) is transferred to a metal pan until there is enough water to come up to the middle of the tubes. The metal rack holding the samples is transferred to the water-containing metal pan and the water is allowed to cool for approximately 10 min to 45°. During hybridization, the extension mix should be prepared. To make enough extension mix for 25 samples, 16.7 $\mu$l 10× extension buffer, 8.3 $\mu$l dNTP-T, 15 $\mu$l [$\alpha$-$^{32}$P]dTTP, 10 $\mu$l water, and 2.5 $\mu$l undiluted reverse transcriptase are mixed by brief vortexing and pulse centrifugation at room temperature. The extension mix is kept on ice until needed. Once the temperature of the water bath has reached 45°, samples are centrifuged briefly at room temperature to concentrate any condensed sample in the tube to the bottom. For the sequencing reactions, 1 $\mu$l of the appropriate 1.5 $\mu M$ stock of ddNTP is added to the appropriate tube at this time. To each sample, 2 $\mu$l extension mix is added and mixed by multiple pipetting up and down. Samples are briefly pulse centrifuged at room temperature and incubated in a 42° oven for 30 min. Next, 1 $\mu$l chase is added to the side of each tube. For the sequencing reactions, 1 $\mu$l of the appropriate 67 $\mu M$ stock of ddNTP is now added to the side of the appropriate tube to supplement the chase reaction. Samples are mixed briefly by vortexing and centrifugation at room temperature and then incubated at 42° for 15 min. To each sample, 120 $\mu$l precipitation mix is added and samples are vortexed thoroughly before incubation at room temperature for 10 min. Samples are centrifuged at room temperature for 10 min and the supernatant is removed carefully using a fine (gel loading) pipette tip on the Pipetman, generally in two steps removing approximately 65 $\mu$l each time. Removal of all ethanol at this stage is very important; however, the pellet will likely be invisible and must not be disturbed. Pellets are dried under vacuum for 5 min and resuspended in 10 $\mu$l primer extension loading buffer with extensive vortexing. Samples are heated to 95° for 2.5 min, and 1.5 $\mu$l is loaded on a 6% acrylamide, 6 $M$ urea, 1$\times$ TBE sequencing gel (60 cm $\times$ 20 cm $\times$ 0.25 mm), which is run at 55 mA for 2.5 hr with an aluminum heat-diffusing plate. The gel is transferred to Whatman (Clifton, NJ) paper and dried at 80° for 30 min. Autoradiography of the gel with Amersham Hyperfilm is typically carried out overnight at room temperature. However, the length of time required for appropriate exposure varies for different primers. Use of an intensifying screen is discouraged because it tends to result in decreased sharpness of the bands.

Interpretation of Cleavage Data

Figure 1 shows a representative autoradiogram of primer extension of the target RNA from a directed hydroxyl radical probing experiment using Fe(II)-S20 complexed to 16S rRNA. Cleavage of the RNA results in the appearance of bands or increased intensity of bands above background bands in the autoradiogram. These new bands correspond to abortive cDNA transcripts ending at the position of cleavage of the RNA template. The autoradiogram in Fig. 1 shows two sequencing lanes (A and G) followed by five experimental lanes that have all been subjected to hydroxyl radical probing. Lane 1 shows the results of probing using wild-type S20 (lacking cysteine residues) that was mock treated with Fe(II)–BABE; thus, any bands appearing in lane 1 cannot be due to directed probing and can be considered as "background" for the other lanes. Bands that are present in all lanes correspond to spontaneous stops by reverse transcriptase; a representative spontaneous stop is indicated in Fig. 1 (K). Lanes 2–5 are four different cysteine-containing S20 mutant proteins that are all Fe(II)–BABE derivatized. Additional bands that are not observed (lane 1) identify positions of directed cleavage; their relative intensities correspond to the

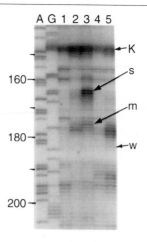

Fig. 1. Directed hydroxyl radical cleavage of 16S rRNA by Fe(II)-derivatized ribosomal protein S20 detected by primer extension. Cleavage of 16S rRNA is shown for the 150–210 region using the 232 primer; (A, G) dideoxy sequencing lanes; (1) mock Fe(II)–BABE-treated wild-type S20; (2) Fe-C14-S20; (3) Fe-C23-S20; (4) Fe-C49-S20; and (5) Fe-C57-S20. Samples in lanes 1–5 were treated with $H_2O_2$ and ascorbate. K indicates a spontaneous stop by reverse transcriptase; s, m, and w show examples of strong, medium, and weak cleavage sites, respectively.

extent of cleavage at each site in 16S rRNA and are correlated with the distance between the RNA target and the position of the Fe(II).[14]

Cleavage sites are scored as strong, medium, or weak based on visual comparison to the intensity of adjacent sequencing lanes as internal controls; typical examples of all three classes of intensity are shown in Fig. 1. Strong cleavages correspond to bands that are twofold or more intense than adjacent sequencing lanes; medium are those of approximately equal intensity; and weak are at least twofold less intense than adjacent sequencing lanes. Calibration experiments show that strong band intensities correspond to a distance between the tethered Fe(II) and the target site of 0 to 22 Å, medium to 12 to 36 Å, and weak to 20 to 44 Å.[14] Thus, the extent of cleavage at a given site can be correlated with its distance from the probe.

Three distinct regions of cleavage of 16S rRNA by Fe(II)-S20 are revealed in Fig. 1. Fe(II)-Cys23-S20 (lane 3) causes cleavage in two regions, most strongly at positions 163–165 and more weakly near position 175. Note that band position in the dideoxy sequencing lanes migrates at a position one nucleotide longer than the corresponding cleavage bands, i.e., a band due to cleavage at nucleotide 163 migrates at the same position as the dideoxy band corresponding to G164 (Fig. 1, lane 3). Fe(II)-Cys14-S20 (Fig. 1, lane 2) more weakly targets the same two regions as Fe(II)-C23-

S20, but the positions and intensities of the cleavage bands are distinctly different from those in lane 3. The cleavage pattern generated by Fe(II)-Cys 57-S20 (lane 5) again differs from those in lanes 2 and 3; Fe(II)-Cys 57-S20 strongly targets nucleotides near position 180 and much more weakly around nucleotides 190–195. No significant cleavage in this region of 16S rRNA is observed from Fe(II)-Cys 49-S20 (lane 4). These data provide information about the proximity of different regions of 16S rRNA relative to different positions on the surface of ribosomal protein S20. Correlating the distances between the tethering positions of the Fe(II) probe and the many RNA target sites yields extensive data regarding the RNA environment around the protein. This approach should also be of utility in mapping the three-dimensional structures of spliceosomes, telomerase, RNase P and MRP, signal recognition particle, and other RNA–protein complexes of biological interest.

## Acknowledgments

The authors thank Claude Meares and co-workers for generous advice and encouragement and for synthesis of BABE. We thank John Diener for critically reading the manuscript. We also acknowledge the contributions of Gabriele Heilek, Simpson Joseph, Kate Lieberman, and Kevin Wilson to the described methodology.

# Section VI

# Cell Biology Methods

# [32] Use of Dimethyl Sulfate to Probe RNA Structure *in Vivo*

*By* SANDRA E. WELLS, JOHN M. X. HUGHES, A. HALLER IGEL, and MANUEL ARES, JR.

## Introduction

Understanding how RNA works requires the coordinated use of diverse experimental approaches. Knowledge of RNA structure and its relationship to function is an essential ingredient for interpreting the biological mechanisms of RNA action. Use of genetic, phylogenetic, biophysical, and computational approaches to divining RNA structure and structural dynamics is greatly enhanced by the application of chemical and enzymatic probes of RNA structure in solution. Among the most versatile chemical probes available for studying RNA and ribonucleoprotein structure is dimethyl sulfate (DMS), which can directly donate a methyl group to specific hydrogen bond-accepting ring nitrogens on A, C, and G residues in RNA. The efficiency of methylation reports the chemical environment of the sensitive ring nitrogens in each base: hydrogen bonding or poor solvent accessibility results in protection from methylation, whereas solvent exposure or an unusual chemical environment may enhance methylation.[1] The efficiency of methylation can be estimated at many positions along the RNA chain by evaluating methylation-dependent stops to primer extension by reverse transcriptase.[2,3] This information provides clues about the environment of individual nucleotides that can be compared to RNA structural models and hypotheses about RNA function.

A major experimental convenience of DMS is its rapid penetration into all compartments of the cell. This feature has allowed probing of RNA structure in a wide variety of cells, including gram-negative[4,5] and gram-positive[6] bacteria, yeast[7] protozoa,[8,9] and plant,[10] including the nucleus,[7]

[1] D. Peatie and W. Gilbert, *Proc. Natl. Acad. Sci. U.S.A.* **77,** 4679 (1980).
[2] T. Inoue and T. Cech, *Proc. Natl. Acad. Sci. U.S.A.* **82,** 648 (1985).
[3] D. Moazed, S. Stern, and H. Noller, *J. Mol. Biol.* **187,** 399 (1986).
[4] S. Climie and J. Friesen, *J. Biol. Chem.* **263,** 15166 (1988).
[5] D. Moazed, J. Robertson, and H. Noller, *Nature* **334,** 362 (1988).
[6] M. Mayford and B. Weisblum, *EMBO J.* **8,** 4307 (1989).
[7] M. Ares and A. H. Igel, *Genes Dev.* **4,** 2132 (1990).
[8] K. Harris, D. Crothers, and E. Ullu, *RNA* **1,** 351 (1995).
[9] A. Zaug and T. Cech, *RNA* **1,** 363 (1995).
[10] J. Senecoff and R. Meagher, *Plant Mol. Biol.* **18,** 219 (1992).

nucleolus,[11] and chloroplasts.[12] Short incubation times at physiological temperatures allow for a quick snapshot of RNA structure *in vivo* with a minimum of perturbation or concern that secondary effects lead to the observed structure. The method can be applied to many cultures simultaneously, facilitating direct determination of the effect of different mutations or treatments on folding of the target RNA. Using primers specific for a number of RNAs, the structure of many RNAs can be determined in the same sample. This article presents methods for the probing of RNA structure in yeast cells using DMS.

Theory

If a particular base in RNA is involved in a secondary or tertiary RNA structure or interacts with a protein or other ligand, it may have altered reactivity with DMS. For example, reactivity toward DMS of N-1 of A and N-3 of C would be reduced when the base is involved in a Watson–Crick base pair, as they accept hydrogen bonds and are not accessible to DMS. A similar situation may occur during interaction of the base with a protein or other ligand. The N-7 of G is also methylated by DMS. Although not involved in Watson–Crick base pairing, N-7 of G residues can be evaluated for unusual base–base interactions, ligand interactions, or special chemical environments as well.

Because DMS penetrates cells readily without the need for extended permeabilization treatments, modification of RNA occurs under nearly *in vivo* conditions. Stopping DMS modification during RNA extraction can pose a significant challenge, however. To quench the DMS reaction, high concentrations of 2-mercaptoethanol are added to react with excess soluble DMS. Aqueous insoluble DMS that may contaminate the cell pellet can be removed by extraction of the culture with water-saturated isoamyl alcohol prior to centrifugation, a treatment that does not reduce the recovery of RNA.

Mapping of the methylated sites can be achieved in the total RNA population using a labeled primer specific for the target RNA. In the case of A and C, the methylated base inhibits reverse transcriptase directly because the methyl group alters the Watson–Crick face of the base. In order to map methylation at N-7 of G, the RNA must be treated with aniline and borohydride to cleave the RNA chain at the methylated Gs prior to reverse transcription. By comparing the pattern of modification-

[11] A. Méreau, R. Fournier, A. Grégoire, A. Mougin, P. Fabrizio, R. Lührmann, and C. Branlant, *J. Mol. Biol.* **273,** 552 (1997).

[12] D. Higgs and D. Stern, personal communication (1998).

dependent reverse transcription stops to dideoxynucleoside triphosphate-generated stops using the same labeled primer, the sites of methylation can be mapped to the RNA sequence. Methylation-dependent stops occur one position before the corresponding dideoxynucleotide stop.

### Practice

The method has the advantages that it is neither technically demanding nor labor intensive, but care and a few pilot experiments for the empirical determination of certain variables are necessary. In outline, the method involves incubating an aliquot of cells in culture with DMS, quenching the reaction and removing the DMS from the cells, extracting total RNA, and mapping the sites of modified nucleotides using a primer-extension assay.

DMS is corrosive and toxic and is a suspected carcinogen. It is volatile and inhalation is hazardous. It is absorbed readily through the skin. For these reasons, DMS should only be handled in the hood or in tighly closed containers. Nonlatex gloves that resist penetration by organic chemicals should be worn and care should be taken to ensure that clothing does not become contaminated. The isoamyl alcohol and 2-mercaptoethanol used in the protocol are also volatile and noxious. Thus, closed containers and proper hood ventilation are necessary for the safe practice of this protocol. Information about DMS is available at http://www.state.nj.us/health/eoh/rtkweb/0768.pdf.

### DMS Treatment of Yeast Cultures

Yeast can be grown in a variety of media, including YEPD or synthetic complete medium[13] at 18–37° depending on the design of the experiment.

1. Take 10-ml aliquots of yeast culture at $A_{600}$ of 0.75–1.5 and place in a 50-ml polypropylene tube in the hood. In order to maintain temperatures, tubes should be prewarmed. Set up a "stop control" tube to evaluate the effectiveness of the DMS quench treatment.
2. Add 200 $\mu$l of a fresh (same day) dilution of DMS (Aldrich, 99%+ grade) in 95% (v/v) ethanol (1:4, v/v). Mix well, but without cavitation. Do not add DMS to the "stop control" sample yet. Cap the tubes tightly.
3. Incubate with shaking for about 2 min if incubation temperature is 30° or higher. At 18° it may take 10 min to achieve a good reaction.
4. Stop the reaction by placing the tube on ice and adding 5 ml of 0.6 M 2-mercaptoethanol and 5 ml of water-saturated isoamyl alcohol.

---

[13] F. Sherman, *Methods Enzymol.* **194**, 3 (1991).

Cap tightly and shake well. *After* the addition of stop solutions to the "stop control" tube, add an equivalent amount of DMS as the experimental tubes received in step 2 and mix again.

5. Centrifuge the cells at 3000g in the cold for 5–10 min in order to pellet the cells and float the DMS-containing isoamyl alcohol phase away from the pellet. Carefully remove the upper isoamyl alcohol phase and the lower aqueous phase from the pellet. The isoamyl alcohol solubilizes any DMS micelles that may pellet with the yeast cells and react with RNA during subsequent manipulations.

6. Suspend the cell pellet in another 5 ml of 0.6 *M* 2-mercaptoethanol and centrifuge again.

7. Extract RNA from the cell pellet.

We usually do at least one stop control reaction during each experiment and have found that confidence in data is enhanced greatly by these controls. Significant DMS reactivity can occur during RNA extraction and generally shows that all A and C residues are accessible. The stop control provides an assurance that the modification pattern observed was generated during the incubation period instead of during the extraction procedure.

*RNA Extraction*

It may be important to begin RNA extraction immediately after stopping the modification reaction. Although frozen cell pellets from untreated yeast cultures can be stored and extracted by the following protocol with satisfactory results, we have not tried to store and extract RNA from frozen, DMS-treated yeast cultures. As with any RNA extraction protocol, use of DEPC-treated water for making buffers, baked glassware or RNase-free plasticware, gloves, and other precautions are necessary.

1. Suspend cell pellets in 250 $\mu$l of AK buffer [1 g triisopropylnaphthalene (ACROS Organics, NJ), 6 g sodium *p*-aminosalicylate (Sigma, St. Louis, MO), 1.17 g sodium chloride, and 6 ml of water-saturated phenol, dissolved in water and brought to a final volume of 100 ml, can be stored frozen for several months] and transfer them rapidly to microcentrifuge tubes containing 0.5 ml of water-saturated phenol preincubated at 65°, followed by vortexing at high speed for 30 sec.

2. Continue incubation of the emulsified phenol–AK buffer mixture at 65° for 30 min, vortexing for 30 sec at 10-min intervals.

3. Cool the mixture on ice for 2 min and centrifuge at 12,000g for 10 min in a microcentrifuge. Transfer the aqueous phase to a fresh tube and reextract the organic phase with 200 $\mu$l AK buffer.

4. Pool the aqueous phases and extract once with phenol/chloroform/ isoamyl alcohol (25:24:1) and once with chloroform/isoamyl alcohol (24:1).

5. Bring the aqueous phases to a final concentration of 0.3 $M$ sodium acetate (pH 5.2), and precipitate the RNA by the addition of 2 volumes of ethanol.

6. Collect the precipitate, rinse with 70% ethanol, dry, and redissolve in water at a concentration of 1 mg RNA per milliliter (determined by absorption at 260 nm). From 10 ml of yeast cells at $A_{600}$ of 1, the usual yield is about 200 $\mu$g of RNA. Take care to observe that the RNA pellet is completely redissolved.

To map the sites of A and C methylation, proceed directly to primer extension (see later). To map the methylation sites of N-7 of G, it is necessary to chemically cleave the RNA at 7-methylguanosine (7mG) residues before primer extension. The chemistry of the cleavage of RNA at 7mG was first reported by Wintermeyer and Zachau,[14] who showed that yeast tRNA[Phe] could be cleaved specifically at a naturally occurring methylated guanosine. The methylated base is reduced with sodium borohydride followed by strand scission of the polynucleotide chain at the reduced residue with aniline. The following method is an adaptation of the guanine-specific cleavage reaction used in chemical sequencing of RNA.[15]

*Borohydride Reduction and Aniline Cleavage at 7-Methylguanosine*

1. To 5 $\mu$g of modified RNA, add 1 $\mu$g carrier *E. coli* tRNA. Ethanol precipitate by the addition of 0.1 volume of 3 $M$ sodium acetate, pH 5.2, and 2.5 volumes of 95% (v/v) ethanol. The addition of carrier 7mG-containing RNA such as tRNA has been shown to be critical for the site-specific cleavage of some RNAs.[16]

2. Resuspend the RNA in 10 $\mu$l 1 $M$ Tris–HCl, pH 8.2. Add 10 $\mu$l freshly prepared 0.2 $M$ NaBH$_4$ (Aldrich, Milwaukee, WI) and incubate for 30 min at 0° in the dark. Stop the reaction by the adding 1 $\mu$l 20% 2-mercaptoethanol, 70 $\mu$l 95% ethanol, and 10 $\mu$l 3 $M$ sodium acetate, pH 5.2. Freeze on dry ice and collect the precipitate by centrifugation at 12,000$g$ for 10 min.

3. Rinse the precipitate with 70% ethanol, dry, and resuspend the pellet in 10 $\mu$l of freshly prepared 1.0 $M$ aniline/acetate buffer, pH 4.5

[14] W. Wintermeyer and H. G. Zachau, *FEBS Lett.* **58,** 306 (1975).
[15] D. Peatie, *Proc. Natl. Acad. Sci. U.S.A.* **76,** 1760 (1979).
[16] V. Zueva, A. Mankin, A. Bogdanov, D. Thurlow, and R. Zimmermann, *FEBS Lett.* **188,** 233 (1985).

[378 $\mu$l doubly distilled $H_2O$, 136.5 $\mu$l glacial acetic acid, and 45.5 $\mu$l aniline (Aldrich 99.5 + %)] and incubate for 20 min at 55° in the dark.

4. Add 100 $\mu$l 0.2 $M$ sodium acetate (pH not adjusted) and extract with 120 $\mu$l water-saturated phenol followed by extraction with 120 $\mu$l chloroform. Precipitate the RNA from the aqueous phase with 95% ethanol, rinse with 70% ethanol, dry, and resuspend in 5–10 $\mu$l doubly distilled $H_2O$.

*Primer Extension to Map Modification Sites*

To map primer-extension stops induced by DMS methylation, it is important to have a primer specific for the target RNA. Reaction of the primer with other RNAs in the cell can obscure signals from the target and complicate interpretation. Detection of signals from low abundance targets may be enhanced by optimizing primer annealing and reverse transcriptase parameters.

1. Kinase the oligonucleotide primer. Mix 2 pmol primer with 1 $\mu$l 10× kinase buffer [0.5 $M$ Tris–Cl, pH 7.5, 0.1 $M$ MgCl$_2$, 50 m$M$ dithiothreitol (DTT), 10 m$M$ EDTA], 4 $\mu$l (40 $\mu$Ci, 3000 Ci/mmol) [$\gamma$-$^{32}$P]rATP, and enough water to bring the volume to 9 $\mu$l. Add 10 units (1 $\mu$l) polynucleotide kinase (New England Biolabs) and incubate at 37° for 30–45 min. Stop by adding 90 $\mu$l of 10 m$M$ Tris–Cl, pH 8.0, and 1 m$M$ EDTA and heating to 65° for 5–10 min. To check that the reaction was successful, spot 1 $\mu$l on a rectangular piece of Whatman (Clifton, NJ) 3MM paper about 1.5 cm from the bottom. Place in a beaker filled to a depth of 1 cm with 0.75 $M$ $K_2PO_4$, pH 3.4, and allow the liquid front to move 5–6 cm. Wrap in plastic wrap and autoradiograph for 2 min. Labeled oligonucleotide remains at the origin, and unincorporated rATP migrates at the front.

2. Set up the primer annealing reactions in a total volume of 7 $\mu$l. To up to 5 $\mu$g RNA, add 0.5 $\mu$l 10× reverse transcriptase buffer (1.25 $M$ Tris–Cl, pH 8.3, and 175 m$M$ KCl) and 0.5–1 $\mu$l of the kinased oligonucleotide. This is ~20 n$M$ stock; final oligonucleotide concentration in the annealing will be 1.5–3 n$M$. Usually this concentration is sufficient for oligonucleotides with good specificity and targets present at about $10^3$ molecules per cell or less, with the lower limit of detection depending on the specific activity of the probe. It is possible to use less RNA and more primer if the target is more abundant and the oligonucleotide is sufficiently specific. Set up annealing reactions on the stop control and on unmodified RNA as well. To generate a sequence ladder, set up four additional annealing reactions with unmodified RNA.

3. Anneal the primer by heating to 95° for 2 min, and bring the reaction to 65° for 5 min and then to the annealing temperature of the oligonucleotide for 30–45 min. The annealing temperature is best determined empirically but can be approximated by adding four times the number of GC base pairs and two times the number of AT and AU base pairs in the hybrid and subtracting 5. Fancier formulas can also be used. Bringing the mixture to the annealing temperature and allowing it to slow cool at least 15° over the course of 45 min usually ensures that efficient and specific annealing will occur. Allowing the temperature to go too low (e.g., placing the annealing reactions on ice) may encourage the formation of hybrids between the oligonucleotide and nontarget RNAs.

4. For reverse transcriptase sequencing reactions, add 0.5 $\mu$l of one of each of the ddNTP (Pharmacia, Piscataway, NJ) stocks to each of the four annealing reactions containing unmodified RNA. The concentration of the ddNTP stock necessary to obtain aesthetically pleasing and informative sequence ladders must be determined empirically and will be influenced by the base composition of the region being sequenced, the particular reverse transcriptase being used, and the quality (effective concentration) of the ddNTP and dNTPs in the reaction. Good starting points for the following reaction using avian myeloblastosis virus (AMV) reverse transcriptase would employ stocks of about 0.75 m$M$ of ddATP or ddGTP and 1–1.5 m$M$ of ddCTP and ddTTP.

5. Assemble a "premix" on ice. Per annealing reaction use 0.5 $\mu$l reverse transcriptase buffer (as defined earlier), 0.5 $\mu$l 0.1 $M$ DTT, 1 $\mu$l 0.1 $M$ MgCl$_2$, 0.5 $\mu$l of a mixture of 2.5 m$M$ each dNTP (Pharmacia), 0.25 $\mu$l of 1 mg/ml actinomycin D (Sigma), and 0.25 $\mu$l of 20 units/$\mu$l AMV reverse transcriptase (Life Sciences, Inc.). To set up $n$ reactions, multiply the just-described volumes by ($n$ + 1), mix, and distribute 3 $\mu$l to each annealing reaction. Incubate at 42° for 30–45 min.

6. Stop the reaction by adding 5 $\mu$l of a solution of 10 $\mu$g/ml RNase A, 30 m$M$ EDTA, and 0.6 $M$ sodium acetate, pH 5.4. Incubate for 5 min at 42°.

7. Remove the RNase by adding 5 $\mu$l of a solution of 0.2% sodium dodecyl sulfate (SDS), 20 $\mu$g/ml proteinase K, and 0.6 $M$ sodium acetate, pH 5.4. Incubate for 10 min at 42°.

8. Add 50 $\mu$l 95% ethanol and freeze on dry ice for 10 min. Add 1 ml 70% ethanol, mix by inversion, and centrifuge at 12,000$g$ for 10 min. Remove ethanol and dry. Resuspend pellet in 1 $\mu$l of 20 $\mu$g/ml proteinase K and 25 m$M$ EDTA. Incubate for 2–5 min at 65°.

9. Add 2 $\mu$l 98% formamide with dyes. Heat for 5 min at 65°. Load on 6–8% acrylamide, 7.5 $M$ urea sequencing gel.

## Examples and Interpretation

### Probing of U2 Small Nuclear RNA in Vivo

Structure-probing data for RNA in solution are often found to be in good agreement with RNA structures determined by other methods. Data obtained by probing *in vivo* can provide support for the biological relevance of structural models based on other data. Probing mutant derivatives of the RNA of interest provides a way to understand the impact of the mutation on RNA structure. As an example, we compare the results of *in vivo* structure probing of wild-type and mutant U2 RNAs in the stem–loop IIa region with the high-resolution nuclear magnetic resonance (NMR) structure of a model oligoribonucleotide of U2 stem–loop IIa obtained by Stallings and Moore[17] (Fig. 1, PDB entry 1U2A).

These mutations were constructed and described in previous studies.[7,18,19] In this experiment, cells expressing only the indicated U2 gene were incubated with DMS as described earlier for 2 min at 30°, the reaction was stopped, and RNA was extracted from the cells and used for primer extension. Wild-type U2 stem IIa loop nucleotides show a characteristic reactivity *in vivo* (Fig. 1A) that agrees well with the NMR structure of the model stem IIa. The hyperreactivity of A60 toward DMS is consistent with its participation in an interaction through its N-6 and N-7 with the N-3 and N-2 of G55 (Fig. 1B), thus holding the DMS reactive N-1 of A60 into solvent. Residues 57–59 are stacked above A60 and project into solvent.[17] A61 does not quite pair with U54 in the NMR structure,[17] which is consistent with its poor protection. Paired residues 62, 52, and 51 are more protected, but not entirely due to alternative conformations of the RNA *in vivo*.[19]

The problem of dynamic RNA structure is not well addressed by the solution structure probing technique because the reactivity observed at any position is the sum of reactivities at that position of all structures found in the population during the incubation time. This means that the most abundant structure will contribute most to the reactivity pattern. The more complex the structural heterogeneity, the more difficult it will be to interpret the results. In this case, a conserved U2 sequence downstream of the loop of stem–loop IIa has the ability to pair with the loop, forming a competing structure that disrupts the stem.[19] Thus signals from the stem–loop structure

[17] S. Stallings and P. Moore, *Structure* **5**, 1173 (1997).
[18] S. Wells and M. Ares, *Mol. Cell Biol.* **14**, 6337 (1994).
[19] M. Zavanelli, J. Britton, A. Igel, and M. Ares, *Mol. Cell Biol.* **14**, 1689 (1994).

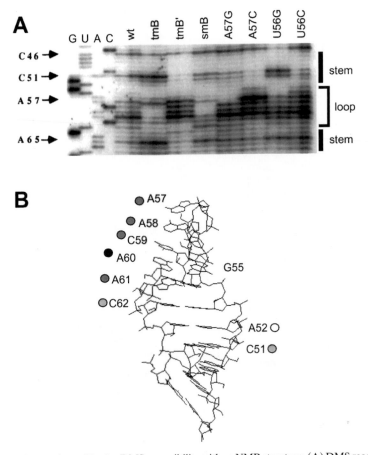

Fig. 1. Comparison of *in vivo* DMS accessibility with an NMR structure. (A) DMS reactivity of wild-type and mutant U2 stem IIa. To determine the structural features associated with specific mutant forms of U2, yeast strains expressing either wild-type or mutant U2 were treated with DMS, and then total RNA was extracted and analyzed by primer extension using an oligonucleotide complementary to U2. The U2 RNA sequence shown in the first four lanes was generated using unmodified RNA as a template by including a different dideoxy-nucleotide in each of four primer-extension reactions. The next lanes show extension products using modified RNA derived from wild-type and seven mutant strains (samples are in duplicate pairs). The mutations include tmB, 58ACA60 to GUU; tmB′, 99UGU to AAC; and smB contains both tmB and tmB′. The single nucleotide changes are as indicated. Note that some of the changes in the reactivity patterns for these mutants result from sequence changes, e.g., changing an A to a G results in a loss of a reactive site rather than protection. The samples were separated by electrophoresis in an 8% polyacrylamide gel containing 7 *M* urea in TBE buffer. Extension products resulting from DMS-induced termination are one nucleotide shorter than the corresponding dideoxynucleotide terminations. (B) Comparison of DMS reactivity *in vivo* with the NMR structure of U2 stem–loop IIa of Stallings and Moore.[17] Residues A52, C51, and C62 are relatively protected (white or light gray dots), whereas A57, A58, C59, and A61 are relatively reactive (dark gray dots). A60 is highly reactive (black dot). The N1 of A60 is held in solvent by the formation of a Sanger type II G–A pair with G55,[17] explaining its hyperreactivity.

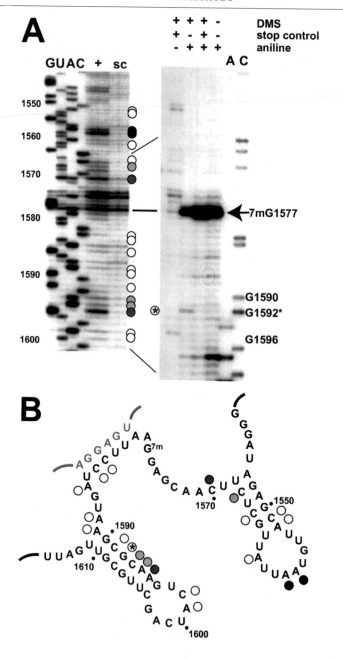

and the competing structure are superimposed. The existence of this alternatively folded form in the pool of U2 molecules in the cell likely explains the incomplete protection of bases in stem IIa.

Mutations in the loop that destabilize stem IIa enhance the formation of the competing structure, as seen by increased reactivity of C51 and A52 in the U56G and tmB mutants (Fig. 1A). The NMR structure places the base moiety of U56 in a space beneath the loop that will not easily accommodate the bulky G residue,[17] and the destabilizing effect on the loop may account for the increased reactivity observed. The increased reactivity of stem bases in the tmB mutant, which replaces residues 58–60 with GUU, may be caused by the loss of the G55-A60 base pair, as well as the stacking interactions above it. Another possibility is that the loop mutations could inhibit the binding of a protein to the loop, destabilizing the stem. This seems less likely considering the effect of tmB' (which destroys the complementary sequence needed to form the alternative structure) and the A57C mutation (which also reduces complementarity between the loop and the downstream sequence). Both of these mutants show increased protection of the stem bases, especially C62. Thus *in vivo,* the structure of wild-type U2 stem–loop IIa in the context of the full-length U2 snRNA is very similar to the model oligonucleotide studied by Stallings and Moore.[17] Furthermore, the effect of mutations on the DMS accessibility of the RNA *in vivo* corresponds to the expected effect of the mutations on the structure.

### Probing DMS Reactivity at N-7 of G Residues in RNA

To demonstrate the usefulness of this method for identifying reactivities at N-7 of G we probed 18S ribosomal RNA. Primer extensions were performed using a primer complementary to nucleotides 1629–1645, just downstream of a conserved region of 18S rRNA (Fig. 2). The reactivity pattern

FIG. 2. Mapping reactive 7-methylguanosine residues in yeast rRNA. (A) Primer-extension reactions using DMS-modified RNA (lane +) or "stop control"-treated RNA (lane sc) either with (right) or without (left) treatment with aniline. Treatment of RNA in the right panel is indicated at the top. Sequencing tracks were generated by primer extension on unmodified RNA with ddGTP (lanes C, left and right panels), ddTTP (lanes A, left and right panels), ddATP (lane U), and ddCTP (lane G). Nucleotide positions within 18S rRNA are indicated on the left. The reactions were performed as described previously except that the amount of RNA in the primer extension was reduced to 2 $\mu$g and the primer was annealed by heating at 95° for 2 min followed by slow cooling to 42°. The G residues marked on the right are positioned to indicate the migration of the expected modification/cleavage-induced stop rather than the dideoxy termination-induced stop. (B) Secondary structure model of the region of yeast 18S rRNA[20] displayed in A. Dots indicate reactivity (black and gray) or protection (white). The hyperreactive N-7 of G1592 is indicated with an asterisk.

obtained is largely consistent with the assignment of helical regions proposed by phylogenetic analysis for this portion of yeast 18S rRNA.[20] Analysis of the pattern of A and C accessibility (Fig. 2A, left-hand side) observed for the regions represented as being single stranded in the model[20] (Fig. 2B), however, requires more information than sequence comparison can provide. Protections and enhancements in these areas may be due to protein, higher order structure, or other unique environments. Reactivities of nucleotides in the homologous region of *E. coli* rRNA have been shown to be enhanced by binding of ribosomal proteins S7, S9, and S19.[21]

To assay the accessibility of the N-7 position of G, the RNA was subjected to reduction with sodium borohydride and aniline-induced cleavage. Primer extensions were performed with the same primer (Fig. 2A, right-hand side). A nearly quantitative stop results from the natural methylation of the G at position 1577 in yeast 18S rRNA[22] in all samples subjected to aniline cleavage (Fig. 2A, arrow, 7mG1577). A stop observed only in the modified and aniline cleaved RNA was observed at G1592 (asterisk, Fig. 2A). The stop at G1592 represents a G residue that is especially available for methylation by DMS at N-7. The reason for the accessibility of this residue is unclear but it may represent a disruption of the helical structure due to protein binding. Corresponding residues in *E. coli* 16S rRNA are also sensitive to N-7 methylation.[23] We have been able to observe hyperreactive G residues using this method, but the reactivity of the typical G residue in RNA seems quite low. Aniline-induced strand scission can produce a background ladder due to a low rate of cleavage at many sites, obscuring subtle protections. It has not been easy to monitor A, C, and G reactivities in the same sample, so we routinely run aniline-treated samples only to determine G accessibility. A report[24] used this approach to detect a hyperreactive G in the HIV Rev-response element RNA expressed in the context of a yeast mRNA.

*Accessibility of Occasional U Residues to DMS in Vivo*

U3 RNA is located in the nucleolus and is also accessible to modification by DMS *in vivo*,[11] as shown in Fig. 3. The region shown includes the most 3′ stem–loop of the U3A structure (A263-U308) and the adjacent short

[20] R. Gutell, B. Weiser, C. Woese, and H. Noller, *Prog. Nucleic Acids Res. Mol. Biol.* **32,** 155 (1985).

[21] S. Stern, T. Powers, M.-L. Changchien, and H. Noller, *Science* **244,** 783 (1989).

[22] V. S. Zueva, A. S. Mankin, A. A. Bogdanor, and L. A. Baratova, *Eur. J. Biochem.* **146,** 679 (1985).

[23] D. Moazed and H. Noller, *J. Mol. Biol.* **211,** 135 (1990).

[24] B. Charpentier, F. Stutz, and M. Rosbash, *J. Mol. Biol.* **266,** 950 (1997).

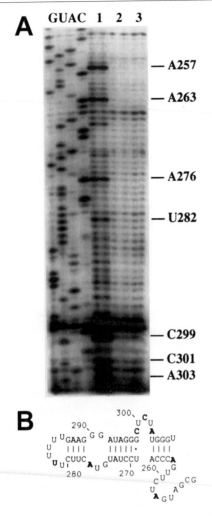

Fig. 3. An unusual U residue in U3 snoRNA is reactive to DMS *in vivo*. (A) Primer-extension assay to map DMS-reactive bases in U3. DMS-modified RNA, lane 1; "stop control"-treated RNA, lane 2; and unmodified RNA, lane 3. Duplicate, independent samples are shown for each treatment. Modified nucleotides are indicated on the right. The primer used is complementary to nucleotides 309–329 of yeast U3A. The RNA sequence shown on the left was generated using the same primer with unmodified RNA and dideoxynucleoside triphosphates as described in the text. (B) Secondary structure model for nucleotides 251–308 of yeast U3A RNA. Modified nucleotides are indicated in bold type.

loop of conserved nucleotides (G251-G262), referred to as "box C," which is required in mammalian U3 for protein binding.[25] The sites of strong DMS modification (Fig. 3A, lane 1, labeled on the right) are indicated in Fig. 3B in bold type. The pattern of modifications and protections corresponds well to the proposed structure of the stem loop. U282, which occurs within a run of six U residues in the terminal loop of the stem, is also specifically modified by DMS. This modification has been observed by others *in vivo*, but not in naked RNA or partially purified U3 snRNPs.[11] Methylation of U and G by DMS is rarely but reproducibly observed, and its chemical basis is not understood.[26] We suggest that the special environment of this residue *in vivo* may stabilize an enol tautomer of certain U residues, rendering them reactive to DMS at the N-3 position. Stabilization of the enol tautomer of G (and methylation at N-1 of G) would also explain the occasional G residue that reacts with DMS and leads to a reverse transcription stop in the absence of aniline treatment.[26]

Troubleshooting

Difficulties encountered in the use of DMS *in vivo* have usually involved either poor modification or inability to stop the modification (e.g., "stop control" reactions that show that modification occurs during RNA extraction). In the case of poor modification, more of a 1:4 dilution of DMS in ethanol should be used per 10 ml of culture, rather than the same amount of a less dilute stock of DMS. Pure DMS has a limited shelf life and it may simply be worth ordering a fresh bottle if modification is poor. A small bottle of fresh DMS every few months rather than a single lifetime supply is recommended.

In the case of modification in the stop control samples, too much DMS is being used, and more dilute stocks (e.g., 1:6, 1:8 of the DMS in ethanol) should be tried until the modification reaction is suppressed in the stop control samples, but is readily detected in the experimental samples.

One problem encountered with primer extension is poor specificity of the oligonucleotide primer. This causes multiple stops that are not interpretable. A specific primer used under optimal reverse transcription conditions for the unmodified RNA target will generate a very strong stop corresponding to the 5′ end of the RNA, as well as several structure-induced stops characteristic of the target. Nontarget RNAs may also produce such stops, but these may be longer than the longest product expected from the target.

[25] S. Baserga, X. Yang, and J. A. Steitz, *EMBO J.* **10**, 2645 (1991).
[26] D. Moazed and H. Noller, *Cell* **47**, 985 (1986).

It is advisable to determine empirically the optimum annealing conditions for each primer.

A second problem is primers that do not label efficiently or do not prime efficiently due to the formation of competing structure within the primer or the target. These problems may require using alternative primers. If several primers on the same target do not produce satisfactory signals, the target may not be abundant enough. Enriching the RNA sample for the target or optimizing the KCl concentration in the reverse transcriptase reaction may help.

A third problem that is usually diagnosed during the evaluation of primer extensions is that of ribonuclease contamination of samples. This is most often seen when a strong stop is observed at a position (or several positions) equivalent to a dideoxyTTP-induced stop at an A residue in the target RNA. The A will almost always be preceded by a U in the sequence. This usually indicates that the target RNA has been hit by a pancreatic-type (RNase A) nuclease that prefers UpA sites for cleavage, producing a strong stop at the cleavage site.

Difficulty in generating a satisfactory sequencing ladder usually can be remedied by adjusting the concentration of the dideoxyNTP stocks used until an appropriate distribution of stops is obtained for each nucleotide.

### Acknowledgments

We thank Uwe von Ahsen for help with aniline cleavage and Sandra Wolin, Carol Dieckmann, Dave Higgs, David Stern, and Dave Brow for sharing *in vivo* DMS experiences with us. This work was supported by NIH Grant GM40478.

## [33] Sensitive and High-Resolution Detection of RNA *in Situ*

*By* Pascal Chartrand, Edouard Bertrand, Robert H. Singer, and Roy M. Long

### Introduction

Unlike other molecular approaches, *in situ* hybridization offers the advantage of studying RNA in the complex environment of the eukaryotic cell. Also, due to their higher spatial resolution, fluorophore-labeled probes are increasingly replacing enzymatic and radioactive methods used previously to detect cellular RNA. For this reason, fluorescence *in situ* hybridization has been applied successfully not only for the simple detection of mRNA,[1] rRNA,[2] tRNA,[3] snRNA,[4] and snoRNA,[5] but also to study the transcription,[6] processing,[7] export,[8] and localization[9,10] of these RNA in several types of eukaryotic cells.

This article provides protocols for fluorescence *in situ* hybridization (FISH) on RNA in yeast and tissue culture cells. These protocols use fluorescent-labeled probes, but are applicable to other detection techniques based on biotin–streptavidin or digoxigenin.

New microscopic designs and methods for image acquisition and analysis are particularly useful when employing fluorescence as a detection method. Algorithms that can remove out-of-focus light and quantitate fluorescent signal[6] have provided powerful means by which to extract information from acquired digitized images. An important consideration in analyzing these images is that decisions on scaling and background threshold settings must be made objectively.[11] While this is not the subject of this article, it will be ultimately the most important component of the data presentation.

Oligonucleotide probes have significant advantages over long probes and some disadvantages. Advantages are that they can be synthesized rapidly from the abundant and growing sequence information in databases. They can be targeted to specific genomic regions, free of repetitive sequences, or the isoform-specific regions of RNA, such as the 3'UTR.[12] Because large amounts of oligonucleotide probes can be made (milligram amounts, enough for thousands of experiments), the probes serve as a consistent reagent over a period of years. Conjugation of the probes with fluorochromes or other reactive groups can be controlled chemically and quality assured. Unlike probes prepared enzymatically, every probe is identical, allowing for better reproducibility in the experiments. Hybridization times are short because of the low complexity and signal-to-noise ratios are higher. A disadvantage is that they are small and hence give a fraction

[1] K. L. Taneja, L. M. Lifshitz, F. S. Fay, and R. H. Singer, *J. Cell Biol.* **119,** 1245 (1992).
[2] I. B. Lazdins, M. Delannoy, and B. Sollner-Webb, *Chromosoma* **105,** 481 (1997).
[3] E. Bertrand, F. Houser-Scott, A. Kendall, R. H. Singer, and D. R. Engelke, *Genes Dev.* **12,** 2463 (1998).
[4] A. G. Matera and D. C. Ward, *J. Cell Biol.* **121,** 715 (1993).
[5] D. A. Samarsky, M. J. Fournier, R. H. Singer, and E. Bertrand, *EMBO J.* **17,** 3747 (1998).
[6] A. Femino, F. S. Fay, K. Fogarty, and R. H. Singer, *Science* **280,** 485 (1998).
[7] G. Zhang, K. L. Taneja, R. H. Singer, and M. R. Green, *Nature* **372,** 809 (1994).
[8] D. A. Amberg, A. L. Goldstein, and C. N. Cole, *Genes Dev.* **6,** 1173 (1992).
[9] A. Ephrussi, L. K. Dickinson, and R. Lehmann, *Cell* **66,** 37 (1991).
[10] R. M. Long, R. H. Singer, X. Meng, I. Gonzalez, K. Nasmyth, and R-P. Jansen, *Science* **277,** 383 (1997).
[11] F. S. Fay, K. L. Taneja, S. Shenoy, L. Lifshitz, and R. H. Singer, *Exp. Cell Res.* **231,** 27 (1997).
[12] E. H. Kislauskis, Z. Li, R. H. Singer, and K. L. Taneja, *J. Cell Biol.* **123,** 165 (1993).

of the signal of a probe containing much more sequence. Although "cocktails" of oligonucleotide probes can be generated, this becomes expensive. Also, they are not applicable if the sequence information is not known. Generally, oligonucleotide probes are more difficult for detecting DNA sequences by FISH, as signal levels are low. As many as 10 probes (500 nucleotides total) will be necessary.[6] In this case, larger probes containing 10 kb of genomic sequence are preferable. Similarly, detection using cRNA probes is also a viable approach in all cases using enzymatically prepared probes. Large probe sizes (>200 nucleotides) have to be removed to avoid a higher background. The key to success is that the iteration of many small probes on the template leads to detection. The following protocols report on preparation and use of this variety of probes. Continuously updated information may be found at www.singerlab.org.

## Preparation of Probes

### Oligonucleotide Probes

DNA oligonucleotide probes are currently preferred for *in situ* hybridization. They are synthesized chemically, which allows the incorporation of amino-modified nucleotides [usually an aminoallyl(dT)] at specific positions in their sequence. Each free amine can then be coupled to a fluorophore after synthesis. The main advantage of direct labeling of probes resides in the high signal-to-noise ratio of these probes compared to indirect detection methods using antibodies. Even if the absolute signal is higher using fluorescent-labeled antibodies, they will contribute to a higher background. Also, these probes are small enough to penetrate easily in the cells (a 50 base oligonucleotide is 50 times smaller than the Fab fragment of an antibody) and can be washed away under mild conditions, compatible with the retention of intracellular RNA, and reduce the background fluorescence signal.

*Design and Synthesis.* Usually, between two and six probes of 50 nucleotides complementary to a given RNA are used to detect this specific RNA in a cell. These probes are designed to have five amino-modified T per oligonucleotide, each separated by a stretch of 10 nucleotides and, if possible, a total G + C content of 50% (if not possible, all the oligonucleotides should be designed to have the same melting temperature). The oligonucleotides can be synthesized on a DNA synthesizer on a low-scale 0.2 $\mu$mol column, which will give enough quantity of oligonucleotides for years. The amino-modified dT used should contain a six-carbon arm between the base and the free amine in order to increase the coupling of the fluorophore.

*Purification.* After synthesis and deprotection, the oligonucleotides are purified on an acrylamide gel in order to remove the lower molecular size oligonucleotides.

1. Set up a 10% acrylamide/urea gel with large wells.
2. Resuspend the dry oligonucleotide in a 9:1 formamide: 1× TBE buffer (100 m$M$ Tris, pH 8.3, 100 m$M$ boric acid, 2 m$M$ EDTA) with dyes. Load and run until the xylene cyanol dye reaches the bottom of the gel.
3. Visualize the DNA under UV. Cut the band, chop up into fine pieces, and place in a 50-ml tube with 25 ml of distilled water. Elute overnight at 37°.
4. Repeat the extraction twice with 15 ml of 10 m$M$ NaCl for 2 hr at room temperature. Lyophilize each extraction and resuspend in 1 ml of distilled water.
5. Load the extract on a Sephadex G-50 column (Pharmacia, Piscataway, NJ) and collect the fractions. Take the OD$_{260}$ of each fraction and pool those containing pure oligonucleotides.

The oligonucleotides can also be purified by reversed-phase chromatography on a Poly-Pak column (Glen Research, Herndon, VA).

*Labeling of Oligonucleotide Probes with Terminal Transferase*

Another approach used for the labeling of the oligonucleotide probes is to incorporate digoxigenin or biotin-labeled nucleotides at the 3′ end of the probes using the terminal transferase enzyme. This approach is less expensive then the incorporation of amino-modified nucleotides during the synthesis of the probes and can give a higher signal because of the utilization of antibodies for the detection. However, the yield of incorporation of modified nucleotides by the terminal transferase can vary among the probes and quantitative measurements are not possible.

1. Purify the oligonucleotide probes by gel electrophoresis.
2. The reaction mix (50 $\mu$l) consists of 25 p$M$ oligonucleotide probe, 0.2 m$M$ digoxigenin-11-dUTP or biotin-16-dUTP, 1 m$M$ CoCl$_2$, 140 m$M$ potassium cacodylate, 30 m$M$ Tris, pH 7.6, 0.1 m$M$ DTT, and 100 U terminal transferase. Incubate at 37° for 1 hr. The terminal transferase and modified dUTP are available from Boehringer-Mannheim (Indianapolis, IN).
3. Purify the probes with two rounds of gel filtration on Micro Bio-Spin columns with Bio-Gel P30 (Bio-Rad, Hercules, CA).

*RNA Probes*

Short RNA probes (50–200 nucleotides) can also be used for *in situ* hybridization. They have the advantage of being less expensive than oligonucleotides probes. However, they give higher background and necessitate more stringent washing conditions. It is very important to design the RNA probes such that no polylinker sequences are present in the resulting transcript. Indeed, the polylinker often contains GC-rich stretches that induce cross-hybridization with ribosomal RNA.[13] These RNA probes are synthesized *in vitro* using the T7, T3, or SP6 RNA polymerase, and an amino-modified nucleotide [usually amino-allyl(U)] is incorporated during the transcription. *In vitro* transcription kits are available from companies such as Ambion (Austin, TX) and can be adapted for the incorporation of modified nucleotides.

*Notes.* The T7 and T3 RNA polymerases incorporate amino-modified nucleotides better than the SP6 RNA polymerase. Digoxigenin-11-UTP, biotin-16-UTP, and fluorescein-12-UTP (all from Boehringer-Mannheim, Indianapolis, IN) can also be incorporated using the same protocol. However, fluorophore-labeled UTPs are incorporated at a lower frequency, which result in a lower specific activity for the probes. Also, these transcription reactions cannot be phenol extracted, as the fluorophore-labeled RNA probe partition with the phenolic phase.

*Reaction Mix*[14]

> 40 m$M$ Tris, pH 7.5
> 6 m$M$ MgCl$_2$
> 2 m$M$ spermidine
> 10 m$M$ NaCl
> 20 m$M$ dithiothreitol
> 1 m$M$ ATP, CTP, GTP, UTP
> 1 m$M$[15] 5-(3-aminoallyl)UTP (Sigma, St. Louis, MO)
> 2 U/$\mu$l RNase inhibitor
> 40 ng/$\mu$l linearized plasmid DNA
> 0.2 U/$\mu$l RNA polymerase

Complete with DEPC-treated H$_2$O. Incubate at 37° for 1–2 hr. Add 1 U of DNase/RNase free and incubate at 37° for 10 min. Purify the RNA product by phenol extraction and ethanol precipitation. The RNA is then resuspended in 1× SSC (0.15 $M$ NaCl, 15 m$M$ sodium citrate, pH 7), and unincorporated nucleotides are removed by two rounds of gel filtration on Micro Bio-Spin columns with Bio-Gel P30 (Bio-Rad, Hercules, CA). The

---

[13] H. Witkiewick, M. E. Bolander, and D. R. Edwards, *BioTechniques* **14**, 458 (1993).
[14] M. R. Jacobson and T. Pederson, *in* "mRNA Formation and Function," p. 341. Academic Press, San Diego, 1997.
[15] If a high ratio of fluorophore coupling is preferred, it is possible to use only aminoallyl-UTP in the reaction instead of a 1:1 mix of UTP:aminoallyl-UTP.

RNA is finally precipitated with ethanol and resuspended in DEPC-treated water.

## Conjugation of Activated Fluorophore to Amino-Modified Probes

Manipulations should be conducted under low luminosity in order to avoid bleaching of the fluorophores. Also, any trace of free primary amine, such as Tris base, should be removed from the nucleic acids, as it will also react with the activated fluorophore.

Dissolve 20 $\mu$g of dried, purified, amino-modified oligonucleotide or RNA in 35 $\mu$l of 0.1 $M$ sodium carbonate buffer, pH 9.0 (use pH 8.8 and shorter incubation time for RNA probes in order to decrease the hydrolysis of the RNA during the incubation). Resuspend the activated fluorophore in 15 $\mu$l dimethyl sulfoxide (DMSO) and add to the oligonucleotide solution (CY3 is water soluble and does not require DMSO). Incubate for 12–16 hr for oligonucleotides and 6–12 hr for RNA probes, in the dark, at room temperature, with occasional vortexing. Commonly used fluorophores are CY3 (Amersham, Arlington Heights, IL) and rhodamine (Molecular Probes, Eugene, OR) as red fluorophores and FITC and Oregon Green 488[16] (both from Molecular Probes, Eugene, OR) as green fluorophores.

Unreacted fluorophores are removed by gel filtration through a Sephadex G-50 column. Fractions containing the labeled oligonucleotide are visualized under a UV lamp and pooled. The specific activity of the probe is calculated by absorption spectroscopy. If several probes are used to detect a single RNA, they can be pooled at a concentration of 1 ng/$\mu$l for each probe and stored at $-20°$ in the dark.

## In Situ Hybridization on Yeast Cells

At this step, it is important that all solutions for fixation, spheroplasting, and in situ hybridization be DEPC-treated or prepared with DEPC-treated distilled water. Also, individuals must wear gloves to avoid RNase contaminations. This protocol is derived from Long and colleagues.[17]

## Preparation of Coverslips

To maintain yeast cells at the surface of the coverslip during hybridization and washing steps, the coverslips must be coated with poly(L-lysine).

---

[16] The coupling of Oregon Green 488 to amino-modified nucleotides is very slow and an incubation time of 48 hr is preferable in this case. We usually do not use this fluorophore with RNA probes because of the higher degradation level of the probes.

[17] R. M Long, D. J. Elliot, F. Stutz, M. Rosbash, and R. H. Singer, *RNA* **1**, 1071 (1995).

Type 1 coverslips (22 × 22 mm, Fisher Scientific, Pittsburgh, PA) are first boiled in 250 ml of 0.1 *N* HCl for 30 min. They are rinsed 10 times with distilled water in a beaker and autoclaved in distilled water. They can be stored at 4° for several months.

Put one coverslip in each well of a six-well tissue culture plate (Becton-Dickinson, Franklin Lakes, NJ) and drop 200 μl of poly(L-lysine) 0.01% (Sigma) on each coverslip. Incubate for 2 min at room temperature, aspirate the excess, and let dry at room temperature (about 2–3 hr). When dry, wash each well three times for 10 min with distilled water, which is removed by aspiration. At the third wash, rest each coverslip on the wall of the wells, with the face treated with poly(L-lysine) on the top, and let dry (do not let the coverslips air dry on the bottom of the wells as they will stick to the plastic).

*Fixation of Yeast Cells*

Yeasts are grown in 50-ml cultures in the appropriate media until they reach early log phase ($OD_{600}$ between 0.2 and 0.4, about $10^8$ cells). Cells are fixed for 45 min at room temperature by directly adding 6.3 ml of 32% formaldehyde (EM grade, Electron Microscopy Sciences, Fort Washington, PA) to the medium. The fixative is removed by three rounds of centrifugation (5 min at 3500 rpm at 4°) and washes with 10 ml of ice-cold buffer B (1.2 *M* sorbitol, 0.1 *M* potassium phosphate, pH 7.5).

*Notes.* For the detection of nuclear RNA, one can use 10 ml of 20% formaldehyde, 50% acetic acid for 10 min. The quality of the formaldehyde is crucial for the preservation and detection of small details by FISH. We strongly suggest using ultrapure, single-usage ampoule-sealed formaldehyde. If not possible, a fresh solution of 40% paraformaldehyde can be prepared by adding 12 g of paraformaldehyde (Sigma) to 22.8 ml of DEPC-treated water with 6 drops of 10 *N* NaOH.

*Spheroplasting*

Cells are resuspended (do not vortex) in 1 ml of buffer B containing 20 m*M* vanadylribonucleoside complex (Gibco-BRL, Gaithersburg, MD), 28 m*M* 2-mercaptoethanol, 0.06 mg/ml phenylmethylsulfonyl fluoride, 5 μg/ml of pepstatin, 5 μg/ml of leupeptin, 5 μg/ml of aprotinin (all from Sigma), and 120 U/ml of RNase inhibitor (Boehringer-Mannheim, Indianapolis, IN) and transferred in a tube containing 0.1 mg of dried oxalyticase (50,000 U/mg, Enzogenetics, Corvallis, OR). Spheroplasting is done by incubating the cells for 8 min at 30° (up to 15 min for strains with more resistant walls). Cells are then centrifuged for 4 min at 3500 rpm at 4° and washed once in ice-cold buffer B. Cells are further resuspended in 750 μl of buffer B, and 100 $\mu$l is added to each of the poly(L-lysine)-coated coverslips in the six-well tissue culture plates. Cells are left to adhere to the coverslips by incubating for 30 min at 4°. Three milliliters of buffer B is then carefully added to each well, removed by suction, and replaced by 5 ml of 70% (v/v) ethanol, which is incubated for at least 15 min at −20° before hybridization. At this stage, the coverslips can be stored a few weeks at −20°.

*Note.* The oxalyticase should be resuspended in 50 m$M$ potassium phosphate, pH 7.5, at 1 mg/ml, aliquoted at 100 $\mu$l (0.1 mg) per tube, and lyophilized. These aliquots can be stored at 4° in a dessicator. To test if the spheroplasting went well, drop 20 $\mu$l of 1% SDS on 100 $\mu$l of spheroplasted cells on a glass slide and mix. Spheroplasts will blow out and a white, flaky precipitate should appear.

### In Situ Hybridization

All the following steps should be done under low luminosity in order to avoid the bleaching of the fluorophores on the probes, especially for fluorescein isothiocyanate (FITC)-labeled probes.

*Preparation of Probes.* For each coverslip used in the hybridization, prepare one tube of probes (we suggest using two coverslip per experiment in order to have a duplicate if one of the coverslip is broken during the manipulations). Add 10 $\mu$l of the 1-ng/$\mu$l probe solution with 4 $\mu$l of a 5-mg/ml solution of 1:1 sonicated salmon sperm DNA : *E. coli* tRNA (Sigma). Lyophylize and resuspend in 12 $\mu$l of 80% formamide (Sigma) and 10 m$M$ sodium phosphate, pH 7.0.

*Hybridization.* The cell-coated coverslips are put in a Coplin jar (Thomas Scientific, Swedesboro, NJ) and rehydrated with two washes of 8 ml of 2× SSC (0.3 $M$ NaCl, 30 m$M$ sodium citrate, pH 7) for 5 min at room temperature. Incubate coverslips in 8 ml of 40% formamide, 2× SSC for 5 min at room temperature. During this incubation, heat the 12-$\mu$l solution of probes at 95° for 3 min and add 12 $\mu$l of 4× SSC (0.75 $M$ NaCl, 75 m$M$ sodium citrate, pH 7), 20 m$M$ vanadylribonucleoside complex, 4 $\mu$g/$\mu$l of RNase-free BSA (Boehringer-Mannheim, Indianapolis, IN), and 50 U of RNase inhibitor. Drop 24 $\mu$l of the probe solution on a Parafilm sheet wrapped around a glass plate (16 × 20 cm) and lay the coverslip on the drop (the surface of the coverslip containing the cells should face the drop. Be careful to avoid air bubbles). Up to 10 coverslips can be placed on these plates. Finally, wrap over a second Parafilm sheet (do not move the coverslips after they have been placed), seal the two Parafilm sheets together, wrap the glass plate in aluminum foil, and incubate at 37° for 3 hr to overnight (a 3-hr incubation gives a lower hybridization signal).

*Notes.* The quality of the formamide is important in order to avoid nonspecific signaling. Keep the formamide stock at 4° and do not store for more than a year. The probes must be resuspended in a solution containing twice the final concentration of formamide, as they will be diluted subsequently. If the hybridization is done against the poly(A) mRNA population with a poly(dT) probe, use a 10% formamide solution instead of 40% at each step of the hybridization. For RNA probes, use 50% formamide.

*Washing.* After the incubation, remove the coverslips from the Parafilm sheet and put them back in the Coplin jar. Wash the coverslips twice with 8 ml of 40% formamide, 2× SSC (preheated at 37°) for 15 min at 37° (for RNA probes, wash at 50°). Wash with 8 ml of 2× SSC, 0.1% Triton X-100 for 15 min at room temperature and then twice with 8 ml of 1× SSC for 15 min at room temperature. Finally, add 8 ml of 1× PBS (phosphate-buffered saline: 1 m$M$ $KH_2PO_4$, 10 m$M$ $Na_2HPO_4$, 140 m$M$ NaCl, 3 m$M$ KCl, pH 7.4) containing 1 ng/ml of DAPI (Molecular Probes, Eugene, OR).

At this step, the coverslips are ready to be mounted on the glass slides (1 mm thick, Gold Seal Products, Highland Park, IL). Drop 10 $\mu$l of mounting medium (see next paragraph) on the slide, lay down the coverslip on the drop (the surface of the coverslip containing the cells should face the drop), and remove excess medium with Kimwipes. Finally, seal the coverslip sides with nail polish. The slides can be stored at −20° for several months without loss of fluorescence.

*Preparation of Mounting Medium.* This medium contains *p*-phenyl-enediamine, which retards the photobleaching of fluorophores. Dissolve 100 mg of *p*-phenylenediamine (Sigma) in 10 ml of 10× PBS (10 m$M$ $KH_2PO_4$, 0.1 $M$ $Na_2HPO_4$, 1.4 $M$ NaCl, 40 m$M$ KCl, pH 7.5) and adjust to pH 8.0 with 0.5 $M$ sodium bicarbonate, pH 9.0 (freshly prepared). Add 90 ml of glycerol and 10 $\mu$l of 1 mg/ml DAPI and keep at −20° wrapped in aluminum foil.

*Note.* We recommend storing in multiple aliquots in order to prevent frequent thawing–freezing of the stock solution. The manipulations should be done in the dark. The color of the solution changes with time, going from yellow to dark brown (which indicates oxidation of the phenylenediamine). Discard if the color is too dark in order to prevent fluorescence artifacts.

## *In Situ* Hybridization on Cultured Cells

### *Preparation of Coverslips*

The coverslips are prepared the same way as in the section on yeast cells. However, if fibroblast cells are used, it is preferable to use gelatin-coated coverslips instead of poly(L-lysine) because the fibroblasts do not adhere well on poly(L-lysine).

To prepare gelatin-coated coverslip, treat the coverslips with HCl and rinse with distilled water (see *in situ* hybridization on yeast cells). Put the coverslips in a solution of 0.5% gelatin (Fisher Scientific, Pittsburgh, PA) and autoclave. Keep at 4°.

Coverslips are placed in a 100-mm petri dish (Fisher Scientific), and cultured cells are grown directly on the coverslips in the appropriate growth medium at 37°, up to 80% confluence.

## Fixation

The cells are washed once in 1× PBS (10 m$M$ Na$_2$HPO$_4$, 1 m$M$ KH$_2$PO$_4$, 137 m$M$ NaCl, 3 m$M$ KCl, pH 7.4) and fixed for 10 min at room temperature in 4% formaldehyde (from a 32% liquid stock, Electron Microscopy Sciences, Eugene, OR), 10% acetic acid, 1× PBS. After two washes with 1× PBS, cells are dehydrated by treatment with 70% ethanol overnight at 4°. Coverslips can be stored for weeks at this stage.

*Notes.* The presence of acetic acid improves detection of nuclear RNA. It can be replaced by a Triton X-100 extraction prior to fixation. Neither of these treatments is necessary when short (50 bases) oligonucleotide probes are used (a 30-min fixation in 4% formaldehyde, 1× PBS is sufficient in this case) or when one wants to detect cytoplasmic RNAs.

## Hybridization

The cells are rehydrated for 5 min, at room temperature, in 2× SSC (0.3 $M$ NaCl, 30 m$M$ sodium citrate, pH 7.0), 50% formamide. Cells are hybridized overnight at 37° (in the dark) in 40 $\mu$l of a solution containing 10% dextran sulfate, 2 m$M$ vanadylribonucleoside complex, 0.02% RNase-free BSA, 40 $\mu$g *E. coli* tRNA, 2× SSC, 50% formamide, and 30 ng of probe. (See the section on yeast for a detailed description of the manipulations.)

## Washing

The cells are washed twice for 30 min at the appropriate stringency: 2× SSC, 50% formamide, 37° for oligonucleotide probes or 0.1× SSC (15 m$M$ NaCl, 1.5 m$M$ sodium citrate, pH 7.0), 50% formamide, 50° for RNA probes. The coverslips are finally mounted on glass slides with mounting medium (see section on yeast cells) and stored at −20°.

*Note.* Probes that are labeled with high specific activity tend to give higher background, but the addition of 0.1% Nonidet P-40 (NP-40) or 0.1% SDS in the washing buffer can diminish this background.

*Simultaneous in Situ Hybridization and Immunofluorescence (Optional)*

This protocol can be used if the simultaneous detection of RNA and protein is required in both yeast[10] and cultured cells.[18] It can also be used for the detection of digoxigenin-labeled probes using antidigoxigenin antibodies.[19] In both cases, the *in situ* hybridization has to be performed first, followed by immunofluorescence.

*Primary Antibody.* Following the last wash with 1× SSC, the coverslips are incubated in 8 ml of 1× PBS, 0.1% BSA, in 8 ml of 1× PBS, 0.1% BSA, 0.1% NP-40, and in 8 ml 1× PBS, 0.1% BSA, all for 5 min at room temperature. The primary antibody is diluted at the appropriate concentration in 1× PBS, 0.1% BSA, 20 m$M$ vanadylribonucleoside complex, and 120 U/ml of RNase inhibitor. Drop 25 $\mu$l of antibody solution on glass plates covered by a Parafilm sheet and lay the coverslips on the drops. Cover with a second sheet of Parafilm, seal, and wrap with aluminum foil. Incubate at 37° for 1–2 hr in the dark.

*Notes.* The 1× PBS, 0.1% BSA solution should be prepared fresh from a 10× PBS DEPC-treated solution and molecular biology grade BSA (Boehringer-Mannheim). The final solution should be filter-sterilized and not autoclaved. Use Triton X-100 instead of NP-40 for mammalian cells. For the detection of digoxigenin-labeled probes, antibodies are available from Boehringer-Mannheim.

*Washing.* Transfer the coverslips in the Coplin jar and wash sequentially for 15 min at room temperature with 8 ml of 1× PBS, 0.1% BSA, followed by 8 ml of 1× PBS, 0.1% BSA, 0.1% NP-40, and finally with 8 ml of 1× PBS, 0.1% BSA.

*Secondary Antibody.* The secondary antibody is diluted at the appropriate concentration in 1× PBS, 0.1% BSA, 20 m$M$ vanadylribonucleoside complex, and 120 U/ml of RNase inhibitor. The coverslips are incubated in the presence of the secondary antibody for 1 hr at room temperature in the dark.

*Notes.* Antibodies conjugated to green fluorophores (e.g., FITC, CY2), red fluorophores (e.g., CY3, Rhodamine, and Texas Red), or blue fluorophore (e.g., AMCA) can be purchased from Jackson ImmunoResearch (West Grove, PA).

*Washing.* Coverslips are subsequently washed once for 15 min at room temperature in 8 ml of 1× PBS, 0.1% BSA, twice in 1× PBS, 0.1% BSA, 0.1% NP-40, and once in 8 ml of 1× PBS, 0.1% BSA. The coverslips are

---

[18] C. Jolly, F. Mongelard, M. Robert-Nicoud, and C. Vourc'h, *J. Histochem. Cytochem,* **45,** 1585 (1997).

[19] P. A. Takizawa, A. Sill, J. R. Swedlow, I. Herskowitz, and R. D. Vale, *Nature* **389,** 90 (1997).

then rinsed in 1× PBS and 1 ng/ml of DAPI and mounted on slides with mounting medium. They can be stored at −20° for several months.

## Microscopy and Imaging

### Microscopes

A good microscope is essential for obtaining good quality results from *in situ* hybridizations. Several manufacturers, e.g., Nikon (Melville, NY), Olympus (Melville, NY), and Zeiss (Thornwood, NY), offer different types of microscopes. Of course, state-of-the-art microscopes are not necessary for most of the FISH and immunofluorescence applications in yeast and cultured cells, and equivalent results are obtained with less expensive microscopes with good optics.

A suitable epifluorescence microscope for *in situ* hybridization work should contain a UV lamp, fluorescence excitation and emission filters, a condenser and filters for Nomarski or phase, and a wide-field objective. Because each of these items is usually sold separately from the microscope, it is possible to set the microscope to your particular needs. Two pieces are essential for good quality microscopy: fluorescence filters and objectives. Even if some fluorophores have similar maximum excitation wavelengths, such as Texas Red (595 nm), Rhodamine Red (570 nm), and CY3 (550 nm), it is important to use the appropriate filter for a given fluorochrome in order to have a maximal light emission from this fluorophore. These filters are available from the microscope manufacturers or from specialized suppliers such as Chroma Technology (Brattleboro, VT).

For objectives, magnification lens of 40 to 100× (for a total magnification of 400–600 to 1000–1500× depending on the eyepiece) are usually appropriate for *in situ* hybridization work in both yeast and cultured cells. It is important to note that the brightness of the fluorescence image is inversely proportional to the magnification of the lens. To increase both the intensity of the excitatory light and the emission fluorescence, it is appropriate to use objectives with high numerical aperture (NA) in oil immersion. The numerical aperture value reflects the angle of the cone of light accessible by the objective; a wider cone will emit and gather more photons. Objectives with higher NA also have the advantage of a higher resolving power. Finally, most objectives are corrected for field curvature aberrations and produce a flat field of observation. These objectives range from plan achromat, plan fluorites to plan apochromats, depending on the quality of the correction (and the price). For an accessible and excellent review on light microscopy, we suggest the web site of Molecular Expressions (www.micro.magnet.fsu.edu).

*Imaging*

After *in situ* hybridizations have been performed, it is important to acquire images of the cells for publication purposes or simply data storage. The main choices for image acquisition technologies are between well-known reflex cameras and charge-coupled device (CCD) cameras. Photomicroscopy with reflex cameras offers a less expensive advantage than CCD cameras (although film costs must be considered). These cameras are usually sold by the microscope manufacturers and use standard black-and-white or color films. Usually, an ASA 400 film is sufficient for good quality images of both cultured and yeast[20] cells. However, several images of the same cells with different exposure times must be acquired and, depending on the signal intensity, exposure times of several seconds are necessary. These long exposure times will increase the bleaching of the samples. Also, the images captured are nonquantitative and have a smaller dynamic range compared to a CCD camera.

CCD cameras use silicon microchips to detect photons and produce a digital readout that can be stored and processed on a computer. These cameras have a very low noise background and are very sensitive, which makes them useful for the detection of low hybridization signals. Exposure times of 100 msec to a few seconds are usually sufficient. Cooled CCD cameras ranging from 8 to 16 bits can be purchased from companies such as Photometrics (Tucson, AZ), Olympus (Melville, NY), or Hamamatsu (Japan). The number of bits reflects the dynamic range (the ability to detect very dim and very bright parts in a single image) and the gray-scale resolution of the camera. The hardware and software must allow the processing and storage of digital images of a few megabytes, which are usually captured. This means that a computer of 32 MB of RAM and several hundred megabytes for storage is necessary for image acquisition. Long-term storage on an optical disk or CD-ROM with a 650MB capacity is recommended. These images can be restored and processed with software such as Adobe Photoshop (Adobe Systems, Mountain View, CA) or NIH Image (NIH, Bethesda, MD).

Conclusion

The use of chemically defined probes for FISH in conjunction with high-resolution optics, sensitive means of image detection, and sophisticated algorithms to extract three-dimensional quantitative information at super-resolution allows the study of the molecular biology of individual cells using

[20] J. R. Pringle, A. E. M. Adams, D. G. Drubin, and B. K. Haarer, *Methods Enzymol.* **191** (1991).

microscopy. The FISH approach allows the conversion of the nucleic acid sequence into physiologically relevant information. New developments are forthcoming in real-time imaging of specific squences in living cells and in more powerful optics and more extensive analysis of images, which will provide us with increasingly abundant information concerning the expression of nucleic acid sequences within cells.

# Author Index

Numbers in parentheses are footnote reference numbers and indicate that an author's work is referred to although the name is not cited in the text.

## A

Abelson, J., 88, 105, 386
Adams, A. E. M., 505
Adams, S. R., 351, 356(15)
Agmon, I., 119
Agrawal, R. K., 119, 251
Agris, P. F., 115
Aldrich, J. V., 442
Alexander, R. W., 118, 122, 127, 130, 133
Allain, F. H., 304
Allen, L. C., 148, 149(11), 158(11)
Allen, P., 204, 238
Allerson, C. R., 167
Allington, A. D., 292
Altman, S., 238, 239, 240(9, 13), 242, 242(9), 244(9), 245(9), 248(9, 13), 249(9, 13, 16), 269(14)
Altschuler, M., 73
Amalric, F., 310, 326(4), 330(27), 331(4), 332
Amberg, D. A., 494
Amontov, S., 203
Anderson, V. E., 6
Angrand, P. O., 365
Ansel-McKinney, P., 425
Anslyn, E., 149
Anthony, D. D., 141
Anthony-Cahill, S., 279
Ares, M., 479, 486, 486(7)
Arruda, I., 407
Aruoma, O. I., 33
Atkins, J. A., 374
Atkins, J. F., 17, 365
Atkinson, B. L., 99
Atkinson, T., 274, 277(32)
Atzberger, A., 297, 331(26), 332, 350(6), 351, 375, 376, 377(3), 380, 383, 399
Auerbach, T., 119
Aurup, H., 195

## B

Ausubel, F. M., 51, 338, 341(17)
Avila, H., 119
Avis, J. M., 304
Axel, R., 371

Bach, M., 392
Bach, R., 20
Bachellerie, J. P., 14
Bai, C., 402
Bain, E. S., 356
Bald, R., 19
Baldwin, R. L., 426
Ban, N., 118, 251
Bantle, J. A., 50
Banu, L., 367
Baratova, L. A., 490
Barofsky, D. F., 442
Baron, C., 367
Barrell, B. G., 71, 180
Barrett, R. W., 269
Barriocanal, J. G., 370, 372(24)
Barta, A., 128
Bartel, D. P., 193, 194, 195, 205, 214(3), 218, 272
Bartel, P., 411
Bartels, H., 119
Barton, J. K., 7, 33
Baserga, S., 492
Bashan, A., 119
Bashkin, J. K., 148
Baskerville, S., 210
Bass, B. L., 48, 49(7, 8), 52(7), 53(7), 57(7), 59(7), 66
Bass, M. B., 407
Bass, S., 334
Bassi, G., 44, 46(12), 47(12)

507

## S

Vlassova, I., 147, 159(1)
Vogel, A., 115
von Ahsen, U., 214, 215, 217(8), 222(8)
von der Eltz, H., 202
von Hippel, P. H., 298, 304(11)
von Jagow, G., 373
von Kitzing, E., 44, 46(12), 47(12)
Vourc'h, C., 503

## W

Wadzack, J., 261
Wakatsuki, S., 136, 441
Walczak, R., 20
Waldsich, C., 229
Wallace, J. C., 140
Wallace, S. T., 214, 215, 217(9), 220(9), 222(9), 223(9), 227(9)
Wallis, M. G., 215, 217(8), 222(8)
Walter, P., 141, 281
Wang, J., 29, 47, 203, 238, 427
Wang, J. F., 175, 189(1)
Wang, R., 118, 120, 122, 128
Wang, S., 297, 335, 349(12), 350(5), 351
Wang, S. Y., 105
Wang, Y., 215, 222(6)
Wang, Y.-H., 33, 34(6), 35(6)
Wang, Z. F., 402, 408(12), 414(12)
Wank, H., 215
Ward, D. C., 494
Ward, W. W., 375
Wassarman, D. A., 22, 85, 87
Watanabe, K., 196
Watkins, N. J., 386
Watts, J. W., 271
Wecker, M., 229, 237(1)
Weeks, K. M., 6, 22, 427
Wei, P., 351
Wei, Z., 148
Weidle, U. H., 365, 366(6), 369(6), 370(6), 371(6), 373(6)
Weidner, M. F., 461
Weinstein, S., 119
Weintraub, H., 203
Weisblum, B., 479
Weiser, B., 181, 186(14), 462, 465, 490
Weiss, M. A., 306, 425
Weitzel, S. E., 298, 304(11)
Weitzmann, C. J., 123, 130

Wekks, K. M., 209
Wells, S. E., 479, 486
Werner, C., 159
Westhof, E., 6, 7, 10, 15(32), 18, 18(32), 19, 19(47, 50), 20, 39, 41(26), 42, 43, 152, 156, 157(40), 160, 161, 161(34), 165, 167, 175(10), 189, 204, 252
White, S. A., 109, 117(15)
White, S. W., 136, 251
Whitfeld, P. R., 53, 55(14)
Whitfield, M. L., 402, 408(12), 414(12)
Wick, C. L., 229
Wickens, M., 297, 331, 332, 350, 355(3), 399, 402, 407, 407(9, 13), 408(9), 414(13), 415(10), 416(10)
Wickstrom, E., 129
Wieder, R., 124
Wiegand, T. W., 237
Wiewiorowski, M., 7, 225
Wiewiorowsky, M., 159
Wigler, M., 371
Wilhelm, J. E., 298, 299(13)
Wilkie, A. M., 23, 25(11), 26(11), 27(11), 32(11)
Will, C. L., 88
Williams, D. M., 195
Williamson, J. R., 252, 290, 291(41), 328, 329(18), 424(7), 425
Willis, M. C., 89, 94, 94(5), 97(5), 98, 98(5), 99
Wills, J. W., 203
Wilm, M., 446
Wilms, C., 114, 115
Wilson, C. B., 203
Wilson, K. S., 10, 119, 461
Wilson, T. M. A., 271
Wimberly, B., 18
Winkeler, K. A., 148
Wintermeyer, W., 19, 158, 483
Wise, J. A., 385
Witherell, G. W., 196, 278, 335, 377
Witkiewick, H., 497
Wittmann, W. G., 460
Wittmann-Liebold, B., 438, 439, 440(3), 441, 442(7)
Witzel, H., 53, 55(14)
Wlodawer, A., 148, 158(13)
Woese, C., 134, 490
Wolff, T., 88
Wolin, S. L., 281

# Subject Index

## A

Adenosine deaminase, *see* Inosine-containing messenger RNA
Affinity chromatography, *see* Small nuclear ribonucleoprotein
Antibiotic–RNA complex
  examples of aptamers, 215
  mechanism of antibiotic action, 214
  recognition motifs, 215
  RNA aptamer characterization
    boundary mapping
      lead(II) acetate cleavage, 223, 225–226
      partial hydrolysis, 222–223
    dimethyl sulfate modification
      hybridization, 227–228
      modification reaction, 226–229
      polyacrylamide gel electrophoresis analysis, 228
      primer extension, 228
      renaturation, 226
    dissociation constant determination, 222
    sequence analysis, 221
    structure characterization by nuclear magnetic resonance, 229
  *in vitro* selection
    affinity chromatography
      antibiotic coupling to Sepharose, 218–220
      buffers, 219
      counterselection, 220
    initial DNA pool design, 217–218
    overview, 215, 217
    polymerase chain reaction, 221
    reverse transcription, 220–221
    transcription, 219, 221
Antitermination, RNA-binding library screening
  components of λ N antitermination complex, 297–299
  doped libraries, 308

$\beta$-galactosidase assays for scoring interactions
  colony color assay
    overview, 300–301
    plating and color development, 301
    reagents, 301
  screening capacity, 304–305
  solution assay, 301–302
  modifications of system, 308
  N–RNA-binding protein fusion, 299, 302–304
  RNA-binding domain library generation
    ligation, 307
    oligonucleotide extraction, 306
    overview, 305–306
    primer extension, 306
  selection of RNA binders from libraries
    false-positive elimination, 307
    nonspecific-positive elimination, 307–308
    primary screen, 307
  two-plasmid reporter system, principles, 298–299
Aptamer, *see* In Vitro selection, RNA aptamers
Arylazides, *see* RNA–RNA cross-linking
Azides, *see* RNA–RNA cross-linking

## B

BABE, *see* 1(*p*-Bromoacetamidobenzyl)-EDTA
Bacteriophage P22, *see* RNA challenge phage system
Bradykinin, RNA ligand selection
  binding reaction, 233
  *N*-bromoacetylbradykinin synthesis, 230
  guanosine 5'-monophosphorothioate
    purification of RNA with thiopropyl-Sepharose, 232–233
    synthesis, 230

secondary structure, 39, 41
end-labeling of RNA, 36–37
enzymes and reagents, 35
overview, 35
polyacrylamide gel electrophoresis,
38–39
principle, 36
probing reaction, 38
purification of labeled RNA, 37
ribonuclease T1 digestion, 37
specificity of RNA cleavage, 33–35
Cross-linking, *see* RNA–protein cross-linking; RNA–RNA cross-linking
1-Cyclohexyl-3-(2-morpholinoethyl)carbo-
diimide metho-*p*-toluene sulfonate
applications of RNA modification, 9
primer extension analysis, 16
RNA modification site, 5

## D

DEPC, *see* Diethyl pyrocarbonate
Diethyl pyrocarbonate
applications of RNA modification, 8–9
direct detection of RNA modification,
13
primer extension analysis, 16
RNA modification site, 5–6
water treatment for RNA studies, 119
Dimethyl sulfate
antibiotic RNA aptamer modification
hybridization, 227–228
modification reaction, 226–229
polyacrylamide gel electrophoresis analysis, 228
primer extension, 228
renaturation, 226
applications of RNA modification, 8–9
direct detection of RNA modification, 12
primer extension analysis, *in vitro*, 16
RNA modification sites, 5, 479–480
*in vivo* probing of RNA structure
advantages, 480–481
borohydride reduction and aniline
cleavage at 7-methylguanosine,
483–484
cell permeability, 479–480
examples
18S ribosomal RNA, 489–490

U2, 486, 489
U3, 490, 492
primer extension analysis
extension reaction, 485
hybridization, 484–485
polyacrylamide gel electrophoresis,
486
primer labeling, 484
RNA extraction, 482–483
safety, 481
theory, 480–481
troubleshooting, 492–493
yeast culture treatment, 481–482
DMS, *see* Dimethyl sulfate

## E

ENU, *see* Ethylnitrosourea
Ethylnitrosourea
applications of RNA modification, 8–9
direct detection of RNA modification, 13
primer extension analysis, 16
RNA modification site, 6

## F

Fast protein liquid chromatography, recombinant 30S ribosomal subunit purification, 452–454
FISH, *see* Fluorescence *in situ* hybridization
Fluorescence *in situ* hybridization, RNA
applications, 493–494, 505–506
cultured cells
coverslip preparation, 501–502
fixation, 502
hybridization, 502
washing, 502
imaging and analysis, 505
microscopes, 504
oligonucleotide probes
advantages and disadvantages, 494–495
design, 495
labeling with terminal transferase, 496
purification, 495–496
synthesis, 495
RNA probes
advantages and disadvantages, 496–497

reverse transcription–polymerase chain reaction, 247
selection progress assessment, 249
transcription, *in vitro*, 241–242, 244–245, 247
Ribonuclease T1
inosine 3'-phosphodiester bond, specific cleavage of messenger RNA
cellular RNA purification, 50–51
control RNA synthesis, 49–50
digestion reaction, 54–55
glyoxal reaction, 54
optimization, 53–54
oxidation of 3'-hydroxyl groups, 51–52
postcleavage processing
cyclic phosphate removal, 55
glyoxal adduct removal, 55–56
principle, 52–53
ladder analysis of modified RNA, 14, 37, 39
Ribosomal RNA
dimethyl sulfate probing *in vivo* of 18S rRNA, 489–490
direct detection of modification or cleavage of 5S rRNA
chemical modification
diethyl pyrocarbonate, 13
dimethyl sulfate, 12
ethylnitrosourea, 13
hydrazine cleavage, 13
lead(II) acetate, 12
sodium borohydride cleavage, 13
denaturation and renatutration, 12
end labeling, 11–12
polyacrylamide gel electrophoresis fractionation, 13–14
iron(II)–EDTA tethered complex for structure probing
anticodon stem–loop analogs, 186–188
environment surrounding 5'-end of phenylalanine transfer RNA, 185
16S rRNA, 183–184
L11–23S rRNA interaction, *in vitro* selection from random RNA fragments
advantages, 252–253, 268
antibiotic interference, 255
binding reaction, 262
binding sites, 253–255
cloning of fragments, 265–266

conservation of interaction between species, 256
cycles of selection, 260–261
minimal binding site identification, 266–267
nitrocellulose filter binding of complexes, 261–262
overview, 253, 260
polymerase chain reaction, 260–261
protein purification, 256–257
RNA pool
construction, 257–259
diversity, 259–260
RNA:protein ratio, 264
selection progress, monitoring
band-shift assay on native gels, 262–264
binding assays, 262–263
transcription, 261
loop E structure elucidation of 5S RNA, 18–20
primer extension analysis of 5S RNA modifications or cleavage
chemical modification reactions, 16
gel electrophoresis, 17
primer annealing and extension, 17
primer selection, 14, 16
Ribosome
assembly, 251–252
iron(II)–EDTA tethered hydroxyl radical probing
advantages, 461–462
buffers, 463–464
equipment and supplies, 464
footprinting S20 on 16S ribosomal RNA, 468
hydroxyl radical probing conditions, 470
iron(II) derivatization of protein, 467
linker–iron(II) complex preparation, 466–467
nuclease contamination prevention, 462
overview, 461
primer extension analysis
extension reaction, 472–473
hybridization, 472
interpretation of data, 473–475
overview, 471
polyacrylamide gel electrophoresis analysis, 473